Communications
in Computer and Information Science 311

Chrisina Jayne   Shigang Yue
Lazaros Iliadis (Eds.)

Engineering Applications of Neural Networks

13th International Conference, EANN 2012
London, UK, September 20-23, 2012
Proceedings

 Springer

Volume Editors

Chrisina Jayne
Coventry University
Priory Street
Coventry CV1 5FB, UK
E-mail: ab1527@coventry.ac.uk

Shigang Yue
University of Lincoln
Lincoln LN6 7TS, UK
E-mail: syue@lincoln.ac.uk

Lazaros Iliadis
University of Thrace
193 Pandazidou St.
68200 N Orestiada, Greece
E-mail: liliadis@fmenr.duth.gr

ISSN 1865-0929
ISBN 978-3-642-32908-1
DOI 10.1007/978-3-642-32909-8
Springer Heidelberg Dordrecht London New York

e-ISSN 1865-0937
e-ISBN 978-3-642-32909-8

Library of Congress Control Number: 2012946639

CR Subject Classification (1998): I.2.6, I.5.1, H.2.8, J.2, J.1, J.3, F.1.1, I.5, I.2, C.2

Typesetting: Camera-ready by author, data conversion by Scientific Publishing Services, Chennai, India

Printed on acid-free paper

Springer is part of Springer Science+Business Media (www.springer.com)

Preface

The First Engineering Applications of Neural Networks (EANN) conference was held in Otaniemi, Finland, in 1995. Since then the EANN conferences have provided academics and industry professionals from across the world with the opportunity to share experiences and to present and demonstrate advances in a wide range of neural network applications.

The 13th EANN 2012 conference was held on the London campus of Coventry University, UK, during September 2012. The primary sponsor for the conference was the International Neural Network Society (INNS). The 13th EANN 2012 attracted delegates from 23 countries across the world: Russia, USA, South Africa, Germany, Italy, UK, Greece, Switzerland, Spain, Brazil, India, Ukraine, France, Poland, Turkey, Chile, Israel, China, Cyprus, Taiwan, Portugal, Belgium, and Finland.

This volume includes the papers that were accepted for presentation at the conference, at the Workshop on Applying Computational Intelligence Techniques in Financial Time Series Forecasting and Trading (ACIFF), and the Workshop on the Computational Intelligence Applications in Bioinformatics (CIAB). The papers demonstrate a variety of applications of neural networks and other computational intelligence approaches to challenging problems relevant to society and the economy. These include areas such as: intelligent transport, environmental engineering, computer security, civil engineering, financial forecasting, virtual learning environments, language interpretation, bioinformatics and general engineering. All papers were subject to a rigorous peer-review process by at least two independent academic referees. EANN accepted approximately 40% of the submitted papers for full length presentation at the conference. The best ten papers were invited to submit extended contributions for inclusion in a special issue of the *Evolving Systems* journal (Springer).

The following keynote speakers were invited and gave lectures on exciting neural network application topics:

1. Nikola Kasabov, Director and Founder, Knowledge Engineering and Discovery Research Institute (KEDRI), Chair of Knowledge Engineering, Auckland University of Technology, Institute for Neuroinformatics - ETH and University of Zurich
2. Danil Prokhorov, President-Elect of INNS, Toyota Research Institute NA, Ann Arbor, Michigan
3. Kevin Warwick, University of Reading, England and Fellow of The Institution of Engineering & Technology (FIET)
4. Richard J. Duro, Grupo Integrado de Ingeniera Escuela Politecnica Superior, Universidade da Coruña

A tutorial on "Fuzzy Networks with Modular Rule Bases" was presented by Alexander Gegov from the University of Portsmouth, UK.

Two workshops were included in the EANN 2012 conference: Applying Computational Intelligence Techniques in Financial Time Series Forecasting and Trading (ACIFF 2012) focused on the scientific areas of computer engineering, finance and operational research; and Computational Intelligence Applications in Bioinformatics (CIAB 2012) focused on problems from the fields of biology, bioinformatics, computational biology, chemical informatics, and bioengineering.

On behalf of the conference Organizing Committee, we would like to thank all those who contributed to the organization of this year's program, and in particular the Program Committee members.

September 2012   Chrisina Jayne
                                                               Shigang Yue

Organization

Organizing Chairs

Chrisina Jayne   Coventry University, UK
Shigang Yue   University of Lincoln, UK

Advisory Chair

Nikola Kasabov

Program Committee Chairs

Chrisina Jayne   Coventry University, UK
Shigang Yue   University of Lincoln UK
Lazaros Iliadis   Democritus University of Thrace, Greece

Program Committee

K.A. Theofilatos	University of Patras, Greece
A. Adamopoulos	University of Thrace, Greece
B. Akhgar	Sheffield Hallam University, UK
A. Andreou	Cyprus University of Technology, Cyprus
P. Angelov	Lancaster University, UK
A. Anjum	University of Derby, UK
R. Bali	Coventry University, UK
G. Beligiannis	University of Western Greece, Greece
I. Bukovsky	Czech Technical University, Czech Republic
F.C. Morabito	University of Reggio Calabria, Italy
C. Christodoulou	Cyprus University, Cyprus
S.D. Likothanassis	University of Patras, Greece
G.D. Magoulas	University of London, UK
K. Dimitrios	Demokritos National Centre for Scientific Research, Greece
F. Doctor	Coventry University, UK
M. Eastwood	Coventry University, UK
M. Elshaw	Coventry University, UK
M. Fiasche	University of Reggio Calabria, Italy
A. Gammerman	Royal Holloway, University of London, UK
A. Gegov	University of Portsmouth, UK

Program Committees ACIFF Workshop

A.S. Karathanasopoulos	London Metropolitan University, UK
G. Sermpinis	University of Glasgow, UK
S.D. Likothanassis	University of Patras, Greece
E.F. Georgopoulos	Technological Educational Institute of Kalamata, Greeece
R. Rosillo	University of Oviedo, Spain
C. Dunis	Horus Partners Wealth Management Group SA
J. Laws	University of Liverpool Management School, UK
A. Andreou	Cyprus University of Technology, Cyprus
G. Beligiannis	University of Western Greece, Greece
H. Papadopoulos	Frederick University, Cyprus
E. Papatheocharous	University of Cyprus, Cyprus
K.A. Theofilatos	University of Patra, Greece
H.J. von Mettenheim	University of Hannover, Germany

Program Committee CIAB Workshop

S.D. Likothanassis	University of Patras, Greece
E.F. Georgopoulos	Technological Educational Institute of Kalamata, Greece
S. Mavroudi	University of Patras, Greece
A. Adamopoulos	University of Thrace, Greece
A. Tsakalidis	University of Patras, Greece
S. Kossida	Academy of Athens, Greece
H. Prez-Snchez	University of Murcia, Spain
A. Gammerman	Royal Holloway, University of London, UK
V.P. Plagianakos	University of Central Greece, Greece
G.D. Magoulas	University of London, UK
P. Kalnis	King Abdullah University of Science and Technology, Saudi Arabia
C. Moschopoulos	Katholieke Universiteit Leuven, Belgium
I. Valavanis	National Hellenic Research Foundation, Greece

Table of Contents

Workshop on Applying Computational Intelligence Techniques in Financial Time Series Forecasting and Trading

Workshop on Computational Intelligence Applications in Bioinformatic

Elimination of a Catastrophic Destruction of a Memory in the Hopfield Model

Iakov Karandashev, Boris Kryzhanovsky, and Leonid Litinskii

CONT SRISA RAS, Moscow, Vavilova St., 44/2, 119333, Russia
Yakov.Karandashev@phystech.edu, {kryzhanov,litin}@mail.ru

Abstract. For the standard Hopfield model a catastrophic destruction of the memory has place when the last is overfull (so called *catastrophic forgetting*). We eliminate the catastrophic forgetting assigning different weights to input patterns. As the weights one can use the frequencies of appearance of the patterns during the learning process. We show that only patterns whose weights are larger than some critical weight would be recognized. The case of the weights that are the terms of a geometric series is studied in details. The theoretical results are in good agreement with computer simulations.

Keywords: Hopfield model, catastrophic forgetting, quasi-Hebbian matrix.

1 Introduction

In the standard Hopfield model one uses the Hebb connection matrix constructed with the aid of M random patterns [1], [2]. For definiteness we suppose that vector-row $\mathbf{x}^\mu = (x_1^\mu, x_2^\mu, ..., x_N^\mu)$ with binary coordinates $x_i^\mu = \pm 1$ is the μ-th pattern, and the number of patterns is equal to M. The Hebb connections have the form:

$$J_{ij} = (1 - \delta_{ij}) \sum_{\mu=1}^{M} x_i^\mu x_j^\mu, \ i, j = 1, ... N .$$

The estimate for the number of random patterns that can be recognized by the Hopfield model is well known: $M_c \approx 0.14 \cdot N$. If the number of patterns written down into connection matrix is larger than M_c, the catastrophe takes place: the network ceases to recognize patterns at all. This destruction of the memory can be explained by the symmetry of the Hebb matrix. Indeed, all patterns entering the Hebb matrix are equivalent. If under some conditions one pattern is recognized, other patterns will be recognized too. The reverse is also true: If under some conditions a pattern is not recognized, any other pattern will not be recognized too. This means that the memory will be destroyed.

The catastrophic forgetting is a troublesome defect of the Hopfield model. Earlier some modifications of the Hebb matrix were proposed to eliminate the memory destruction. As the result of such modifications an unlimited number of random patterns can be fearlessly written down into matrix elements one by one. However,

C. Jayne, S. Yue, and L. Iliadis (Eds.): EANN 2012, CCIS 311, pp. 1–10, 2012.

the memory of the network is restricted. If as previously the maximum number of recognized patterns is denoted by M_c, for the models discussed in [3]-[6] $M_c \approx 0.05 \cdot N$. The general weak-point of all these models is as follows: only last patterns written down in the network are recognized. In other words, the memory of a network is formed by patterns whose order numbers satisfy the inequality $M - M_c \leq \mu \leq M$. Patterns with order numbers less than $M - M_c$ are excluded from the memory irretrievably.

We succeeded in eliminating of the catastrophic memory overfull in the Hopfield model. In our approach a weight is assigned to each pattern. Different weights allow us to individualize conditions of recognition of different patterns and this is why the catastrophic forgetting can be eliminated.

In previous papers [7], [8] with the aid of statistical physics methods the main equation (3) for the Hopfield model with the quasi-Hebbian matrix

$$J_{ij} = (1 - \delta_{ij}) \sum_{\mu=1}^{M} r_\mu x_i^\mu x_j^\mu \qquad (1)$$

was obtained. The weights r_μ are positive and put in decreasing order: $r_1 \geq r_2 \geq \ldots \geq \ldots \geq 0$. In [7] for a special distribution of the weights we succeeded in solution of the main equation. In this paper we examined the case when only one coefficient differed from all other that were identically equal: $r_1 = \tau$, $r_2 = r_3 = \ldots = r_M = 1$. Theoretical results were confirmed by computer simulations.

In this paper we give the solution of the main equation in the general case. The main result is as follows. For every weights distribution $\{r_\mu\}$ there is such a critical value r_c that only patterns whose weights are greater than r_c will be recognized by the network. Other patterns are not recognized. The case of the weights, which are the terms of a decreasing geometric series $r_\mu = q^\mu$ ($q < 1$) is discussed in details. For these weights the number of the recognized patterns is $\sim 0.05 \cdot N$. The results are confirmed by computer simulations.

Note, for the first time the quasi-Hebbian connection matrix (1) was discussed many years ago [9], [10]. For such a matrix the system of transcendent equations was obtained in [9] in the replica symmetry approximation. However, the authors failed to transform this system into the main equation (3), and when solving the system they made an error. In [10] only the standard Hopfield model ($r_\mu \equiv 1$) was examined.

2   Different Weights

2.1   Main Equation

Statistical physics methods allow one to obtain equations for the overlap m_k of the pattern $\mathbf{x}^k$ with a nearest fixed point $\tilde{\mathbf{x}}^k$: $m_k = \sum_{i=1}^{N} x_i^k \tilde{x}_i^k / N$. After solving these equations it is possible to understand under which conditions the overlap of the k-th pattern with the nearest fixed point is of the order of 1. In other words: when the fixed point coincides (or nearly coincides) with the k-th input pattern. This means that the k-th pattern is recognized by the network. If $m_k \sim 0$, the k-th pattern is not recognized by the network.

Repeating step by step calculations performed in [1], [2] for the standard Hopfield model, and as the last step tending the temperature to zero, we obtain the system of equations for the overlap m_k:

$$m_k = \mathrm{erf}\left(\frac{r_k m_k}{\sqrt{2}\sigma}\right), \quad \sigma^2 = \frac{1}{N}\sum_{\mu\neq k}^{M}\frac{r_\mu^2}{(1-C\cdot r_\mu)^2}, \quad C = \frac{1}{\sigma}\sqrt{\frac{2}{\pi}}\exp\left[-\left(\frac{r_k m_k}{\sqrt{2}\sigma}\right)^2\right]. \tag{2}$$

Let us introduce an auxiliary variable $y = r_k m_k / \sqrt{2}\sigma$. Then excluding σ and C from the system (2), we obtain the main equation:

$$\frac{\gamma^2}{\alpha} = \frac{1}{M-1}\sum_{\mu\neq k}^{M}\left(\frac{r_\mu}{r_k\varphi - r_\mu}\right)^2. \tag{3}$$

It is supposed that M and N are very large: $M,N \gg 1$. $\alpha = M/N$ is the load parameter, $\gamma = \gamma(y)$ and $\varphi = \varphi(y)$ are monotonic functions of $y \geq 0$:

$$\gamma(y) = \sqrt{\frac{2}{\pi}}e^{-y^2}, \quad \varphi(y) = \frac{\sqrt{\pi}}{2}\frac{\mathrm{erf}(y)}{y}e^{y^2}.$$

When y_0 is the solution of Eq.(3), the overlap of the k-th pattern is $m_k = \mathrm{erf}(y_0)$.

In the case of the standard Hopfield model ($r_\mu \equiv 1$) the system (2) reduces to the well-known system (see Eqs. (2.71)-(2.73) in [2]), and Eq. (3) turns into the well-known equation obtained in [1].

For unequal weights and the temperature $T > 0$ the analog of the system (2) was obtained long ago in [9]. However, when using the zero-temperature limit the authors mistakenly assumed that $C = 0$ and $\sigma^2 = \sum_{\mu\neq k}^{M} r_\mu^2 / N$. Solving the remained equation for m_k they obtained that at a moment when the patterns ceased to be recognized the overlap $m_k = 0$. In other words, the transition from the case of patterns recognition to the case when the patterns were not recognized was the phase transition of the second kind. This is not true. We show that there is a breakdown of the solution of the main equation. This means that the phase transition of the first kind takes place. We are the first, who obtain the equation (3). The solution of this equation is discussed in the next item.

2.2   Algorithm of Solution

Let us transform Eq.(3) dividing the left hand and right hand sides by M. The main equation takes form:

$$N = \sum_{\mu \neq k}^{M} f_\mu^{(k)}(y),$$

(4)

where $f_\mu^{(k)}$ are the functions of γ, φ and r_μ:

$$f_\mu^{(k)}(y) = \left(\frac{t_\mu^{(k)}}{\gamma(\varphi - t_\mu^{(k)})} \right)^2, \quad t_\mu^{(k)} = \frac{r_\mu}{r_k}, \quad \mu(\neq k) = 1,...,M.$$

The values $t_\mu^{(k)}$ are arranged in decreasing order. Note, the first $k-1$ of these values are larger than 1, and the other ones are less than 1:

$$t_1^{(k)} > t_2^{(k)} > ... > t_{k-1}^{(k)} > 1 > t_{k+1}^{(k)} > t_{k+2}^{(k)} > ....$$

(5)

The r.h.s. of Eq. (4) is the sum of functions $f_\mu^{(k)}(y)$. It is easy to see that when $y \to \infty$ the denominator $\gamma(\varphi - t_\mu^{(k)})$ of any function $f_\mu^{(k)}(y)$ tends to 0: at the infinity each function $f_\mu^{(k)}(y)$ increases unrestrictedly. Consequently, at the infinity the r.h.s. of Eq.(4) increases unrestrictedly too.

The behavior of the function $f_\mu^{(k)}(y)$ for finite values of its argument depends on its number μ. If $t_\mu^{(k)} < 1$, the function $f_\mu^{(k)}(y)$ is everywhere continuous and limited, because for $y > 0$ the function $\varphi(y) > 1$. However, if $t_\mu^{(k)} > 1$, the function $f_\mu^{(k)}(y)$ has a singular point. In this case the denominator of the function $f_\mu^{(k)}(y)$ is equal to zero for some value $y_\mu^{(k)}$ of its argument: $\varphi\left(y_\mu^{(k)}\right) = t_\mu^{(k)} \Leftrightarrow y_\mu^{(k)} = \varphi^{-1}\left(t_\mu^{(k)}\right)$, where φ^{-1} is the reciprocal function with regard to φ. We see that for every $t_\mu^{(k)} > 1$ the function $f_\mu^{(k)}(y)$ has the discontinuity of the second kind in the point $y_\mu^{(k)}$. Since in the series (5) the first $k-1$ values of $t_\mu^{(k)}$ are greater than 1, it is easy to understand that the r.h.s. of Eq.(4) has the discontinuities of the second kind in the points $y_1^{(k)} > y_2^{(k)} > ... > y_{k-1}^{(k)}$. At the infinity the r.h.s. of Eq.(4) increases unrestrictedly.

For simplicity let us go to reciprocal quantities in Eq.(4):

$$\frac{1}{N} = F_k(y), \quad \text{where} \quad F_k(y) = \left(\sum_{\mu \neq k}^{M} f_\mu^{(k)}(y) \right)^{-1}.$$

(6)

The nonnegative function $F_k(y)$ in the r.h.s. of Eq.(6) is equal to zero in the points $y_1^{(k)} > y_2^{(k)} > ... > y_{k-1}^{(k)}$. At the infinity it tends to zero. In Fig.1 for the weights in the form of the terms of the harmonic series, $r_\mu = 1/\mu$, the behavior of the functions

$F_3(y)$, $F_6(y)$ and $F_9(y)$ is shown. To the right of the rightmost zero $y_1^{(k)}$ every function $F_k(y)$ at first increases, and after reaching its maximal value the function $F_k(y)$ decreases monotonically. Let $y_c^{(k)}$ be the coordinate of the rightmost *maximum* of $F_k(y)$. The value of $F_k(y_c^{(k)})$ determines the critical characteristics related to the recognition of the k-th pattern. Let us explain what it means.

Generally speaking, Eq.(6) has for every k several solutions, that respond to intersections of the function $F_k(y)$ with the straight line that is parallel to abscissa axis at the height $1/N$ (see Fig.1). These solutions correspond to stationary points of the saddle-point equation [1], [2], [10]. However, only one intersection corresponding to the minimum of the free energy is important. Its coordinate is *to the right of the rightmost maximum* $y_c^{(k)}$. Other solutions of Eq.(6) can be omitted.

For a fixed value N at the height $1/N$ let us draw the line that is parallel to abscissa. For patterns whose numbers k are not so large, the graph of the function $F_k(y)$ necessarily intersects this straight line in the point that is to the right of its rightmost maximum $y_c^{(k)}$. (In Fig. 1 there are such intersections have the graphs of the functions $F_3(y)$ and $F_6(y)$.) Let us increase k little by little. We see that when k increases the local maximum $F_k(y_c^{(k)})$ decreases steadily. Consequently, sooner or later we reach such a value k_m that the curve $F_{k_m}(y)$ still has the intersection with the straight line that is to the right of its rightmost maximum, but the curve $F_{k_m+1}(y)$ has no such intersection. (In Fig.1 the curve $F_9(y)$ has no intersection with the straight line that is to the right of its rightmost maximum, and the curve $F_8(y)$, it is not shown, has such an intersection.)

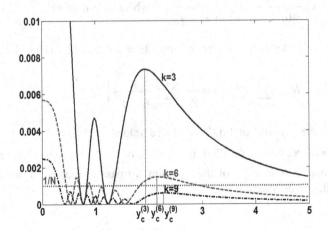

Fig. 1. The behavior of different functions $F_k(y)$ (6) when the weights are equal to $r_\mu = 1/\mu$: $k = 3, 6, 9$, $y_c^{(k)}$ are the local maximums of functions $F_k(y)$, and straight line corresponds to $1/N = 0.001$ (see the body of the text)

For given dimensionality N the pattern with the number $k_m = k_m(N)$ is the last in whose vicinity there is a fixed point: this is the last pattern recognized by the network. For $k < k_m$ Eq.(6) has a solution in the region $y > y_c^{(k)}$ too. These patterns will also be recognized. On the contrary, for $k > k_m$ Eq.(6) has no solutions in the region $y > y_c^{(k)}$. Consequently, the patterns with such numbers will not be recognized. We see that the overlap m_{k_m} of the last recognized pattern is always sufficiently larger than zero.

Let r_c be the weight corresponding to the pattern with the number k_m: $r_c = r_{k_m}$. The obtained result means that only the patterns, whose weights are not less than the critical value r_c will be recognized by the network: $r_k \geq r_c$. Patterns whose weights are less than the critical value r_c are not recognized in spite of the fact that they are written down in the quasi-Hebbian connection matrix. In other words, the memory of the network is limited, but the catastrophic forgetting does not occur.

Our analysis is true for arbitrary weights r_μ. In the next Section we examine a specific distribution of the weights.

3   Weights as the Terms of Geometric Series

Let us discuss in details the case of the weights in the form $r_\mu = q^\mu$, where $q \in (0,1)$. These weights were discussed earlier in [5] and [6]. We assume that in Eq.(4) the first value of the summation index is equal to zero and the first weight is equal to 1: $r_0 = 1$. Then

$$f_\mu^{(k)}(y) \sim \frac{q^{2\mu}}{(q^k \varphi - q^\mu)^2} = \left(\frac{q^\mu}{s_k - q^\mu} \right)^2 \text{, where } s_k = q^k \varphi(y).$$

Let the number of patterns be equal to infinity: $M = \infty$. Then Eq.(4) has the form

$$N = \sum_{\mu=0 \neq k}^{\infty} f_\mu^{(k)}(y) = \frac{1}{\gamma^2} \sum_{\mu=0 \neq k}^{\infty} \left(\frac{q^\mu}{s_k - q^\mu} \right)^2. \tag{7}$$

Our interest is the solution of Eq.(7) for large values of the argument y when the inequality $s_k = q^k \varphi(y) > 1$ is fulfilled. In the r.h.s. of Eq.(7) we replace summation by integration. Then in both sides of the equation we pass to reciprocal quantities and obtain the analogue of equation (6):

$$\frac{1}{N} = \frac{\gamma^2 (\varphi-1)^2}{(\varphi-1)^2 \Phi_k(y) - 1} \text{, where } \Phi_k(y) = \frac{\ln \left(\dfrac{s_k - 1}{s_k} \right) + \dfrac{1}{s_k - 1}}{|\ln q|} \text{, } s_k = q^k \varphi > 1. \tag{8}$$

When solving Eq.(8) numerically we can find the maximal value of the number of the pattern k_m which is recognized by the network yet: $k_m = k_m(N,q)$. We need to find such value of the parameter q that corresponds to the maximal memory of the network. In other words, we are looking for the optimal value of the parameter q for which the value of k_m is maximal: $\tilde{k}(N) = \max_q k_m(N,q)$. It is evident that this optimal q have to exist; $q_m = q_m(N)$ denotes its value.

On the left panel of Fig.2 the dependence of the ratio $k_m(N,q)/N$ on q for three dimensionalities N is shown. We see that all curves have distinct points of maximum, but the value of all maximums is the same:

$$\lim_{N \to \infty} \tilde{k}(N)/N \approx 0.05. \qquad (9)$$

In other words, the maximal number of patterns that can be memorized by the network is $M_c \approx 0.05 \cdot N$. It is almost three times less than the storage capacity for the standard Hopfield model, but the catastrophic overfull of the memory does not occur. Let us list the optimal values $q_m(N)$ for different N: $q_m = 0.992, 0.9992$ and 0.99992 for $N = 1000, 10000$ and 100000, respectively.

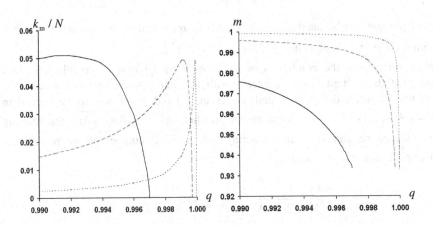

Fig. 2. For three dimensionalities N we show: on the left panel the dependence k_m/N on q; on the right panel synchronous values of the last recognized pattern overlaps with the nearest fixed point. On both panels the solid line corresponds to the dimensionality $N=1000$, the dashed line corresponds to $N=10000$, the point line corresponds to $N=100000$.

When the value of q becomes larger than q_m, the number of recognized patterns decreases. It is clear that as far as the parameter q tends to 1, our model more and more resemble the standard Hopfield model for which the dimensionality N is finite and the number of the patterns M is infinitely large. It is evident that when $q=1$ the network memory is destroyed. It turns out that the destruction of the memory occurs even before q becomes equal to 1. Let q_c denotes the critical value when only the first pattern is recognized: for $q > q_c$, the network ceases to recognize patterns at all. The more N the more critical value q_c becomes closer to 1. It may be shown that the following estimate for q_c is true: $q_c = 1 - \delta$, where $\delta \approx 1/0.329 \cdot N$. The fitting shows that the estimate $q_m \approx 1 - 2.75 \cdot \delta$ is true.

For the pattern with the number k_m on the right panel of Fig.2 the dependence of the overlap on q is shown. In the point of the solution "breakdown" q_c (when the patterns cease to be recognized) the overlaps have approximately the same values $m_c \approx 0.933$.

We can move up in analytical calculations if for $q \to 1$ we simplify the r.h.s. of Eq.(8) for large values of y (let us say, for $y \geq 2$):

$$\frac{1}{N} = g^2 \frac{2q^{2k} \, |\ln q|}{1 - 2q^{2k} \, |\ln q|}, \quad \text{where } g = \frac{erf \, y}{\sqrt{2} y} \approx \frac{1}{\sqrt{2} y}. \qquad (10)$$

Equation (10) allows us to write down the explicit expression for k :

$$k = \frac{\ln[2(Ng^2 + 1) \, |\ln q|]}{2 \, |\ln q|}. \qquad (11)$$

This expression can be maximize with regard to q. When $q_0 \approx 1 - e/2(Ng^2 + 1)$ the maximal value of k is equal to $k_0 \approx Ng^2 / e \approx N/2ey^2$.

Now we require the conditions of *the perfect* recognition to be fulfilled. That means that the pattern has to coincide with the fixed point. In other words, the difference between the pattern and the nearest fixed point has to be less than $1 : m = erf(y) > 1 - 1/N$. We obtain that in this case $y \approx \sqrt{\ln(N/4)}$. Substituting y into the expression for k_0 and denoting the maximal number of the pattern for perfect recognition as k_p, we have:

$$\frac{k_p}{N} \approx \frac{1}{2e \ln(N/4)} \xrightarrow{N \to \infty} 0. \qquad (12)$$

In other words, the requirement of the perfect recognition substantially decreases the storage capacity of the network (compare expressions (12) and (9)). The same is true for the standard Hopfield model.

In Fig.3 three graphs corresponding to the dimensionality $N = 1000$ are shown. The solid line is the dependence of the value k_m on q. The dashed line shows the dependence of the value k_p on q. This dependence is calculated by substitution the expression $y \approx \sqrt{\ln(N/4)}$ in equation (11). Markers in the figure show the numerical results for the number of perfectly recognized patterns. The results of computer simulations are averaged over 500 random matrices. On the whole the agreement with the theory is quite good.

We have investigated in the details the networks whose weights are equal to the terms of the harmonic series as well as to the terms of the arithmetical progression [10]. In the first case the memory of the network is not too large: $k_m \sim \sqrt{N / \ln N}$. In the second case for the optimal value of the common difference the memory of the network is extensively large $k_m \sim 0.1N$. These theoretical results are confirmed by computer simulations [10].

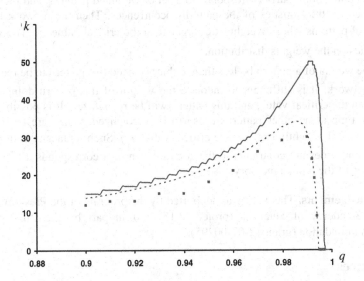

Fig. 3. For $N = 1000$ the dependence of the critical numbers of the patterns on q is shown. The solid line shows the maximal number of the recognized patterns k_m , the dotted line shows the number of the perfectly recognized patterns k_p ; markers are the experimental numbers of the perfectly recognized patterns.

4   Conclusions

The catastrophic forgetting is a troublesome defect of the standard Hopfield model. Indeed, let us imagine a robot whose memory is based on the Hopfield model. It is natural to think that his memory is steadily filled up. When the robot sees an image, it is written additionally to its memory. Catastrophic forgetting means that when the number of stored patterns exceeds M_c , the memory is completely destructed. Everything that was previously accumulated in the memory would be forgotten.

This behavior is contrary to the common sense. In the brain there are mechanisms allowing it to accumulate the obtained information continuously. If the Hopfield model is assumed as an artificial model of the human memory, it has to work even if new information is written down continuously.

The method of eliminating of catastrophic overfull of the memory, proposed in our paper, seems to be very effective. In our approach it is not necessary to separate the learning and the working stages. One can learn the network even during the working stage. Each pattern that occurs in the "field of vision" of the network modifies matrix elements according to the standard Hebbian rule. If this pattern is the same as the one written down previously, its weight increases by 1. If the pattern is new, it is written down into the connection matrix with the weight equals to 1. As the result, the obtained connection matrix corresponds to a set of weighted patterns, and the weights are defined by the statistics of the patterns occurrences. Then the network memory consists of patterns whose weights are larger than the critical value r_c. The value of r_c depends on the weights distribution.

Let the weight of a pattern be less than r_c but we need this pattern to be recognized by the network. It is sufficient to increase the weight of this pattern doing it to be larger than the critical value, and this pattern will be recognized. It is possible that at the same time some other patterns cease to be recognized. (They are those whose weights are only slightly exceeds the critical value r_c). Such replacement of patterns by other ones does not contradict to the common sense. It corresponds to the general conception of the human memory.

Acknowledgements. The work was supported by the program of the Presidium of the Russian Academy of Sciences (project 2.15) and in part by the Russian Basic Research Foundation (grant 12-07-00295).

References

1. Amit, D., Gutfreund, H., Sompolinsky, H.: Statistical Mechanics of Neural Networks Near Saturation. Annals of Physics 173, 30–67 (1987)
2. Hertz, J., Krogh, A., Palmer, R.: Introduction to the Theory of Neural Computation. Addison-Wesley, Massachusetts (1991)
3. Parisi, G.: A memory which forgets. Journal of Physics A 19, L617–L620 (1986)
4. Nadal, J.P., Toulouse, G., Changeux, J.P., Dehaene, S.: Networks of Formal Neurons and Memory Palimpsest. Europhysics Letters 1(10), 535–542 (1986)
5. van Hemmen, J.L., Keller, G., Kuhn, R.: Forgetful Memories. Europhysics Letters 5(7), 663–668 (1988)
6. Sandberg, A., Ekeberg, O., Lansner, A.: An incremental Bayesian learning rule for palimpsest memory (preprint)
7. Karandashev, Y., Kryzhanovsky, B., Litinskii, L.: Local Minima of a Quadratic Binary Functional with a Quasi-Hebbian Connection Matrix. In: Diamantaras, K., Duch, W., Iliadis, L.S. (eds.) ICANN 2010, Part III. LNCS, vol. 6354, pp. 41–51. Springer, Heidelberg (2010)
8. Karandashev, I., Kryzhanovsky, B., Litinskii, L.: Weighted Patterns as a Tool for Improving the Hopfield Model. Phys. Rev. E 85, 041925 (2012)
9. van Hemmen, J.L., Zagrebnov, V.A.: Storing extensively many weighted patterns in a saturated neural network. J. of Phys. A 20, 3989–3999 (1987)
10. van Hemmen, J.L., Kuhn, R.: Collective Phenomena in Neural Networks. In: Domany, E., van Hemmen, J.L., Shulten, K. (eds.) Models of Neural Networks, pp. 1–105. Springer, Berlin (1992)

An Operational Riverflow Prediction System in Helmand River, Afghanistan Using Artificial Neural Networks

Bernard Hsieh and Mark Jourdan

US Army Engineer Research and Development Center, Vicksburg, Mississippi, 39180, USA
hsiehb@wes.army.mil

Abstract. This study uses historical flow record to establish an operational riverflow prediction model in Helmand River using artificial neural networks (ANNs). The tool developed for this research demonstrates that the ANN model produces results with a very short turn-around time and with good accuracy. This river system used for this demonstration is quite complex and contains uncertainties associated with the historical record. These uncertainties include downstream flow rates that are not always higher than the combined upper stream values and only one continuously operating stream gage in the headwaters. With these characteristics, improvements in the hydrologic predictions are achieved by using a best additional gage search and a two-layered ANN strategy. Despite the gains demonstrated in this research, better simulation accuracy can be achieved by constructing a new knowledge base using more recent information on the hydrologic/hydraulic condition changes that have occurred since the available period for 1979. Follow-on research can also include developing extrapolation procedures for desired project events outside the range of the historical data and predictive error correction analysis.

Keywords: Riverflow prediction, Helmand River, neural networks, prediction improvement analysis.

1 Introduction

When performing an operational riverflow prediction of for a military site; rapid response and high accuracy are required. One such method for delivering answers is a system-based approach, such as Artificial Neural Networks (ANNs). The ANNs model is a mathematical or computational model formulated as a network of simple units, each having local memory. A "signal" is transmitted through the network layers and its value is increased or decreased according to its relevance to the final output. Patterns of signal weights and parameter biases are analyzed to find the optimal pattern that best fits the input parameters and results. Once the network has been trained (calibrated) to simulate the best response to the input data, the configuration of the network is fixed and a validation process is conducted to evaluate the performance of the ANNs as a predictive tool. This research project employs a commercial ANN software tool, NeuroSolution [1], to perform the computations. Three algorithms;

C. Jayne, S. Yue, and L. Iliadis (Eds.): EANN 2012, CCIS 311, pp. 11–20, 2012.
© Springer-Verlag Berlin Heidelberg 2012

namely Multi-layer Feed Forward Neural Networks (MLPs), Jordan and Elman Recurrent Neural Networks (JERs), and Time-Lagged Recurrent Neural Networks (TLRNs) are considered to test the prediction system. The detailed theoretical development for the algorithms can be found on Haykin [2] and Principe, et al, [3].

The most widely used methodologies for water level (stage) forecasting use either a conceptual structure with different levels of physical information, a stochastic structure or a combination of both. These approaches, which became wide spread in the 1990's, started in the 1960s. Since 2000, new types of data-driven models, based on artificial intelligence and soft computing techniques have been more frequently applied. Currently, ANNs are one of the most widely used techniques in the forecasting field (Hsu et al., [4], and Thimuralaiah and Deo [5]). Most applications based on these models consider the discharge as the forecasting variable (Imrie, et al. [6], Dawson et al. [7], Moradkhany et al. [8], Hsieh and Bartos [9], and Kisi,[10]), primarily because of historical contiguity with the classes of conceptual and physical based rainfall-runoff models. Such an approach requires the knowledge of the rating curve in the cross section of interest to parameterize the model. However, the knowledge of the stage is required within the framework of a flood warning system and thus the rating curve has to be used also to transform the forecasted flows into stages.

Several previous research studies identified the use of a special design to perform the pre-processing of the input variables as the first stage to build the flow forecasting system. Moradkhaqni et al. [8] explored the applicability of a Self Organizing Radial Basis (SORB) function to one-step ahead forecasting of daily stream flow. The architecture employed consisted of SOFM as an unsupervised training scheme for data clustering, which correspondingly provides the parameters required for the Gaussian functions in RBF neural network. Spread of the Gaussian functions extracted from SOFM seemed to be tunable, and tuning was done in parallel to training the RBF network. A flow region specific flow forecasting approach that identified improvements of prediction for typical high flows was developed by Sivapragasam and Liong [11]. Attributes were decided based on the underlying hydrological process of the flow region and the model was implemented by the Support Vector Machine (SVM) learning algorithm. This research was applied to Tryggevaelde Catchment for 1- and 3-lead days and promising results were obtained, particularly for high flow in a 3-lead day model.

The Helmand River Basin (Figure 1) is the largest river basin in Afghanistan and the river stretches for 1,150 km. The Helmand River Basin is a desert environment with rivers fed by melting snow from high mountains and infrequent storms. Great fluctuations in stream flow, from flood to drought, can occur annually. The purpose of this study is to identify the best river flow prediction model, from the upstream Kajakai reservoir gauge to downstream locations, for providing the guidance of military operation plans. We also developed a daily operational tool for performing predictions using a laptop or desktop computer. River discharge measurements began in Afghanistan in the mid-1940s at a few sites (USGS). Measurements were discontinued soon after the Soviet invasion of Afghanistan in 1979. Discharge for five downstream gauges – Lashkargeh, Darweshan, Marlakan, Burjak, and Khabgah with two tributaries, the Musa Qala and Arghndab Rivers – were collected during this period. However, only Kajaki and Burjak gauges, as well as the two tributaries, had sufficiently complete data. The knowledge base for developing the ANNs model included daily discharge measurements from 10/01/1953 to 09/30/1979. Missing data were filled by using ANNs algorithms with high correlation relationships.

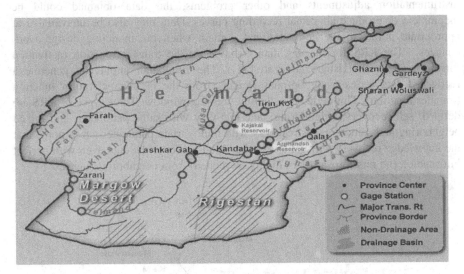

Fig. 1. Helmand River Basin and its flow gauges

2   Knowledge Base Development for Helmand River Flow Discharge and Basic Data Analysis

To develop a good data-driven model, a knowledge base with minimum uncertainty is a critical factor to make the first successful step. A schematic drawing representing the flow gauges in the Helmand River and two tributaries (Musa Qala River and Arghastan River,) used for building this operational flow prediction model, is shown in Figure 2. Other than those regular USGS flow gauges, there are two artificial points: "K+M," approximately representing the total flow combined by Kajakai and Musa Qala gauges, and "L+Q," representing the total flow merged by Lashkargah and Qala-i-Bust gauges. The reason to set up these two control points is to check how these two tributaries will contribute downstream flow from historical events since there is only one upstream gauge (Kajaki) currently collecting the measurement.

2.1   Knowledge Recovery System

In the historical records for these 8 gauges, about 40% of data from the main stem of Helmand River (Figure 5) is missing. Those missing windows have to be filled before performing the analysis. The knowledge recovery system KRS (Hsieh and Pratt [11]) has been applied to several tidal and riverine ERDC projects. KRS deals with activities that are beyond the capabilities of normal data recovery systems. It covers most of the information that is associated with knowledge oriented problems, and is not limited to numerical problems. Almost every management decision, particularly for a complex and dynamic system, requires the application of a mathematical model and the validation of the model with numerous field measurements. However, due to instrumentation adjustments and other problems, the data obtained could be incomplete or produce abnormal recording curves. Data may also be unavailable at appropriate points in the computational domain when the modeling design work changes. The KRS for missing data is based on the transfer function (activation function) approach (Hsieh and Pratt [11]). The simulated output can generate a recovered data series using optimal weights from the best-fit activation function (transfer function). This series is called the missing window. Three types of KRS are defined: self-recovery, neighboring gage recovery with same parameter, and mixed neighboring and remote gages recovery with multivariate parameters. It is noted that the knowledge recovery will not always be able to recover the data even with perfect information (no missing values or highly correlated) from neighboring gages. .

Fig. 2. A schematic of flow gauges for building Helmend River operational flow Prediction Model

2.2   KRS for Helmand River Flow Data

To develop the completed knowledge base, the missing window (6/1/57 – 9/30/72) for Lashkargah flow gauge is first applied by KRS. Instead of using either the Kajakai flow or Darweshan flow to recover this missing window, due to a stronger bonding structure it is better to use both flow gauges as inputs and Lashkargah flow as output during the period between 10/01/72 and 09/30/1979 to build a transfer function (ANNs model). Three algorithms – MLPs, TLRNs, and JERNs (defined as the first paragraph in the introduction section) – are used to compare which one has best generalization. The following system parameters for filling above mentioned missing window are the examples in which to use TLRNs.

System Structure: 2 inputs – 1 output system
Training Exemplars =2550; Total Iterations = 2500
Hidden Layer = 1
    Activation Function = TanhAxon
    Learning Rate = Momentum; Step Size= 1.0; Momentum = 0.70
Output Layer
    Activation Function = TanhAxon
    Learning Rate = Momentum; Step Size= 0.1; Momentum = 0.70
Memory = GammaAxon Function; Depth in Samples = 10

This training (TLRNs) process is examined by MSE (Mean Square Error) versus Epoch through each iteration and training result are very good (correlation coefficient (CC) = 0.97). The comparison of training results among three selected algorithms and the measure of statistical parameters – MSE, NMSE (Normalized Mean Square Error), MAE (Mean Average Error), Min Abs E (Minimum Absolute Error), Max Abs E (Maximum Absolute Error), and CC (Table 1) indicate the MLPs obtain a slightly better performance, but the TLRNs present clearer time-delay response from upstream to downstream flow transport phenomenon. The TLRNs are selected as main algorithm for conducting the remaining study. After this training process has been done, the missing window can be simulated by using a set of weights for Malakan flow gauge as Figure 3. The developed KRS procedures can be applied to other missing windows with the least uncertainty involved. The blue line represents the best estimates for missing windows while the pink line shows existing data points. The same procedures are used to simulate the remaining missing windows for other downstream gauges.

Table 1. Performance comparison for three training algorithms (MLP, TLRNs, and JERNs)

Performance/Networks	MLPs	JERNs	TLRNs
MSE	1030	1085	1776
NMSE	0.034	0.036	0.059
MAE	21.56	21.80	28.24
Min Abs E	0.006	0.014	0.051
Max Abs E	474.95	503.74	495.90
Corre. Coeff.	0.982	0.981	0.976

3   An Operational Riverflow Prediction System in Helmand River

The main goal for this project is to develop an operational downstream riverflow prediction data-driven model when only the most upper stream Kajaki reservoir flow is given. In this section, three operational models are presented. The first is for only the relationship between the Kajakai reservoir flow and the operational downstream location. The other two also consider to the impacts of tributary inflows.

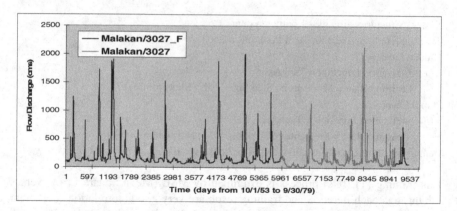

Fig. 3. Modeled and measured flow for the Malakhan gauge. The pink portion is the measured data, the blue is the data obtained from the KRS.

According the historical record after KRS processing, the best available data covers the riverflow from Kajaki reservoir to 5 downstream gauges and 2 tributaries during the period of 10/1/53 to 09/30/79 (historical knowledge base). These data are used to perform this operational model development. To test the operational model, the newly collected data below Kajakai from 10/01/09 to 04/23/10 is adopted. The historical knowledge base is divided into a training set (10/01/53 – 09/30/73) and a verification set (10/01/73 – 09/30/79). The neural networks architecture is TLRs with 1x5 (one input – five outputs) structure. The correlation coefficients for training are between 0.95 (Lashkargen) and 0.89 (Char Burjak). The test results are between CC (Correlation Coefficient) 0.96 for Lashkargen (Figure 4) and CC 0.87 for Khwabgan. The operational flow prediction (203 days) is then simulated as Figure 5. This simulation only takes seconds (20 seconds) to be done once the operational prediction model has been trained. The total processing time, including model training, new data transfer, and simulation mode, should be less than around 10 minutes. For the future use, the prediction model does not need to be trained again if the prediction horizon is short (e.g. within a year) or if no significant hydrologic/hydraulic conditions have changed.

As it has been stated in the previous section, two tributaries (Musa Qala River and Arghastan River) merge into the upper main stem of the Helmand River. To examine the impact of the tributary flows on the downstream river flows, the model is adapted to include two inflow data set scenarios. The second scenario deals with two additional artificial points ("K+M" and "L+Q"), the third scenario only takes these two additional points as well as three downstream gauges. Thus, the second scenario has 1x7structure, and the third scenario remains as 1x5 structure. Comparison of correlation coefficients among these three scenarios is summarized as Table 2. Except the gauge Char Burjak, which reduces the correlation coefficient for training of the third scenario runs, no significant variations are found for other conditions. More detailed investigations, such as comparing the overall time history variation as well as other statistical parameters for the performance, may better identify more significant correlation impacts.

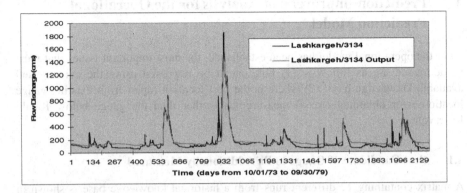

Fig. 4. Test results for the operational flow prediction model (gauge Lashkargen) from 10/01/73 to 09/30/79

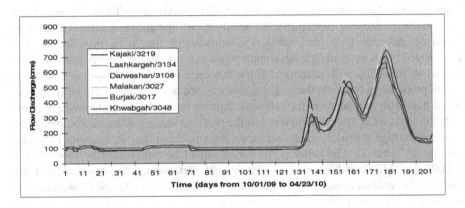

Fig. 5. Results of operational riverflow prediction model during the period from 10/01/09 to 04/23/10 for 5 downstream gauges when the flow below Kajakai Reservoir is provided

Table 2. Comparison of correlation coefficients among three prediction model scenarios

	Training 1	Test 1	Training 2	Test 2	Training 3	Test 3
K+M	N/A	N/A	0.970	0.972	0.973	0.976
Lashkargeh	0.949	0.962	0.948	0.960	N/A	N/A
L+Q	N/A	N/A	0.900	0.901	0.888	0.892
Darweshan	0.914	0.876	0.910	0.876	N/A	N/A
Malakan	0.929	0.870	0.928	0.871	0.931	0.871
Burjak	0.890	0.855	0.887	0.854	0.848	0.848
Khwabgah	0.912	0.870	0.909	0.868	0.908	0.861

4   Prediction Improvement Analysis for the Operational Prediction Model

After the operational model has been established, the most important issues that must be addressed are as follows: (1) how can it be improved using the existing and available knowledge base; (2) where is the best location (apart from existing gauge locations) for obtaining stream measurements other than the gauge below Kajaki Reservoir.

4.1   Best Candidate of Gauge to Be Additionally Measured

A matrix containing 12 different runs from a historical knowledge base is shown in Table 3. This table indicates which gauge is used as input (use symbol I) and which gauge is represented as output (use symbol O). From the design, it consists of one 1x7, five 2x6, one 1x5, and five 2x4 runs. The objective is to see the overall performance from selected statistical parameters (MSE, NMSE, MAE, and CC) and to select the best possible candidate to improve the reliability of operational model. The idea is based on the concept of the boundary condition that has to be provided from a generalized numerical model development during the model set-up process. The prediction reliability will be reduced if the influence of the input signal is dampened due to propagation loss over the long distance traveled to reach the output location. It is noted that these runs perform the full-scale training (entire length of knowledge base) with the same set of system parameters as the previous section. After the completion of runs the average statistical parameters are computed from the output series (The last column of Table 3). The better performance should be represented by lower MSE, NMSE, and MAE, and higher CC. The overall performance strongly indicates the gauge at Darweshan is the best candidate to be selected as an additional measurement location. This is shown by the high CC in run 9.

Table 3. Average corresponding outputs statistical parameters for 12 designed runs

Run	Kaj.	K+M	Las.	L+Q	Dar.	Mal.	Bur.	Khw.	CC
1	I	O	O	O	O	O	O	O	0.916
2	I	O	O	O	I	O	O	O	0.967
3	I	O	I	O	O	O	O	O	0.943
4	I	O	O	O	O	I	O	O	0.957
5	I	I	O	O	O	O	O	O	0.934
6	I	O	O	I	O	O	O	O	0.963
7	I	N/A	O	N/A	O	O	O	O	0.911
8	I	N/A	I	N/A	O	O	O	O	0.944
9	I	N/A	O	N/A	I	O	O	O	0.970
10	I	N/A	O	N/A	O	I	O	O	0.952
11	I	N/A	O	N/A	O	O	I	O	0.964
12	I	N/A	O	N/A	O	O	O	I	0.951

4.2   A Two-Layered Operational Flow Prediction Model

With the above two examinations (best additional measurement gauge and error correction analysis), a two-layered ANN model is proposed to improve the present capability of the prediction model. The first layer uses a 1x1 structure (Kajakai reservoir flow as input and Darweshan gauge flow as output) to simulate the prediction flows at Darweshan. Then it again uses Kajakai Reservoir flow as well as the simulated Darweshan gauge flow as inputs to predict the flow at the other four gauges. The ANN structure for the second layer of the prediction model is 2x4. Figures 6 shows the training results for downstream gauge at Lashkargeh The performance improvement between original 1x5 and 1x1 – 2x4 combination is summarized as Table 4. Although the performance for the prediction of Darweshan can be neglected, the rest of gauges gain significant improvement in predictive ability. This test demonstrates how the two-layered operational flow prediction model can generate an increase in predictive capacity.

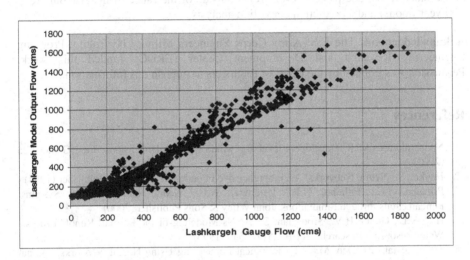

Fig. 6. Results of test performance for a 1x1 – 2x4 ANNs model (Lashkargen gauge)

Table 4. Performance comparison between original 1 x 5 systems and alternative 2 x 4 approach

Gauges	MSE (1 x 5)	CC (1 x 5)	MSE (2 x 4)	CC (2 x 4)
Lashkargeh	3255	0.949	1512	0.976
Darweshan	9964	0.914	10040	0.913
Malakan	7487	0.929	927	0.992
Burjak	12184	0.890	5470	0.953
Khwabgah	9580	0.912	4329	0.961

5   Conclusions

This study uses historical flow records and ANNs to construct an operational riverflow prediction model in Helmand River, Afghanistan. The developed tool demonstrates this model has a very short turn-around run time as well as good accuracy. However, some significant uncertainties from the historical record show the impact of tributary flow rates on downstream flow rates as well as channel transmission losses – the downstream flow rates are not always higher than the upper stream flow rates. In addition, there is only one upper stream gauge currently being operated. To improve the predictive reliability, some two external factors need to be considered, adding a measurement location and improving the modeling by a proposed two-layered ANN model. Since the performance of data-driven modeling relies heavily on the quality of the information, it is critical to construct a new knowledge base using more recent information due to hydrologic/hydraulic condition changes beyond the existing historical record. The extrapolation procedures should be extended to include project-level events outside of the range of historic data (e.g. 100-year floods) and a prediction correction analysis.

Acknowledgement. The U.S. Army Corps Engineers Military Hydrology Program, Engineering Research and Development Center (ERDC) funded this work. Permission was granted by the Chief of Engineers to publish this information.

References

1. Neurosolutions v5.0: Developers level for window. NeuroDimensions, Inc., Gainesville (2003)
2. Haykin, S.: Neural Networks: A Comprehensive Foundation. Macmillan Publisher (1994)
3. Principe, J.C., Euliano, N.R., Lefebvre, W.C.: Neural and Adaptive Systems: Fundamentals through Simulations. John Wiley & Sons Publisher (2000)
4. Hsu, K.L., Gupta, H.V., Srooshian, S.: ANNs Modeling of the Rainfall-Runoff Process. Water Resources Research 31(10), 2517–2530 (1995)
5. Thimuralaiah, K., Deo, M.C.: Hydrological Forecasting Using Neural Networks. Journal of Hydrologic Engineering 5(2), 180–189 (2000)
6. Imrie, C.E., Durucan, S., Korre, A.: River Flow Prediction Using ANNs: Generalizations beyond the Calibration Range. Journal of Hydrology 233, 138–153 (2000)
7. Dawson, C.W., Harpham, C., Wilby, R.L., Chen, Y.: Evaluation of ANNs Technique in the River Yangtze, China. Hydrology, Earth System Science 6, 6129–6626 (2002)
8. Hsieh, B., Bartos, C.: Riverflow/River Stage Prediction for Military Applications Using ANNs Modeling, ERDC/CHL Technical Report, TR-00-16 (2000)
9. Kisi, O.: River Flow Forecasting and Estimation Using Different Artificial Neural Network Techniques. Hydrology Research 39(1), 27–40 (2008)
10. Sivaprgasam, C., Loing, S.-Y.: Flow Categorization Model for Improving Forecasting. Nordic Hydrology 36(1), 37–48 (2005)
11. Hsieh, B., Pratt, T.: Field Data Recovery in Tidal System Using ANNs, ERDC/CHL, CHTN-IV-38 (2001)

Optimization of Fuzzy Inference System Field Classifiers Using Genetic Algorithms and Simulated Annealing

Pretesh B. Patel and Tshilidzi Marwala

Faculty of Engineering and the Built Environment, University of Johannesburg,
P.O. Box 524, Auckland Park, 200, Johannesburg, South Africa
{preteshp,tmarwala}@uj.ac.za

Abstract. A classification system that would aid businesses in selecting calls for analysis would improve the call recording selection process. This would assist in developing good automated self service applications. This paper details such a classification system for a pay beneficiary application. Fuzzy Inference System (FIS) classifiers were created. These classifiers were optimized using Genetic Algorithm (GA) and Simulated Annealing (SA). GA and SA performance in FIS classifier optimization were compared. Good results were achieved. In regards to computational efficiency, SA outperformed GA. When optimizing the FIS 'Say account' and 'Say confirmation' classifiers, GA is the preferred technique. Similarly, SA is the preferred method in FIS 'Say amount' and 'Select beneficiary' classifier optimization. GA and SA optimized FIS field classifier outperformed previously developed FIS field classifiers.

Keywords: Classification, fuzzy inference system, interactive voice response, optimization, genetic algorithm, simulated annealing.

1 Introduction

Large and small businesses are adopting Interactive Voice Recognition (IVR) technology within their call centers due to the benefits of cost reduction and increased productivity. In order to support the growth of the adoption rate, callers must be satisfied to use these systems.

Currently, call centres receive thousands of calls daily. In order to evaluate caller behaviour within IVR applications, many call centres select random call recordings for analysis. The aim of this research is to develop a classification system, using computational intelligent techniques that could aid businesses in identifying calls to further analyze. It is anticipated this would improve the call recording selection method used.

IVR system is an automated telephony system that interacts with callers to gather relevant information to route calls to the appropriate destinations [1]. These systems have also been used to automate certain business transactions such as account

C. Jayne, S. Yue, and L. Iliadis (Eds.): EANN 2012, CCIS 311, pp. 21–30, 2012.

transfers. The inputs to the IVR system can be voice, Dual Tone Multi-Frequency (DTMF) key-pad selection or both. IVR systems can provide appropriate responses in the form of voice, fax, callback, e-mails and other media [1]. Generally, an IVR system solution consists of telephony equipment, software applications, databases and supporting infrastructure.

Voice Extensible Markup Language (VXML) is a speech industry adopted standard. The IVR application defined in VXML are voice-based dialog scripts comprising of form and dialog elements. These elements are used to group input and output sections together. Field elements are used to obtain, interpret and react to user input information [2]. Field elements are found in form elements. Further information on field elements can found in [2].

Recent research illustrated that Artificial Neural Network (ANN) and the Support Vector Machine (SVM) techniques outperformed the Fuzzy Inference System (FIS) technique in call classification [3]. This research examines the use of Genetic Algorithm (GA), and Simulated Annealing (SA) to optimize the membership functions and set of fuzzy inference rules of the FIS model.

GA is an evolutionary heuristic that employs population-based search method. The algorithm conducts an exploration of the search space to identify optimum solutions by utilizing a form of direct random search process. The mechanism of natural evolution is an inspiration to GA. It is an effective global search method with good applications across different industries [4][5].

SA is an iterative approach that continuously updates a single candidate solution to reach a termination condition. SA is a member of the probabilistic hill-climbing class of algorithms. These algorithms dynamically alter the probability of accepting inferior cost or fitness values [6]. This approach has also been used in a range of applications across various industries [7][8].

The following section examines the classification system and the data used in this implementation. FIS classifiers explanation follows. Thereafter, the GA and SA optimization of the FIS classifiers are described. The paper ends with a comparison of the results of the optimization methods and the selection of the appropriate technique.

2 Classification System

The objective of the classification system is to identify calls to be used to determine areas of improvement within the IVR applications. In the proposed system, field classifiers are trained on data extracted from IVR log event files. When a call is placed to the IVR system, the platform generates these files. Data such as call begin, form enter, automatic speech recognition events and call disconnect information are written to the log files [9].

Field classifier inputs and outputs are shown in Table 1. The inputs characterize the caller experience at a field within an IVR application. Interaction classes, which summarize the caller behaviour, form the outputs of the classifiers.

Table 1. Field classifier inputs and outputs

Inputs	Outputs	Output interaction class
Confidence	Field performance	Good, acceptable, investigate, bad
No matches	Field transfer reason	Difficulty, no transfer, unknown
No inputs	Field caller disconnect reason	Difficulty, no transfer, unknown
Max speech timeouts	Field difficulty attempt	Attempt 1, attempt 2, attempt 3
Barge-ins	Field duration	High, medium, low
Caller disconnect	Field recognition level	High, medium, low
Transfer to CSA	Experienced caller	True, false
DTMF transfer		
Duration		
System error		
Confirmed		

Confidence input, a percentage value, indicates the IVR speech recognition probability. IVR applications accommodate a finite number of caller responses to automated questions. When caller responses are not within the specified answers, no match events occur. These events assist in identifying calls where the caller misunderstood the self service prompt and unique responses that the IVR application did not accommodate. As a result, identifying these types of calls assists in improving the IVR application field recognition coverage. In this research, the IVR application caters for 3 no match events per field. Thereafter, if the caller is still not successful in completing the field, the call is transferred to a Customer Service Agent (CSA).

At times, the caller may remain silent in response to a prompt. No input events represent these occurrences. Identifying calls with these events assists in determining whether the prompt is confusing for majority of callers thus resulting in the caller not responding. Similar to the no match events, the IVR application provides 3 no input events per field. After the third no input event, the caller is transferred to a CSA.

When caller response duration is larger than the allocated timeout period of a field, a maximum speech timeout event occurs. Identifying calls comprising of these events will assist in evaluating the timeout period of fields. The IVR application accommodates 3 maximum speech timeout events per field. Thereafter, the call is also transferred to a CSA.

A caller may interrupt the application prompt play. The barge-in inputs represent these events. Caller disconnects and transfers to CSA inputs indicate instances when caller terminates the call and when a call is transferred from the IVR application to a CSA, respectively. During a call, when a caller selects the hash (#) key, the input mode is changed. DTMF transfer input represents these instances. Identifying calls comprising of these events, assist in determining the various levels of difficulty the callers experience at the different fields within the IVR application.

Duration and system error inputs indicate the total time spent and system errors experienced in a field, respectively. When the speech recognition is low, a confirmation prompt to verify the caller intentions is played. The confirmation input represents the caller response to the confirmation prompt. This input is also used to indicate whether or not the prompt is played to the caller.

The classification system provides a summarized interpretation of caller behaviour at a field. This achieved by defining output interaction classes. Table 1 illustrates these classifier outputs. Field performance interaction class categorizes caller behaviour into good, acceptable, investigate or bad. Field durations and recognition level outputs illustrates 3 categories of performance, low, medium and high.

Field transfer and caller disconnect reason outputs indicate whether or not transfer to CSA and caller disconnect events occurred, respectively. These outputs also provide possible explanation for the transfer or caller disconnect event. Field difficulty attempt output computes the number of no match, no input and maximum speech timeout events that were experienced. Thus, these classifier outputs provide further detail of caller behaviour and assists in selecting various types of calls to further analyze. The experienced caller output uses caller difficulty inputs such as no match events, caller end status, duration and recognition information to classify whether or not the caller is a regular user and familiar with the application call flow. This output assists in determining the usage of the IVR application.

A 3 digit binary word notation is used to provide the no match, no input and maximum speech timeout data to the classifiers. A bit binary word notation is employed to provide the barge-in, caller disconnect, transfer to CSA, DTMF transfer and system error inputs. The confirmed input is presented to the classifier using a 2 digit binary word notation. The classifier interaction outputs were interpreted using a similar binary notation scheme.

In order to precondition the data for classifier inputs, the confidence and duration inputs were normalized. Training, validation and test sets are used to ensure over-fitting and under-fitting were avoided. The training dataset is used to teach the classifiers to determine general classification groups in the data. Validation and test datasets are used to evaluate the classifier on unseen data and to confirm the classification capability of the developed models, respectively.

This paper is concerned with the development of field classifiers. At the 'Say account' field element in an IVR pay beneficiary telephone banking application, the caller is asked to respond with the account name or number. When a call enters the 'Say amount' field, the caller is requested to say the actual amount to pay a beneficiary. The beneficiary is selected within the 'Select beneficiary' field. At the 'Say confirmation' field, the recognized information is read to the caller. The caller is also asked to approve the transaction. Due to the varying responses required at the fields with the IVR application, caller behaviour is unique per field. Therefore, field classifiers specifically trained on data relevant to the field are developed.

3   Fuzzy Inference System

FIS, based on fuzzy logic, is a technique that computes outputs based on the fuzzy inference rules and present inputs. FIS use fuzzification, inference and defuzzification processes [10]. Refer to [10] for further information about these processes.

Numerical data clustering establishes the basis of many classification and system modeling applications. The objective of clustering is to locate natural groupings in a set of inputs. Fuzzy inference rules computed using data clustering are specifically tailored to the data. As a result, when compared to FIS developed without clustering; this is an advantage [11].

In this research, the optimal number and form of fuzzy inference rules are determined using subtractive clustering [12]. A subtractive clustering implementation is dependent on parameters such as the cluster radius, squash factor and reject parameters [13]. When using a small cluster radius values, many cluster centers in the data are created. Similarly, when using a large cluster radius values, few cluster centers are produced.

The squash factor parameter is used to specify the squash potential for outlying points to be considered as a part of a specific cluster. In this research, a squash factor parameter equal to 1.25 is used. This ensures only clusters far apart are found. Accept and reject ratios are used to set the potential above and below which a data point can be accepted or rejected as a cluster center, respectively. In this research, accept and reject radios set to 0.9 and 0.7 is used, respectively. This ensures only data points with strong potentials for being a cluster center are accepted.

Caller behaviour classification is a multi-dimensional problem. GA and SA are used to determine the optimal FIS classifier. These techniques are used to identify the optimal cluster radius values for each input. GA and SA employed an evaluation function that mapped the radius values to the validation and test dataset accuracy values of the inference system. The minimum value of these accuracies determined fitness of the individual. The accuracy of the classifier is determined utilizing a confusion matrix that identifies the number of true and false classifications produced by the FIS developed. This is then utilized to calculate the true accuracy of the classifiers using equations in [3]. The sections to follow examine the GA and SA implementations.

4   Genetic Algorithm

GA manages, maintain and manipulate populations of solutions using a survival of the fittest strategy to identify an optimal solution [14]. However, similar to biological evolution, inferior individuals can also survive and reproduce.

The chromosome representation scheme determines the manner in which the classification problem is structured in the GA. This research used a binary and real coded chromosome representation. GA utilizes a probabilistic selection function to determine the individuals that will proceed onto the next generation. In this research, normalized geometric ranking and tournament selection functions are used within GA implementations. The types of crossover and mutation genetic operators employed are dependent on the chromosome encoding used within the GA [15]. The binary coded GA used binary mutation and simple cross over genetic operators. The real coded GA utilized non-uniform mutation and arithmetic cross over genetic operators [16].

GA advances from generation to generation, until a termination criterion such as maximum number of generations reached or lack of improvement of the best solution for a number of generations is achieved. In this research, the numbers of individuals within the population were varied. GA produced 25 generations of these populations with population sizes of 30, 100, 200, 300, 400 and 500. Table 2 illustrates the results of the most accurate GA implementations.

Table 2. Results of most accurate GA implementations. In this table, size, executions and ranking represent population size, evaluation function executions and normalized geometric ranking, respectively.

FIS classifier	Encoding	Selection function	Size	Accuracy(%)	Executions
'Say account'	binary	Ranking	300	95.660	7500
	real		400	95.649	580
	binary	Tournament	30	95.660	750
	real		300	95.649	490
'Say amount'	binary	Ranking	30	93.164	750
	real		30	93.164	215
	binary	Tournament	30	93.164	750
	real		30	93.164	222
'Say beneficiary'	binary	Ranking	100	95.777	2500
	real		100	95.680	280
	binary	Tournament	100	95.747	2500
	real		100	95.680	290
'Say confirmation'	binary	Ranking	30	91.482	750
	real		30	91.482	216
	binary	Tournament	30	91.482	750
	real		30	91.482	220

In the FIS 'Say account' field classifier optimization, the binary and the real coded GA converged to a solution value of 95.660% and 95.649%, respectively. However, during FIS 'Say account' field classifier optimization, the real coded GA using tournament selection function executed the evaluation function the least. Additionally, the binary coded GA that employed the tournament selection function achieved the solution value of 95.660% with the smallest population size of 30.

When optimizing FIS 'Say amount' and 'Say confirmation' field classifiers, both encodings converged to a solution value of 93.164% and 91.482%, respectively. The binary coded GA produced a solution value of 95.777% in 'Select beneficiary' field classifier optimization.

In all classifier optimizations considered, the real coded GA executed the evaluation function the least. Additionally, a population size of 30 is most suitable for FIS 'Say amount' and 'Say confirmation' field classifier optimization. A population size of 100 is appropriate for FIS 'Select beneficiary' classifier optimization.

5   Simulated Annealing

SA is derived from the annealing process used to produce an optimal metal crystal structure [17]. In order to allow inputs to assume a wide range of values, SA utilizes a high initial temperature value. At iterations, the range of inputs is restricted by gradually decreasing the temperature using an annealing schedule.

SA begins by computing the fitness value, using an evaluation function, of a randomly chosen point. A new point is selected from a random number generator. When comparing the fitness values of the new point and the previously selected point, the new point is chosen if its fitness value is better. When the new point fitness value is worse than the previously selected point, the new point is accepted based on a computed acceptance function probability. A new point is always chosen based on the best point. At high iteration values, the probabilities of large deviations from the best point and the probabilities of acceptance decrease. Therefore, SA initially ensures the exploration of distant points. However, as the temperature is reduced utilizing an annealing schedule, the algorithm does not generate or accept distant points.

In this research, standard Boltzmann, exponential and fast annealing schedules were considered [18]. Additionally, the number of iterations is also varied from 1 to 100. Initial temperature values of 1 and 100 were also considered. Table 3 illustrates the results of the SA implementation.

The initial temperature of 1 yielded good results for 'Say account', 'Select beneficiary' and 'Say confirmation' FIS field classifier optimization. This is also true for all annealing schedules considered. When optimizing the 'Say amount' field classifier, both initial temperatures considered achieved good results with accuracy values of 93.164%. These annealing schedules executed the evaluation function the same number of times.

All annealing schedule achieved the same accuracies of 93.164% and 91.482% for 'Say amount' and 'Say confirmation' FIS field classifier optimization, respectively. 'Say account' field classifier optimization illustrated that the exponential annealing schedule achieved the largest accuracy and executed the evaluation function the least. The exponential annealing schedule executed the evaluation function the least in 'Select beneficiary' field classifier optimization. Both Boltzmann and exponential annealing schedules executed the evaluation function the least in 'Say confirmation' field classifier. When optimizing FIS 'Say amount' and 'Say confirmation' field classifier, the numbers of iterations used by all annealing schedules were low.

The initial temperature of 100 did not produce the most accurate FIS in majority of the SA implementations. This can be attributed to initial temperature of 100 resulting in the exploration of points very far from the optimal value. As a result, SA does not converge to the optimal value as this temperature is reduced.

Table 3. Results of most accurate SA implementations. In this table, annealing, initial temp and executions represent annealing schedule, initial temperature and evaluation function executions, respectively.

FIS classifier	Iterations	Annealing	Initial Temp	Accuracy(%)	Executions
'Say account'	86	Boltzmann	1	95.659	88
	78	Exponential	1	95.660	80
	99	Fast	1	95.659	101
'Say amount'	1	Boltzmann	1 and 100	93.164	3
	1	Exponential	1 and 100	93.164	3
	1	Fast	1 and 100	93.164	3
'Say beneficiary'	100	Boltzmann	1	95.699	102
	9	Exponential	1	95.680	11
	67	Fast	1	95.684	69
'Say confirmation'	4	Boltzmann	1	91.482	6
	4	Exponential	1	91.482	6
	5	Fast	1	91.482	7

6   Comparison of Optimization Results

GA and SA are compared in terms of computational efficiency and quality of solution. In regards to this classification problem, computational efficiency is the number of evaluation function executions the algorithms used to converge to an optimal solution. In all FIS field classifier optimizations, SA converged to an optimal solution using the least number of evaluation function executions. As a result, SA is most computationally efficient. Quality of solution is the confirmation that the optimal FIS field classifier yielded by the optimization techniques is truly the most optimal solution. This is confirmed by computing accuracy, sensitivity and specificity values as defined in [3]. Table 4 illustrates these performance measures.

When presented with the test dataset, the FIS 'Say account' and 'Say confirmation' field classifiers produced by GA yielded the better accuracy value. The sensitivity and specificity values confirm this. In FIS 'Say amount' field classifier optimization, GA and SA yielded the same results for validation as well as test datasets. When optimizing 'Select Beneficiary' field classifier, GA produced the better classifier accuracy value for validation dataset. However, SA yielded the better classifier accuracy value for test dataset. The difference between the validation and test dataset accuracy values yielded by the FIS 'Select beneficiary' classifier optimized using both techniques is 0.078 and 0.064, respectively. Therefore, in regards to quality of solution, SA outperforms GA for 'Select beneficiary'.

It is evident from the high sensitivity and specificity values that the FIS field classifiers developed possess high positive and negative classification rates, respectively. However, the number of evaluation function executions used by GA and SA differ. As illustrated in Table 2 and 3, SA executes the evaluation function less than GA. Also, the field classifiers optimized using SA are less than 0.05% accurate than classifiers developed using GA. Therefore, in this research, when computational efficiency is a priority, SA would be preferred for all field classifier optimizations. Additionally, acceptable classification accuracy values can be achieved.

Table 4. Performance measures of optimization algorithms considered. In this table, annealing and initial temp represent annealing schedule function and initial temperature, respectively.

FIS classifier	Algorithm	Accuracy (%)		Sensitivity (%)		Specificity (%)	
		Validation	Test	Validation	Test	Validation	Test
'Say account'	GA	95.660	96.641	94.452	96.062	96.884	97.224
	SA	95.660	96.624	94.452	96.038	96.884	97.213
	Previous	80.680	80.770	72.000	66.150	99.430	98.630
'Say amount'	GA	93.962	93.164	91.561	89.800	96.425	96.654
	SA	93.962	93.164	91.561	89.800	96.425	96.654
	Previous	82.650	82.540	70.220	68.470	98.600	99.510
'Say beneficiary'	GA	95.777	95.951	93.333	93.963	98.285	97.980
	SA	95.699	96.015	93.109	93.654	98.361	98.437
	Previous	78.430	77.820	61.170	62.730	99.000	99.080
'Say confirmation'	GA	91.482	92.147	85.815	86.901	97.524	97.711
	SA	91.482	92.136	85.815	86.879	97.524	97.711
	Previous	79.510	79.470	71.380	65.760	99.580	96.040

However, if small accuracy improvements are important in this classification system, GA is preferred technique for FIS 'Say account' and 'Say confirmation' field classifiers. Similarly, SA is the preferred technique for FIS 'Say amount' and 'Say beneficiary' classifier optimization. Research conducted has concluded that SA is more effective than GA [19] and vice versa [8]. Therefore, the performance of these techniques is very dependent on the optimization problem under consideration.

Results of previous FIS field classifiers developed are also provided in Table 4 [3]. These field classifiers used the same cluster radius for each input or dimension. As illustrated in Table 4, GA and SA has provided a significant improvement on the accuracy of FIS field classifiers.

7   Conclusion

GA and SA techniques considered in this research produced highly accurate FIS field classifiers. In order to generate inference rules specifically tailored to the data, subtractive clustering is used. The evaluation function utilized by GA and SA is an error function that maps the cluster radii to the FIS classifier accuracy values of validation and test datasets.

Binary and real coded GA that employed normalized geometric and tournament selection functions were used to compute optimal cluster radius values. The population sizes used within GA were varied. In FIS 'Say account' and 'Select beneficiary' field classifier optimization, binary coded GA outperformed real coded GA. Both GA encodings produced the same FIS 'Say amount' and 'Say confirmation' classifier results. Real coded GA executed the evaluation function the least.

In this research, Boltzmann, exponential and fast annealing schedules as well as initial temperature values of 1 and 100 were considered. The numbers of iterations used by SA were also varied. When optimizing majority of the FIS field classifiers, an initial temperature value of 1 achieved better results. Good results were achieved by exponential and Boltzmann annealing schedules in FIS 'Say account' and 'Select beneficiary' classifier optimizations, respectively. 'Say confirmation' field classifier optimization determined that exponential and Boltzmann annealing schedules yielded the same results. The same results were obtained by all annealing schedules and initial temperature values in FIS 'Say amount' classifier optimization.

When computational efficiency is a priority, SA is the preferred technique for all FIS field classifier optimizations. SA will achieve acceptable classification accuracy values. However, when small accuracy improvements are of greater importance than computational efficiency, GA is preferred technique for FIS 'Say account' and 'Say confirmation' field classifiers. Similarly, SA is the preferred optimization method for FIS 'Say amount' and 'Select beneficiary' field classifiers.

When comparing this research to previous research conducted, the optimization of FIS has yielded improvements in the accuracy values of the field classifiers.

References

1. Nichols, C.: The Move from IVR to Speech – Why This is the Right Time to Make the Move to Speech Applications in Customer-Facing Operations. Intervoice (2006)
2. VoiceXML 2.0/VoiceXML 2.1 Reference, http://developer.voicegenie.com/voicexml2tagref.php?tag=st_field&display=standardtags
3. Patel, P.B., Marwala, T.: Caller Behaviour Classification Using Computational Intelligence Methods. Int. Journal of Neural Systems 20(1), 87–93 (2010)
4. Panduro, M.A., Brizuela, C.A., Balderas, L.I., Acosta, D.A.: A comparison of genetic algorithms, particle swarm optimization and the differential evolution method for the design of scannable circular antenna arrays. Progress in Electromagnetics Research B 13, 171–186 (2009)
5. Jones, K.O.: Comparison of genetic algorithms and particle swarm optimization for fermentation feed profile determination. In: Int. Conf. on Computer Systems and Technologies, pp. IIIB.8-1–IIIB.8-7 (2006)
6. Romeo, F., Sangiovanni-Vincentelli, A.: Probabilistic Hill Climbing Algorithms: Properties and Applications. In: Proceedings of the 1985 Chapel Hill Conference on VLSI, pp. 393–417 (1985)
7. Ethni, S.A., Zahawi, B., Giaouris, D., Acarnley, P.P.: Comparison of Particle Swarm and Simulated Annealing Algorithms for Induction Motor Fault Identification. In: 7th IEEE International Conference on Industrial Informatics, INDIN 2009, pp. 470–474 (2009)
8. Manikas, T.W., Cain, J.T.: Genetic Algorithms vs. Simulated Annealing: A Comparison of Approaches for Solving the Circuit Partitioning Problem. Technical report, University of Pittsburgh (1996)
9. VoiceGenie Technologies Inc.: VoiceGenie 7 Tools User's Guide. VoiceGenie Technologies Inc. (2005)
10. Siler, W., Buckley, J.J.: Fuzzy Expert Systems and Fuzzy Reasoning. John Wiley & Sons, New Jersey (2004)
11. Elwakdy, A.M., Elsehely, B.E., Eltokhy, C.M., Elhennawy, D.A.: Speech recognition using a wavelet transform to establish fuzzy inference system through subtractive clustering and neural network (ANFIS). Int. Journal of Circuits, Systems and Signal Processing 2, 264–273 (2008)
12. Yen, J., Wang, L.: Constructing optimal fuzzy models using statistical information criteria. Journal of Intelligent and Fuzzy Systems: Applications in Engineering and Technology 7, 185–201 (1999)
13. Akbulut, S., Hasiloglub, A.S., Pamukcu, S.: Data generation for shear modulus and damping ratio in reinforced sands using adaptive neuro-fuzzy inference system. Soil Dynamics and Earthquake Engineering 24, 805–814 (2004)
14. Michalewicz, Z.: Genetic Algorithms + Data Structures = Evolution Programs. Springer, Heidelberg (1996)
15. Booker, L.B., Goldberg, D.E., Holland, J.H.: Classifier systems and genetic algorithm. Artificial Intelligence 40(1-3), 235–282 (1989)
16. Houck, C.R., Joines, J.A., Kay, M.G.: A genetic algorithm for function optimization: a Matlab implementation. Technical Report, North Carolina State University (1995)
17. van Laarhoven, P., Aarts, E.: Simulated Annealing: Theory and Applications (Mathematics and its applications). Kluwer Academic Publishers, Springer, Heidelberg (1987)
18. Ingber, L.: Adaptive simulated annealing (ASA): Lessons learned. Control and Cybernetics 25(1), 33–54 (1996)
19. Lahtinen, J., Myllymaki, P., Tirri, H.: Empirical comparison of stochastic algorithms. In: Proc. of the Second Nordic Workshop on Genetic Algorithms and their Applications, pp. 45–59 (1996)

Information Theoretic Self-organised Adaptation in Reservoirs for Temporal Memory Tasks

Sakyasingha Dasgupta, Florentin Wörgötter, and Poramate Manoonpong

Bernstein Center for Computational Neuroscience, Georg-August-Universität,
Göttingen, Germany
{dasgupta,worgott,poramate}@physik3.gwdg.de

Abstract. Recurrent neural networks of the Reservoir Computing (RC) type have been found useful in various time-series processing tasks with inherent non-linearity and requirements of temporal memory. Here with the aim to obtain extended temporal memory in generic delayed response tasks, we combine a generalised intrinsic plasticity mechanism with an information storage based neuron leak adaptation rule in a self-organised manner. This results in adaptation of neuron local memory in terms of leakage along with inherent homeostatic stability. Experimental results on two benchmark tasks confirm the extended performance of this system as compared to a static RC and RC with only intrinsic plasticity. Furthermore, we demonstrate the ability of the system to solve long temporal memory tasks via a simulated T-shaped maze navigation scenario.

Keywords: Recurrent neural networks, Self-adaptation, Information theory, Intrinsic plasticity.

1 Introduction

Reservoir Computing (RC) is a powerful paradigm for the design, analysis and training of recurrent neural networks [3]. The RC framework has been utilized for mathematical modeling of biological neural networks as well as applications for non-linear time-series modeling, robotic applications and understanding the dynamics of memory in large recurrent networks in general. Traditionally the reservoir is randomly constructed with only the output connections trained with a regression function. Although both spiking and analog neurons have been explored previously, here we focus on the Echo-state network (ESN) type [2] using sigmoid leaky integrator neurons.

Although the generic RC shows impressive performance for many tasks, the fixed random connections and variations in parameters like spectral radius, leak-rate and number of neurons can lead to significant variations in performance. Approaches based on Intrinsic Plasticity (IP) [1] can help to improve such generic reservoirs. IP uses an information theoretic approach for information maximization at an individual neuron level in a self-organized manner. The IP performance

C. Jayne, S. Yue, and L. Iliadis (Eds.): EANN 2012, CCIS 311, pp. 31–40, 2012.

significantly depends on the type of transfer function, degree of sparsity required and the use of different probability distributions. However the conventional IP method is still outperformed by specific network connectivities like permutation matrices, in terms of the memory capacity performance [8].

We overcome this here, by first utilizing a new IP method [4] based on a Weibull distribution for information maximization. This is then combined with an adaptation rule for the individual neuron leak-rate based on the local information storage measure [7]. Transfer entropy is another measure for such an adaptation rule. However conventionaly this is more difficult to compute, and as it also maximizes input to output information transfer, it is difficult to combine with an IP rule. We achieve such a combination in a self-organized way to guide the individual units for both, maximizing their information and their memory based on the available input. Subsequently through two standard benchmark tasks and a simulated robot navigation task, we show that our adapted network has better performance and memory capacity as compared to static and only IP adapted reservoirs. All three tasks involve a high degree of non-linearity and requirement of adaptable temporal memory. Specifically in robotics and engineering control tasks with nonlinear dynamics and variational inputs (in the time domain), our adaptation technique can show significant performance.

2 Self-adaptive Reservoir Framework

Here we present the description of the internal reservoir network dynamics and briefly explain (i) The local neuron memory adaptation based on information storage measure (ii) The self-organised adaptation of reservoir neurons inspired by intrinsic plasticity. This is based on maximization of the input-output mutual information of each neuron considering a Weibull distribution as the expected output distribution. Subsequently, we combine both mechanisms for a comprehensive adaptive framework.

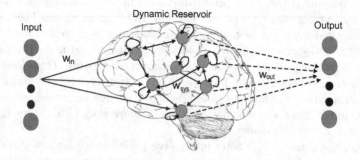

Fig. 1. The Reservoir network architecture, showing the flow of information from input to reservoir to output units. Typically only the output connections are trained. W_{in} and W_{sys} are set randomly.

2.1   Network Description

The recurrent network model is depicted in Fig. 1 as an abstract model of the mammalian neo-cortex. The firing activity of the network at discrete time t is described by the internal state activation vector $\mathbf{x}(t)$. Neural units are connected via weighted synaptic connections $\mathbf{W}$. Specifically $\mathbf{W}_{in}$ are the $K \times N$ connections from the input neurons K to the reservoir neurons N, $\mathbf{W}_{out}$ are the $N \times L$ connections from the reservoir neurons to the output neurons L and $\mathbf{W}_{sys}$ represents the weight matrix for the internal $N \times N$ connections.

The state dynamics at a particular instant of time for an individual unit is given by:

$$\mathbf{x}(t+1) = (\mathbf{I} - \mathbf{\Lambda})\boldsymbol{\theta}(t) + \mathbf{\Lambda}[\mathbf{W}_{sys}\boldsymbol{\theta}(t) + \mathbf{W}_{in}\mathbf{v}(t)], \tag{1}$$

$$\mathbf{y}(t) = \mathbf{W}_{out}\mathbf{x}(t), \tag{2}$$

$$\lambda_i = \frac{1}{T_c}(\frac{1}{1+\alpha_i}), \tag{3}$$

where $\mathbf{x}(t) = (x_1(t), x_2(t), ..., x_N(t))^T$ is the N dimentional vector of internal neural activation, and $\mathbf{v}(t) = (v_1(t), v_2(t), ..., v_K(t))^T$ is the K-dimensional time-dependent input that drives the recurrent network. $\mathbf{y}(t) = (y_1(t), y_2(t), .., y_L(t))$ is the output vector of the network. $\mathbf{\Lambda} = (\lambda_1, \lambda_2, ..., \lambda_N)^T$ is a vector of individual leak decay rates of reservoir neurons with global time constant $T_c > 0$, while $\alpha_i \in \{0, 1, 2, ..., 9\}$ is the leak control parameter. It is normally adjusted manually and kept fixed, but here it is determined by the local information storage based adaptation rule for each reservoir neuron (Section 2.2).

The ouput firing rate of neurons is given by the vector $\boldsymbol{\theta} = (\theta_1, \theta_2, ..., \theta_N)^T$.

$$\theta_i(t) = \text{fermi}(a_i x_i(t) + b_i). \tag{4}$$

A simple transformation from the fermi-dirac distribution function to the tanh transfer function can be done using:

$$\tanh(x) = 2\theta\text{fermi}(2x) - 1. \tag{5}$$

Here b_i acts as the individual neuron bias values, while a_i governs the slope of the firing rate curve. We adapt these parameters according to IP learning mechanism, presented in Section 2.3.

The output weights $\mathbf{W}_{out}$ are computed as the ridge regression weights of the teacher outputs $\mathbf{d}(t)$ on the reservoir states $\mathbf{x}(t)$. The objective of training is to find a set of output weights such that the summed squared error between the desired output and the actual network output $\mathbf{y}(t)$ are minimised. This is done by changing the weights incrementaly in the direction of the error gradient. This could also be done in a one-shot manner using the recursive least squared algorithm [11].

2.2   Neuron Memory Adaptation : Information Storage

An information theoretic measure named *local information storage* refers to the amount of information in the previous state of the neuron that is relevant in predicting its future state. It measures the amount of information stored in the current state of the neuron, which provides either positive or negative information towards its next state. Specially, the instantaneous information storage for a variable x is the local (or un-averaged) mutual information between its semi-infinite past $x_t^{(k)} = \{x_{t-k+1}, ..., x_{t-1}, x_t\}$ and its next state x_{t+1} at the time step $t+1$ calculated for finite-k estimations. Hence, the local information storage is defined for every spatio-temporal point within the network (dynamic reservoir). The local unaveraged information storage can take both positive as well as negative values, while the Average (active) information storage $A_x = \langle a_x(i,t) \rangle$ is always positive and bounded by $\log_2 N$ bits. N is the network size. The local information storage for an internal neuron state x_i is given by:

$$a_x(i, t+1) = \lim_{k \to \infty} \log_2 \left(\frac{P(x_{i,t}^{(k)}, x_{i,t+1})}{P(x_{i,t}^{(k)})p(x_{i,t+1})} \right). \tag{6}$$

Where $a_x(i, t+1, k)$ respresents finite-k estimates. In case of neurons with a certain degree of leakage (applied after the non-linearity) as introduced in [2] for Echo-state networks (a type of RC), the leakage rate (λ) tells how much a single neuron depends on its input, as compared to the influence of its own previous activity. Since λ varies between 0 and 1, $(1-\lambda)$ can be viewed as a local neuron memory term. Where in, lower the value of λ, stronger the influence of the previous level of activation as compared to current input to the neuron, or high local memory and vice versa.

Using epochs(ϕ) with finite history length $k = 8$, the active information storage measure at each neuron adapts the leak control parameter α_i as follows :

$$\alpha_i = \begin{cases} \alpha_i + 1 & \text{if } A_x(i, \phi) - A_x(i, \phi - 1) > \epsilon \\ \alpha_i - 1 & \text{if } A_x(i, \phi) - A_x(i, \phi - 1) < \epsilon, \end{cases} \tag{7}$$

where $\epsilon = \frac{1}{2}\log_2 N$ and $0 < \alpha_i < 9$.

After each epoch, α_i and λ_i are adjusted and these values are used for the subsequent epoch. Once all the training samples are exhausted the pre-training of reservoir is completed and λ_i is fixed.

2.3   Generic Intrinsic Plasticity

Homeostatic regulation by way of intrinsic plasticity is viewed as a mechanism for the biological neuron to modify its firing activity to match the input stimulus distribution. In [6] a model of intrinsic plasticity based on changes to a neurons non-linear activation function was introduced. A gradient rule for direct minimization of the Kullback-Leibler divergence between the neurons current firing-rate distribution and maximum entropy (fixed mean), exponential output distribution was presented. Subsequently in [1] an IP rule for the hyperbolic tangent transfer function with a Gaussian output distribution (fixed variance maximum entropy distribution) was derived. During testing the adapted reservoir dynamics, it was observed that for tasks requiring linear responses (NARMA) the Gaussian distribution performs best. However on non-linear tasks, the exponential distribution gave a better performance. In this work, with the aim to obtain sparser output codes with increased signal to noise ratio for a stable working memory task, we implement the learning rule for IP with a Weibull output distribution. The shape parameter of the Weibull distribution can be tweaked to account for various shapes of the transfer function (equation 4). With the aim for a high kurtosis number and hence more sparser output codes, we choose the shape parameter $K = 1.5$. This model was very recently introduced in [4]. However the application of this rule in the reservoir computing framework and its effect on the network performace for standard benchmark tasks had not been studied so far.

The probability distribution of the two-parameter Weibull random variable θ is given as follows:

$$f_{weib}(\theta; \beta, k) = \begin{cases} \frac{k}{\beta}(\frac{\theta}{\beta})^{k-1} exp - \frac{\theta}{\beta}^k & \text{if } \theta \geq 0 \\ 0 & \text{if } \theta < 0 \end{cases} \tag{8}$$

The parameters $k > 0$ and $\beta > 0$ control the shape and scale of the distribution respectively. Between $k = 1$ and $k = 2$, the weibull distribution interpolates between the exponential distribution and Rayleigh distribution. Specifically for $k = 5$, we obtain an almost normal distribution. Due to this generalization capability it serves best to model the actual firing rate distribution and also account for different types of neuron non-linearities. The neuron firing rate parameters a and b of equation (4) are calculated by minimising the Kullbeck-Leibler divergence between the real output distribution f_θ and the desired distribution f_{weib} with fixed mean firing rate $\beta = 0.2$. Here the Kullbeck-Leibler divergence is given by:

$$D_{KL}(f_\theta, f_{weib}) = \int f_\theta(\theta) log\left(\frac{f_\theta(\theta)}{f_{weib}(\theta)}\right) d\theta$$

$$= -H(\theta) + \frac{1}{\beta^k}E(\theta^k) - (k-1)E(log(\theta)) - log(\frac{k}{\beta^k}) \tag{9}$$

Where, $f_\theta(\theta) = \frac{f_x(x)}{\frac{d\theta}{dx}}$. for a single neuron with input x and output θ

Differentiating D_{KL} with respect to a and b (not shown here) we get the resulting online stochastic gradient descent rule for calculating a and b with the learning rate η and time step h as:

$$\Delta a = \eta\left[\frac{1}{a} + kx(h) - (k+1)x(h)\theta(h) - \frac{k}{\beta^k}x(h)\theta(h)^k(1 - \theta(h))\right], \tag{10}$$

$$\Delta b = \eta\left[-\theta(h) + k(1 - \theta(h)) - \frac{k}{\beta^k}\theta(h)^k(1 - \theta(h))\right]. \tag{11}$$

IP tries to optimize the neurons information content with respect to the incoming input signal. In contrast the neural local memory adaptation rule tries to modulate the leakage based on a quantification of the extent of influence, the past activity of a neuron has on it's activity in the next time step. We therefore combine IP learning with the neuron memory adaptation rule in series. The leakage adaptation is carried out after the intrinsic adaptation of the neuron non-linearity. This combination leads to a single self-adaptive framework that controls the memory of each neuron based on the input to the entire network.

3   Experiments

To show the performance of our self-adapted RC, we test it on two benchmark tests, namely the NARMA-30 time series modeling task, and a delayed 3-bit parity task. We choose these two tasks taking into consideration the inherent non-linearity and requirements of extended temporal memory. Finally we use a simulated delayed response task of robot navigation though a T-junction. This task shows the potential application of our network for solving real robotic tasks with long delay period between memory storage and retreival.

3.1   Experimental Setup

For all experiments the internal reservoir network was constructed using N=400 leaky integrator neurons initialised with a 10% sparse connectivity. Internal reservoir weights W_{sys} were drawn from the uniform distribution over [-1,1] and were subsequently scaled to a spectral radius of 0.95. Input weights and output feedback weights(if present) can be randomnly generated in general. The firing rate parameters are initialised as $a = 1$ and $b = 0$ with the learning rate for the stochastic gradient descent algorithim fixed at $\eta = 0.0008$. Weibull IP and leak adaptation were carried out in 10 epochs in order to determine the optimal parameters a, b and λ for each neuron. Performance evaluation was done after the neuron leak and transfer function parameters have been fixed. For the delayed 3-bit parity memory task the setup consisted of a single input neuron, the internal reservoir and 800 output units. The simulated robotic task was performed using 6 input neurons (number of sensors) and 2 output neurons (number of actuators) with the internal reservoir size fixed to the original value.

Dynamic System Modelling with 30th Order NARMA: Its dynamics is given by:

$$y(t + 1) = 0.2y(t) + 0.004y(t) \sum_{i=0}^{29} y(t - i) + 1.5v(t - 29)v(t) + 0.001 \qquad (12)$$

Here $y(t)$ is the output of the system at time 't'. $v(t)$ is the input to the system at time 't', and is uniformly drawn from the interval [0,0.5]. The system was trained to output $y(t)$ based on $v(t)$. The task in general is quite complex considering that the current system output depends on both the input and the previous outputs. Consequently we use feedback connections (W_{back}) from the output neurons to the internal neurons with Equation (1) modified to: $x(t+1) = (I - \Lambda)\theta(t) + \Lambda[W_{sys}\theta(t) + W_{in}v(t) + W_{back}y(t)]$, and Equation (2) modified to: $y(t+1) = W_{out}[x(t), y(t)]$. The task requires extended temporal memory. The training, validation and tesing were carried out using 1000, 2000 and 3000 time steps respectively. Five fold cross-validation was used with the training set. Here the first 50 steps was used to warm up the reservoir and was not considered for the training error measure. Performance is evaluated using the normalised root mean squared error:

$$NRMSE = \sqrt{\frac{\langle(d(t) - y(t))^2\rangle}{\langle(d(t) - <y(t)>)^2\rangle}} \tag{13}$$

Delayed 3-bit Parity Task: The delayed 3-bit parity task functions over input sequences τ time steps long. The input consists of a temporal signal $v(t)$ drawn uniformly from the interval [-0.5,0.5]. The desired output signal was calculated as the PARITY $(v(t-\tau), v(t-\tau-1), v(t-\tau-2)$ for increasing time delays of $0 \le \tau \le 800$. Since the parity function (XOR) is not linearly seperable, this task is quite complex and requires long memory capabilities. We evaluated the memory capacity of the network as the amount of variance of the delayed input signal recoverable from the optimally trained output units summed over all delays. For a given input signal delayed by k time steps, the net memory capcity is given by:

$$MC = \sum_k MC_k = \sum_k \frac{cov^2(y(t-k), d(t))}{var(y(t))var(d(t))} \tag{14}$$

where cov and var denote covariance and variance operations, respectively.

Simulated Robot T-maze Navigation: In order to demonstrate the temporal memory capacity of our system, we employ a four wheeled mobile robot (NIMM4 Fig. 2(a)) to solve a delayed response task. The primary task of the robot was to drive from the starting position untill the T-junction. Then it should either take a left or right turn depending on whether a spherical ball(Yellow) appeared to its right or left before the T-junction. Hence, here the robot has to learn both the reactive behaviarol task of turning at the T-junction as well as remembing the event of a colored ball being shown before. Therefore conventional methods like landmarks to identify the T-junction is not needed. Moreover, to demonstrate the generalization capability of the system to longer time delays, we divided the task into two mazes of different lengths. Maze B requires a longer temporal memory as compared to maze A. NIMM4 consists of four infrared sensors (two on the left and right respectively), a relative distance sensor, a relative angle of deviation sensor and four actuators to control the desired turning and speed. The experiment consists of data-set aquisition, training of our adapted RC and offline testing. During the first phase using the simulator, we manually controlled the robot movement through the maze using simple keyboard instructions and recorded the sensor and actuator values. We recorded 80 examples in total with different initial starting positions. 40% of these were used for trainng and 60% for offline testing. After the first phase, the self-adapted RC was trained using imitation learning on the collected data with the actuator values from manual control as desired output. Finally we peformed offline testing using the remaining set of recorded data. Simulations were carried out using the C++ based LPZRobot simulator.

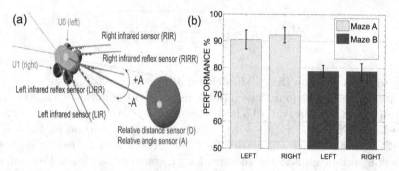

Fig. 2. (a) Model of the simulated robot NIMM4 showing the sensors (LIR,RIR,LIRR,RIRR) and actuators (U0,U1). The red ball in front of the robot represents its goal. (b) Performance of the robot in the two maze tasks (Maze B longer than Maze A) measured in terms of percentage of correct times the robot took the proper trajectory (left/right turn at T-junction). 5% noise is considered on sensors.

3.2   Experimental Results

In Table 1. we summarize the results of our self-adaptive RC with information maximization based on a Weibull distribuition as compared to the performace obtained by a static RC and RC with Gaussian distribution based intrinsic plasticity [1]. Our network outperforms the other two models, both in terms of lowest normalised root mean squared error (0.362) for the 30th order NARMA task, as well as an extended average memory capacity of 47.173 for the delayed 3-bit parity task. Non-normal networks (e.g. a simple delay like network) have been shown to theoretically allow extensive memory [9] which is not possible for arbitary recurrent networks. However our self-adaptive RC network shows considerable increase in the memory capacity, which was previously shown to improve only in case of specifically selected network connections (permutation matrices).

Figure 3(left,centre) shows screenshots of the robot performing the maze navigation task and successfully making the correct turn at the T-junction.

Table 1. Normalised root mean squared error (NRMSE) and average memory capacity performance for the NARMA-30 and 3-bit parity tasks, comparing the basic RC (ESN) model, the RC model with a intrinsic plasticity method using Gaussian probability density and our self-adapted RC (SRC) network using Weibull probability density

Dataset	Measure	ESN	RC with IP(GAUSS)	SRC with IP(Weib)
NARMA-30	NRMSE	0.484	0.453	**0.362**
	Std. Dev.	0.043	0.067	0.037
3-bit Parity	Memory Cap.(MC)	30.362	32.271	**47.173**
	Std. Dev.	1.793	1.282	1.831

The turn depends on the prior input appearing while driving along the corridor. Our adapted RC is able to successfully learn this task and use only the sensor data to drive along the corridor and negotiate the correct turn. The offline testing results in the form of the percentage of correct turns from the total test set for both mazes are shown in Fig. 2(b). In case of the shorter maze A (smaller memory retention period) we achieve average performance of 92.25% with a standard deviation of 2.88. A good generalization capability for the longer maze B is also observed with the average performace of 78.75% with a standard deviation of 3.11, both for right turn. This is quite high as compared to previous results obtained by [5] for a similar task with static Echo-state network. Furthemore in Fig.3 (right) one can see that our adapted reservoir network clearly outperforms a static RC for the same task. Note that if additional sensors were availabe to the robot, this could further improve the performace, owing to the availability of additional information.

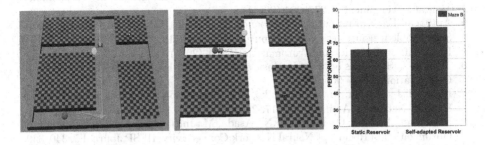

Fig. 3. (left) Screenshots of the robot successfully navigating through the large maze B. Yellow ball is cue to turn right at the T junction, Red ball marks its goal. (centre) Screenshot of the robot navigating through small maze A. (right) Performance on the maze B task after 80 trials for static reservoir vs our self-adapted reservoir. Our network outperforms by 10%. 5% noise is considered on sensors.

4   Conclusion

We have presented and evaluated an adaptation rule for the reservoir computing network that successfully combines intrinsic plasticity using a Weibull output distribution with a neuron leak adjustment rule based on local information storage measure. The neuron leak rate governs the degree of influence of local memory, while intrinsic plasticity ensure information maximization at each neuron output. The evaluated performance on the two benchmark tasks and the robotic simulation demonstrates a reduction in performance error along with an increased memory capacity, of our adapted network as compared to basic RC setups. Future works include using online testing of our network on more complex navigation scenarios. This will require longer working memory and fast adaptation of reservoir time scale. We will also apply this network to an actual hexapod robot AMOSII [10] for enabling memory guided behaviour. Moreover, possible hierarchical arrangement of such adapted reservoirs for both short-term and long term memory (different time scales) within a single framework will be evaluated.

Acknowledgements. This research was supported by the Emmy Noether Program (DFG, MA4464/3-1) and the Bernstein Center for Computational Neuroscience II Göttingen (01GQ1005A, project D1).

References

1. Schrauwen, B., Wardermann, M., Verstraeten, D., Steil, J.J., Stroobandt, D.: Improving Reservoirs using Intrinsic Plasticity. Neurocomputing 71, 1159–1171 (2008)
2. Jaeger, H., Lukosevicius, M., Popovici, D., Siewert, U.: Optimization and Applications of Echo State Networks with Leaky-integrator Neurons. Neural Networks 20, 335–352 (2007)
3. Lukoševičius, M., Jaeger, H.: Reservoir Computing Approaches to Recurrent Neural Network Training. Computer Science Review 3, 127–149 (2009)
4. Li, C.: A model of Neuronal Intrinsic Plasticity. IEEE Trans. on Autonomous Mental Development 3, 277–284 (2011)
5. Antonelo, E., Schrauwen, B., Stroobandt, D.: Mobile Robot Control in the Road Sign Problem using Reservoir Computing Networks. In: Proceedings of the IEEE Int. Conf. on Robotics and Automation, ICRA (2008)
6. Triesch, J.: Synergies between Intrinsic and Synaptic Plasticity Mechanisms. Neural Computation 4, 885–909 (2007)
7. Lizier, T.J., Pritam, M., Prokopenko, M.: Information Dynamics in Small-world Boolean Networks. Artificial Life 17, 293–314 (2011)
8. Boedecker, J., Obst, O., Mayer, M.N., Asada, M.: Initialization and Self-Organized Optimization of Recurrent Neural Network Connectivity. HFSP Journal 5, 340–349 (2009)
9. Ganguli, S., Dongsung, H., Sompolinsky, H.: Memory Traces in Dynamical Systems. PNAS 105, 18970–18975 (2008)
10. Steingrube, S., Timme, M., Wörgötter, F., Manoonpong, P.: Self-Organized Adaptation of a Simple Neural Circuit Enables Complex Robot Behaviour. Nature Physics 6, 224–230 (2010)
11. Jaeger, H.: Adaptive Nonlinear System Identification with Echo State Networks. In: Advances in Neural Information Processing Systems, vol. 15 (2003)

Fuzzy-Logic Inference for Early Detection
of Sleep Onset in Car Driver

Mario Malcangi[1,*] and Salvatore Smirne[2]

[1] DICo - Dipartimento di Informatica e Comunicazione,
[2] Dipartimento di Tecnologie per la Salute
Università degli Studi di Milano
Milano - Italy
{mario.malcangi,salvatore.smirne}@unimi.it

Abstract. Heart rate variability (HRV) is an important sign because it reflects the activity of the autonomic nervous system (ANS), which controls most of the physiological activity of the subjects, including sleep. The balance between the sympathetic and parasympathetic branches of the nervous system is an effective indicator of heart rhythm and, indirectly, heart rhythm is related to a patient's state of wakefulness or sleep. In this paper we present a research that models a fuzzy logic inference engine for early detection of the onset of sleep in people driving a car or a public transportation vehicle. ANS activity reflected in the HRV signal is measured by electrocardiogram (ECG). Power spectrum density (PSD) is computed from the HRV signal and ANS frequency activity is then measured. Crisp measurements such as very low, low, and high HRV and low-to-high frequency ratio variability are fuzzified and evaluated by a set of fuzzy-logic rules that make inferences about the onset of sleep in automobile drivers. An experimental test environment has been developed to evaluate this method and its effectiveness.

Keywords: onset sleep, heart rate variability, power spectrum density, fuzzy logic, autonomic nervous system.

1 Introduction

Falling asleep at the wheel is a cause of very dangerous accidents, so many methods have been investigated to find a practical solution for early detection of the onset of sleep to achieve an higher level of safety in private and public transportation systems.

To detect sleep onset early, continuous monitoring of the driver's physiological state needs to be carried out. The electrocardiogram (ECG) carries most of the information about physiological status.

Capturing an ECG is a complex task that requires the accurate application of several electrodes to the patient. This may be feasible in a clinical context, but is not practical for a person driving a car. A non-invasive method for capturing ECG signal needs to

---

* Corresponding author.

C. Jayne, S. Yue, and L. Iliadis (Eds.): EANN 2012, CCIS 311, pp. 41–50, 2012.

be developed for practical application to monitoring a driver's physiological status in the automobile environment [1] [2].

Other physiological information can be monitored along with the ECG, such as eye movements, breathing rate, muscular tone, and arm movements. This additional information can also be captured with non-invasive methods, so that a very robust and effective set of features can be measured.

Such information is highly fuzzy, so a smart inference system is needed to infer about sleep onset. The fuzzy-logic-based inference system can be very effective if several physiological features concur in the decision, but a practical system for early detection of oncoming driver sleep onset can also be based also on EEG measurements alone. This is because there is enough information in the EEG signal directly related to sleep-wake control [3] [4].

Sleep is a physiological state characterized by variations in the activity of the autonomic nervous system that is reflected in heart rate and its variability (HRV). The power spectral density (PSD) of heart rate varies with the change from wakefulness to sleep [5] [6]. The low-to-high frequency ratio is a valid indicator of such change because it reflects the balancing action of the sympathetic nervous system and parasympathetic nervous system branches of the autonomic nervous system.

When the activity of the sympathetic nervous system increases, the parasympathetic nervous system diminishes its activity, causing an acceleration of cardiac rhythm (shorter beat intervals). Cardiac rhythm deceleration is caused by low activity of the sympathetic nervous system and increased parasympathetic nervous system activity, producing a deceleration of the heart rhythm (longer beat intervals).

The PSD of HRV (Fig. 1) signal shows that sympathetic activity is associated with the low frequency range (0.04–0.15 Hz) while parasympathetic activity is associated with the higher frequency range (0.15–0.4 Hz). Because the frequency ranges of sympathetic and parasympathetic activity are distinct, it is possible to separate the sympathetic and parasympathetic contributions.

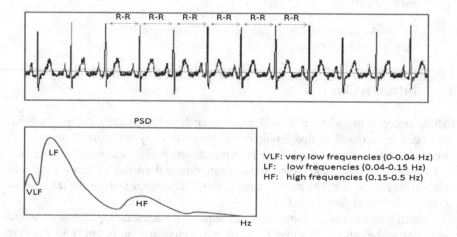

Fig. 1. Very low, low and high frequencies in power spectrum density (PSD), computed from the heart rate variability signal (HRV)

The PSD analysis of beat-to-beat HRV provides a useful means for understanding when sleep is setting in. Sleep and wakefulness are directly related to the autonomous nervous system [7]. In the awake state, the low-frequency spectral component (sympathetic modulation activity) was significantly higher and the high-frequency spectral components (parasympathetic modulation activity) significantly lower. Conversely, in the asleep state, the low-frequency spectral component (sympathetic modulation activity) was significantly lower and the high frequency spectral components (parasympathetic modulation activity) significantly higher. If we consider the balance of low frequency versus high frequency in a person's PSD, it is possible to predict the onset of falling asleep.

A set of experiments demonstrates that, when a person tries to resist falling asleep, the LF/HF ratio of PDS computed from the HRV signal increases significantly a few minutes before becoming significantly lower during the sleep stage. Like a reaction to falling asleep, it causes high activity of the sympathetic system while the parasympathetic system decreases its activity.

Making inferences about physiological status from the HRV signal is very difficult because of the high degree of variability and the presence of artifacts. Softcomputing methods can be very effective for inferring in such a context [8].

There are several methods [9] [10] [11] for performing predictions with artificial neural networks (ANN). Mager [12] utilizes Kohonen's self-organizing map (SOM) to provide a method of clustering subjects with similar features. This method, applied to the problem of detecting oncoming sleep early, allows artifacts to be filtered and the variability component of noise combined with the primary HRV signal to be smoothed.

The drawback is that ANNs are very difficult to train for HRV of normal subjects who fall asleep at the wheel, because it is difficult to detect precisely the time when the event happens.

An alternative approach   uses fuzzy decision logic [13] [14]   to model the oncoming onset of sleep. Such an approach is effective because it allows use of the membership function to model data features and of a sleep-disease specialist's ability to interpret the PSD LF/HF ratio dynamics as an index of oncoming onset of sleep.

2   System Framework

The whole system consists of a signal acquisition and preprocessing subsystem, a feature extraction subsystem, and a fuzzy-based decision logic module (Fig. 2).

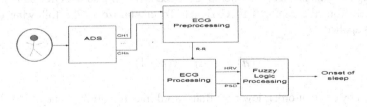

Fig. 2. System architecture consists of an Analog-to-Digital Subsystem (ADS), an ECG preprocessing subsystem, an ECG processing subsystem, and a fuzzy logic processing engine

ECG signal acquisition is not a simple task because noise and artifacts are very strong (Fig. 3). Good signal acquisition can be assured by a high-quality, analog-to-digital subsystem (ADS), specifically designed for ECG signals.

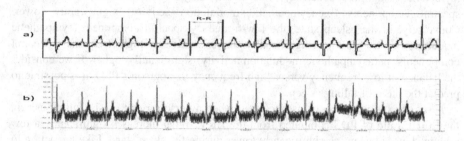

Fig. 3. ECG signal acquired at thorax level (a) and at hand level (b)

The ECG signal is sampled at 500 samples per second (SPS) with a depth of 24 bits.

A set of signal-processing algorithms was applied to the acquired ECG signal to remove noise and artifacts, so that the QRS complex can be detected (Fig. 4). The ECG is filtered to remove 60-Hz noise, baseline fluctuations, and muscle noise.

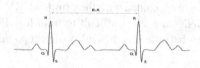

Fig. 4. The QRS complex and R-R interval

Baseline oscillations are removed using a zero phase fourth-order, high-pass filter (1-Hz cutoff frequency).

2.1   ECG Processing

To compute the HRV signal, the heartbeat needs to be extracted from the acquired ECG signal. This is bandpass filtered (centered at 17 Hz), so the QRS complex will be extracted from the captured ECG signal. To emphasize it, the following derivative filter is applied:

$$y(n) = x(n) - x(n-1) \tag{1}$$

followed by an eight-order, low-pass Butterword filter (cutoff frequency at 30 Hz) [15].

The QRS complex is now ready to be thresholded and measured for peak-to-peak period (R-R interval). This is done by squaring the sample values and passing them through the following moving average filter [15]:

$$y(n) = \frac{1}{N} \sum_{i=0}^{N-1} x(n-i) \tag{2}$$

2.2   HRV and PSD Computation

HRV is computed from the R-R intervals. These are measured and collected as a series of times. Because it is an irregular interval-time sequence, it needs to be converted into a uniformly sampled time-spaced sequence.

PSD distribution is then computed so that measurements on the following three frequency bands can be carried out:

- very low frequencies (0-0.04 Hz)
- low frequencies (0.04-0.15 Hz)
- high frequencies (0.15-0.5 Hz)

The low-to-high frequency ratio is also computed.

2.3   Fuzzy Decision Logic

A fuzzy logic engine (Fig. 5) was tuned to make inferences about sleep-onset events. The fuzzy engine makes epoch-by-epoch (20 or 60 seconds per epoch) inferences. HRV and PSD features are fed to the engine in a fuzzified form.

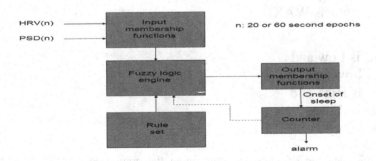

Fig. 5. Fuzzy-logic decision engine tuned to predict the onset of sleep in drivers

To fuzzify such features, a set of membership functions are derived from the distribution of the crisp values in the respective measurement domains (Fig. 6).

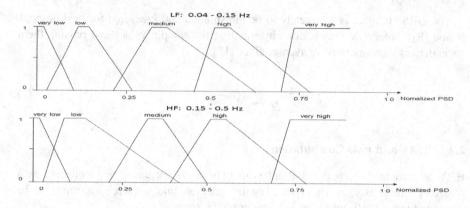

Fig. 6. Membership functions to fuzzify input features

A set of rules has been defined and tuned manually to achieve the best performance for the decision logic. The fuzzy rules looks like this:

if HRV(n) is Low and
    LF(n) is Medium Low and
        HF(n) is Medium High and
        LF/HF is Medium
    then the epoch is ONSET_SLEEP

. . .

if HRV(n) is High and
    LF(n) is High and
        HF(n) is Low and
        LF/HF is High
    then the epoch is WAKE

. . .

if HRV(n) is Low and
    LF(n) is Low and
        HF(n) is High and
        LF/HF is Low
    then the epoch is SLEEP

These three rules are the strongest in determining the output for ONSET_SLEEP, WAKE, and SLEEP states. There are more variants of these rules are in the rule set, each generated during tuning to correct for false detectios that have occurred due to noise and artifacts.

The output of the fuzzy-logic engine consists of a set of singleton membership functions (Fig. 7). The "center of gravity" algorithm is applied to defuzzify the final decision:

$$crisp\_output = \frac{\sum (fuzzy\_output) \; x \; (singleton\_position)}{\sum fuzzy\_output}$$

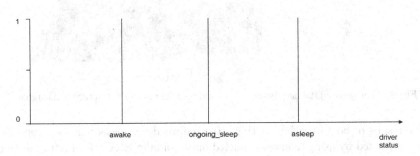

Fig. 7. Singleton membership function to defuzzify the final decision

An output module counts how many times a short epoch (20 seconds) has been classified as ONSET_SLEEP and how many long epochs (60 seconds) are classified as SLEEP. Such counts are used as a feedback (memory) to the inference engine, as well as to integrate the epoch-by-epoch outputs of the fuzzy-logic engine.

3   Experimental Results

A MATLAB-based application (Fig. 8) was developed to conduct experimental tests. This environment was connected to an ECG acquisition board (Fig. 9) so that bioelectrical signals are captured at the thorax or arm level.

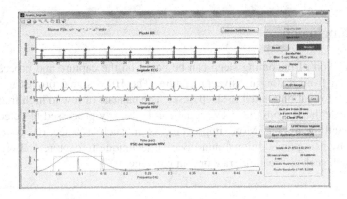

Fig. 8. MATLAB-based application developed to host the experiments

Two sets of experimental tests were carried out. The first set used ECGs acquired in a clinical context (thorax). The second refers to ECGs acquired in a field context (hands). This signal is noisy and less well-defined in QRS complex However, after more rules are added, sleep-onset detection showed the same detection rate as clinically collected data (90% true detection on a set of 10 analyzed ECG).

Fig. 9. Analog-to-digital acquisition board and cable connector to capture EEG signal

The results of both experiments (Fig. 10) confirmed our thesis that the sleep onset can be predicted by using features extracted only from the HRV signal (clinical data). The experiments also confirmed that sleep-onset detection may be successfully based on the analysis of an ECG signal captured from the driver's hands.

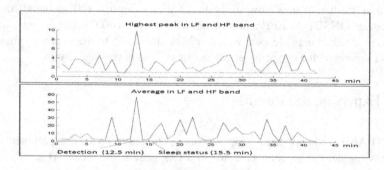

Fig. 10. One early detected sleep onset

4   Conclusion

Early detection of oncoming sleep can be based on the capture and processing of an ECG signal from a non-invasive procedure. The HRV signal proves to be a good carrier of information related to sleep onset. The balance of low to high frequencies of the PSD calculated from the HRV signal can be fuzzily processed to detect the early signs of falling asleep. More improvements in such a detection system can be achieved by using an additional non-invasive technique to capture and process some other signal, such as breathing rate or arm movement.

Compared to non-HRV measurement-based methods, the proposed method has several advantages, mainly that it is non-invasive and based on neurological information rather than on visual signs (eyes movements) or gestures (head movements).

The fuzzy-logic method proves to be the most appropriate way to make inferences based on HRV information, because its main features are relative power reading at different frequencies. These features can easily mapped onto membership functions and compiled into fuzzy rules by using an expert's knowledge, i.e. that of a physician who is expert in sleep disorders).

References

1. Dorfman Furman, G., Baharav, A., Cahan, C., Akselrod, S.: Early Detection of Falling Asleep at the Wheel: a Heart Rate Variability Approach. Computers in Cardiology 35, 1109–1112 (2008)
2. Zocchi, C., Giusti, A., Adami, A., Scaramellini, F., Rovetta, A.: Biorobotic System for Increasing Automotive Safety. In: 12th IFToMM World Congress, Besançon (France), June18-21 (2007)
3. Estrada, E., Nazeran, H.: EEG and HRV Signal Features for Automatic Sleep Staging and Apnea Detection. In: 20th International Conference on Electronics, Communications and Computer (CONIELECOMP), February 22-24, pp. 142–147 (2010)
4. Lewicke, A.T., Sazonov, E.S., Schuckers, S.A.C.: Sleep-wake Identification in Infants: Heart Rate Variability Compared to Actigraphy. In: Proceedings of the 26th Annual International Conference of the IEEE EMBS, San Francisco, CA, USA, September 1-5, pp. 442–445 (2004)
5. Manis, G., Nikolopoulos, S., Alexandridi, A.: Prediction Techniques and HRV Analysis. In: MEDICON 2004, Naples, Italy, July 31-August 5 (2004)
6. Rajendra Acharya, U., Paul Joseph, K., Kannathal, N., Lim, C.M., Suri, J.S.: Heart Rate Variability: a Review. Med. Bio. Eng. Comput. 44, 1031–1051 (2006)
7. Tohara, T., Katayama, M., Takajyo, A., Inoue, K., Shirakawa, S., Kitado, M., Takahashi, T., Nishimur, Y.: Time Frequency Analysis of Biological Signal During Sleep. In: SICE Annual Conference, September 17-20, pp. 1925–1929. Kagawa University, Japan (2007)
8. Ranganathan, G., Rangarajan, R., Bindhu, V.: Evaluation of ECG Signal for Mental Stress Assessment Using Fuzzy Technique. International Journal of Soft Computing and Engineering (IJSCE) 1(4), 195–201 (2011)
9. Ranganathan, G., Rangarajan, R., Bindhu, V.: Signal Processing of Heart Rate Variability Using Wavelet Transform for Mental Stress Measurement. Journal of Theoretical and Applied Information Technology 11(2), 124–129 (2010)
10. Lewicke, A., Shuckers, S.: Sleep versus Wake Classification from Heart Rate Variability Using Computational Intelligence: Consideration of Rejection in Classification Models. IEEE Transactions on Biomedical Engineering 55(1), 108–117 (2008)
11. Mager, D.E., Merritt, M.M., Kasturi, J., Witkin, L.R., Urdiqui-Macdonald, M., Sollers, J.I., Evans, M.K., Zonderman, A.B., Abernethy, D.R., Thayer, J.F.: Kullback–Leibler Clustering of Continuous Wavelet Transform Measures of Heart Rate Variability. Biomed. Sci. Instrum. 40, 337–342 (2004)

12. Alexandridi, A., Stylios, C., Manis, G.: Neural Networks and Fuzzy Logic Approximation and Prediction for HRV Analysis. In: Hybrid Systems and their Implementation on Smart Adaptive Systems, Oulu, Finland (2003)

13. Dzitac, S., Popper, L., Secui, C.D., Vesselenyi, T., Moga, I.: Fuzzy Algorithm for Human Drowsiness Detection Devices. SIC 19(4), 419–426 (2010)

14. Catania, D., Malcangi, M., Rossi, M.: Improving the Automatic Identification of Crackling Respiratory Sounds Using Fuzzy Logic. International Journal of Computational Intelligence 2(1), 8–14 (2004)

15. Azevedo de Carvalho, J.L., Ferreira da Rocha, A., Assis de Oliveira Nascimento, F., Souza Neto, J., Junqueira Jr., L.F.: Development of a Matlab Software for Analysis of Heart Rate Variability. In: 6th International Conference on Signal Processing Proceedings, pp. 1488–1491 (2002)

Object-Oriented Neurofuzzy Modeling and Control of a Binary Distillation Column by Using MODELICA

Javier Fernandez de Canete[1,*], Alfonso Garcia-Cerezo[1], Inmaculada Garcia-Moral[1], Pablo del Saz[1], and Ernesto Ochoa[2]

[1] System Engineering and Automation, C/ Doctor Ortiz Ramos s/n,29071 Malaga, Spain
[2] Acenture S.L., Av Lopez-Peñalver 28, 29590 Malaga, Spain
`{canete,alfonso.garcia,gmoral,pab.saz}@isa.uma.es,`
`ernesto.ochoa@hotmail.com`

Abstract. Neurofuzzy networks offer an alternative approach both for the identification and the control of nonlinear processes in process engineering. The lack of software tools for the design of controllers based on hybrid neural networks and fuzzy models is particularly pronounced in this field. MODELICA is an oriented-object environment widely used which allows system-level developers to perform rapid prototyping and testing. Such programming environment offers an intuitive approach to both adaptive modeling and control in a great variety of engineering disciplines. In this paper we have developed an oriented-object model of binary distillation column with nonlinear dynamics, and an ANFIS (Adaptive-Network-based Fuzzy Inference System) neurofuzzy scheme has been applied to derive both an identification model and a adaptive controller to regulate distillation composition. The results obtained demonstrate the effectiveness of the neurofuzzy control scheme when the plant's dynamics is given by a set of nonlinear differential algebraic equations (DAE).

Keywords: Object-oriented, DAE equations, MODELICA, Neurofuzzy Modeling, Neurofuzzy Control, ANFIS structure, Distillation Column.

1 Introduction

Neural and fuzzy applications have been successfully applied to the chemical engineering processes [1], and several control strategies have been reported in literature for the distillation plant modeling and control tasks [2]. Recent years have seen a rapidly growing number of neurofuzzy control applications [3]. Beside this, several software products are currently available to help with neurofuzzy problems.

An ANFIS system (Adaptive Neuro Fuzzy Inference System) [4] is a kind of adaptive network in which each node performs a particular function of the incoming signals, with parameters updated according to given training data and a gradient-descent learning procedure. This hybrid architecture has been applied mostly to the control of nonlinear single input single output (SISO) nonlinear systems [5], while

---

* Corresponding author.

C. Jayne, S. Yue, and L. Iliadis (Eds.): EANN 2012, CCIS 311, pp. 51–60, 2012.
© Springer-Verlag Berlin Heidelberg 2012

application to general multiple inputs multiple outputs (MIMO) control problems rely both on decoupling control to produce a set of SISO controllers or else designing a direct multivariable controller .

Distillation columns constitute a major part of most chemical engineering plants and remains as the most important separation technique in chemical process industries around the world [7]. Improved distillation control can have a significant impact on reducing energy consumption, improving product quality and protecting environmental resources. However, both distillation modeling and control are difficult tasks because the plant behavior is usually nonlinear, non-stationary, interactive, and is subject to constraints and disturbances [8]-[9]. Therefore, the use of adaptive techniques as the neurofuzzy approach here used can contribute to the optimization of the distillation plant operation.

There are a considerable number of specialized and general-purpose modeling software available for systems simulation [10] that can generally be classed as structure-oriented and equation-oriented approaches. A general purpose simulation environment that is often used for physical system modeling is MATLAB-SIMULINK$_{TM}$ [11]. However, this software follows a causal modeling approach and requires explicit coding of mathematical model equations or representation of systems in a graphical notation such as block diagrams, which is quite different from common representation of physical knowledge.

The object-oriented environment MODELICA$_{TM}$ [12]-[13] is ideally suited as an architectural description language for complex systems described by set of DAE, which allows the system, subsystem, or component levels of a whole physical system to be described in increasing detail by using a non-causal approach. The MODELICA$_{TM}$ environment has been used for a long time in different fields of engineering [14]-[15].

In this paper we describe the application of an adaptive network based fuzzy inference system (ANFIS) to both the adaptive modelling and the control of a binary distillation column by using a oriented-object strategy under the MODELICA$_{TM}$ environment. Results obtained during composition control operation demonstrate the effectiveness of the neurofuzzy control scheme even in the case of the plant dynamics is described by a set of nonlinear differential algebraic equations.

2   Process Description

The distillation column used in this study is designed to separate a binary mixture of ethanol and water (Fig. 1), which enters the feed tray as a continuous stream with flow rate F and composition x_F (this tray N_F located between the rectifying and the stripping section), obtaining both a distillate product stream D with composition x_D and a bottom product stream B with composition x_B. The column consists of $N = 40$ bubble cap trays. The overhead vapor is totally condensed in a water cooled condenser (tray $N+1$) which is open at atmospheric pressure. The process inputs that are available for control purposes are the heat input to the boiler Q and the reflux flow rate L (L_{N+1}). Liquid heights in the column bottom and the receiver drum (tray 0) dynamics are not considered for control since flow dynamics are significantly faster than composition dynamics and pressure control is not necessary since the condenser is opened to atmospheric pressure.

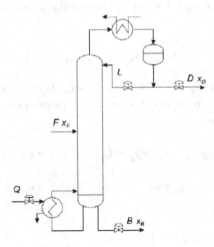

Fig. 1. Schematic of the binary distillation column, condenser and boiler, showing the input mixed flowrate F together with the distillate and the bottom product flowrates D and B

2.1   Mathematical Model of the Distillation Column

The model of the distillation column used throughout the paper is developed starting from [16], composed by the mass, component mass and enthalpy balance equations, and it has been used as basis to implement a MODELICA$_{TM}$ simulation model which describes the nonlinear column dynamics as a set of DAE multivariable system with two inputs (Q, L) and two outputs (x_D, x_B).

The mass balance equations that determine the change of the molar holdup n_l of the trays are given by

$$\dot{n}_l = V_{l-1} - V_l + L_{l+1} - L_l \tag{1}$$

only valid for $l = 1 \dots N_F - 1, N_F + 1, \dots N$ trays, where V_l and L_l are the molar vapour flux and the liquid flux leaving the lth tray respectively. For the feed tray N_F the equation is given by

$$\dot{n}_{N_F} = V_{N_F-1} - V_{N_F} + L_{N_F+1} - L_{N_F} + F \tag{2}$$

and the corresponding equations for the condenser and the reboiler are

$$\dot{n}_{N+1} = V_N - L_{N+1} - D \tag{3}$$

$$\dot{n}_0 = L_1 - V_0 - B \tag{4}$$

The composition balance equation for the most volatile component will follow the relation

$$(n_l \dot{x}_l) = V_{l-1}y_{l-1} - V_l y_l + L_{l+1}x_{l+1} - L_l x_l \tag{5}$$

valid for $l = 1 \dots N_F - 1, N_F + 1, \dots N$ where x_l is the liquid concentration of the most volatile component and y_l is the composition of the vapour flow out of the lth tray. For the feed tray we will have

$$(n_{N_F} \dot{x}_{N_F}) = V_{N_F-1}y_{N_F-1} - V_{N_F}y_{N_F} + L_{N_F+1}x_{N_F+1} - L_{N_F}x_{N_F} + Fx_F \tag{6}$$

where x_{N_F} corresponds to x_F, while for the reboiler and the condenser we have respectively

$$(n_0 \dot{x}_0) = -V_0 y_0 + L_1 x_1 - B x_0 \tag{7}$$

$$(n_{N+1} \dot{x}_{N+1}) = V_N y_N - L_{N+1}x_{N+1} - D x_{N+1} \tag{8}$$

where x_{N+1} and x_0 correspond to x_D and x_B respectively.

In order to derive the enthalpy balance equations we will assume that the condenser pressure is fixed to the outside atmospheric pressure, and that the pressure in the trays can be calculated under the assumption of constant pressure drop from tray to tray $P_l = P_{l+1} + \Delta P_l$, and the tray temperatures are implicitly defined by the assumption that the total pressure is equal to the sum of the partial pressures $P_l = x_l P_1^s(T_l) + (1 - x_l)P_2^s(T_l)$, where T_l holds for the temperature of the lth tray, and where the partial pressures $P_k^s, k = 1,2$ can be evaluated as $P_k^s(T) = \exp(A_k - \frac{B_k}{T+C_k})$, with different values of the constants according to the liquid we are evaluating.

The non-ideality of the trays and other unmodelled effects are accounted in the tray efficiencies α_l. With these tray efficiencies we can calculate the composition of the vapour flow out of the lth tray as a linear combination of the ideal vapour composition on the tray and the incoming vapour composition of the tray below by $y_l = \alpha_l \frac{P_1^s(T_l)}{P_l} x_l + (1 - \alpha_l)y_{l-1}$, starting with $y_0 = \frac{P_0^s(T_0)}{P_0} x_0$.

We will formulate then the enthalpy balances in order to determine the vapour streams that leave each tray as follows

$$(n_l \dot{h}_l^L) + n_l \left(\frac{\partial h_l^L}{\partial x_l} \dot{x}_l + \frac{\partial h_l^L}{\partial T_l} \dot{T}_l \right) = V_{l-1}h_{l-1}^V - V_l h_l^V + L_{l+1}h_{l+1}^l - L_l h_l^l \tag{9}$$

valid for $l = 1 \dots N_F - 1, N_F + 1, \dots N$, where h_l^k is the enthalpy with k indexing to the phase (L refers to liquid and V to vapour). For the feed tray and the condenser we will have

$$\left(n_{N_F}\dot{h}^L_{N_F}\right) + n_{N_F}\left(\frac{\partial h^L_{N_F}}{\partial x_{N_F}}\dot{x}_{N_F} + \frac{\partial h^L_{N_F}}{\partial T_{N_F}}\dot{T}_{N_F}\right) =$$

$$V_{N_F-1}h^V_{N_F-1} - V_{N_F}h^V_{N_F} + L_{N_F+1}h^l_{N_F+1} - L_{N_F}h^l_{N_F} + Fh^L_F$$
(10)

$$\left(n_0\dot{h}^L_l\right) + n_0\left(\frac{\partial h^L_0}{\partial x_0}\dot{x}_0 + \frac{\partial h^L_0}{\partial T_0}\dot{T}_0\right) = Q - Q_{loss} - V_0h^V_0 + L_1h^l_1 - Bh^l_0$$
(11)

where the enthalpies h^l_l and h^V_l are given by $h^l_l = x_l h^L_{l1}(T) + (1 - x_l)h^L_{l1}(T)$ and $h^V_l = y_l h^V_{l1}(T, P) + (1 - y_l)h^V_{l1}(T, P)$.

2.2   MODELICA Model of the Distillation Column

MODELICA$_{TM}$ is a modeling language that supports both high-level modeling using pre-prepared complex model components in various libraries and detailed modeling by differential algebraic equations (DAE). The graphical diagram layer and the textual layer can be efficiently combined. The basic construct in MODELICA$_{TM}$ is class. There are several types of classes with several restrictions: class, model, block, type, etc.

The concepts in oriented-object modeling follow in many aspects to oriented-object programming (e.g., inheritance). From a modeler point of view, oriented-object means that one can build a model similar to a real system by taking the components and connecting them into a model. It is very important that oriented-object environments enable the so-called acausal modeling approach, so that modelers can concentrate on physical laws and then translation process transforms all the model equations into a set of vertically and horizontally sorted equations. More precisely, the translation process transforms an oriented-object model into a flat set of equations, constants, variables and function definitions. After the flattening, all of the equations are topologically sorted according to the data-flow dependencies between the equations.

The most important difference with regard to the traditional block-oriented simulation tools is in the different way of connecting components. So, a special-purpose class connector as an interface defines the variables of the model shared with other (sub)models, without prejudicing any kind of computational order. In this way the connections can be, besides inheritance concepts, thought of as one of the key features of oriented-object modeling, enabling effective model reuse.

For the model implementation in MODELICA$_{TM}$ an appropriate modeling environment is needed. We chose Dymola$_{TM}$ [17], which has a long tradition and good maintenance. The basic idea of implementation in MODELICA$_{TM}$ is to decompose the described system into components that are as simple as possible and then to start from the bottom up, connecting basic components (classes) into more complicated classes, until the top-level model is achieved.

The whole system integrating column trays, reboiler and condenser was simulated using the object oriented language MODELICA$_{TM}$ according to the block diagram shown in Fig. 2. For this purpose, several types of connectors have been used ranging from heat flow (Q), liquid flow (L) to vapour flow (V) among others.

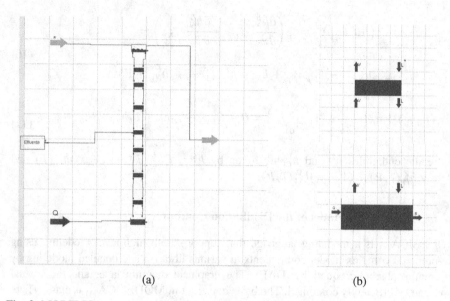

(a)   (b)

Fig. 2. MODELICA$_{TM}$ scheme of the binary distillation column (a), showing details of trays, boiler, condenser and effluent blocks (b)

It is important to highlight that the behaviour of the column's trays (excepting the feed tray) was described by a similar set of equations, so we only have modelled just one block and reused it (inheritance).

3   NeuroFuzzy Identification and Control

An ANFIS system is a kind of adaptive network in which each node performs a particular function of the incoming signals, with parameters updated according to given training data and a gradient-descent learning procedure. This hybrid architecture has been applied to the modeling and control of multiple-input single-output (MISO) systems [4].

The architecture of the ANFIS is constituted by five layers (Fig. 3). If we consider for simplicity two inputs x and y and two outputs f_1 and f_2 for a first-order Sugeno fuzzy model, with A_i and B_j being the linguistic label associated with x and y respectively, every node in layer 1 represents a bell-shaped membership function $\mu_{A_i}(x)$ or $\mu_{B_i}(y)$ with variable membership parameters. Usually we choose the bell-shaped functions. Nodes of layer 2 output the firing strength defined as the product $\omega_{ji} = \mu_{A_i}(x)\mu_{B_i}(y)$, where the set of nodes in this layer are grouped for each output j. A normalization process is computed in layer 3 giving the normalized $\bar{\omega}_{ji}$, and the Sugeno-type consequent of each rule with variable parameters p_i, q_i and r_i is implemented in layer 4 yielding f_i as the output of the single summation node $f_i = \sum_i \bar{\omega}_{ji} (p_i x + q_i y + r_i)$, and finally the single node of layer 5 computes de overall output as a summation of all incoming signals.

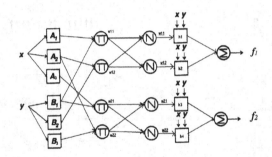

Fig. 3. Architecture of the ANFIS structure

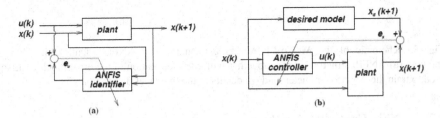

Fig. 4. Schematic of the ANFIS identification structure (a) and ANFIS model reference adaptive controller (b)

Prior to the design of the ANFIS neuro-fuzzy controller an identification process is accomplished in order to predict the plant dynamics. The MODELICA$_{TM}$ simulated binary distillation plant model described formerly has been used to obtain representative data for the training process, and an ANFIS neural network has been used as an identification model of the distillation column dynamics (Fig. 4). Once the ANFIS identifier is trained, a model reference adaptive control scheme is used to train the ANFIS controller by using the jacobian of the ANFIS identifier to propagate backwards the plant output error to the ANFIS controller [18].

Both ANFIS structures have been designed under the MODELICA$_{TM}$ simulation environment. We have assumed that all state variables were measurable and thus available. This assumption is often unrealistic, since usually the only data available are manipulated input-controlled output measurements.

3.1 ANFIS Identification Structure

The ANFIS inference model considered is constituted by two inputs Q and R (percentage of reflux valve opening) and two outputs x_D and x_B and is defined by 9 if-then Sugeno fuzzy rules with 3 membership functions associated to each input variable. A corresponding membership function's parameter and consequent's parameter sets have been tuned by using the backpropagation of the plant's output error in composition starting from a set of I/O plant operation data constituted by 2500 points. A schematic diagram of this structure has been implemented in MODELICA$_{TM}$ (Fig. 5).

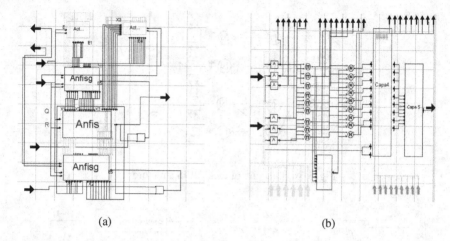

(a)   (b)

Fig. 5. Schematic of the MODELICA$_{TM}$ implementation of ANFIS identification block, showing (a) ANFIS structure (Anfis) together with the propagation of the error block (Anfisg) used to train the ANFIS parameter and (b) details of the five layer disposition

3.2   ANFIS Controller Structure

The ANFIS neurofuzzy controller receives as inputs both the actual composition error and its integral for each of two compositions x_D and x_B, and produces as outputs the heat flow Q and the percentage of reflux valve opening R, each one through the use of two distinct ANFIS structures in parallel (Fig. 6).

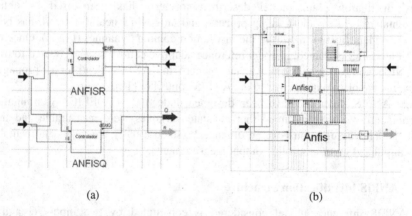

(a)   (b)

Fig. 6. Schematic of the MODELICA$_{TM}$ implementation of neurofuzzy controller, showing the (a) two ANFIS control structures in parallel (ANFISR and ANFISQ), being each one (b) similar to the ones defined previously

The ANFIS inference model considered is constituted by two inputs Q and R (percentage of reflux valve opening). The results obtained show the validity of the MODELICA$_{TM}$ simulation model as compared with the results reported by [16].

A set of 6 Sugeno fuzzy rules with 3 membership functions associated to each input variable has been used to define each ANFIS controller structure, which has been tuned by using the ANFIS model previously defined and the same set of I/O plant operation data already gathered. It has been used a double PI controller (one for each manipulated variable) previously tuned to estimate the number of Sugeno rules and its parameter values as initial condition at the training stage.

4   Results

In order to demonstrate the performance of the neurofuzzy control approach, a MODELICA$_{TM}$ simulation model of the distillation column has been operated to separate a mixture of ethanol and water under different operational conditions (F, x_F, $L(R)$ and Q variable) starting from a 50% composition in alcohol for each tray as initial condition. The validity of the MODELICA$_{TM}$ simulation model has been contrasted with the results reported by [16].

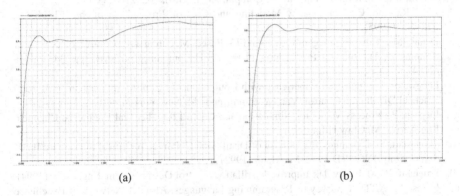

(a)   (b)

Fig. 7. Distillation column response to (a) changes in x_D set point value from 90 to 95% and for a 30% feed composition (b) 1000 W/min flow heat to the boiler, reflux rate opening of 90%, and molar flow rate of 2 mol / s and a feed composition of 0.3 in alcohol.

In Fig. 7.a it is shown the response for set point changes in ethanol top composition x_D obtained under the MODELICA$_{TM}$ neurofuzzy controller, while in Fig. 7.b it is demonstrated the ability of the proposed approach to maintain the desired x_D set point when a disturbance is presented as a change in the feed composition x_F.

5   Conclusions and Future Works

In this paper it has been described the application of an adaptive network fuzzy inference system (ANFIS) to both the adaptive modeling and the control of a binary distillation column by using a oriented-object strategy under the MODELICA$_{TM}$ environment. Neurofuzzy modeling and control modules have been programmed into the MODELICA$_{TM}$ simulation environment due to its modularity and adaptability, which have made possible the inheritance and reuse of previously defined structures. The results obtained demonstrate the validity of the proposed approach to control the distillation column when both set-point changes and disturbances are present.

The ease of programming using $MODELICA_{TM}$ contrasts with that of $SIMULINK_{TM}$, mainly due to the object-oriented features it exhibits $MODELICA_{TM}$ besides the DAE representation which enables the non-causal approach as opposed to the causal one which uses $SIMULINK_{TM}$. This modeling approach is not unique to chemical engineering processes, and can be easily extended to other fields.

Future works are directed to the stability and robustness analysis of the neuro-controlled distillation plant when modeling errors are considered.

References

1. Bulsari, A.: Neural Networks for Chemical Engineers. Elsevier, Amsterdam (1995)
2. Hussain, M.A.: Review of the Applications of Neural Networks in Chemical Process Control. Simulation and online Implementations. Artif. Intel. Eng. 13, 55–68 (1999)
3. Engin, J., Kuvulmaz, J., Omurlu, E.: Fuzzy control of an ANFIS model representing a nonlinear liquid level system. Neural Comput. Appl. 13(3), 202–210 (2004)
4. Jang, R., Sun, C.: Neuro-fuzzy Modeling and Control. Proceedings of the IEEE 83(3), 378–405 (1995)
5. Denai, M., Palis, F., Zeghbib, A.: ANFIS Based Modeling and Control of Nonlinear Systems: A Tutorial. In: Proceedings of the IEEE Conference on SMC, pp. 3433–3438 (2004)
6. Leegwater, H.: Industrial Experience with Double Quality Control. In: Luyben, W.L. (ed.) Practical Distillation Control. Van Nostrand Reinhold, New York (1992)
7. Luyben, P.: Process Modeling, Simulation and Control for Chemical Engineers. Chemical Eng. Series. McGraw Hill (1990)
8. Skogestad, S.: Dynamics and Control of Distillation Columns. A Critical Survey. In: IFAC Symposium DYCORD 1992, Maryland (1992)
9. Fruehauf, P., Mahoney, D.: Improve Distillation Control Design. Chem. Eng. Prog. (1994)
10. Burns, A.: ARTIST Survey of Programming Languages. ARTIST Network of Excellence on Embedded Systems Design (2008)
11. MATLAB (2010), http://www.mathworks.com/
12. Mattsson, S.E., Elmquist, H., Otter, M.: Physical System Modelling with Modelica. Control Eng. Pract. 6, 501–510 (1998)
13. Bruck, D., Elmqvist, H., Olsson, H., Matteson, S.E.: Dymola for Multi-engineering Modelling and Simulation. In: Proceedings of the 2nd Int. Modelica Conference, Germany, pp. 551–558 (2002)
14. Tiller, M.: Introduction to Physical Modeling with Modelica. Kluwer Ac. Press (2001)
15. Fritzson, P.: Principles of Object-oriented Modeling and Simulation with Modelica 2.1. Wiley-IEEE Press (2003)
16. Diehl, M., Uslu, I., Findeisen, R.: Real-time optimization for large scale processes: Nonlinear predictive control of a high purity distillation column. In: On Line Optimization of Large Scale Systems: State of the Art. Springer (2001)
17. Zauner, G., Leitner, D., Breitenecker, F.: Modeling Structural Dynamic Systems in MODELICA/Dymola, MODELICA/Mosilab and AnyLogic. In: Proceedings of the International Workshop on Equation-Based Oriented Languages and Tool (2007)
18. Norgaard, M., Ravn, O., Poulsen, N.K., Hansen, L.K.: Neural Networks for Modeling and Control of Dynamic Systems. Springer (2000)

Evolving an Indoor Robotic Localization System Based on Wireless Networks

Gustavo Pessin[1,3], Fernando S. Osório[1], Jefferson R. Souza[1], Fausto G. Costa[1], Jó Ueyama[1], Denis F. Wolf[1], Torsten Braun[2], and Patrícia A. Vargas[3]

[1] Institute of Mathematics and Computer Science (ICMC)
University of São Paulo (USP) - São Carlos, SP, Brazil
{pessin,fosorio,jrsouza,fausto,joueyama,denis}@icmc.usp.br
[2] Institute of Computer Science and Applied Mathematics
University of Bern - Bern, Switzerland
braun@iam.unibe.ch
[3] School of Mathematical and Computer Sciences (MACS)
Heriot-Watt University - Edinburgh, UK
p.a.vargas@hw.ac.uk

Abstract. This work addresses the evolution of an Artificial Neural Network (ANN) to assist in the problem of indoor robotic localization. We investigate the design and building of an autonomous localization system based on information gathered from Wireless Networks (WN). The paper focuses on the evolved ANN which provides the position of one robot in a space, as in a Cartesian plane, corroborating with the Evolutionary Robotic research area and showing its practical viability. The proposed system was tested on several experiments, evaluating not only the impact of different evolutionary computation parameters but also the role of the transfer functions on the evolution of the ANN. Results show that slight variations in the parameters lead to huge differences on the evolution process and therefore in the accuracy of the robot position.

1 Introduction

Mobile robot navigation is one of the most fundamental and challenging directions in mobile robot's research field and it has received great attention in recent years [6]. Intelligent navigation often depends on mapping schemes which turns out on depending on the localization scheme. For indoor or outdoor environments the mapping and localization schemes have their own features. Thereby, localization is a key problem in mobile robotics and it plays a pivotal role in various successful mobile robot systems [13].

This paper describes an investigation on the evolution of a system for indoor localization, which learns the position of one robot based on information gathered from WNs. The signal strength of several wireless nodes are used as input in the ANN in order to measure the position of one robot in an indoor space. The evolution of the ANN, detailed in section 2.1 is done using Particle Swarm Optimization [1,3]. We show the complete hardware and software architecture

C. Jayne, S. Yue, and L. Iliadis (Eds.): EANN 2012, CCIS 311, pp. 61–70, 2012.

for the robotic system developed so far. Our main focus in this work is to report findings which corroborate the use of evolutionary computation techniques to create autonomous intelligent robots [5].

In indoor spaces, sensors like lasers and cameras might be used for pose estimation [10], but they require landmarks (or maps) in the environment and a fair amount of computation to process complex algorithms. These sensors also have a limited field of vision, which makes harder the localization task. In the case of video cameras, the variation of light is also a serious issue. Another commonly used sensor is the encoder, which provides odometry. Odometry is an useful source of information in some cases [8] but it has an incremental error that usually invalidates its use in real systems.

Wireless Networks (WNs) are widely available in indoor environments and might allow efficient global localization while requiring relatively low computing resources. Other advantages of this technology are that it may provide high degrees of scalability and robustness. However, the inherent instability of the wireless signal does not allow it to be used directly for accurate position estimation. One machine learning technique that could reduce the instability of the signals of the WN is the Artificial Neural Networks; due to its capacity to learn from examples, as well as the generalization and adaptation of the outputs [9].

In [2], it has been shown that obtaining an absolute performance in localization, by means of a WN, depends on the environmental configuration. This means that different approaches are required for different environments, such as using different kinds of signals and filters. Evaluations in large indoor areas (like a building) present specific difficulties not always the same as in small indoor areas (like a room). These difficulties are related to the problem of attenuation and reflection of the signals on the walls and the different sources of interferences. The use of WNs addressing localization inside a building can be seen in [4,7]. Another approach for localization uses Wireless Sensor Network (WSN) where a large number of small sensors are used to pick up information from the environment. The information acquired by the sensors can be regarded as a fingerprint [12]. The drawback of the canonical WSN approach is that, as it requires a huge number of resources, it could make the system more expensive.

This paper has the following structure: Section 2 outlines the methodology that is employed to set up and to evaluate the experiments. Section 3 describes all the evaluations that have been carried out. The final section makes some concluding comments and examines the future perspectives of this area of research.

2   Methodology

The indoor localization system uses an evolved ANN[1]. The inputs of the ANN are signals strength measurements from WNs (802.11b/g) received by the robot[2] from 8 statically positioned Access Points (AP) as shown in Fig. 1.

---

[1] Source-code and data files used to evolve the ANN are available in goo.gl/vfXN2
[2] Although we have used the humanoid robot NAO, the proposed methodology may be applied to any kind of device with wireless capabilities.

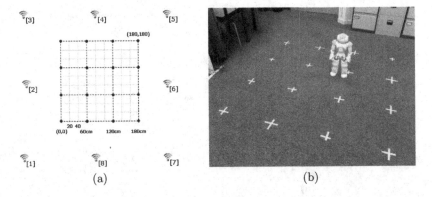

Fig. 1. (a) Graphical representation of the working area – It represents an area of 180 cm by 180 cm. (b) Picture of the working area with the robot, similar to what is represented in Figure 1(a). Each small cross are placed 60 cm long.

The evolution of the ANN is carried out using data collected by the robot. We use the robot inside the working area (Fig. 1(b)) and collected 3 minutes readings (i.e. ≈180 readings) at each marked point. With a displacement of 60 cm, mapping out a plane of 180 cm by 180 cm, it means 16 points to read resulting in ≈2880 readings altogether.

The signal obtained from the WN is the Received Signal Strength Indication (RSSI). This value is obtained with the aid of the GNU/Linux command *iwlist* (used as *iwlist <interface> scanning*). As we use the *iwlist* command, there is no need to establish a connection (or login) with different specific networks. The scan of the networks, without a connection, provides enough information for this evaluation. Without a connection, the system becomes easier to use, more lightweight and flexible. Furthermore, as the robot NAO has an operating system based on GNU/Linux, this approach may be generalized to any other GNU/Linux based system.

Our approach relies on the ANN learning and generalization capabilities in an attempt to reduce the effect of unstable data (due to signal strength oscillation), and increase the accuracy of the position estimation of the robot. However, as the values obtained from the reading of the APs are quite unstable, we improve the learning capability using the noise filter proposed in [11]. The behaviour of the noise filter can seen in Fig. 2(a), where two lines represent scanning of one AP in a period of time. The red line shows the raw value and the black line shows how the median filter removes some of the noise. Although it generates a delay of 8 seconds in acquiring the new position it was shown in [11] that the accuracy turns out to be widely better.

The number of neurons in the input layer is equivalent to the number of available APs – as we use 8 APs, the inputs of the ANN use one neuron for each network signal. The order is important, and hence, AP 1 was linked to neuron 1, AP 2 with neuron 2 and so on. The outputs of the network are two values, i.e. the coordinates (x, y). We measure the output errors by using the Euclidean

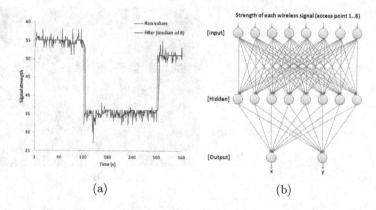

(a)   (b)

Fig. 2. (a) Sample of filter behaviour – The red line shows the raw value from one of the access points. The black line shows how the median filter removes some of the noise. (b) Example of ANN topology.

distance, as shown in Eq. 1. The value d is the error (distance, in centimetres), (x_1, y_1) are the expected values from the ANN validation set and (x_2, y_2) are the obtained value while using the ANN.

$$d = \sqrt{(x_2 - x_1)^2 + (y_2 - y_1)^2} \tag{1}$$

2.1 Evolutionary Localization

As we seek to evaluate ANN characteristics and also the evolution process, we started with a simple ANN topology as it can be seen in Fig. 2(b). We evaluate changes in the ANN topology and in the role of the transfer function. Furthermore, as the evolution is carried out using Particle Swarm Optimization (PSO) [1,3] we evaluate several aspects that influence the search efficiency in the PSO. These aspects are related to confidence models, neighbourhood topology, and inertia among others.

We use PSO to evolve two different structures. In the first one, we use the PSO to evolve just the ANN weights. In the second, we evolve the ANN weights plus the slope of the transfer function. As an example, the ANN in Fig. 2(b) has 80 connections plus 10 weights for bias, hence, the PSO particle has 90 positions (i.e. it is an vector with 90 positions). For the slope, we consider its use just in the hidden layer, and it is the same value for all neurons. Hence, it adds only one more value to the PSO particle. The evolutionary process considerers 2/3 of the dataset as training data and 1/3 as validation data. The ANNs are evolved for 10k cycles (generations) and all presented results are from the validation dataset.

The PSO is a stochastic optimization technique, inspired by social behaviour of bird flocking and fish schooling [1,3]. The optimization process occurs in two different ways simultaneously: through cooperation (group learning) and competition (individual learning) among particles (individuals) from a swarm (population). It shares many concepts with evolutionary computation techniques such as Genetic Algorithms (GA), where there is an initial population (where each individual represents a possible solution) and a fitness function (whose value represents how far an individual is to an expected solution). However, unlike GA, PSO has no explicit concepts of evolution operators such as crossover or mutation. In the PSO, there is a swarm of randomly created particles. On each algorithm iteration, each particle is updated following: (i) the best population fitness; (ii) the best fitness found by the particle (considering past generations of the particle). Each particle has a position x (or a position vector) and a velocity v (or velocity vector). The position represents a solution for the problem and the velocity defines the particles displacement direction weight.

New particle's position is given by Eq. 2. Where x_k^i is the position of particle i at instant k and v_k^i is particle's i velocity at k moment. Particle's velocity is updated according to Eq. 3.

$$x_{k+1}^i = x_k^i + v_{k+1}^i \qquad (2)$$

$$v_{k+1}^i = w \cdot v_k^i + c_1 \cdot r_1(pbest - x_k^i) + c_2 \cdot r_2(gbest - x_k^i) \qquad (3)$$

On Eq. 3, v_k^i is the particle actual velocity, w represents a particle inertia parameter, $pbest$ is the best position among all positions found by the individual (particle best), $gbest$ is the best position among all positions found by the group (group best), c_1 and c_2 are trust parameters, r_1 and r_2 are random numbers between 0 and 1. Parameters (w, c_1, c_2, r_1 e r_2) are detailed below.

The velocity is the optimization's process guide parameter [3] and reflects both particle's individual knowledge and group knowledge. Individual knowledge is known as *Cognitive Component* while group knowledge is known as *Social Component*. Velocity consists of a three-term sum: (i) Previous speed: utilized as a displacement direction memory and can be seen as a parameter that avoids drastic direction changes; (ii) Cognitive Component: directs the individual to their best position found so far (i.e. memory of the particle); (iii) Social Component: directs the individual to the best particle in the group.

Parameters c_1 and c_2 (confidence or trust) are used to define individual or social tendency importance. Default PSO works with static and equal trust values ($c_1 = c_2$), which means that the group experience and the individual experience are equally important (called *Full Model*). When parameter c_1 is zero and parameter c_2 is higher than zero, PSO uses only group information (called *Social Model*). When parameter c_2 is zero and parameter c_1 is higher than zero, PSO uses only particle's information, disregarding group experience (called *Cognitive Model*). Random value introduction (r_1 and r_2) on velocity adjustment allows PSO to explore on a better way the search space [3]. Inertia parameter aims to balance local or global search. As the value approximates to 1.0, search gets close to global while lower values allow local search. Usually this value is between 0.4 and 0.9. Some authors suggest its linear decay, but they warn that it is not always the best solution. Most parameters are problem-dependent [3,14].

3 Results

This section describes the four steps that were carried out to perform the evaluations, considering changes in the ANN and PSO techniques. As a first step we sought to evaluate the impact of using different range values in the initialization of PSO's velocity and position. Furthermore we evaluate how the number of generations and the swarm size impact the evolution.

Table 1(a) shows the evaluation set related to swarm size and the number of generations. Table 1(b) shows the evaluation set for weights of connections (i.e. position) and velocity parameters in the PSO.

Table 1. First evaluation set performed with the PSO

(a)				(b)		
Generations	Swarm Size			Velocity	Position	
	200	500	1000		{-2.0;2.0} {-5.0;5.0} {-20.0;20.0}	
500	E1	E2	E3	{-2.0;2.0}	A1 A2 A3	
1000	E4	E5	E6	{-5.0;5.0}	A4 A5 A6	
2000	E7	E8	E9	{-20.0;20.0}	A7 A8 A9	

We performed 25 runs for each parameter set, considering different random seeds on each initialization. The results can be seen in Fig. 3. We can see in Fig. 3(a) that more generations and bigger swarm size provides better results (E9). However, even using swarm size equal to 1,000 and number of generations equal to 2,000 the evolution process was not reaching learning stabilization (the learning curve still have slightly improvements).

In Fig. 3(b) we can see that the two sets that obtained the best (lower error) results considering the range of position and velocity are A8 and A2. Both use positions between {-5.0;5.0} but have different velocities range. We can see that although A8 has the lowest minimum error it has the biggest dispersion among all. The A2 set has the lowest median and a minimum error close to A8. Although, considering the graph scale (2 cm), all results are not much different. Thereafter, to the next steps, we maintain the set A2 in the PSO. However, given that the evolution process was not reaching complete stabilization, we extrapolate E9 with 10k generations instead of 2k.

As a second step, we wanted to understand the behaviour of the PSO evolving the ANN considering confidence models, the inertia and the role of the transfer function. Table 2 shows the evaluated parameter set. Linear and Logistic are the two types of transfer function used in the ANN hidden layer.

We can see that results (Fig. 4) using the *Cognitive Model* or the *Social Model* were worse than using the *Full Model*. The four best sets are {Fi3, Fi5, Fo3, Fo5}. We can see that the sets with Logistic Transfer Function obtained best results near to 15 cm while the sets with Linear Transfer Function obtained best results near to 45 cm. In these sets, we can also see that when inertia was used as 0.3 we have better results than using inertia equal to 0.5 or 0.7. It makes sense due to the fact that lower the inertia better the local search (fine tuning). Hence, the best parameter set which we maintain to next steps is Fo3.

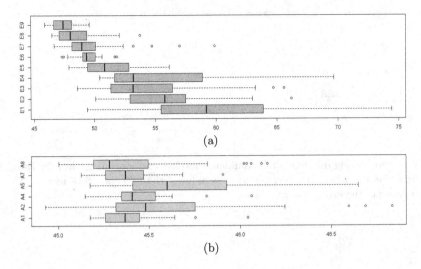

Fig. 3. (a) Results using different swarm size and number of generation (Table 1(a)). (b) Results using different initialization in PSO position and velocity (Table 1(b)). The graphs are in different scales. The x axes represent the error in centimetres.

Table 2. Second evaluation set performed with the PSO (inertia and confidence models) and the ANN (transfer function)

Inertia	Full		Social		Cognitive	
	Linear	Logistic	Linear	Logistic	Linear	Logistic
0.3	Fi3	Fo3	Si3	So3	Ci3	Co3
0.5	Fi5	Fo5	Si5	So5	Ci5	Co5
0.7	Fi7	Fo7	Si7	So7	Ci7	Co7

Fo3 set uses Logistic Transfer Function with a slope of 0.02. We also had a different PSO evolving the slope, however, the results when we leave the PSO to evolve the slope were similar to the use of the pre-defined value. Of course, the finding of slope equal to 0.02 was not trivial as it was encountered analysing the output of the sum of the hidden layers. Such a situation may encourage the use of the PSO in order to find the slope of the Transfer Function. We also performed some evaluations taking into account linear decay of the inertia value, but the results were not better than the currently configuration.

In the third step, we sought to evaluate the impact of using different PSO neighbourhood topologies (NT). PSO NT are related to the choice of the best particle to follow. In the star model, all particles follow the best among all. Nevertheless, it may sometimes be more susceptible to local minima. Others NT may be less susceptible to local minima, where the particle does not to follow the global best but the best of some subpopulation. It, in general, also increases the diversity. We have evaluated 4 different types of NT: (i) the star model; (ii) two subpopulations – i.e. 2 groups of neighbours; (iii) four subpopulations – i.e. 4 groups of neighbours; and (iv) four subpopulations but being the best particle to follow the average among the two best in the subpopulations. Upon these results

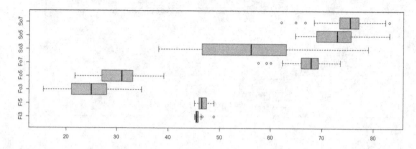

Fig. 4. Results using different PSO confidence models and inertia, as show in Table 2. We show the sets where the average error were lower than 80 cm − 8 of 18 sets. The x axis represents the error in centimetres.

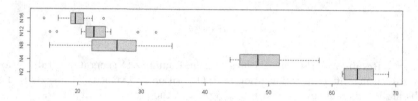

Fig. 5. Results using different number of neurons in the ANN hidden layer. The graphs are in different scales. The x axes represent the error in centimetres.

we performed a statistical test to verify its significance. Using t-test among all sets they are accepted as equivalent (using 95% of confidence). It means that, for this problem, the use of different NT does not substantially affect the results.

The final step was the evaluation of the number of neurons in the hidden layer. We evaluate the use of 2, 4, 8, 12 and 16 neurons. Results can be seen in Fig. 5. We can see that using 2 and 4 neurons, the results are quite bad. Using 8 and 12 neurons they have good minimum errors but high dispersion. The ANN with 16 neurons presents the lowest error and the more homogeneous results. Statistical analyses (t-test) upon N12 and N16 shows that, with 95% of confidence, they are not accepted as equivalent (p-value of 0.013), i.e., the use of 16 neurons does improve the system.

After the evaluation of the number of neurons in the hidden layer, we performed a new evaluation considering the best set found so far but running the evolutionary process for 100k generations instead of 10k. Results can be seen in Fig. 6. The data shown in Fig. 6(a) and 6(b) presents an average error and standard deviation, respectively, of (14.6, 18.4) and (10.2, 14.4), i.e. running the PSO for 100k generations allowed us to decrease the average error in more ≈30%. We can see that Fig. 6(a) has less results in the first class (0 to 5 cm) and a bigger tail than Fig. 6(b). Statistical evaluation using the Mann-Whitney test (as it is a non normal distribution) showed p-value equal to $1.517e^{-11}$ (not accepted as equivalent using 95% of confidence). Fig. 7 shows a section

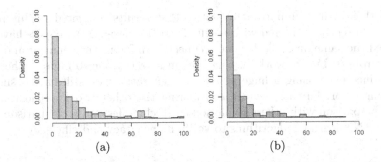

Fig. 6. Histogram of the localization error using the best acquired ANN. (a) Using 10k generation in the PSO. (b) Using 100k generation in the PSO. The x axes represent the error in centimetres.

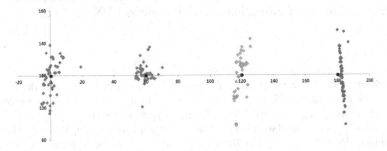

Fig. 7. Section of the plane (Fig. 1(a)) with expected and obtained values for 4 coordinates. The black dots are the expected value; diamonds show the obtained values. Each colour represents a different position. Axes x and y are in centimetres.

of the plane (Fig. 1(a)) with expected and obtained values for 4 positions, using the best acquired ANN. For all positions in the evaluated plane, 86% of the errors are below 20 cm.

4   Conclusion and Future Work

In this paper we have shown an investigation addressing the evolution and the use of an ANN to assist in the problem of indoor localization by using data gathered from WNs. Upon the data obtained from the WN we employed a median noise filter as shown in [11]. We evaluated several PSO and ANN settings. Results showed that the use of transfer function in the ANN along with the PSO Full model allowed us to decrease the average error from ≈45 cm to ≈15 cm. Also, we might see that the use PSO neighbourhood topology did not allow us to have any significant improvement. Finally, the system could be improved using larger number of neurons in the hidden layer and larger number of generations, leading to an average error of ≈10 cm.

Nevertheless, it is important to notice that results presented in this paper cannot be directly compared with results from [11] because the papers have not employed the same data, as they were collect in different environment and with different robots. Future work may include an investigation to improve the local search, since even using a huge number of generations we still have a slightly decreasing error. Further, we are considering the other two approaches, that is, the comparison with others Evolutionary Techniques and a comparison with classical ANN learning algorithms to verify its accuracy and efficiency.

Acknowledgments. The authors would like to acknowledge the financial support granted by CNPq and FAPESP to the INCT-SEC (National Institute of Science and Technology – Critical Embedded Systems – Brazil), processes ID 573963/2008-8 and 08/57870-9. Also, we would like to acknowledge the Capes Foundation, Ministry of Education of Brazil, process BEX 4202-11-2.

References

1. Eberhart, R.C., Kennedy, J., Shi, Y.: Swarm Intelligence. M. Kaufmann (2001)
2. Elnahrawy, E., Li, X., Martin, R.: The limits of localization using signal strength: a comparative study. In: IEEE SECON, pp. 406–414 (2004)
3. Engelbrecht, A.P.: Fundamentals of Comp. Swarm Intelligence. Wiley (2005)
4. Espinace, P., Soto, A., Torres-Torriti, M.: Real-time robot localization in indoor environments using structural information. In: IEEE LARS (2008)
5. Fogel, D.: Evolutionary Computation: Toward a New Philosophy of Machine Intelligence. IEEE Press Series on Computational Intelligence (2006)
6. Fu, S., Hou, Z., Yang, G.: An indoor navigation system for autonomous mobile robot using wsn. In: Networking, Sensing and Control, pp. 227–232 (2009)
7. Ladd, A., Bekris, K., Rudys, A., Wallach, D., Kavraki, L.: On the feasibility of using wireless ethernet for indoor localization. IEEE Trans. on Robotics and Automation 20(3), 555–559 (2004)
8. Martinelli, A.: The odometry error of a mobile robot with a synchronous drive system. IEEE Trans. on Robotics and Automation 18(3), 399–405 (2002)
9. Mitchell, T.M.: Machine Learning. McGraw-Hill (1997)
10. Napier, A., Sibley, G., Newman, P.: Real-time bounded-error pose estimation for road vehicles using vision. In: IEEE Conf. on Intelligent Transp. Systems (2010)
11. Pessin, G., Osório, F.S., Ueyama, J., Souza, J.R., Wolf, D.F., Braun, T., Vargas, P.A.: Evaluating the impact of the number of access points in mobile robots localization using artificial neural networks. In: Proc. of the 5th International Conference on Communication System Software and Middleware, pp. 10:1–10:9 (2011)
12. Robles, J., Deicke, M., Lehnert, R.: 3d fingerprint-based localization for wireless sensor networks. In: Positioning Navigation and Communication, WPNC (2010)
13. Thrun, S., Fox, D., Burgard, W., Dellaert, F.: Robust monte carlo localization for mobile robots. Artificial Intelligence 128(1-2), 99–141 (2001)
14. Yao, X.: Evolving artificial neural networks. Proc. of IEEE 87(9), 1423–1447 (1999)

Employing ANN That Estimate Ozone in a Short-Term Scale When Monitoring Stations Malfunction

Antonios Papaleonidas and Lazaros Iliadis

Democritus University of Thrace , Department of Forestry & Management of the Environment
& Natural Resources, 193 Pandazidou st., 68200 N Orestiada, Greece
papaleon@sch.gr, liliadis@fmenr.duth.gr

Abstract. This paper describes the design, development and application, of an intelligent system (operating dynamically in an iterative manner) capable of short term forecasting the concentration of dangerous air pollutants in major urban centers. This effort is the first phase of the partial fulfillment of a wider research project that is related to the development of a real time multi agent network serving the same purpose. Short term forecasting of air pollutants is necessary for the proper feed of the real time multi agent system, when one or more sensors are damaged or malfunctioning. From this point of view the potential outcome of this research is very useful towards real time air pollution monitoring. A vast volume of actual data vectors are combined from several measurement stations located in the center of Athens. The final target is the continuous estimation of Ozone (O_3) in the historical city center, considering the effect of primitive pollutants and meteorological conditions from neighboring stations. A group comprising of hundreds artificial neural networks has been developed, capable of estimating effectively the concentration of O_3 at a specific temporal point and also after 1, 2, 3 and 6 hours.

1 Introduction

Due to the importance of the air pollution problem in major urban centers, the optimization of forecasting and depicting mechanisms should be a primary target of the civil protection authorities.

The basic task accomplished by this research is the effective short term estimation of O_3 in the historical center of Athens at a specific moment and then successively in a horizon of (h+1) hours ($0<=h<=5$) with the employment of neural modeling. The most important design aspect that gives more merit to this effort is the concurrent combination of actual real time meteorological and air pollution data, from surrounding stations that have a serious effect in the situation inside the ring of the city center.

In the literature, there are some cases of Artificial Neural Networks (ANN) towards long term estimation and modeling of air quality, and O_3 concentration (Wahab and Alawi, 2002), (Paschalidou et al., 2007), (Iliadis et al., 2007), (Ozcan et al., 2007), (Ozdemir et al., 2008), (Inal, 2010). Also there are papers that deal with the hourly estimation of O_3 (Paoli, 2011). However all of these research efforts resulted in the

C. Jayne, S. Yue, and L. Iliadis (Eds.): EANN 2012, CCIS 311, pp. 71–80, 2012.

employment of an optimal ANN which had the best output. More over they examined the problem with data related to specific periods of time and seasons (e.g. summer or spring) and not for long temporal periods.

1.1   Innovation of This Research

The innovation of the approach presented here is that it aims in overcoming the three basic limitations of the existing methods. The first limitation is the potential incapability of the ANN to perform properly in the case of lack of data or lack of "accurate" data vectors, due to unexpected potential malfunction of the sensors. The possibility for such a problem is quite high, despite the credibility of modern measurement stations, due to their service requirements and also due to random events. This is a major problem especially when ANN are a part of an intelligent system that makes short term estimations of vulnerability, or when the data sent by the simple sensors feed a common central monitoring system.

The following table 1 provide evidence of the actual extend of the problem. It presents the high percentage of measurements' loss for five air pollution stations in the wider area of Athens during the time period 2003-2004 (MinEnv, 2012). The areas corresponding to the codes of table 1 will be explained later in this chapter.

Table 1. Percentage of data loss in measurements' stations in Athens 2003-2004

	2003		**2004**	
Expected number of hourly meteorological and air pollution measurements	8760		8784	
Station code	Actual number of measurements	Percentage % of lost values	Actual number of measurements	Percentage % of lost values
AGP	8408	**4,02**	7684	**12,52**
ATH	2292	**73,84**	6097	**30,59**
ARI	8678	**0,94**	8397	**4,41**
GEO	3538	**59,61**	5514	**37,23**
PAT	6632	**24,29**	7474	**14,91**

It is quite impressive, that in several cases the data loss appears to be as high as 74%.

A second problem that motivated this research is the difficulty of existing forecasting models to adjust in serious temporal changes of the respective data. In major urban centers like Athens the transfer of air pollutants from one area to the other through the changes of the wind direction are really important for the determination of the O_3 concentrations. From the following figure 1 it is clear that in 2010 the measurements of "Patision" station have been influenced by 27% from pollutants, transferred by north to north east winds, whereas this number for south to south west winds was almost as high as 20% (MinEnv, 2011).

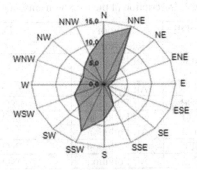

Fig. 1. Wind direction distribution for the "Patision" station for year 2010

Indeed, data coming from stations like "Amarusio" or "Lykovrish" which are located in the north, would offer better results from data coming from the southern part of Athens like "Piraeus" because the frequency of north wind is higher. However in cases of strong southern wind the ANN would have a problem of adjusting itself because it would have been trained with data coming from the northern stations.On the other hand extremely complicated ANN that combine data from all stations do not offer good levels of convergence as it will be shown later. Finally the need for a model that covers the whole year is high, because the problem is not met only in the summer but it is also high in winter (eg in February) and in spring (March) when we have also very high ozone concentrations (MinEnv, 2011).

This research effort is innovative because it faces the problem under all of the above described perspectives and it contributes towards a more rational model development.

1.2   Area of Research

The following table 2 presents all of the stations that were used in this study. Some of them are performing meteorological measurements whereas others measure air pollutants' levels. Data are provided by air pollution department of Ministry of Environment, Energy & Climate Change (MinEnv, 2012). More specifically mean hourly values gathered in 24hours basis, concerning 8 air pollution stations and 2 meteorological stations for a period starting at the day of establishment for each station up to 31/12/2010 for all stations.

The station of "Athinas" is one of the oldest and it is located in the heart of the city. Looking at the data related to the "Athinas" station, in the 23.31% of the measurements at least one of the primitive pollutants is missing (MinEnv, 2012). The actual high level of missing values of this station and its location in the center of the city made this spot ideal to be selected as the point of study and application. The value -9990,00 has been recorded when a measurement station is malfunctioning or if there is a missing value for any other reason. These cases were not taken into consideration by the model in order to avoid noise.

Table 2. Description of the measurements' stations

ID	Station's name	CODE	Type of Measurement	Data measured
1	Ag. Paraskevi	AGP	Air Pollution	NO, NO_2
2	Amarusio	MAR	Air Pollution	NO, NO_2, CO
3	Peristeri	PER	Air Pollution	NO, NO_2, CO, SO_2
4	Pathsion	PAT	Air Pollution	NO, NO_2, CO, SO_2
5	Aristotelous	ARI	Air Pollution	NO, NO_2
6	Geoponikis	GEO	Air Pollution	NO, NO_2, CO, SO_2
7	Piraeus 1	PIR	Air Pollution	NO, NO_2, CO, SO_2
8	N Smyrnh	SMY	Air Pollution	NO, NO_2, CO, SO_2
9	Penteli	PEN	Meteorological	Temp, Wspeed, SunTime, Wdirection, Radiation, RH
10	Thiseion	THI	Meteorological	Temp, Wspeed, SunTime, Wdirection, Illumin, RH
	Athinas	ATH	Air Pollution	NO, NO_2, CO, SO_2, O_3

Image 1. Location of the measurement stations in the wider area of Athens

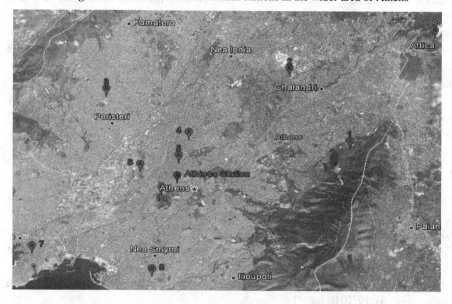

2   Developing the Neural Network

2.1   Determining the Layers' Structure

The data vectors of all pollutant measuring stations (hourly values) were combined in every possible way, in order to form the input vectors of the developed neural networks. Thus, ANN have been developed, having as many input neurons as the aggregation of the number of the independent parameters related to 2 stations, 3 stations, 4, 5, 6, 7 and 8, till all potential combinations between air pollution stations are done (for details on the considered features see table 3). Totally 247 ANN were developed after following the described approach. The input parameters related to each one of the 247 ANN were combined with the features of each of the 2 meteorological stations, in order to form the input layers of additional networks. The last ANN produced, had an input layer comprising of $n+m$ input neurons where n and m are the number of independent parameters of the two meteorological stations respectively.

Thus, a total number of 247X3=741 ANN were developed and they were assigned a vast volume of data vectors. All of the networks were multi layer feed forward and they employed back propagation optimization (Kecman, 2001) (Haykin,1999) (Jones, 2005).

The following table 3 presents a small sample of the stations considered by some of the networks. Obviously not all ANN have the same input layers. Additionally, every input vector contains also the number of the month, the day of the week (in a scale from 1 to 7) and the exact time of the current measurement (in a scale from 1 to 24) in its first four fields. This was scheduled based on the annual reports of air quality of Ministry of Environment (MinEnv, 2011) (MinEnv, 2010). Also, except from the current pollutant concentration values, hourly values of each one of the previous 4 hours and hourly meteorological values of the previous 7 hours were considered. This approach was determined after several trial and error experiments. Tests with average daily values did not offer good results at all.

Concluding it must be clarified that the number of input neurons varies from 85 to 235 depending on the case and on the combination.

Table 3. Sample of the stations considered by a sample of the developed ANN

Network Id	Stations considered by the network
1	THI, ARI, PAT
2	THI, ARI, GEO
3	THI, ARI, SMY
....	....
479	PAT, ARI, PAT, GEO, PIR, MAR, AGP
480	PAT, ARI, PAT, GEO, SMY, AGP, PER
....	....
741	THI, PAT, ARI, PAT, GEO, SMY, PIR, MAR, AGP, PER

The output layer in all ANN that were developed, comprised of 5 output neurons, related with the hourly O_3 concentration for the temporal points of 0, 1, 2, 3, 6 hours. The determination of the number of hidden neurons for each ANN was done based on the indicative use of the following equation 1, that relates the size of the training set to the number of input and output neurons and to the potential level of noise in the data (Taylor, 2006).

$$N_i = \frac{Number\ of\ training\ datapairs}{(Number\ of\ input\ neurons\ + Number\ of\ output\ neurons) * noise\ factor} \quad (1)$$

The noise factor in this case was assigned the value 1, since the data were checked and the abnormal values were not considered. The size of the used training sets (training datapairs) varied from 11,300 to 21,400 data vectors, thus the potential number of hidden neurons could be roughly estimated by the following relation R1 (where M_I is the number of input neurons).

$$\frac{11300}{(M_I+5)} \prec N_i \prec \frac{21400}{(M_I+5)} \tag{R1}$$

Considering the above equation and the minimum and maximum number of input neurons off all possible ANN (85 to 235) we can turn R1 to the following relation R2.

$$\frac{11300}{235+5} < N_i < \frac{21400}{85+5} => 47 < N_i < 237 \tag{R2}$$

Given the above relation, and the minimum and maximum number of input neurons off all possible ANN three weights were applied (1, 0.75, 0.5) that relate the size of the input vector with the number of its neurons for every ANN. Thus, three alternative networks were developed for each weight respectively and the total number of developed networks was as high as 2,223.

Due to the extremely high volume of data vectors and due to the vast number of ANN developed, a piece of software entitled *ann_devel.m* was developed in Matlab R2011b that automatically created all of the data files used for the training and testing of the networks, it determined and implemented the architecture of the networks and it stored the actual output results.

2.2   Training and Assessment of the ANN

The construction of the networks was done automatically by running the custom implemented script *ann_devel.m* (that was discussed before) under the MATLAB platform. A percentage as high as 60% of the data was used in training 20% was kept for validation and 20% was used in the testing phase. More specifically, training was performed with data from 1985 to 2006 whereas validation and testing was done using data vectors related to years 2007, 2008 and 2009. The number of input records provided to the ANN ranged from 18,834 to 93,768. The vast volume of training vectors and the rational number of iterations performed in training are in favor of the generalization ability of the developed model.

The following table 4 presents the number of cases with the best estimated values for each weight factor used.

Table 4. Number of best estimations for each weight factor

weight factor	Number of ANN with the best performance	Number of cases with the best estimation of values for all input vectors and for all of the ANN
0,5	108	2,588,563
0,75	216	2,575,653
1	417	2,634,863

Although a weight equal to1 seems to offer a higher number of ANN with the best performance, this is not the case when we examine the number of cases with the best estimated output values for all input vectors and for all ANN. This is clearly shown on table 5. That is why for every case and for every network we have used trial and error and we have finally applied the number of hidden neurons that offer the best results. The obtained networks were assessed based on the coefficient of determination R^2 and table 5 presents an indicative ranking. Another assessment was based on the number of times that every network offered the best results compared to the rest. An indicative view of this analysis is shown in table 6.

Table 5. Ranking of the ANN based on the coefficient of determination

Net ID	Ranking	R^2	Factors
394	1	0.791032	PEN, ARI, PAT, SMY, AGP
691	2	0.784287	THI, PEN, ARI, PAT, SMY, MAR, AGP
395	3	0.782871	PEN, ARI, PAT, SMY, MAR
....	....	....	
740	525	0.657559	THI, PEN, PAT, GEO, SMY, PIR, MAR, AGP, PER
741	546	.0651572	All stations
....	....	....	

An important finding was that large and complicated ANN did not have good ranking in both assessments although they are using data from a large number of stations so it was expected to be able to predict best values under any circumstances. Afterwards, the selection of the best model was done based in two heuristics. Based on the first rule of thumb, the network with id 394 was considered the one with the best performance.

According to the second heuristic approach, in every estimation step we have used the network that had offered the best results in the previous step. This means that the *ann_devel.m* script was calculating the output of the 741 ANN and that every time for the next estimation it was using the network that previously had the best performance.

Table 6. Ranking based on the number of best estimations

Net ID	Ranking	Number of best estimated cases for the moment 0 h	Factors
259	1	1435	PEN, PAT, AGP
253	2	1207	PEN, ARI, AGP
280	3	1125	PEN, ARI, AGP, PAT
....	....	....	
740	740	1	THI, PEN, PAT, GEO, SMY, PIR, MAR, AGP, PER
735	741	1	

This principle was applied for all of the 5 output features. We have applied this rule of thumb, considering that the change of the values of the independent parameters between two successive measurements (hourly values) depict also the change in the environmental conditions. These two hypotheses were evaluated based on four characteristics, namely: R for every output, RMSE for every estimated value,

Error histograms and the number of cases that either of them satisfies. Data for the year 2012 (and the 8593 data vectors available) were used in this process.

3 Assessment of the Two Heuristic Approaches

Totally 4236 outputs were obtained, after running ANN #394 with approximately 52% of the expected measurements missing due to malfunctioning of sensors. The metrics used were R^2 and RMSE (Root Mean Square Error) shown in the following equation 2 (Iliadis, 2007) where N is the number of data records (training or testing) y_j is the actual value and $\hat{y}_j$ is the estimated value by the system (in training or in testing). Table 7 presents the performance of ANN#394.

$$RMSE = \sqrt{\frac{\sum\limits_{j=1}^{N}\left(y_j - \hat{y}_j\right)^2}{N}} \tag{2}$$

Table 7. Performance of network 394

Output	R^2	MSE (RMSE)
0 hours	0,6971	438,982 (20,951)
1 hours	0,6559	411,373 (20,282)
2 hours	0,5975	437,698 (20,921)
3 hours	0,5395	469,357 (22,279)
6 hours	0,4278	554,953 (23,557)

After the execution of the program that performs the second rule of thumb, we obtained 8571=365X24 outputs. In this case only 0.26% of the expected measurements corresponded to missing or unreliable data. Table 8 presents the performance of the networks that were developed following the second approach, for 0, 1, 2, 3, 6 hours, where the time spot 0, corresponds to the data gathered at 00.00 hours of 1/1/2010.

Table 8. Performance of networks developed when following the second heuristic

Output	R	R^2	MSE (RMSE)
0 – current estimation	0,9146	0,8365	139,712 (11,820)
1 hour prediction	0,9057	0,8203	154,735 (12,439)
2 hour prediction	0,8909	0,7937	173,413 (13,169)
3 hour prediction	0,8754	0,7663	193,302 (13,903)
6 hour prediction	0,8441	0,7125	230,007 (15,166)

The following figures 2a and 2b present the error histogram and R for the 0 hour (current estimation) and 6 hour prediction outputs.

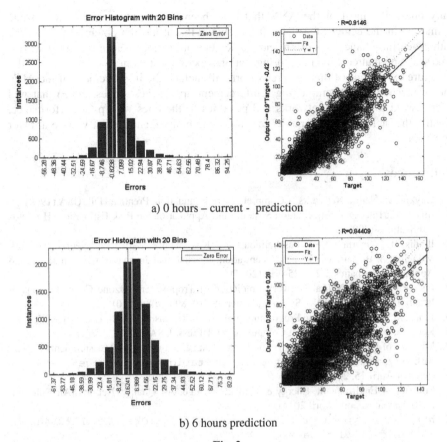

a) 0 hours – current - prediction

b) 6 hours prediction

Fig. 2.

The second approach does not always offer the best choice of the optimal network but it offers much better results than the selection of a unique "Save Best" ANN and as it can be shown to table 8 and figures 2a and 2b although R^2 falls from 0,8365 to 0,7125 and RMSE raises from 11,820 to 15,166 the dispersion of the errors shows no outliers or significant abnormalities.

4   Conclusions and Future Work

The model that was developed here proves its ability to function under all circumstances for the year 2010, when the percentage of good function of the stations was as high as 99.74%. The good function index of the measurements' stations was not considered in other research effort so far, since the availability of reliable data was believed to be certain. This has been proven wrong by the data of table 1. This model has the advantage that it is not influenced by a potential malfunction of monitoring stations, since it exploits the other available ANN to function. Its results are very encouraging and moreover it functions based on hourly values and not on average daily ones, like most of the ANN that have been developed so far for the Ozone estimation in the center of Athens. It has shown that it has the potential to offer the authorities the chance to design and implement protection measures almost in real time scale, no matter if the monitoring stations operate well or not.

Future extensions of this research effort will include the improvement of the ANN choice mechanism, by employing reinforcement approaches (Jones, 2005) that will offer reward to the best networks and penalties to the ones with poor performance. Finally, the inclusion if this model to a real time multi agent system will be a major step ahead.

References

1. Haykin, S.: Neural Networks A Comprehensive Foundation. Prentice Hall, USA (1999)
2. Iliadis, L.: Intelligent Information Systems and Applications in Risk Estimation. Hrodotos Publications (2007)
3. Iliadis, L., Spartalis, S., Paschalidou, A., Kassomenos: Artificial Neural Network Modeling of the surface Ozone concentration. International Journal of Computational and Applied Mathematics 2(2), 125–138 (2007)
4. Inal, F.: Artificial Neural Network Prediction of Tropospheric Ozone Concentrations in Istanbul, Turkey. CLEAN – Soil, Air, Water 38(10), 897–908 (2010)
5. Jones, M.T.: AI Application Programming, 2nd edn. Thomson Delmar Learning (2005)
6. Kecman, V.: Learning and Soft Computing. MIT Press, USA (2001)
7. Ministry of Environment, Energy & Climate Change, Air Pollution Measurements (2012), http://www.ypeka.gr/Default.aspx?tabid=495&language=el-GR (valid at April 15, 2012)
8. Ministry of Environment, Energy & Climate Change, Air Quality, Reports, Air Pollution 2010 Annual Report (April 2011), http://www.ypeka.gr/LinkClick.aspx?fileticket=5mmiFX%2b4n2M%3d&tabid=490&language=el-GR
9. Ministry of Environment, Energy & Climate Change, Air Quality, Reports, Air Pollution 2009 Annual Report (2010), http://www.ypeka.gr/LinkClick.aspx?fileticket=NIpIqbQg
10. Ozcan, H.K., Bilgili, E., Sahin, U., Bayat, C.: Modeling of trophospheric ozone concentrations using genetically trained multi-level cellular neural networks. Advances in Atmospheric Sciences 24(5), 907–914 (2007)
11. Ozdemir, H., Demir, G., Altay, G., Albayrak, S., Bayat, C.: Environmental Engineering Science 25(9), 1249–1254 (2008)
12. Paoli, C.: A Neural Network model forecasting for prediction of hourly ozone concentration in Corsica. In: Proceedings IEEE of the 10th International Conference on Environment and Electrical Engineering, EEEIC (2011)
13. Paschalidou, A., Iliadis, L., Kassomenos, P., Bezirtzoglou, C.: Neural Modeling of the Tropospheric Ozone concentrations in an Urban Site. In: Proceedings of the 10th International Conference Engineering Applications of Neural Networks, pp. 436–445 (2007)
14. Brian, T.J.: Methods and procedures for the verification and validation of artificial neural networks. Springer Publications (2006)
15. Wahab, A.-S.A., Al-Alawi, S.M.: Assessment and prediction of tropospheric ozone concentration levels using artificial neural networks. Environmental Modeling & Software 17, 219–228 (2002)

An Ontology Based Approach to Designing Adaptive Lesson Plans in Military Training Simulators

D. Vijay Rao[1], Ravi Shankar[2], Lazaros Iliadis[3], and V.V.S. Sarma[4]

[1] Institute for Systems Studies and Analyses
Defence Research and Development Organisation,
Metcalfe House, Delhi 110054, India
doctor.rao.cs@gmail.com
[2] Department of Management Studies
Indian Institute of Technology, Hauz Khas, New Delhi 110016, India
ravi1@dms.iitd.ac.in
[3] Department of Forestry & Management of the Environment and Natural Resources
Democritus University of Thrace, GR-68200, Orestiada, Hellas
liliadis@fmenr.duth.gr
[4] Department of Computer Science and Automation
Indian Institute of Science, Bangalore 560 012, India
vvs@csa.iisc.ernet.in

Abstract. Several classes of simulators have been designed in the military domain for training and operational analysis. Joint Operations Simulation System (JOpsSS) is a virtual warfare analysis system that has been developed for planning, operational analysis and evaluating joint operations. A major concern in the design and development of these simulators is the training lesson plans for trainees with different backgrounds. The design of intelligent lesson plans that are adaptable to the varied needs of the trainees is a challenging task. In this paper, we propose an ontology based design of training, learning and evaluation agents that has been used to design intelligent training systems. This approach has proved to be very effective in modelling complex and adaptive warfare scenarios. An ontology that represents the military domain concepts, context and data is used to represent and store the knowledge-base required for intelligent training simulators and designing intelligent lesson plans. The Instructor agent assesses each trainee from the past credentials for the level of lesson plans and builds a *concept- graph*. This is dynamically adapted to suite the level of trainee based on the responses and bridges the gap between the expected and actual competency level.

Keywords: Training simulators, Adaptive Lesson Plans, Knowledge Representation, Ontology, Training Effectiveness.

1 Introduction

Recent concerns (and the many ill-effects that military forces are suffering around the world) on the damages caused to the environment and bio-diversity by military

C. Jayne, S. Yue, and L. Iliadis (Eds.): EANN 2012, CCIS 311, pp. 81–93, 2012.

operations have prompted scientists to consider virtual-reality based simulators that give the same training experience as the real-world systems [1],[2]. With rapid advances in technology and increasingly complex defence systems in operation, substantial effort and resources are spent on training for their effective usage. In order to improve the efficiency, effectiveness, usage and safety of training, organizations and user agencies are investing heavily into developing intelligent computer-based training simulators. Agent based systems, that are loosely coupled but highly cooperative and collaborative in achieving the tasks and goals have proved to be very successful in design and development of virtual warfare simulators [1], [2], [3], [4], [5]. An important success criterion for designing such simulators is their fidelity to emulate the real-world training conditions. This is achieved by designing the lesson plans that caters to the training needs of the users. Traditionally, simulators are designed with pre-defined lesson plans where the rules, narratives and environments are created during development, as static elements with which a dynamic player will interact. While this approach ensures robustness and easy testability, the trainees easily anticipate the lessons after some cycles of learning and overcome the learning quickly. As often observed in many cases, if trainees can predict certain outcomes, their progress can be achieved by repeatedly exploiting a successfully strategy. This is prominent in domain dependent games and scenario-based exercises where the domain knowledge and experience of the players is important. An important and crucial requirement to the design of intelligent training simulators is, thus, the representation of domain knowledge and lesson plans that are dynamically adaptable to the various training needs of the users [7],[11],[12]. Design of intelligent lesson plans that are adaptable to the varied needs of the trainees and explicitly representing the domain knowledge that is sharable and reused across different classes of users with different levels of abstraction and resolutions is a challenging task. In this paper, we propose an ontology based design of training, learning and evaluation agents that has been used to develop intelligent lesson plans in military training simulators. The paper is organized as follows: We describe the issues in designing intelligent mission planning simulators, followed by ontology to design the lesson plans and their adaptivity to suit the various classes of users and trainers. This is followed by a case study for joint mission planning operations, and concludes the paper with a discussion of the results obtained and directions for future work.

2   Learning Theories and Instructional Design

In designing and developing simulators, several important factors such as the learning theories, cognition, behaviors and psychology need to be considered. The development of theories about how people learn began with Aristotle, Socrates and Plato. In more recent years, instructional design techniques have been developed for those learning theories that assess in the development of learning experiences. These instructional design techniques are the basis for the e-learning authoring software tools. The following three learning theories have been used to design the content and instruction in the Learning Management System in the virtual warfare training simulator [7], [22], [23].

a) *Behaviourism*: Focuses on repeating a new behavioral pattern until it becomes automatic. The emphasis is on the response to stimulus with little emphasis on the thought processes occurring in the mind. This is used to design the instructions where the doctrines of warfare play an important role. For example, Understanding the capabilities of the different aircraft and weapons.

b) *Cognitivism*: is similar to behaviorism in that it stresses repetition, but it also emphasizes the cognitive structures through which human process and store information. For example, use of smart weapons vs conventional dumb bombs in different situations.

c) *Constructivism*: Takes a completely different approach to learning. It states that knowledge is constructed through an active process of personal experience guided by the learner himself. For example, a military goal is identified and the trainees have to construct a concept paper on how to achieve the goals.

There are two basic types of learning environments: synchronous and asynchronous:

- **Synchronous**: Synchronous learning environment is one in which an instructor teaches a somewhat traditional class but the instructor and students are online simultaneously and communicate directly with each other. Software tools for synchronous e-learning include audio-conferencing, video conferencing, and virtual whiteboards that enable both instructor and students to share knowledge.

- **Asynchronous**: In an asynchronous learning environment, the instructor only interacts with the student intermittently and not in real-time. Asynchronous learning is supported by such technologies as online discussion groups, email, and online courses and can further be classified as Traditional asynchronous e-learning, Scenario-based e-learning, Simulation-based e-learning and Game-based e-learning [7].

In this work, we focus on the design of asynchronous, scenario-based, simulation-based and game-based learning simulators where the lesson plans for the trainees are dynamically composed form the learning objects after considering the competency level and gap analysis of the trainee. The learning objects are stored in a repository and the Instructor agent is designed to generate the lesson plan from the ontology concepts stored in a knowledge base after appropriately reasoning from the ontology and fill the competency gaps.

3   Intelligent Mission Planning Simulators

Several classes of simulators have been designed in the military domain for training, analysis, to generate strategic scenarios for forecasting, creating what-if scenarios and evaluating effectiveness of military operations and procedures [3].

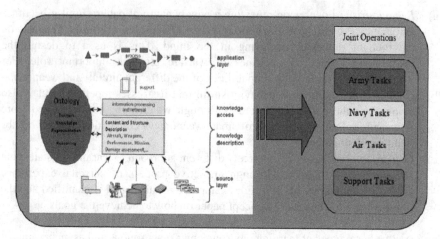

Fig. 1. Architecture of the LMS for dynamically composing joint operations lesson plans

Agent-oriented system development aims to simplify the construction of complex systems by introducing a natural abstraction layer on top of the object-oriented paradigm composed of autonomous interacting actors [6]. It has emerged as a powerful modeling technique that is more realistic for today's dynamic warfare scenarios than the traditional models which were deterministic, stochastic or based on differential equations. These approaches provide a very simple and intuitive framework for modeling warfare and are very limited when it comes to representing the complex interactions of real-world combat because of their high degree of aggregation, multi-resolution modeling and varying attrition rate factors. The effects of random individual agent behavior and of the resulting interactions of agents are phenomenon that traditional equation-based models simply cannot capture. Fig. 2(a),(b) shows the agent based architecture of a virtual warfare training simulator [2],[4].

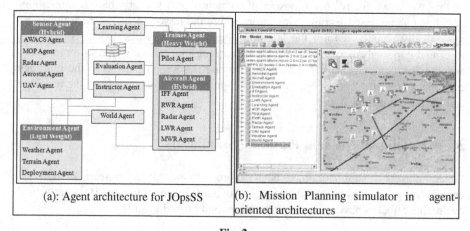

| (a): Agent architecture for JOpsSS | (b): Mission Planning simulator in agent-oriented architectures |

Fig. 2.

Using this simulator, users from army, navy and air force are trained in different aspects of warfare. Such requirements demand that the content should be dynamic and adaptable to the varying needs of users. Joint Operations Simulation System (JOpsSS) is a virtual warfare analysis simulator that has been developed for planning, analysis and evaluating joint operations of the Armed forces (army, navy and air-force) and other support organizations in order to meet the objectives set by the instructors for training. While the individual armed forces wargames have been used successfully for training and operational analysis, joint operations seem to be much more complex as the domain knowledge required for decision making far exceeds the individual games. A mission objective (goal) set by the instructor is designed within a contextual setting and also describing the scenario and settings within which the training is imparted and the trainees are assessed. The lesson plans are designed using all the four types of learning depending upon the nature of lessons and training to be imparted. The lesson plans are designed based on the domain knowledge that is explicitly represented by ontology of the warfare resources, aircraft, weapons, performance characteristics, constraints, weather, and terrain information. The lesson plans are dynamically adapted by asking relevant questions on the concepts of learning from the ontology and reasoning based on the answers to change the lesson plans accordingly. The goals are decomposed as tasks, and sub-tasks in a hierarchical manner, indicating the roles of the armed services, support organizations and people who would be collaborating to collectively achieve the objective (Fig. 4). The sequence and timing diagrams of the tasks are generated and these are associated with the resource constraints and resolution of conflicts. The assessment of the trainees is done by evaluating the plans made by the trainees to meet the goals. The Learning Management sub-system (LMS) in this simulator architecture (JOpsSS) is responsible for planning the lessons for the trainees, storing and updating the contents, evaluate the trainees and also learn from the behavior of the trainees for further lesson planning. The LMS consists of three prominent agents: Instructor agent, Learning agent and Evaluation agent. The Instructor agent is composed of a Lesson Planner that identifies a goal for the trainees, composes the lesson plan from the learning objects and given to all the trainees. The trainees decompose the task into a number of independent tasks that are to be achieved by each of the teams, in order to achieve the objectives of the goal [5].

The Instructor agent updates the state of a lesson plan and creates a scenario that is based upon the information received from weather, terrain and deployment agent and provides an information service to the world agent after its own process of reasoning. This information is then used by other agents such as Manual Observation Post (MOP), Pilot, Unmanned Air Vehicle (UAV), Identification Friend/Foe (IFF), Radar Warning Receiver (RWR), Missile Warning Receiver (MWR), Laser Warning Receiver (LWR), Mission Planning, Sensor Performance, Target Acquisition and Damage Assessment and Computation (Fig. 2 (a),(b)).

4   An Ontology Based Approach for Designing Lesson Plans

Three main challenges in designing reusable learning objects are (i) *intelligence*; (ii) *sharable*; and (iii) *dynamic*. This is overcome by developing semantic metadata for providing intelligence to learning objects; developing content packaging for enhancing the sharability of learning objects and developing learning object repository with ontologies and Semantic Web technologies for making learning objects more dynamic [14],[19]. To meet these challenges the following methodological steps are followed to design and develop the online environment of learning object repository.

• *Stage 1*: To develop a metadata framework which integrates the most suitable metadata as well as proposed pedagogical and military metadata elements that can be applied to a variety of learning objects.
• *Stage 2*: To apply a content packaging standard that packages learning objects together so they can be exported to and retrieved from various learning management systems.
• *Stage 3*: To identify the ontology (i.e. a common vocabulary of terms and concepts) for construction education and to develop a Semantic Web environment that will increase sharability of objects within construction domains.

Ontologies are specifications of the conceptualisation and corresponding vocabulary used to describe a domain [13],[19]. It is an explicit description of a domain and defines a common vocabulary as a shared understanding. It defines the basic concepts and their relationships in a domain as machine understandable definitions. The concept *Aircraft* is understood by its Type, Role, Armament that the aircraft carries, its Range and Combat potential. An example of SU 30 MKI aircraft is *Fighter aircraft*, *Multi-Role*, *RVVAE*, and *1000Kms* and *High Combat potential*. The concept *Ship* in the Navy is defined by its classification as a Warship, Passenger ship or Cargo ship. A Navy Warship is identified based on its size, function, and role such as *Aircraft Carrier*, *Destroyer*, *Frigate*, *Corvette*, and *Amphibious Ships*. The sub-concept of *Aircraft Carrier* itself can be further classified based on *Types* of *Aircraft* onboard, *Range*, *Missile Type* and *Number of Missiles* it can carry, *Onboard Sensors* and *Performance*, *Lethality* of the missiles, and so on. Such a classification is the basis of the military ontology that built for all the military resources. This ontology forms the knowledge base that the training simulators use to generate the lesson plans. We design a military ontology consisting of a formal and declarative representation which includes the vocabulary (or names) for referring to the terms of army, navy and airforce and the logical statements that describe what the terms are, how they are related to each other, and how they can or cannot be related to each other (Fig.3). Ontology therefore provides a vocabulary for representing and communicating knowledge about some aspect of military training and a set of relationships that hold among the terms in that vocabulary [15],[16],[17]. The main purpose of ontology is, however, not to specify the vocabulary relating to an area of interest but to capture the underlying conceptualisations. Noy and McGuinness [9] have identified five reasons for the development of an ontology:

• to share common understanding of the structure of information amongst people or software agents;
• to enable reuse of domain knowledge;
• to make domain assumptions explicit;
• to separate domain knowledge from the operational knowledge;
• to analyse domain knowledge.

A generic organisational structure of learning objects for the domain of pedagogy design based on the premise that ontologies can help people to better share knowledge and to demonstrate the usefulness of the proposed resource organisational structure for pedagogy design. The ontology links between user needs and characteristics of the learning material and enable the agents to discover learning objects that are decentralised across the network and internet.

In the design of the JOpsSS, military domain knowledge is represented and stored as ontology in Protégé. (Fig.4). Protégé is a freely available, open-source platform that provides a suite of tools to construct domain models and knowledge-based applications that use ontologies. At its core, Protégé implements a rich set of knowledge-modeling structures and actions that support the creation, visualization, and manipulation of ontologies in various representation formats (including the Web Ontology Language, OWL and Resource Description Framework (RDF)). Protégé can be customized to provide domain-friendly support for creating knowledge models and entering data. Further, Protégé can be extended by way of a plug-in architecture and a Java-based Application Programming Interface (API) for building knowledge-based tools and applications.

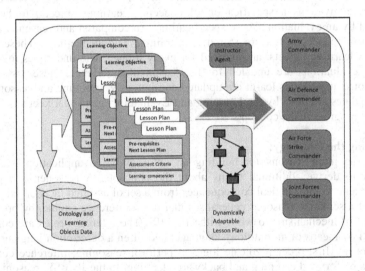

Fig. 3. Ontology based Instructor agent to dynamically plan adaptable lessons based on competency gaps

Protégé can load OWL/RDF ontologies, edit and visualise classes and properties; execute reasoners such as description logic classifiers and edit OWL individuals for SemanticWeb. Protégé is widely used for modelling of simple applications to high-tech, high-powered applications [8],[9],[10]. It also offers support to ontology libraries and OWL language.

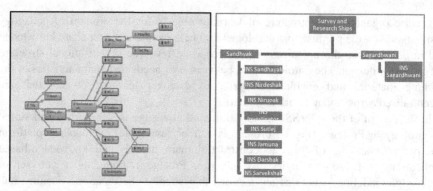

Fig. 4. Ontology of lesson plan of *Aircraft* and *Survey and Research Ships* concept in Protégé

Fig. 4(a) shows a screenshot of the Protégé which shows OWL classes of the military ontology that was designed and developed for the JOpsSS. While metadata of learning objects describe the artifacts of learning objects that are shared by diverse domains, an ontology represents a knowledge domain that shares the relationships of learning objects within a specific context [20],[21]. The training needs of every trainee are given as a specification of concepts (learning objects) to be taught, followed by an evaluation. These specifications are compared and reasoned from the ontology of concepts that are stored in a training knowledge-base. These lessons, stored as learning objects, are updated for every trainee who undergoes the training and the evaluations are updated for the lesson plans. Hence, these lesson plans, evaluations, and lessons learnt are updated continuously with time and lessons taken. These updates are done by the Learning and Evaluation agents that keeps track of the trainees' progress in the concepts.

Reasoning the Ontology:
One of the main reasons for building an ontology-based application is to use a reasoner to derive additional truths about the concepts. A reasoner is a piece of software able to infer logical consequences from a set of asserted facts or axioms. The notion of a semantic reasoner generalises that of an inference engine, by providing a richer set of mechanisms to work with [10],[19]. The inference rules are commonly specified by means of an ontology language, and often a description language. Many reasoners use first-order predicate logic to perform reasoning; inference commonly proceeds by forward chaining and backward chaining. In the JOpsSS, reasoning helps in formulating questions for testing the understanding of related concepts. This is used to evaluate the competency of trainees, lessons planned and a gap analysis.

Adaptive Lesson Plans Using Game Trees:
The adaptability of the designing a lesson plan depends on the background and training needs of different users. The lesson plans are selected by two main cognitive criteria: memory and learning. The competency level required by the lesson plan is compared with the competency level of the trainee. The gap is reduced by reasoning the ontology concepts and choosing the lesson plans that are represented in a *concept-map* and implemented as a *concept-graph*. The trainee is now switched to a new lesson that matches with the competency of the lessons.

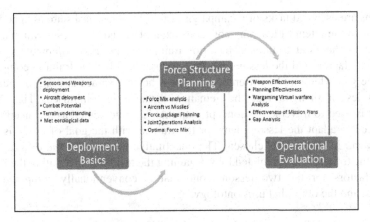

Fig. 5. An adaptable sequence of a *Lesson Plan* for Trainee 2

The lesson plans for the Army, Navy and Air Force and Joint operations which demand inter-disciplinary domain knowledge are organized as a concept graph. The starting node for the trainee is identified based on the competency level and based on the various preliminary questions, the Instructor agent reasons from the ontology *concept-graph* and composes the new lesson plan by traversing it to reduce the competency gap.

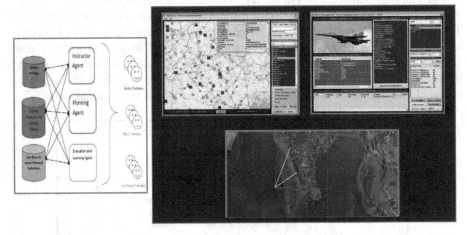

Fig. 6. Architecture of agents in JOpsSS and screens depicting the Army, Air and Naval tasks for trainees

5   Application Results of a Training Exercise

Consider a training exercise for military operations in which the trainees from different branches of specializations with different skills, prior training and field operations are assigned tasks of a campaign (Table 1). These tasks are assigned to the trainees with the intent of teaching concepts, examples, and field cases which are then evaluated in the field training. The prior training is used to compute the trainee competency factor, and the lesson plan initially assigned has the training competency level. The gap which is the difference between the two values is used to decide the switched lesson plan so that the semantic distance is minimised. The military ontology is used to traverse the concept-map that is implemented as a concept graph, and is used to adapt the lesson plans for the trainee with the goal of minimising the semantic gap in the lessons chosen. The quantitative answers for the different tasks given to the trainees are calculated by wargaming the tasks and generating the mission success factors for the two lessons: one that is conventionally computed using databases, and the other that uses ontology.

(a) *Trainee 1*: To understand and evolve different strategies to gather Location based Intelligence necessary as pre-curser to destroy the target (Fig. 6).
(b) *Trainee 2*: To understand the concepts in Mission Planning and Air Tasking operations (Fig. 7).

The mission success factor for Trainee 1 increased from 7.2 to 9.3 and from 5.3 to 9.8 on running the JOpsSS wargame by using the military ontology, reasoning and dynamically adapting the lesson plans to suit the training requirements of the trainee. The lessons plans for all the trainees are shown summarized in Table 1.

The ontology requirements found an importance in military simulators mainly because of the Joint Warfare operations that are introduced in the course of training. These values may or may not have increased as much with the individual service wargames. This gives an intuitive indication of synergy in joint wargames that demand a much greater understanding of the warfare concepts and applying them in joint missions that surpass the boundaries of individual war games. The inherent notion of quantifying the synergistic effects is being explored in a separate work.

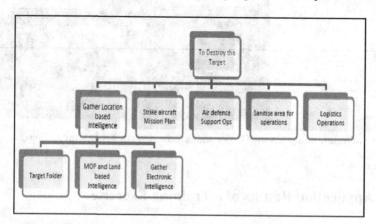

Fig. 7. T-001 Game Scenario Lesson Plans generated for Trainee 1

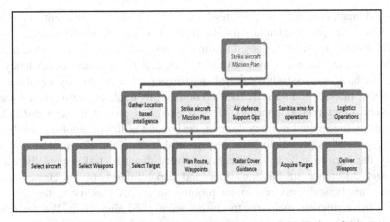

Fig. 8. T-002 Game Scenario Lesson Plans generated for Trainee 2

Table 1. Fuzzy antecedents (factors) to generate training lesson plans

Mission_ ID	Trainee	Service	Training and Field Operatio ns Skills	Lesson Plan (Task)	Trainee Compet- ency Factor	Training Compet- ency Factor	Compet- ency Gap	Mission Success Factor (1-10) Conventional	Mission Success Factor (1-10) Ontology and adaptive lessons
#001	T-001	Intelligence	Low	ESM and Location based Intelligence gathering	High	High	Very Low	7.2	9.3
#002	T-002	AirForce	High	Strike and Air-Air Combat	High	High	Very Low	5.3	9.8
#003	T-003	Navy	High	Anti-Submarine Warfare	Low	Medium	Medium	4.6	6.4
#004	T-004	Logistics	Low	Mobilization of Logistics and Road Move Plan	Low	Very High	Very High	6.8	7.4
#005	T-005	Army	Very High	Air defence Guns Deployment	Medium	High	Medium	7.2	9.3

6   Conclusions and Discussion

We present a novel approach to the design and development of intelligent training simulators and the design of adaptable lesson plan using ontology. In this paper, we propose an ontology based design of training, learning and evaluation agents that has been used to design intelligent training systems. This approach has proved to be very effective in modelling complex and adaptive warfare scenarios in an agent-oriented framework. An ontology that represents the military domain concepts, context and data is used to represent and store the knowledge-base required for intelligent training simulators and designing intelligent lesson plans. A Joint Operations warfare simulation system is used as a case study to demonstrate the applicability of the military domain concepts using Protégé software. The Instructor agent assesses each trainee from the past credentials for the level of lesson plans and builds a *concept-graph*. This is dynamically adapted to suite the level of trainee based on the responses and bridges the semantic gap between the expected and actual competency level. While we have successfully designed and implemented a military ontology of the concepts of interest in a specific focused domain, its implementation on a large scale for several other related concepts of field exercises proved to be a difficult task. However, in the limited military domain, design and utility of ontology for designing training simulators proved to be a very fruitful investment. However, the ability to classify the various military resources is a very daunting task especially when the roles and difference in functionalities and capabilities between classes becomes blurred. This typically poses a major problem in crisply classifying the entities and hence a belongingness measures to classes are a good alternative. We are presently exploring the possibility fuzzy ontology and reasoning in classifying representing relationships in the concepts to build the military ontology.

References

1. Vijay Rao, D., Iliadis, L., Papaleonidas, A., Spartalis, S.: Modelling Environmental Factors and Effects in Virtual Warfare Simulators by using a Multi Agent approach. International Journal of Artificial Intelligence (to appear, 2012) (ISSN 0974-0635)
2. Vijay Rao, D., Iliadis, L., Spartalis, S.: A Neuro-Fuzzy Hybridization Approach to Model Weather Operations in a Virtual Warfare Analysis System. In: Iliadis, L., Jayne, C. (eds.) EANN/AIAI 2011, Part I. IFIP AICT, vol. 363, pp. 111–121. Springer, Heidelberg (2011)
3. Vijay Rao, D.: The Design of Air Warfare Simulation System. Technical report, Institute for Systems Studies and Analyses (2011)
4. Vijay Rao, D., Kaur, J.: A Fuzzy Rule-Based Approach to Design Game Rules in a Mission Planning and Evaluation System. In: Papadopoulos, H., Andreou, A.S., Bramer, M. (eds.) AIAI 2010. IFIP AICT, vol. 339, pp. 53–61. Springer, Heidelberg (2010)
5. Vijay Rao, D., Saha, B.: An Agent oriented Approach to Developing Intelligent Training Simulators. In: SISO Euro-SIW Conference, Ontario, Canada (June 2010)
6. Taher, J., Zomaya, A.Y.: Artificial Neural Networks. In: Zomaya, A.Y. (ed.) Handbook of Nature-Inspired and Innovative Computing, Integrating Classical Models with Emerging Technologies, pp. 147–186. Springer, USA (2006)
7. Woelk, D.: E-Learning Technology for improving Business performance and Lifelong Learning. In: Singh, M.P. (ed.) The Practical Handbook of Internet Computing. Chapman & Hall, USA (2005)
8. http://protege.stanford.edu/doc/users.html#papers
9. http://www-ksl-svc.stanford.edu:5915/doc/frame-editor/what-is-an-ontology.html
10. http://www.cs.man.ac.uk/~sattler/reasoners.html
11. Uschold, M., Gruninger, M.: Ontologies: Principles, Methods and Applications. Knowledge Engineering Review 11(2) (1996)
12. Gomez-Perez, A., Corcho, O., Fernandez-Lopez, M.: Ontological Engineering: with examples from the areas of Knowledge Management, e-Commerce and the Semantic Web. Springer (2004)

13. Neches, R., Fikes, R., Finin, T., Gruber, T., Patil, R., Senator, T., Swartout, W.: Enabling Technology for Knowledge Sharing. AI Magazine (Fall 1991)
14. Fisher, M., Sheth, A.: Semantic Enterprise Content Management. In: Singh, M.P. (ed.) The Practical Handbook of Internet Computing. Chapman & Hall, USA (2005)
15. Naval Operations Analysis, prepared by the Operations Committee, Naval Science Department, US Naval Academy, Annapolis, Maryland (1968)
16. Brewin, R.L.: A Review of the Concept of Military Worth and its application in Military Decision making, Master's Thesis in Management, U.S. Naval Postgraduate School (1964)
17. Dockery, J.T., Woodcock, A.E.R. (eds.): The Military Landscape: Mathematical Models of Combat. Woodhead Publishing, Cambridge (1993)
18. Meliza, L.L., Goldberg, S.L., Lampton, D.R.: After Action Review in Simulation-Based Training, RTO-TR-HFM-121-Part-II, Technical report
19. Raju, P., Ahmed, V.: Enabling Technologies for developing Next Generation Learning Object Repository For Construction. Automation in Construction 22 (2012)
20. Berry, J., Benlamri, R., Atif, Y.: Ontology-Based Framework For Context-Aware Mobile Learning. In: IWCMC 2006, Canada (2006)
21. Sabou, M., Wroeb, C., Goble, C., Stuckenschmidt, H.: Learning domain ontologies for Semantic Web service descriptions. Journal of Web Semantics 3 (2005)
22. Brusilovsky, P., Wolfgang, N.: Adaptive Hypermedia and Adaptive Web. In: Singh, M.P. (ed.) The Practical Handbook of Internet Computing. Chapman & Hall, USA (2005)
23. Aparicio IV, M., Singh, M.P.: Concepts and Practice of Personalisation. In: Singh, M.P. (ed.) The Practical Handbook of Internet Computing. Chapman & Hall, USA (2005)

A Continuous-Time Model
of Analogue K-Winners-Take-All Neural Circuit

Pavlo V. Tymoshchuk[1,2]

[1] EIT Department, Lodz University,
65, Narutovich Str., Lodz, 90-131, Poland
pawtym@uni.lodz.pl
[2] CAD Department, L'viv Polytechnic National University,
12, S. Bandera Str., L'viv, 79013, Ukraine
pautym@polynet.lviv.ua

Abstract. A continuous-time model of analogue K-winners-take-all (KWTA) neural circuit which is capable to extraction the K largest from any finite value N unknown distinct inputs, where $1 \leq K < N$, is presented. The model is described by one state equation with discontinuous right-hand side and output equation. A corresponding functional block diagram of the model is given as N feedforward and one feedback hardlimiting neurons, which is used to determine the dynamic shift of inputs. The model combines such properties as high accuracy and convergence speed, low computational and hardware implementation complexity, and independency on initial conditions. Simulation examples demonstrating the model performance are provided.

Keywords: Continuous-time model, State equation, Functional block-diagram, Hardlimiting neuron, Analogue K-winners-take-all neural circuit.

1 Introduction

It is known that K-winners-take all (KWTA) neural networks realize selection of K largest from N inputs, where $1 \leq K < N$. When K is equal to unity, the KWTA network is the winner-takes all (WTA) one, that finds the maximum from a set of N inputs [1].

KWTA neural networks have many applications, in particular, choosing K largest elements from a longer list being a fundamental operation in data and signal processing, in decision making, in pattern recognition, as well as in competitive learning and sorting [2] - [4]. The KWTA networks are used in telecommunications [5] and vision systems [6], for solving problems of filtering [7], decoding [8], image processing [9], clustering [10]. The KWTA operation is applied in machine learning, mobile robot navigation, feature extraction [11], [12]. The KWTA networks can be used as the basic building blocks in computer-based medical diagnostics, in browsing and information retrieval, in data mining and analysis, in financial prediction, as a teaching aid, in analyzing surveys and in questionnaires on diverse items, etc. [13].

C. Jayne, S. Yue, and L. Iliadis (Eds.): EANN 2012, CCIS 311, pp. 94–103, 2012.
© Springer-Verlag Berlin Heidelberg 2012

Analogue (or continuous-time) KWTA neural networks compared to discrete-time analogs are capable of providing stable performance in a wider parameter change range and in a wider diapason of varying a convergence speed [14]. Many different analogue neural networks have been proposed to solve the KWTA problem [1], [3], [15] - [17]. In particular, a continuous-time model of KWTA neural circuit which is capable of identifying the K-winning from N neurons, where $1 \le K < N$, whose input signals are larger than those of the remaining N - K neurons, was proposed in [16]. A convergence analysis of the model state variable trajectories to the KWTA operation is presented in [18]. A discrete-time version of the model and functional block-scheme of corresponding digital neural circuit have been proposed in [19].

In this paper, a modification of continuous-time model of analogue KWTA neural circuit proposed in [16] is derived and simulated. In contrast to the predecessor which exploits signum activation functions, the present model uses more simple step activation functions. A residual function of the model is derived by simplifying the residual function of the previous model. The model state equation which is a generalization of state equation of the existing model allows to avoid a dependency of its solutions of initial conditions. Since the operation of the corresponding circuit is independent of the initial states therefore, such a circuit does not require periodical resetting, special hardware and any additional processing time for this mode. This makes it possible simplify the model and corresponding hardware implementation, and to increase the speed of signal processing. It is shown by computer simulations that the model convergence speed to the KWTA operation is close to that of one of the most fast Hopfield type analogue KWTA neural networks while a computational and hardware implementation complexity of the model is less than the complexity of this network. The hardware implementation complexity of the model is close to that of one of the simplest continuous-time KWTA models whereas the convergence speed to the KWTA operation of the model is less than that of this comparable model.

2   A Model of Analogue KWTA Circuit

Let us be given the input vector $\mathbf{a} = (a_{n_1}, a_{n_2}, \cdots, a_{n_N})^T$, $1 < N < \infty$ with unknown finite value elements while the inputs are assumed to be located in the known range $[a_{min}, a_{max}]$, where the numbers a_{min} and a_{max} represent the minimal and the maximal possible values of inputs, respectively with $a_{max} - a_{min} = A$. Suppose the inputs are distinct and arranged in a descending order of magnitude satisfying the inequalities

$$\infty > a_{n_1} > a_{n_2} > \cdots > a_{n_N} > -\infty, \tag{1}$$

where $n_1, n_2, ..., n_N$ are numbers of the first largest input, the second largest input and so on up to Nth largest input inclusive. Let us design a model of analogue neural circuit that identifies the K largest of these inputs, which are referred to as the winners, where $1 \le K < N$ is a positive integer.

Let us assume that the designed model should process the input vector **a** to obtain, after a finite convergence time, such an output vector $\mathbf{b} = (b_{n_1}, b_{n_2} \cdots, b_{n_N})^T$ that

$$b_{n_i} > 0, i \in 1, 2, \cdots, K; b_{n_j} < 0, j \in K+1, K+2, \cdots, N. \tag{2}$$

Preprocessing an input vector a by subtracting from all its components the value a_{min} yields preprocessed inputs

$$\infty > c_{n_1} > c_{n_2} > \cdots > c_{n_N} > 0, \tag{3}$$

where $c_{n_k} = a_{n_k} - a_{min}$, $k = 1, 2, ..., N$. Let us present the outputs of the model designed herein by

$$b_{n_i} = c_{n_i} - x > 0, i \in 1, 2, \cdots, K;$$
$$b_{n_j} = c_{n_j} - x < 0, j \in K+1, K+2, \cdots, N, \tag{4}$$

where x is a scalar dynamic shift of inputs [16].

In order to design a model of an analogue KWTA neural circuit let us construct a procedure of finding a value of x satisfying the conditions (4). For this purpose, let us assume that there exists a certain time instant t^* when a variable x takes on a steady state value $x = x^*$ that satisfies (4) and keeps it thereafter. To stop a computational process at the instant t^* let us formulate a condition which will control a number of positive outputs at each time instant during the computational process. To this end, the following residual function presented in [16] can be used:

$$R(x) = 2K - N - \sum_{k=1}^{N} \beta_k, \tag{5}$$

where $\beta_k = \begin{cases} 1, & \text{if } c_{n_k} - x > 0; \\ 0, & \text{if } c_{n_k} - x = 0; \\ -1, & \text{if } c_{n_k} - x < 0 \end{cases}$ is a signum (hard limiting) function. Let us

simplify the function (5) to the following form:

$$E(x) = K - \sum_{k=1}^{N} S_k(x), \tag{6}$$

where

$$S_k(x) = \begin{cases} 1, & \text{if } c_{n_k} - x > 0; \\ 0, & \text{if } c_{n_k} - x \leq 0 \end{cases} \tag{7}$$

is a step function, and a sum $\sum_{k=1}^{N} S_k(x)$ determines the number of positive outputs. As it can be seen, the function E(x)=0 if the quantity of positive outputs is equal to K. Therefore, an equality E(x)=0 can be used to identify a necessary number of largest inputs K.

Let us take into account that $x \in [0, A]$. Now we design a continuous-time trajectory $x(t)$ which can go through the whole range $[0, A]$ and reach the value $x(t^*)$ satisfying the equality $E(x) = 0$ from any initial condition $x_0 \in [0, A]$. Suppose that a trajectory $x(t)$ is a solution of the corresponding differential equation. Let us update $x(t)$ by an exponential function. For this purpose we use the following state equation presented in [16]:

$$\frac{dx}{dt} = -\mu x; \quad x_0 = 1, \tag{8}$$

where $\mu = \begin{cases} 0, & \text{if} \quad R(x) = 0; \\ \alpha, & \text{if} \quad R(x) \neq 0, \end{cases}$ α is a constant parameter (or decaying coefficient)

which can be used to control a convergence speed of state variable trajectories to the KWTA operation. As one can see, the state variable trajectories of (8) are dependent on initial value of state variable x_0. In order to avoid this dependency let us generalize equation (8) to the following form:

$$\dot{x} = -\alpha \begin{cases} x, & \text{if} \quad E(x) > 0; \\ 0, & \text{if} \quad E(x) = 0; \\ x - A, & \text{if} \quad E(x) < 0. \end{cases} \tag{9}$$

It is easy to see that the state variable of differential equation (9) can accept any finite initial value $0 \leq x_0 \leq A$. The mathematical model of analogue KWTA neural circuit can be given by state equation (9) and output equation

$$b_{n_k} = c_{n_k} - x, \, k = 1, 2, \ldots, N. \tag{10}$$

If $E(x) > 0$, then according to (6) $\sum_{k=1}^{N} S_k(x) < K$ and the dynamic shift $x(t)$ should be decreased. On the contrary, if $E(x) < 0$, then $x(t)$ must be increased. In the steady state, when $\sum_{k=1}^{N} S_k(x) = K$, $x(t)$ should not be changed further. To identify K largest inputs, the state equation (9) must progressively provide a proper shift x that gradually approaches and finally falls into the range between the (K+1)st and the Kth maximum values of inputs, i.e. $c_{K+1} \leq x(t^*) < c_K$. Once a shift x is in the range between c_{K+1} and c_K, the outputs (10) exactly provide the KWTA operation. A rigorous analysis of global convergence of the model state variable trajectories to the KWTA operation exceeds the maximal permissible volume of this paper. Such the analysis can be performed under the condition $\alpha > 0$ like as it was fulfilled for the previous model in [18]. The results of computer simulations demonstrating a convergence of the model state variable trajectories to the KWTA operation are given below.

Note that the model (9), (10) can be also used in the case of time-varying inputs $a_{n_k}(t)$, $k = 1,2,...,N$ if the module of speed change of such inputs is much less than that of state variable x during transients. In other words, in this case condition

$$\left|da_{n_k}/dt\right| << \left|dx/dt\right|, \tag{11}$$

$k = 1,2,...,N$ should be satisfied for each $t < t^*$. As it can be seen from (9) in order to match the condition (11), the value of parameter α should be chosen large enough.

The functional block-diagram of an analogue KWTA neural circuit built based on the model described by state equation (9) and output equation (10) is shown in Fig. 1. The diagram contains inputs $a_1...a_N$, summers $\sum$, an inverting integrator I with a gain α, external sources of constant signals K, x_0, A, a_{min}, blocks $S_1,...,S_N$ of step functions $S_k(x)$, $k = 1,2,...,N$, outputs $b_1...b_N$, and variable structure functions

$$S_{N+1}(x) = \begin{cases} 0, & \text{if } E(x) \geq 0; \\ x-A, & \text{if } E(x) < 0 \end{cases} \text{ and } S_{N+2}(x) = \begin{cases} x, & \text{if } E(x) > 0; \\ 0, & \text{if } E(x) \leq 0. \end{cases}$$

As one can see, from an analogue hardware implementation complexity point of view, the circuit contains $N+2$ summers, $N+2$ switches, one integrator and four sources of constant signals (or three sources of constant signals if $x_0 = 0$). For comparison, one of the first continuous time KWTA neural networks of Hopfield type proposed in [1] can be implemented in analogue hardware using three multipliers, $N+4$ summers, $N+1$ sigmoid limiters, N integrators and six sources of constant signals. It is easy to see that the presented architecture is simpler than this network from implementation point of view. An implementation of one of the simplest comparable models of analogue KWTA neural network, recently proposed in [15], requires N+1 summers, N switches, one integrator and one source of constant signals. Thus, the hardware implementation complexity of the proposed model is close to that of this comparable model.

From a computational complexity point of view the presented model in each updating cycle needs a consecutive performing of N+5 additions/subtractions, two logic operations, one amplification and one integrating operation. The KWTA network presented in [1] requires for this purpose a consecutive fulfilling of N+2 multiplications, $N+3$ additions/subtractions, one sigmoid function operations and one integrating operation. As it is known, a multiplication requires much larger processing time than an addition/subtraction. Therefore, it is observable that the computational complexity of an updating cycle of the proposed model is less than that of this network. The model presented in [15] needs N+2 additions/subtractions, one logic operation, one amplification and one integrating operation in each updating cycle. Thus, the computational complexity of an updating cycle of the presented model is close to the computational complexity of this comparable model.

A resolution of the proposed model is theoretically infinite and it does not depend on its parameter values. In other words, if inputs are distinct, then the model can always identify them in accordance with the KWTA property (2). Since the model is capable of correctly processing any finite value distinct inputs, its resolution is the same as in other comparable neural networks with the same property [1], [15].

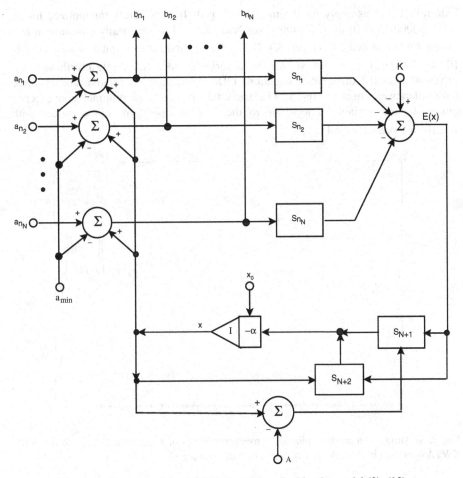

Fig. 1. Architecture of the KWTA circuit described by the model (9), (10)

Since in contrast to the predecessor the present model can operate correctly with any initial condition $0 \leq x_0 \leq A$, therefore the KWTA circuit implemented on the base of this model does not require a periodical resetting for repetitive processing of input sets, corresponding analogue supervisory circuit as well as spending additional processing time. This simplifies the hardware and increases the speed of signal processing.

3   Computer Simulation Results

In order to illustrate the theoretical results presented in this paper, let us consider two examples with corresponding computer simulations which demonstrate the processing of inputs by the herein proposed continuous-time model of analogue KWTA neural circuit. Let us use for this purpose corresponding program codes of MATLAB language and a 1.81 GHz desktop PC.

Example 1. Let us apply for the model (9), (10) 100 uniformly randomized inputs $a \in [-10000, 10000]$ of increasing sizes N=2, 3, ..., 100, uniformly randomized and rounded towards nearest integer $K \in \{1,2,...,N-1\}$, and random initial states $x_0 \in [-10000, 10000]$, i. e. A=20000. We use a variable order Adams-Bashforth-Moulton solver of non-stiff differential equations ODE113 having set relative and absolute error tolerances equal to 1e-15. The maximal, average and minimal convergence times of state variable trajectories to the KWTA operation in the model with $\alpha = 10^6$ are presented in Fig. 2.

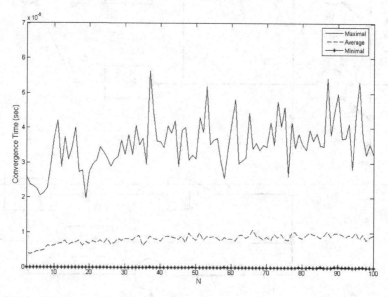

Fig. 2. Maximal, average and minimal convergence times of state variable trajectories to the KWTA operation in the KWTA model (9), (10) in Example 1.

Let us compare the model (9), (10) performance with that of one of the most fast model of analogue KWTA neural networks of Hopfield type presented in [17]. We consider the transient behaviors of the KWTA network model (20) and KWTA model (31), (32) with piecewise-constant activation functions presented in Fig. 5 – Fig. 7 and Fig. 8 – Fig. 11 correspondingly in [17]. Let us compare these behaviors with the transient dynamics of the proposed model depicted above in Fig. 2. It is not hard to see, that all these models demonstrate close convergence time of state variable trajectories to the KWTA operation. However, the model (20), and the model (31), (32) proposed in [17] are more complex and contain restrictions on its parameter values.

We compare the model (9), (10) performance with that of one of the simplest models of analogue KWTA neural networks proposed in [15] which contains the Heaviside step activation function. We analyze the state variable transient behaviors shown above in Fig. 2 and that in the kWTA model proposed in [15] and depicted in Fig. 6 of this paper. As it can be seen, the convergence time of state variable trajectories to the KWTA operation in the model (9), (10) is less by three orders on average.

Example 2. In order to analyze the model performance for time-varying inputs, let us consider the simulation presented below (adopted from [17]). We simulate the model behavior in the case of a set of the following four continuous-time sinusoidal signals: $a_{n_k}(t) = 10\sin[2\pi(t + 0.2(k-1))]$ (k=1,2,3,4), i.e. $N = 4$ and K=2. Since such inputs are time-varying, the corresponding KWTA problem is also time-varying. To reduce the computational time, let us use the finite-difference equation with a sampling period $\Delta t = 0.001$ instead of differential equation (9) in order to realize the model iteratively. The four inputs, transient state variable and the four outputs of the model are presented in Fig. 3, in which $\alpha = 20$, $A = 10$ and $x_0 = 0$. The simulation results show that the KWTA model is capable of determining the two largest inputs from the time-varying signals. Note that a correct performance is achieved with the value of parameter α being by one order less than that of the model presented in [17].

Thus, according to computer simulation results, the convergence time of state variable trajectories to the KWTA operation in the proposed model is close to that of the models of analogue KWTA networks of Hopfield type. However, a hardware implementation complexity of the model is less than that in these networks. On the other hand, the convergence speed in the presented model is higher than that of one of the simplest continuous-time KWTA models. The simulation results demonstrate a good matching of theoretical derivations and show that the proposed model is capable of effectively identifying not only the largest time-constant inputs but also the maximal time-varying signals.

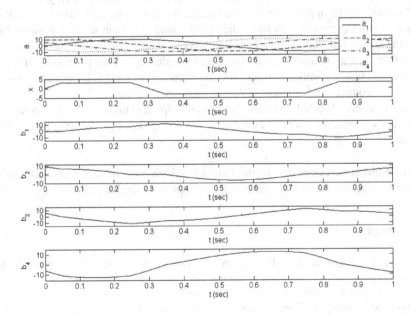

Fig. 3. Inputs, state variable and output signals of the KWTA model (9), (10) in Example 2.

4   Concluding Remarks

This paper presents a continuous-time model and a corresponding functional block-diagram of an analogue K-winners-take-all neural circuit designed based on the input signal dynamic shifting approach. The model is capable of selecting K maximal among any finite value N unknown distinct inputs, where $1 \leq K < N$. The model residual function and activation functions are simpler than such functions in the previous continuous-time KWTA model. Computer simulations show that a convergence time of the model state variable trajectories to the KWTA operation is close to that of one of the fastest continuous-time KWTA networks of Hopfield type. A hardware implementation complexity of the proposed model is less than that of this network. The hardware implementation complexity of the proposed model is close to that of one of the simplest analogue KWTA models. According to computer simulations, the convergence speed of state variable trajectories to the KWTA operation in the presented model is less than that in this comparable model.

In contrast to the previous model, the operation of the proposed model is independent of the initial states. Therefore, the analogue neural circuit implemented on the base of this model does not require a periodical resetting, the corresponding hardware and spending extra processing time for this mode which is useful for real time signal processing. Since the presented model does not contain limitations on its parameter values and it is simpler than analogue KWTA neural networks of Hopfield type therefore the model and the corresponding architecture of analogue KWTA neural circuit can be applied for rising a functioning precision and simplification of analogue sorting networks, order-statistics filters, analogue fault-tolerant systems [4]. Since the presented model has a higher speed of processing inputs than other comparable analogs therefore it can be used for reducing a data processing time on the basis of faster parallel sorting, for acceleration a digital image and speech processing, coding and digital TV by rank-order filtering [7], [15]. Further investigations are directed towards the model implementation with an up-to-date hardware, its generalization on the case of time-varying signal processing, and various applications.

Acknowledgments. The author would like to thank each of the reviewers and editors for their helpful comments and suggestions on improving the paper.

References

1. Majani, E., Erlanson, R., Abu-Mostafa, Y.: On the K-winners-take-all network. In: Touretzky, D.S. (ed.) Advances in Neural Information Processing Systems, vol. 1, pp. 634–642. Morgan Kaufmann Publishers Inc., San Francisco (1989)
2. Atkins, M.: Sorting by Hopfield nets. In: IEEE Int. Joint Conf. on Neural Networks, pp. 65–68. IEEE Press (1982)
3. Urahama, K., Nagao, T.: K-winners-take-all circuit with 0(N) complexity. IEEE Trans. on Neural Networks 6, 776–778 (1995)

4. Kwon, T.M., Zervakis, M.: KWTA networks and their applications. Multidimensional Syst. and Signal Processing 6, 333–346 (1995)
5. Binh, L.N., Chong, H.C.: A neural-network contention controller for packet switching networks. IEEE Trans. on Neural Networks 6, 1402–1410 (1995)
6. Itti, L., Koch, C., Niebur, E.: A model of saliency-based visual attention for rapid scene analysis. IEEE Trans. on Pattern Analysis and Machine Intelligence 20, 1254–1259 (1998)
7. Cilingiroglu, U., Dake, T.L.E.: Rank-order filter design with a sampled-analog multiple-winners-take-all core. IEEE Journal on Solid-State Circuits 37, 978–984 (2002)
8. Erlanson, R., Abu-Mostafa, Y.: Analog neural networks as decoders. In: Touretzky, D.S. (ed.) Advances in Neural Information Processing Systems, vol. 1, pp. 585–588. Morgan Kaufmann Publishers Inc., San Francisco (1991)
9. Fish, A., Akselrod, D., Yadid-Pecht, O.: High precision image centroid computation via an adaptive k-winner-take-all circuit in conjunction with a dynamic element matching algorithm for star tracking applications. Analog Integrated Circuits and Signal Processing 39, 251–266 (2004)
10. Jain, B.J., Wysotzki, F.: Central clustering of attributed graphs. Machine Learning 56, 169–207 (2004)
11. Liu, S., Wang, J.: A simplified dual neural network for quadratic programming with its KWTA application. IEEE Trans. on Neural Networks 17(6), 1500–1510 (2006)
12. DeSouza, G.N., Zak, A.C.: Vision for mobile robot navigation: a survey. IEEE Trans. on Pattern Analysis and Machine Intelligence 24, 237–267 (2002)
13. Graupe, D.: Principles of Artificial Neural Networks, 2nd edn. World Sci. Publisher, Singapore (2007)
14. Cichocki, A., Unbehauen, R.: Neural Networks for Optimization and Signal Processing. John Wiley & Sons, Chichester (1993)
15. Wang, J.: Analysis and design of a k-winners-take-all model with a single state variable and the Heaviside step activation function. IEEE Trans. on Neural Networks 9, 1496–1506 (2010)
16. Tymoshchuk, P.V.: A dynamic K-winners take all analog neural circuit. In: IVth IEEE Int. Conf. "Perspective Technologies and Methods in MEMS Design", pp. 13–18. IEEE Press, L'viv (2008)
17. Liu, Q., Wang, J.: Two k-winners-take-all networks with discontinuous activation functions. Neural Networks 21, 406–413 (2008)
18. Tymoshchuk, P.V.: Stability analysis of continuous-time model of K-Winners-Take-All neural circuit. In: XVI Ukrainian-Polish Conf. "CAD in Machinery Design. Implementation and Educational Problems", pp. 29–30. IEEE Press, L'viv (2008)
19. Tymoshchuk, P.V.: A discrete-time dynamic K-winners-take-all neural circuit. Neurocomputing 72, 3191–3202 (2009)

Network Intrusion Detection System Using Data Mining

Lídio Mauro Lima de Campos, Roberto Célio Limão de Oliveira,
and Mauro Roisenberg

Universidade Federal do Pará - UFPA
Av. dos Universitários s/n - Jaderlandia - Castanhal- PA - Brasil Cep: 68746-360
{lidio,limao,mauro}@ufpa.br
http://www.campuscastanhal.ufpa.br

Abstract. The aim of this study is to simulate a network traffic analyzer that is part of an Intrusion Detection System - IDS, the main focus of research is data mining and for this type of application the steps that precede the data mining : data preparation (possibly involving cleaning data, data transformations, selecting subsets of records, data normalization) are considered fundamental for a good performance of the classifiers during the data mining stage. In this context, this paper discusses and presents as a contribution not only the classifiers that were used in the problem of intrusion detection, but also the initial stage of data preparation. Therefore, we tested the performance of three classifiers on the KDDCUP'99 benchmark intrusion detection dataset and selected the best classifiers. We initially tested a Decision Tree and a Neural Network using this dataset, suggesting improvements by reducing the number of attributes from 42 to 27 considering only two classes of detection, normal and intrusion. Finally, we tested the Decision Tree and Bayesian Network classifiers considering five classes of attack: Normal, DOS, U2R, R2L and Probing. The experimental results proved that the algorithms used achieved high detection rates (DR) and significant reduction of false positives (FP) for different types of network intrusions using limited computational resources.

Keywords: Datamining, Network Intrusion Detection System, Decision Tree, Neural Network, Bayesian Network.

1 Introduction

With the enormous growth of computer networks usage and the huge increase in the number of applications running on top of it, network security is becoming increasingly more important. As shown in [2], all computer systems suffer from security vulnerability whose solution is not only technically difficult but also very expensive to be solved by manufacturers. Therefore, the role of Intrusion Detection Systems (IDSs), as special purpose devices to detect anomalies and attacks in the network, has become more and more important. KDDCUP'99 dataset is widely used as one of the few publicly available data sets for network-based

C. Jayne, S. Yue, and L. Iliadis (Eds.): EANN 2012, CCIS 311, pp. 104–113, 2012.

anomaly detection systems. It has been used as the main intrusion detection dataset for both training and testing [1] different Intrusion Detection schemes. However your research shows that there are some inherent problems in the KD-DCUP'99 dataset[1] that must be corrected before performing any experiment.

Many researchers are devoted to study methodologies to project (IDSs), [3] employed 21 learned machines (7 learners, namely J48 decision tree learning [4], Naive Bayes [5], NBTree [6], Random Forest [7], Random Tree [8], Multi-layer Perceptron [9], and Support Vector Machine (SVM) [10] from the Weka [11] collection to learn the overall behavior of the KDDCUP'99 data set), each trained 3 times with different train sets to label the records of the entire KDD train and test sets, which provided 21 labels for each record. Surprisingly, about 98% of the records in the train set and 86% of the records in the test set were correctly classified with all the 21 learners. Moreover, each dataset record was annotated with a #successfulPrediction value, which was initialized to zero. Once the KDD set had provided the correct label for each record, they compared each record predicted label given by a specific learner with actual label, where #successfulPrediction was incremented by one by one if a match was found. Through this process, the number of learners capable of correctly labeling that given record was calculated. The highest value for #successfulPrediction was 21, which conveys the fact that all learners were able predict that record label. Once conducted a statistical analysis on this data set and proposed a new data set,NSL-KDD,which consists of selected records of the complete KDDCUP99 [13] dataset and does not suffer from any of mentioned shortcomings.

[12]Proposed a new learning algorithm for adaptive network intrusion detection using Naive Bayesian classifier and decision tree, which performs balance detections and keeps false positives at an acceptable level for different types of network attacks, thus eliminating redundant attributes as well as contradictory examples from training data that make the detection model complex. Panda and Patra [14] used Naive Bayes for anomaly detection and achieved detection rate of 95%. Faroun and Boukelif [15] used Neural Networks with K-mean clustering and showed detection rate of 92%. Gaddam and Phoha [16] proposed a method to cascade clustering and decision tree for classifying anomalous and normal data. We used the dataset KDDCUP'99 in our research as proposed by [3] and then we proposed some improvements changes in the dataset through preprocessing to reduce the number of attributes from 42 to 27. Using the modified dataset, a study was conducted on the problem of Intrusion Detection using data mining, Initially, we tested a Decision Tree and a Neural network using this dataset, suggesting improvements in it, by reducing the number of attributes from 42 to 27 and considering only two detection classes normal and intrusion. Following the simulation we discuss some of the improvements in the work of [3]. Then, using the original KDDCUP'99 Dataset [13], we solved the same problem using two classifiers (Decision Tree and Bayesian networks) considering five classes of detection: Normal, DOS,R2L,U2R and Probing. In section 2 we presented the considerations about KDDCUP'99 Dataset are presented, in section 3 the concepts about Intrusion Detection Systems are discussed, in section 4 we presented

the Description of the used algorithms, in section 5 a description and discussion of the experiments, and finally the conclusions on section 6.

2   Considerations about the KDDCUP'99 Dataset

The KDDCUP'99 dataset was used in the 3rd International Knowledge Discovery and Data Mining Tools Competition for building a network intrusion detector. In 1998, DARPA intrusion detection evaluation program, a simulated environment was set up to acquire raw TCP/IP dump data for a local-area network (LAN) by the MIT Lincoln Lab to compare the performance of various intrusion detection methods. The KDDCUP'99 dataset contest uses a version of DARPA'98 dataset[12]. DARPA98 is about 4 gigabytes of compressed raw (binary) tcpdump data of 7 weeks of network traffic, which can be processed into about 5 million connection records, each with about 100 bytes. The two weeks of test data have around 2 million connection records. KDD training dataset consists of approximately 4.900.000 single connection vectors each of which contains 41 features and is labeled as either normal or an attack, with exactly one specific attack type [3]. Attack types were divided into 4 main categories as follow: **i. Probing Attack** is an attempt to gather information about a network of computers for the apparent purpose of circumventing its security controls. **ii. Denial of Service (DOS)** Denial of Service (DOS) is a class of attacks where an attacker makes some computing or memory resource too busy or too full to handle legitimate requests, denying legitimate users access to a machine. **iii. User to root (U2R)** is a class of exploit in which the attacker starts out with access to a normal user account on the system (perhaps gained by sniffing passwords, a dictionary attack, or social engineering) and is able to exploit some vulnerability to gain root access to the system. **iv. Remote to user (R2L)** This attack happens when an attacker sends packets to a machine over a network that exploits the machines vulnerability to gain local access as a user illegally. There are different types of R2U attacks; the most common attack in this class is done by using social engineering. In the KDDCUP'99 dataset these attacks (DoS, U2R, R2L, and probe) are divided into 22 different attacks types that are tabulated in Table 1. Not only they refer to the specific case of KDD-CUP'99 Dataset, there are lots of known computer system attack classifications and taxonomies, some of them have been analyzed in this research [19].

Table 1. Different Types of attacks in KDDCUP'99 Dataset

Attack Classes	22 Types of Attacks
DoS	back,land,neptune,pod,smurt,teardrop
R2L	ftp-write,guess-passwd,imap,multihop,phf,spy,warezclient,warezmaster
U2R	buffer-overflow,perl,loadmodule,rootkit
Probing	ipsweep,nmap,portsweep,satan

2.1   Inherent Problems of KDDCUP'99 DataSet

The total number of records in the original labeled training dataset is 972.781 for Normal, 41.102 for Probe, 3.883.370 for DoS, 52 for U2R, and 1.126 for R2L attack classes. One of the most important deficiencies in the KDD data set is the huge number of redundant records, which causes the learning algorithms to be biased towards the frequent records, and thus prevent them from learning unfrequented records which are usually more harmful to networks such as, U2R and R2L attacks. Besides, the existence of these repeated records in the test set will lead to biased evaluation results by the methods with better detection rates on the frequent records. We addressed this matter by removing all the repeated records on both KDD train and test set, and kept only one copy of each record. Tables 2 and 3 show the statistics of repeated records on the KDD train and test sets, respectively.

Table 2. Statistics of Redundant Records in the KDD Train Set [3]

	Original Records	Distinct Records	Reduction Rate
Attacks	3.925.650	262.178	93.32%
Normal	972.781	812.814	16.44%
Total	4.898.431	1.074.992	78.05%

Table 3. Statistics of Redundant Records in the KDD Test Set [3]

	Original Records	Distinct Records	Reduction Rate
Attacks	250.436	29.378	88.26%
Normal	60.591	47.911	20.92%
Total	311.027	77.289	75.15%

3   Intrusion Detection Overview

Intrusion detection (ID) is a type of security management system for computers and networks. An ID system gathers and analyzes information from various areas within a computer or a network to identify possible security breaches, which include both intrusions (attacks from outside the organization) and misuse (attacks from within the organization). A network based IDS (NIDS) monitor and analyze network traffics, and use multiple sensors for detecting intrusions from internal and external networks [17]. IDS analyze the information gathered by the sensors, and return a synthesis of the input of the sensors to system administrator or intrusion prevention system. System administrator carries out the prescriptions controlled by the IDS. Today, data mining has become an indispensable tool for analyzing the input of the sensors in IDS. Ideally, IDS should have an attack detection rate (DR) of 100% along with false positive (FP) of 0%. Nevertheless, in practice this is really hard to achieve. The most

Table 4. Parameters for performance estimation of IDS[2]

Parameters	Definition
True Positive (TP) or Detection Rate (DR)	Attack occur and alarm raised
False Positive (FP)	No attack but alarm raised
True Negative (TN)	No attack and no alarm
False Negative (FN)	Attack occur but no alarm

important parameters involved in the performance estimation of IDS are shown in Table 4.

Detection rate (DR) and false positive (FP) are used to estimate the performance of IDS [18] which are given as bellow:

$$DR = \frac{Total\_Detected\_Attacks}{Total\_Attacks} * 100 \tag{1}$$

$$FP = \frac{Total\_Misclassified\_Process}{Total\_Normal\_Process} * 100 \tag{2}$$

4   Description of the Used Algorithms

An Artificial Neural Network (ANN) is an information processing paradigm that is inspired by the way biological nervous systems, such as the brain, process information. It is composed of a large number of highly interconnected processing elements (neurons) working in unison to solve specific problems. ANNs, like people, learn by example. An ANN is configured for a specific application, such as pattern recognition or data classification, through a learning process. Learning in biological systems involves adjustments to the synaptic connections that exist between the neurons. This is true of ANNs as well. In this study we use multilayer neural network (MLP) employing backpropagation algorithm.

Decision trees are a classic way to represent information from a machine learning algorithm, and offer a fast and powerful way to express structures in data. The J48-WEKA algorithm used to draw a Decision Tree. The same is a version of an earlier algorithm developed by J. Ross Quinlan, the very popular C4.5.

Bayesian networks (BNs), belong to the family of probabilistic graphical models . A Bayesian network, or belief network, shows conditional probability and causality relationships between variables. The probability of an event occurring given that another event has already occurred is called a conditional probability. The probabilistic model is described qualitatively by a directed acyclic graph. The vertices of the graph, which represent variables, are called nodes. The nodes are represented as circles containing the variable name. The connections between the nodes are called arcs or edges. The edges are drawn as arrows between the nodes, and represent dependence between the variables.

5   Methodology and Experiments

Nowadays data mining has become an indispensable tool for analizing the input of used sensors in IDS. The objective of this research is to simulate a network traffic analyzer that is part of an IDS, as described in the abstract, to do this we tested the performance of three classifiers by employing the KDDCUP99 dataset and selected the best classifiers based on the parameters described in Table 4. The methodology used in this study used the data mining steps that consists of three stages: (1) the initial exploration - this stage usually starts with data preparation which may involve cleaning data, data transformations, selecting subsets of records and in case of data sets with large numbers of variables ("fields") - performing some preliminary feature selection operations to bring the number of variables to a manageable range (depending on the statistical methods which are being considered) (2) model building or pattern identification with validation/verification - this stage involves considering various models and choosing the best one based on their predictive performance and (3) deployment - that final stage involves using the model selected as best in the previous stage and applying it to new data in order to generate predictions or estimates of the expected outcome.

In the initial experiments, we used the modified KDDCUP'99 dataset, proposed by [3]. However some modifications were initially made by reducing the numbers of attributes from 42 to 27, for the following reasons: using statistics of software "Weka", some attributes that had unique value were eliminated, among them "num_outbound_cmds" and "is_host_login." Additionally, we eliminated attributes with high correlation coefficient, it was considered attributes strongly correlated those with correlation coefficients greater than or equal to 0.8. Our aim was to make the selection of attributes instead of synthesis, reason why we eliminated these attributes[1]. Highly correlated attributes influence each other and bring little information, as a result it is not interesting to maintain them in the data set, and so we used PCA (Principal Components Analysis) available in the "Weka". The following attributes were removed : sensor_rate, same_srv_rate, srv_serror_rate, st_host_srv_serror_rate, rerror_rate, srv_rerror_rate, srv_count. It is important to mention that Data mining is a step in the KDD process that consists of applying data analysis and discovery algorithms that produce a particular enumeration of patterns (or models) over the data. We focus primarily on the statistical approach to model fitting, which tends to be the most widely used basis for practical data mining applications given the typical presence of uncertainty in real world data generating processes, in other words the modifications made to the dataset reflect real world conditions.

Some attributes were selected and normalized: wrong_fragment, count, duration num_failed_logins, num_compromised,dst_host_srv_rerror_rate,num_file_creations,num_access_files and dst_host_count, these values were normalized with the values assumed in the interval [0,1]. Normalization is necessary in order to provide the data the same order of magnitude. Without this procedure some quantities could have existed quantities which would be more important than others. .Once the changes were made the "dataset" provided by [16] now has 27 attributes, as a result we obtained some improvements in relation to the performance of the classifiers, decision tree and neural network used in the study of [3] whose simulation results are shown on Table 5. The first classifier used was a decision tree(algorithm J48 from "Weka") to conduct training and testing. The algorithm J48 is an implementation of the C4.5 algorithm in java. In Table 5, it is clear that by using the test data and decision tree, we obtained a detection rate of 99.4% for normal connections and 91.1% for intrusion and false positives of 8.9% and 6%. During the simulations with a neural network we used the following parameters 23 neurons in the input layer, two in the intermediate layer, an one in the output layer, learning rate 0.3, momentum 0.2, sigmoid activation function for all neurons, 50000 epochs, For the test data we obtained a detection of 95% for normal connections, 92.3% for intrusion and 7.7% of false positives for normal connections and 5% for intrusion. The total number of instances correctly classified by the decision tree, was 95.12% and in the work of [3] was 93.82% by the neural network was 93.47% and in the work of [3] was 92.26%, thus reducing attributes according to the techniques previously shown which favored a better performance of the classifiers with respect to the results obtained by the work of [3]. In the experiments described above we used the "modified dataset" proposed by[3] considering only two classification classes : normal and intrusion. The dataset proposed in [3] is suitable, however the changes made in this study show that the results obtained by the classifiers are better. We made additional experiments, using the dataset proposed by [13], adopting the five classes of attack: Normal, DOS, Probing, R2L and U2R. Some modifications were made in the dataset, before carrying out the next step, data mining : we eliminated the single-valued attributes num_outbound_cmds and is_host_login. After selecting these attributes we normalize the following attributes: wrong_fragment, num_failed _logins, num_compromised, num_file_creations, num_access_files, count, dst_host_count and duration, these values were normalized with the values assumed in the interval [0,1]. Following, there was a balance between the classes, selecting records in a manner inversely proportional to the occurrences in accordance with Table 7. While doing this process, we encountered two invalid records in the KDD test set, number 136.489 and 136.497. These two records contain an invalid value, ICMP, as their service feature. Therefore, we removed them from the KDD test set. After the changes were made in the dataset, two classifiers were used in the simulations , Decision Tree algorithm (J48) and the Bayesian Network whose results are shown on Table 6. The decision tree scored better than the Bayesian network, especially in the classification of instances belonging to the class of R2L attacks, the decision tree correctly classified 95.2% of the records of this class, since the Bayesian network, only managed to correctly classify 69.3%. As for the other classes (Normal, Probe, DOS, U2R), the performance of both classifiers was similar. Comparing the results presented in this study, Table 6, with the work of [12] which used a hybrid system employing the algorithm ID3 with a Naive Bayes classifier, the results of [12] were better, except for the detection of false positive of R2L class. The following values were obtained by [12] for detection rate (DR%) and false positive (FP%): normal 99.72% and 0.06%, probe 99.25% and 0.39% Dos 99.75% and 0.04%, U2R 99.20% and 0.11%, R2L 99.26% and 6.81%. However, the results obtained in the second experiment, Table 6, are acceptable. For R2L class we achieved better

Table 5. Results for test using J48 Decision Tree and Neural Network MLP, given the dataset provided by [3] and the reduction of attributes from 42 to 27

Classifier	Normal	Intrusion
Decision Tree (DR%)	99.4%	91.1%
Decision Tree (FP%)	8.9%	6%
Neural Network (DR%)	95%	92.3%
Neural Network (FP%)	7.7%	5%

Table 6. Results for the test dataset [3], considering five classes

Classifier	Normal	Probe	Dos	R2L	U2R
Decision Tree (DR%)	98.9%	98.3%	99.7%	95.2%	93.9%
Decision Tree (FP%)	0.04%	0.02%	0.03%	0.01%	0.01%
Bayesian Network (DR%)	99.1%	93.5%	98.7%	69.3%	90.03%
Bayesian Network (FP%)	0.13%	0.05%	0.02%	0.14%	0.06%

Table 7. Number of Records by class in the Kddcup99 (10%) Dataset[13], with proposed reductions for train and test

Class attack	Dataset (10%)	Train	Test
Normal	97294	12607	2887
Denial of Service (DOS)	391458	36929	9607
Remote to User (R2L)	1113	911	202
User to Root (U2R)	51	31	21
Probing	4106	1247	293
Total	494022	51816	13020

results, such as 0.01 for false positive (FP%) and [12] obtained 6.81%. while the DoS attack type appears in 79% of the connections, the U2R and R2L attack types only appears in 0.01% and 0.225% of the records respectively. And these attacks types are more difficult to predict and and the more costly if missed.

Some rules extracted from decision tree used in the simulations of the second experiment, using the J48 algorithm, are illustrated below . The first rule associated with attacks of type "probe" corresponds to the detection of open ports and services on a live server used during an attack. The third rule concerns the "scans" performed on multiple hosts looking for open ports (e.g. TCP port = 1433). The rules associated with attacks of type R2L are characteristic of access to e-mail box and operations of download and upload files. The last two rules, which identify the type U2R attacks identify hidden files trying to evade antivirus programs running on the client machine overloading the buffer.

RULE 1 (DOS) - IF((flag=REJ OR flag=RSTO OR flag=SO) AND land LESS THAN 0.5) THEN label = neptune.

RULE 2 (Probe) - IF (count LESS THAN 3.5 AND (service=eco_i OR service=ecr_i) THEN label=ipsweep.

RULE 3 (R2L) - IF (service=pop_3 OR service=telnet) THEN label=guess
_passwd.IF((service=pop_3 OR service=telnet) AND (num_failed_logins LESS
THAN 0.5) AND (flag!=REJ OR flag!=RSTO) AND (service=http OR ser-
vice=login)) THEN label=ftp_write.

RULE 4 (U2R) - IF (dst_bytes LESS THAN 665.5) THEN label=rootkit. IF
(dst_bytes GREATER THAN 665.5) THEN label=Buffer_overflow

6   Conclusions

In the models proposed in this study some simplifications were considered:
"Probing" is not necessarily a type of attack except if the number of iterations
exceeds a specific threshold. Similarly a packet that causes a buffer overflow is
not necessarily an attack. Traffic collectors such as TCP DUMP that is used in
DARPAS'98 are easy to be overwhelmed and drop packets in heavy traffic, were
not checked the possibilities of packets dropped. The "dataset" proposed by [3]
which consists of selected records of the "dataset" has unique advantages, as
it does not include redundant or duplicate records which could bias the results
obtained by the classifiers. However, the experiments in this study showed that
it is possible to reduce the number of attributes from 42 to 27, improving the
performance of the decision tree classifiers and neural network in accordance
with the results shown in Table 5 and compared to the work of [3].

In experiments performed with the original dataset [13] and modified according to Table 7 and Table 6, we conclude that the initial stages of the process
of knowledge discovery in databases: data selection, pre-processing and transformation are essential for the data mining. The procedures used in the second
experiment: attribute selection, data normalization, allowed to obtain satisfactory results shown in Table 6 that are within acceptable standards in accordance
with the results presented in [12]. The results obtained by [12] were higher than
those shown in Table 6, because the use of hybrid systems proposed by [12].
However [12] did not provide the decision rules obtained by decision tree, what
is important in the process of knowledge discovery in database.

Tests were performed with a group of classifiers, where the best classifiers
for both cases were the decision tree (J48 algorithm) and Neural Network. The
Bayesian network was not a good classifier for five classes of attacks. We believe
that the methodology presented in this study may help researchers to compare
different methods of intrusion detection.

References

1. Stolfo, S.J., et al.: KDD cup 1999 data set. KDD repository. University of California, Irvine, http://kdd.ics.uci.edu
2. Landwehr, C.E., Bull, A.R., McDermott, J.P., Choi, W.S.: A taxonomy of computer
   program security flaws. ACM Comput. Surv. 26(3), 211–254 (1994)
3. Tavallaee, M., Bagheri, E., Lu, W., Ghorbani, A.: A Detailed Analysis of the KDD
   CUP 99 Data Set. Submitted to Second IEEE Symposium on Computational Intelligence for Security and Defense Applications, CISDA (2009)

4. Quinlan, J.: C4.5: Programs for Machine Learning. Morgan Kaufmann (1993)
5. John, G., Langley, P.: Estimating continuous distributions in Bayesian classifiers. In: Proceedings of the Eleventh Conference on Uncertainty in Artificial Intelligence, pp. 338–345 (1995)
6. Kohavi, R.: Scaling up the accuracy of naive-Bayes classifiers: A decision-tree hybrid. In: Proceedings of the Second International Conference on Knowledge Discovery and Data Mining, vol. 7 (1996)
7. Breiman, L.: Random Forests. Machine Learning 45(1), 5–32 (2001)
8. Aldous, D.: The continuum random tree. I. The Annals of Probability, 1–28 (1991)
9. Ruck, D., Rogers, S., Kabrisky, M., Oxley, M., Suter, B.: The multilayer perceptron as an approximation to a Bayes optimaldiscriminant function. IEEE Transactions on Neural Networks 1(4), 296–298 (1990)
10. Chang, C., Lin, C.: LIBSVM: a library for support vector machines (2001), Software available at http://www.csie.ntu.edu.tw/~cjlin/libsvm
11. Waikato environment for knowledge analysis (weka) version 3.5.7 (June 2008), http://www.cs.waikato.ac.nz/ml/weka/
12. Farid, D.M., Harbi, N., Rahman, M.Z.: Combining naive Bayes and Decision Tree for adaptative Intrusion Detection. International Journal of Network Security & Its Applications (IJNSA) 2(2) (April 2010)
13. KDD Cup 1999 (October 2007), http://kdd.ics.uci.edu/datasets/kddcup99/kddcup99.html
14. Panda, M., Patra, M.R.: Network intrusion detection using naive bayes. IJCSNS (2006)
15. Faroun, K.M., Boukelif, A.: Neural network learning improvement using k-means clustering algorithm to detect network intrusions. IJCI (2006)
16. Gaddam, S.R., Phoha, V.V., Balagani, K.S.: Means+id3 a novel method for supervised anomaly detection by cascading k-means clustering and id3 decision tree learning methods. IEEE Transactions on Knowledge and Data Engineering (2007)
17. Wasniowski, R.: Multi-sensor agent-based intrusion detection system. In: Proc. of the 2nd Annual Conference on Information Security, Kennesaw, Georgia, pp. 100–103 (2005)
18. Chen, R.C., Chen, S.P.: Intrusion detection using a hybrid support vector machine based on entropy and TF-IDF. International Journal of Innovative Computing, Information, and Control (IJICIC) 4(2), 413–424 (2008)
19. Alvarez, G., Petrovic, S.: A new taxonomy of web attacks suitable for efficient encoding. Computers and Security 22(5), 435–449 (2003)

A Near Linear Algorithm for Testing Linear Separability in Two Dimensions

Sylvain Contassot-Vivier[1] and David Elizondo[2]

[1] Université de Lorraine, Loria, UMR 7503, Nancy, France
http://www.loria.fr/~contasss/homeE.html
[2] Centre for Computational Intelligence, De Montfort University,
Leicester, United Kingdom
http://www.dmu.ac.uk/faculties/technology/staff/staff_david_elizondo.jsp

Abstract. We present a near linear algorithm for determining the linear separability of two sets of points in a two-dimensional space. That algorithm does not only detects the linear separability but also computes separation information. When the sets are linearly separable, the algorithm provides a description of a separation hyperplane. For non linearly separable cases, the algorithm indicates a negative answer and provides a hyperplane of partial separation that could be useful in the building of some classification systems.

Keywords: Classification, linear separability, 2D geometry.

1 Introduction

Linear separability is an important problem in the sense that it is underlying to numerous other problems, especially in the area of classification. However, the problem of linear separation is that all the known algorithms have high minimal complexities (greater than $\mathcal{O}(n)$) to determine the separability and/or find a separation direction. This is mainly due to the fact that either they stay at an abstract level without using geometrical aspects, or they use too specific and costly geometrical properties such as the convex hull.

In this paper, we propose an efficient algorithm (with complexity close to $\mathcal{O}(n)$) that determines the linear separability of two data sets in a two dimensional space. Moreover, when the sets are separable, the algorithm provides a separation direction, that is to say, the normal to a separation line, together with a threshold allowing for the distinction of the two sets along that direction.

According to the high speed of that algorithm, it could be useful in many classification processes based on a succession of linear separations. It could also be useful for feature reduction/selection by choosing a minimal subset of features where the sets are linearly separable, or to provide different separators for different subsets of features.

2 State of the Art

There are several existing algorithms to test the linear separability of two sets of points in a given space and to determine a separating hyperplane. The most

C. Jayne, S. Yue, and L. Iliadis (Eds.): EANN 2012, CCIS 311, pp. 114–124, 2012.
© Springer-Verlag Berlin Heidelberg 2012

common one is the perceptron [13], whose complexity order is given by the Novikoff's theorem [11]. More complex techniques like SVMs may also be used but, as they provide better quality results, their complexity is even higher. In the same way, there are convex and linear programming techniques. There exist also faster methods but whose results are not ensured in all cases, like the quadrant solution proposed in [6,5]. Another approach, inspired from the quadrant algorithm is to compute the set of points $X - Y$ and to determine whether the origin of the space is included in the convex hull of that set or not. However, that method is at least in $\mathcal{O}(\|X\|.\|Y\|)$ which is more expensive than some classical algorithms and the algorithm we propose in this paper. The following paragraph presents a brief list of the main methods for testing linear separability together with their complexities in two dimensions (for further details, see the more complete survey by D. Elizondo [4]).

2.1 Methods for Testing Linear Separability

The methods for testing linear separability between two classes can be divided into five groups:

- **Solving systems of linear equations.** These methods include: the Fourier-Kuhn elimination algorithm [9], and the Simplex algorithm [1]. The original classification problem is represented as a set of constrained linear equations. If the two classes are LS, the two algorithms provide a solution to these equations.
- **Computational geometry techniques.** The principal methods include the convex hull algorithm and the class of linear separability method [15]. If two classes are LS, the intersection of the convex hulls of the set of points that represent the two classes is empty. The class of linear separability method consists in characterising the set of points P of $\mathbb{R}^d$ by which it passes a hyperplane that linearly separates two sets of points X and Y.
- **Neural networks techniques.** The perceptron learning algorithm [10,14] is the most widely used neural network based method for testing linear separability. If the two classes are LS, the perceptron algorithm is guaranteed to converge after a finite number of steps, and will find a hyperplane that separates them. However, if the sets are not separable this method will not converge.
- **Quadratic programming.** These methods find a hyperplane that linearly separates two classes by solving a quadratic optimisation problem. This is the case for the SVM [2,3,7,12].
- **The Fisher Linear Discriminant method.** This method [8] tries to find a linear combination of input variables, $w \times x$ that maximises the average separation of the projections of the points belonging to the two classes C_1 and C_2 while minimizing the within class variance of the projections of those points. However, the resulting hyperplane does not necessarily separates the classes.

Other original methods exist, like the Tarski elimination algorithm [16]. However, they have larger complexities than most of the algorithms listed above.

If we consider the particular case of $\mathbb{R}^2$, we obtain the complexities given in Table 1. According to all these algorithms, our original contribution is the design and implementation of a linear separability determination algorithm whose complexity is near linear. So, it is faster than all those algorithms in most cases.

Table 1. Computational complexities of classical methods for testing the linear separability for n points in $\mathbb{R}^2$

Method	Complexity
Fourier Kuhn	$\mathcal{O}(n^4)$
Simplex	$>> \mathcal{O}(n^2)$
Quick Hull	$\mathcal{O}(n.log(n))$
Class of Linear Separability	$\mathcal{O}(n^2)$
Perceptron Algorithm	$\geq \mathcal{O}(n^2)$ for LS sets and ∞ for non LS sets

3   Theoretical Aspect

The design of our determination algorithm is based on several theoretical results that are valid, for some of them, in n dimensions. However, according to the scope of this paper, they are all expressed in two dimensions.

Theorem 1. *Let A and B, two sets of points of $\mathbb{R}^2$. A and B are lineraly separable iff $\exists v \in \mathbb{R}^2$, a direction in $\mathbb{R}^2$, such that:*

$$\forall a \in A, \forall b \in B, \quad a.v > b.v$$
$$or \ \forall a \in A, \forall b \in B, \quad a.v < b.v \tag{1}$$

Together with this theorem, the following lemmas and definition are the fundamental basis of our linear separability determination algorithm.

Lemma 1. *Let A and B, two linearly separable sets of $\mathbb{R}^2$. Then, there exists at least one couple $(a_s, b_s) \in A \times B$ such that a normal vector n_s to the vector $[a_s b_s]$ verifies:*

$$\forall a \in A, \forall b \in B, \quad a.n_s \geq b.n_s$$
$$or \ \forall a \in A, \forall b \in B, \quad a.n_s \leq b.n_s \tag{2}$$

Definition 1. *A tangential line of a set A is any line T, with normal n_T, that contains at least one point a_T of A such that:*

$$\forall a \in A, \quad a.n_T \geq a_T.n_T$$
$$or \ \forall a \in A, \quad a.n_T \leq a_T.n_T \tag{3}$$

The following lemma is the last theoretical basis that is used in our algorithm.

Lemma 2. *Let T be a tangential line common to sets A and B with normal n_T and contact points $a_T \in A$ and $b_T \in B$ such that:*

$$a_T.n_T = b_T.n_T \tag{4}$$

$$\forall a \in A, \ b \in B, \quad (a_T - a).n_T \times (b_T - b).n_T \leq 0 \qquad (5)$$

Let's denote by A_T and B_T the respective subsets of A and B that lie on line T.
   If A_T and B_T are linearly separable then the sets A and B are also linearly separable.

The Lemma 1 directly comes from two simple facts. The former is that the convex hull of a set is geometrically determined by the points in that set, and the latter is that two non-intersecting sets (linearly separable) share at least one common *tangential line*. In our polyhedral context, the notion of tangential line of a set A is any line that contains at least one point of A while all the other points of A lie on the same side of the line. This is depicted in Figure 1 and formally described in Definition 1.

   It can be noticed that there may be an infinite number of tangential lines going through one same point of a given set.

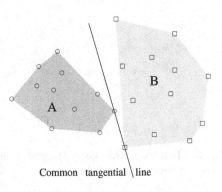

Fig. 1. Tangential line common to two sets A and B in two dimensions

The justification of Lemma 2 comes from the fact that if the points in A_T and B_T are linearly separable, then it is possible to find a slight non-null rotation of T such that:

 – exclusively some points of A_T (or B_T symmetrically) lie on the rotated line
 – all the points of B (symmetrically A) are on the opposite side of the points of A (symmetrically B) according to the rotated line

thus providing a line of separation. The main configurations of A_T and B_T sets in two-dimensions are depicted in Figure 2 together with the LS property of the sets A and B. It can be seen that for the three LS configurations, it is possible to perform a slight rotation of the tangential line in order to keep on the line only some points from the same set (A or B), whereas this is not possible in the last case on the bottom-right, due to the spacial overlapping of the sets A_T and B_T.

   Finally, the approach we propose in this paper to determine the linear separability of two sets A and B in two dimensions consists in finding a tangential line T common to both sets A and B, whose respective contact points fulfill Lemma 1 and such that the corresponding subsets A_T and B_T respectively of A and B on T are linearly separable.

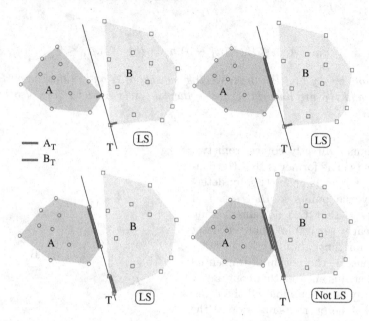

Fig. 2. Main configurations of tangential line common to sets A and B in two dimensions and their respective LS nature

4   Algorithm

The global scheme of our algorithm can be decomposed in several parts:

1- Finding a common tangent to the two sets A and B such that A and B are on opposite sides of the tangent
2- Testing whether the subsets of points of A and B that are on the tangent are linearly separable
3- Deducing the separation line (if it exists) and its associated normal
4- Determining the separation threshold along the normal

4.1   Finding a Common Tangent

Finding a common tangent consists in finding a couple of points $(a_s, b_s) \in A \times B$ that verifies Lemma 1. As stated before, such a couple does not exist in most cases of non linearly separable sets. The particular case of the existence of such a couple when the sets are not linearly separable is taken into account in the following step of the process. So, in this part we focus only on the process used to find whether such a couple exists or not and to determine one when it exists.

The finding process is an iterative process that modifies a direction vector n in order to find one for which the projections of sets A and B (scalar products) have an overlapping interval reduced to a single value. That common value corresponds to the projections of points a_s and b_s along the direction n, which is normal to $[a_s b_s]$ by construction. So, the algorithmic scheme is as follows:

Algorithm 1. Common tangent finding algorithm

---

n : current direction vector for the scalar products
t : current tangent
$indMaxA$: index of the point of A having the largest scalar product with n
$psMaxA$: scalar product of $A(indMaxA)$ and n
$indMinB$: index of the point of B having the smallest scalar product with n
$psMinB$: scalar product of $B(indMinB)$ and n
$history$: list of couples of points defining t that have already been encountered
$cycle$: Boolean indicating that a cycle has been found in the process

initialize n (with gravity centers of A and B)
repeat
    Compute the scalar products of points of A and B with n and stores the max of
    A in $(indMaxA, psMaxA)$ and the min of B in $(indMinB, psMinB)$
    // In the above treatment, the orientation of n is taken such that
    // it maximizes the difference $psMinB - psMaxA$
    if $psMinB < psMaxA$ **then**
        if $(indMaxA, indMinB) \in history$ **then**
            $cycle \leftarrow$ true   // The sets A and B are not linearly separable
        else
            insert $(indMaxA, indMinB)$ into $history$
            $t \leftarrow$ unity vector going from point $A(indMaxA)$ to point $B(indMinB)$
            $n \leftarrow$ normal vector to t
        end if
    end if
until cycle = true or $psMaxA \leq psMinB$

Our algorithm relies on the fact that when there is no common tangent to the sets A and B, the process reaches a cycle. Figures 3 and 4 respectively depict the process in the two cases. Another important remark related to the termination of that process is when $psMaxA$ is strictly smaller than $psMinB$. In this case, the linear separability is directly obtained and steps 2 and 3 can be skipped in the main algorithmic scheme.

4.2 Separability of Points on the Tangent

This part of the algorithm is quite simple as it consists in determining whether two sets of points that lie on the tangent are overlapping or not. So, it reduces to a very simple case of linear separation determination in one dimension as the considered points are aligned by construction. This can be achieved straightforwardly by computing the minimal and maximal positions of both sets of points that lie on the tangent and checking if the obtained segments are overlapping or not. Some representative examples are drawn in blue and green in Figure 2.

4.3 Final Separation Line and Normal

This step consists in deducing one separation line and its associated normal from the common tangent. This is achieved by applying a slight rotation to the

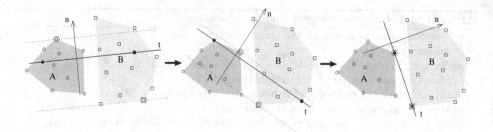

Fig. 3. Example of a tangent finding in 3 iterations. Black points are the successive a_s and b_s. Blue outlines denote the max of A and min of B along vector n.

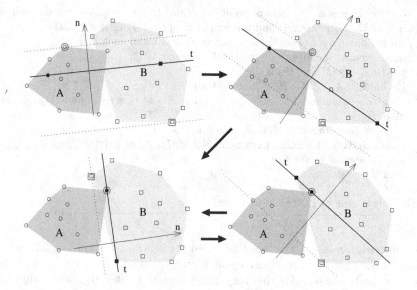

Fig. 4. Example of cycle reaching for non linearly separable sets

common tangent. In fact, that rotation is performed by adding to the current tangent a very small orthogonal vector in the pertinent orientation. Then, the resulting vector is normalized and its normal is deduced. So, there are two issues in this step. The former is how to determine the rotation direction, i.e. the orientation of the added orthogonal vector? and the latter is how to determine the rotation amplitude, i.e. the length of the added orthogonal vector?

Rotation direction: in the previous steps, the orientation of the normal is systematically chosen to put A under B along that direction. Thus, if we choose the orientation of the common tangent so that the couple $(tangent, normal)$ is direct, the rotation direction can be directly deduced from the relative order of A_T and B_T along the tangent. Indeed, if A_T is under B_T, the rotation must be to the right (that is to say in the opposite direction to the normal), otherwise the rotation must be to the left (same direction as the normal).

Rotation amplitude: it corresponds to the length of the orthogonal vector that is added to the tangent. This length is upper bounded by the relative positions of sets A and B on either sides of the tangent. An upper bound can be deduced by the ratio of the length of the projection interval of both sets A and B over the tangent, and the minimal distance to the tangent of all the points not on it.

Finally, by computing the projection interval of both sets A and B over the tangent $(Pint)$, and the minimal distance to the tangent line $(Dmin)$ of all the points but those on it, the separation line s is obtained by the following combination of the tangent t and its normal n:

$$s = Pint.t + rd.Dmin.n \qquad (6)$$

where rd is the rotation orientation (either 1 or -1).

As mentioned before, the orientation of the normal is always
computed in order to get A under B along that direction, in
the previous steps. This has the
practical advantage of implicitly
providing the separation convention that points of A will always
be under the threshold along the
normal direction whereas points of
B will be over it.

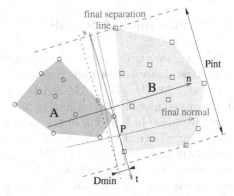

Fig. 5. Computation of s and deduction of the separation line and normal

4.4 Separation Threshold

The separation threshold is the value that separates the projections of A and B along the normal. It is obtained by computing the scalar product of the normal n with a point P (the pivot) that is located between the two sets along the normal. Hence, the scalar product of n with any point of A and B will be either strictly under the threshold or strictly over it. That pivot P can be taken as the middle point of the segment formed by the maximal point of A along n, and the minimal point of B. This requires a last series of scalar product computations of n with all the points of A and B. Figure 5 depicts the elements implied in the tangent rotation and translation to obtain the final separation line and normal.

Finally we obtain a complete linear classifier (a vector and its associated threshold) that allows us to completely separate the two sets A and B by simply projecting the points over the normal and compare them to the threshold.

5 Complexity

Concerning the global computational complexity of our algorithm, the first step is the more complex one and the most expensive. In fact, the complexities of the other steps are at most in $\mathcal{O}(n)$ (where $n = ||A|| + ||B||$) whereas the complexity of the first step is in $\mathcal{O}(\alpha.n)$ whose coefficient α corresponds to the number of iterations to get the tangent (or a cycle). That number is difficult to estimate as the number of required iterations does not directly depend on the number of points in A and B but on their positions. However, a worst case can be exhibited that consists of two sets forming two kinds of hyperbolas. In such a case, the number of iterations to get the tangent could be $\mathcal{O}(n)$, thus implying a global complexity of $\mathcal{O}(n^2)$. Fortunately, a particular choice of initial direction with the two gravity centers avoids such behavior and reduces the number of iterations much under n. In the following section, a statistical analysis is performed on the execution times of the algorithm in order to estimate its complexity range.

6   Experimental Results

The program implementing our algorithm is written in standard C++. The statistical analysis has been performed on a laptop with an Intel i7-2720QM CPU at 2.20GHz, 8Gb RAM and Linux 64bits with 3.2 kernel. However, we do not focus on absolute performances here but on the global shape of the execution time evolution in function of the data sets sizes. In the following experiments, the reported execution times are a mean of 20 executions. Since the standard deviations and the differences between max and min times stay very small in all cases, they are not included, for clarity sake, in Figure 6 that presents the executions times in function of the number of points for typical cases of sets.

A first important feature shown by those results is that the global behavior of the algorithm is merely the same whatever the configurations of sets A and B. In addition, linear bounds have been included to point out the overall linear tendency of the algorithm. A comparison between linear regressions of the results

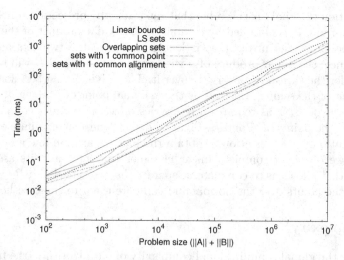

Fig. 6. Execution times in function of the problem size for typical cases

and $x.log_2(x)$ regressions have also confirmed that the curves are closer to the linear case. Moreover, it has been observed that the number of iterations in the first step of the algorithm does not follow the progression of the number of points and stays very small (between 2 and 6), confirming the limited influence of the sets sizes over the convergence speed.

7 Conclusion

An algorithm has been proposed to determine the linear separability of two sets of points in two dimensions. When the sets are separable, the algorithm provides a linear classifier (vector and threshold) that separates the two sets. A complexity analysis together with experimental results have shown that the algorithm has a near linear complexity. The natural following of this work will be to extend our algorithm to deal with higher dimensional spaces while preserving its near linear complexity.

References

1. Bazaraa, M.S., Jarvis, J.J.: Linear Programming and Network Flow. John Wiley and Sons, London (1977)
2. Boser, B.E., Guyon, I.M., Vapnik, V.N.: A training algorithm for optimal margin classifiers. In: Proceedings of the Fifth Annual Workshop on Computational Learning Theory, COLT 1992, pp. 144–152. ACM, New York (1992), http://doi.acm.org/10.1145/130385.130401
3. Cortes, C., Vapnik, V.: Support-vector network. Machine Learning 20, 273–297 (1995)
4. Elizondo, D.A.: The linear separability problem: some testing methods. IEEE Transactions on Neural Networks 17(2), 330–344 (2006), http://doi.ieeecomputersociety.org/10.1109/TNN.2005.860871
5. Elizondo, D.A.: Artificial neural networks, theory and applications (2008), French HDR
6. Elizondo, D.A., Ortiz-de-Lazcano-Lobato, J.M., Birkenhead, R.: A Novel and Efficient Method for Testing Non Linear Separability. In: de Sá, J.M., Alexandre, L.A., Duch, W., Mandic, D.P. (eds.) ICANN 2007, Part I. LNCS, vol. 4668, pp. 737–746. Springer, Heidelberg (2007), http://dx.doi.org/10.1007/978-3-540-74690-4_75
7. Ferreira, L., Kaszkurewicz, E., Bhaya, A.: Solving systems of linear equations via gradient systems with discontinuous righthand sides: application to ls-svm. IEEE Transactions on Neural Networks 16, 501–505 (2005)
8. Fisher, R.A.: The use of multiple measurements in taxonomic problems. Annual Eugenics 7(II), 179–188 (1936)
9. Fourier, J.B.J.: Solution d'une question pariculière du calcul des inégalités. In: Oeuvres II, pp. 317–328 (1826)
10. McCulloch, W., Pitts, W.: A logical calculus of the ideas imminent in nervous activity. Bulletin of Mathematical Biophysics 5, 115–133 (1943)
11. Novikoff, A.: On convergence proofs on perceptrons. In: Symposium on the Mathematical Theory of Automata, vol. XII, pp. 615–622 (1962)

12. Pang, S., Kim, D., Bang, S.Y.: Membership authentication using svm classification tree generated by membership-based lle data partition. IEEE Transactions on Neural Networks 16, 436–446 (2005)
13. Rosenblatt, F.: The perceptron: A probabilistic model for information storage in the brain. Psychological Review 65, 386–408 (1958)
14. Rosenblatt, F.: Principles of Neurodynamics. Spartan, Washington, D.C. (1962)
15. Tajine, M., Elizondo, D.: New methods for testing linear separability. Neurocomputing 47(1-4), 295–322 (2002)
16. Tarski, A.: A decision method for elementary algebra and geometry. Tech. rep., University of California Press, Berkeley and Los Angeles (1951)

A Training Algorithm for Locally Recurrent NN Based on Explicit Gradient of Error in Fault Detection Problems

Sara Carcangiu, Augusto Montisci, and Patrizia Boi

University of Cagliari, Department of Electrical and Electronic Engineering
Piazza d'Armi 09123 Cagliari, Italy

Abstract. In this work a diagnostic approach for nonlinear systems is presented. The diagnosis is performed resorting to a neural predictor of the output of the system, and by using the error prediction as a feature for the diagnosis. A locally recurrent neural network is used as predictor, after it has been trained on a reference behavior of the system. In order to model the system under test a novel training algorithm that uses an explicit calculation of the cost function gradient is proposed. The residuals of the prediction are affected by the deviation of the parameters from their nominal values. In this way, by a simple statistical analysis of the residuals, we can perform a diagnosis of the system. The Rössler hyperchaotic system is used as benchmark problem in order to validate the diagnostic neural approach proposed.

Keywords: Locally recurrent neural networks, nonlinear systems diagnosis, gradient-based training.

1 Introduction

Different approaches for fault detection using mathematical models have been developed in the last 20 years. See, e.g., [1] [2]. One of the most common approaches to fault diagnosis is based on models. The basic idea is to compare the operating conditions of the system under test with that calculated by the model. The difference between said two signals is called residual, and throughout a suitable threshold can be used to perform the diagnosis. The residual is calculated by means of analytical methods such as observers, parameter estimation methods or parity equations [2] [3], or through artificial intelligence techniques, such as neural networks [4] [5] if an accurate mathematical model of the system is not available. The main advantages of neural networks are the ability to cope with nonlinearities and the possibility to train the diagnostic system with historical data without requiring a precise knowledge of the process [6] [7]. In the case of diagnosis of dynamic systems, dynamic neural networks are used to catch the behavior of the system [8] [9]. The calculus of the residual of dynamic systems in general is difficult to accomplish, because the behavior depends on the particular initial state of the system, so that the model which generates the reference

C. Jayne, S. Yue, and L. Iliadis (Eds.): EANN 2012, CCIS 311, pp. 125–134, 2012.

signal should be able to predict the behavior of the system even for scenarios never experimented before. In this case modeling the systems often becomes an identification problem, in which locally recurrent neural networks showed to be quite suitable in several kinds of dynamic systems [7] [10] [11] [12]. Several authors resorted to the training algorithms proposed in [7] and [10], but no new substantial evolution has been presented in literature after that works, excepted some proposals to assume a second order minimization method rather than the classical gradient descent method [13]. On the other hand, some issues hold in the use of such algorithms, in particular because the formulation of the error function to minimize during the training is implicitly dependent on the previous samples. Furthermore, one cannot directly control the stability of the training. In [14] the issue of the stability of such kind of networks has been widely treated. In the present work, a new training algorithm is presented, which at the same time overcomes the problems of training and stability, in this way allowing one to extend the applicability of the method. Said algorithm is prompted by the Binet's formula (1843) which calculates the generic term of the Fibonacci's sequence as an explicit function of the golden ratio. The same formulation allows us to express the impulse response of an IIR filter as a linear combination of power of poles of the filter, then it is possible to explicitly calculate the gradient of the error with respect to the parameters and to take under control the stability. This paper describes a model-based diagnostic approach designed using artificial neural networks. In fact a dynamic neural network can be trained to predict the next sample of the input to the diagnosed process at normal operating conditions and then to generate residuals. Fault detection can be performed if a parameter deviation in the system under test determines a change in the dynamics of the diagnostic signal that appreciably affects the residual.

The organization of the paper is as follows. In Section 2 locally recurrent neural networks are introduced. In Section 3, a training method for neural networks with locally recurrent neurons is presented. Such neural networks are used as signal predictor of the process at normal operating conditions. In Section 4, a method of diagnosis model-based is presented. Section 5 reports the experimental results referring to the benchmark of Rössler system.

2   Neural Model

The artificial neural network used to model the system behavior belongs to the class of so-called locally recurrent globally feed-forward networks [7]. Its structure is similar to a multi-layer perceptron where neurons are organized in layers, but dynamic properties are achieved using neurons with internal feedbacks. Each neuron is an infinite impulse response (IIR) filter, which gives rise to an Auto-Regressive-Moving-Average (ARMA) filter. This structure is a generalized version of the neuron with an activation feedback model [14]. The block structure of the $k - th$ neuron considered is presented in Fig. 1.(a). Dynamics are introduced into the neuron in such a way that the neuron activation depends on its internal states. This is done by introducing a linear dynamic system (the IIR filter) into the neuron structure.

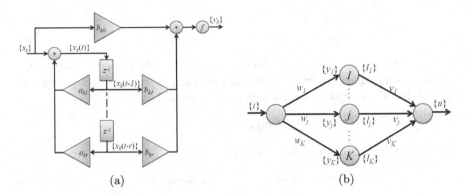

Fig. 1. Neural model: (a) Structure of the $k - th$ neuron with IIR filter, (b) Neural network structure

The input of the $k - th$ neuron $s_k(t)$ is passed to the IIR filter of the order r. Here, the filters under consideration are linear dynamic systems of different orders. The states of the $k - th$ neuron in the network can be described by the following state equation:

$$x_k(t) = \underline{a}_k \cdot \underline{x}_k(t) + s_k(t) \tag{1}$$

where $\underline{a}_k = [a_{k1}, a_{k2}, \ldots, a_{kr}]$ is the weights vector of the AR part of the IIR filter and $\underline{x}_k(t) = [x_k(t-1), x_k(t-2), \ldots, x_k(t-r)]^T$ is the state vector. Finally the neuron output is described by:

$$y_k(t) = f(\underline{b}_k \cdot \underline{x}_k(t) + b_{k0} \cdot s_k(t)) \tag{2}$$

where $f(\cdot)$ is the activation function, $\underline{b}_k = [b_{k1}, b_{k2}, \ldots, b_{kr}]$ is the vector of feed-forward filter parameters, b_{k0} is the weight of the input.

The locally recurrent networks possess many advantages over classical recurrent networks [6] [7], but it is a difficult task training them because of the implicit representation of the error, then it is impossible calculating the gradient of the cost function to apply a gradient-based training algorithm. In this paper a new formulation is proposed that overcomes this limitation. The reasoning applied here to the locally recurrent neural networks represents the extension of the Binet's formula for the calculation of the Fibonacci's sequence terms. In fact, said sequence can be viewed as the impulse response of an ARMA filter of the second order having unitary recursive weights. More specifically, our aim is to calculate the impulsive response of the single neuron, so that it becomes possible to explicitly calculate the output of the neuron by means of the convolution product and then calculate the error with respect to the desired output.

First of all, let us consider the scalar state variable represented by $\{x\}$ (see Fig. 1.(a)). The calculation of the sequence $\{x\}$ is performed by the equation (1). In particular, when the input of the neuron is a unitary impulse, the (1) gives us the impulse response $\{h\}$:

$$h(1) = 1$$
$$h(t) = a_1 \cdot h(t-1) + \cdots + a_r \cdot h(t-r) \tag{3}$$

The derivative of this expression cannot be calculated with respect to coefficients a_i because they affect the value of all the past samples of the state. Let us now suppose that there exist two constants c and z such that:

$$h(t) = c \cdot z^t \tag{4}$$

By substituting this expression in (3), we obtain:

$$c \cdot z^t = a_1 \cdot c \cdot z^{t-1} + \cdots + a_r \cdot c \cdot z^{t-r} \tag{5}$$

from which:

$$c \cdot z^{t-r} \cdot (z^r - a_1 \cdot z^{r-1} - \cdots - a_{r-1} \cdot z - a_r) = 0 \tag{6}$$

The (6) has two trivial solutions: $c = 0$ and $z = 0$, which cannot generate the sequence $h(t)$. Furthermore, we have r solutions corresponding to the zeros of the polynomial between parentheses:

$$z^r - a_1 z^{r-1} - \cdots - a_{r-1} z - a_r = 0 \tag{7}$$

Each solutions of (7) gives rise to a succession of the form (3), but none of them furnishes exactly such succession, because they have only one degree of freedom. Nevertheless, as (3) is linear, a linear combination of general solutions is itself a general solution of (7), therefore, by combining the r solutions of the form (4) deriving from the solutions of (7), we can obtain a general solution with r degrees of freedom

$$h(t) = c_1 \cdot z_1^t + \cdots + c_r \cdot z_r^t. \tag{8}$$

The r constants c_i in (8) can be determined by imposing the equality of the first r terms of the succession given by (3) and by solving the corresponding following system, where the unknowns are the constants c_i:

$$\begin{cases} h(1) = c_1 \cdot z_1 + \cdots + c_r \cdot z_r \\ h(2) = c_1 \cdot z_1^2 + \cdots + c_r \cdot z_r^2 \\ \vdots \\ h(r) = c_1 \cdot z_1^r + \cdots + c_r \cdot z_r^r \end{cases} \tag{9}$$

Denoting with $*$ the convolution product, the (2) can be so re-written:

$$y_k(t) = f\left((\underline{b_k} * \{s_k\} * \{h_k\})_t + b_{k0} \cdot s_k(t) \right) \tag{10}$$

where the subscript t indicates the $t-th$ term of the convolution product within parenthesis.

In the (10) the parameters of the neuron explicitly appear in the output of the neuron, which allows one to calculate the derivative of the output with respect to said parameters.

2.1   Neural Network Scheme

In Fig. 1.(b) the structure of the neural network assumed in this work is shown. For sake of simplicity, we have a single-input single-output network, only one hidden layer, where dynamic of the network is concentrated, and a linear activation function is assigned to the output neuron. In the following of the paper we will refer to said neural structure, so that a lake of generality will be unavoidable, but such a treatment has the advantage of the simplicity and matches the exigencies of the paper.

In the scheme of Fig. 1.(b), $\{i\}$ represents the input succession, $\underline{w} = [w_1 \ldots w_K]^T$ is the weights vector of the links between the input neuron and the hidden layer, K is the number of hidden neuron, $\{\underline{s}\}$ and $\{\underline{y}\}$ are respectively the input and the output vector successions of the hidden layer, $\underline{v} = [v_1 \ldots v_K]^T$ is the weights vector of the links between the hidden layer and the output neuron, $\{u\}$ is the output succession. On the basis of these definitions and the equation (10) we can explicitly express the output of the neural network to the time instant t as a function of the input and the parameters of the network:

$$u(t) = f\left[\left[\underline{b} * (\underline{w} \cdot \{i\}) * \{\underline{h}\}\right]_t + \underline{b}_0 \cdot \left[\underline{w} \cdot i(t)\right]\right] \cdot \underline{v} \tag{11}$$

where $\underline{b}$ is the collection of all the forward weights in the K hidden neurons, $\{\underline{h}\}$ is the set of the impulsive responses, $\underline{b}_0$ is the vector of adynamic weight in the hidden neurons (see Fig. 1.(a)).

3   Training Algorithm

The neural network is trained to iteratively forecast the next sample for a given stream of N samples. The classical gradient descent method is used as training algorithm, minimizing the mean squared error of the training set. Let us define the following cost function:

$$J = \frac{1}{2N} \sum_{t=1}^{N} [u(t) - i(t+1)]^2 \tag{12}$$

3.1   Evaluation of the Gradient

In order to perform the minimization of J we need to define the derivative with respect to each parameter of the network. In the following, said derivatives are reported. In order to calculate the derivatives, for a given set of network parameters, all the intermediate values, for the time instant $t = 1, \ldots, N$ have to be evaluated. This allows us to simplify the expression of the derivatives. By defining $\varepsilon(t) = u(t) - i(t+1)$, the derivatives can be so written:

1. Derivative with respect to the weight v_k, $k = 1, \ldots, K$:

$$\frac{\partial J}{\partial v_k} = \frac{1}{N} \sum_{t=1}^{N} \varepsilon(t) \cdot y_k(t) \tag{13}$$

2. Derivative with respect to the weight b_{k0}, $k = 1, \ldots, K$:

$$\frac{\partial J}{\partial b_{k0}} = \frac{1}{N} \sum_{t=1}^{N} \varepsilon(t) \cdot \frac{\partial u(t)}{\partial b_{k0}} = \frac{1}{N} \sum_{t=1}^{N} \varepsilon(t) \cdot v_k \cdot \frac{\partial y_k(t)}{\partial b_{k0}}$$

$$= \frac{1}{N} \sum_{t=1}^{N} \varepsilon(t) \cdot v_k \cdot w_k \cdot f'(y_k(t)) \cdot i(t) \tag{14}$$

3. Derivative with respect to b_{ks}, $k = 1, \ldots, K$, $s = 1, \ldots, r$. This calculation is a bit more difficult than the previous ones, because a convolution product is involved.

$$\frac{\partial J}{\partial b_{ks}} = \frac{1}{N} \sum_{t=1}^{N} \varepsilon(t) \cdot \frac{\partial u(t)}{\partial b_{ks}} = \frac{1}{N} \sum_{t=1}^{N} \varepsilon(t) \cdot v_k \cdot \frac{\partial y_k(t)}{\partial b_{ks}}$$

$$= \frac{1}{N} \sum_{t=1}^{N} \varepsilon(t) \cdot v_k \cdot f'(y_k(t)) \frac{\partial[(\underline{b}_k * \{s_k\} * \{h_k\})_t + b_{k0} \cdot s_k(t)]}{\partial b_{ks}} \tag{15}$$

$$= \frac{1}{N} \sum_{t=1}^{N} \varepsilon(t) \cdot v_k \cdot f'(y_k(t))(\{s_k\} * \{h_k\})_{t-s}$$

where $f'(\cdot)$ indicates the derivative of the activation function, $(g)_t$ indicates the $t - th$ sample of g and the subscripts that do not fulfill the constraint correspond to a null value of g.

4. Derivative with respect to a_{ks}, $k = 1, \ldots, K$, $s = 1, \ldots, r$. This derivation needs some preliminary remarks. Such coefficients affect the impulse response of the recursive part of the neuron (see Fig. 1.(a)), so that the derivation chain requires one more passage than the previous derivative. Furthermore, the impulse response h is given by a linear combination of powers of the zeros (see eq. (8)), therefore in order to take under control the stability of the network it is preferable to modify directly the zeros rather than the parameters a_{ks} [14]. Once the zeros have been set, the parameters a_{ks} can be easily determined and then we can calculate the constants c_i with the method described above. As a consequence, the derivatives will be calculated with respect to the zeros z_i rather than to the feedback coefficients a_{ks}, and the dependence of both a_{ks} and c_i on z_i is neglected in calculating the gradient.

$$\frac{\partial J}{\partial z_{ks}} = \frac{1}{N} \sum_{t=1}^{N} \varepsilon(t) \cdot \frac{\partial u(t)}{\partial z_{ks}} = \frac{1}{N} \sum_{t=1}^{N} \varepsilon(t) \cdot v_k \cdot \frac{\partial y_k(t)}{\partial z_{ks}}$$

$$= \frac{1}{N} \sum_{t=1}^{N} \varepsilon(t) \cdot v_k \cdot f'(y_k(t)) \frac{\partial[(\underline{b}_k * \{s_k\} * \{h_k\})_t + b_{k0} \cdot s_k(t)]}{\partial z_{ks}}$$

$$= \frac{1}{N} \sum_{t=1}^{N} \varepsilon(t) \cdot v_k \cdot f'(y_k(t)) \frac{\partial[(\underline{b}_k * \{s_k\} * \{\sum_{m=1}^{r} c_m \cdot z_m^p\}_{p<t})_t]}{\partial z_{ks}} \tag{16}$$

$$= \frac{1}{N} \sum_{t=1}^{N} \varepsilon(t) \cdot v_k \cdot f'(y_k(t)) \sum_{p=1}^{t-1} (\{s_k\} * \{b_k\})_{t-p} \cdot p \cdot c_s \cdot z_s^{p-1}$$

5. Derivative with respect to the weight w_k. This weight affect the output of the neuron through the input s_k to the $k-th$ hidden neuron. We will obtain:

$$\frac{\partial J}{\partial w_k} = \frac{1}{N} \sum_{t=1}^{N} \varepsilon(t) \cdot \frac{\partial u(t)}{\partial w_k} = \frac{1}{N} \sum_{t=1}^{N} \varepsilon(t) \cdot v_k \cdot \frac{\partial y_k(t)}{\partial w_k}$$

$$= \frac{1}{N} \sum_{t=1}^{N} \varepsilon(t) \cdot v_k \cdot f'\big(y_k(t)\big) \frac{\partial [(\underline{b}_k * \{s_k\} * \{h_k\})_t + b_{k0} \cdot s_k(t)]}{\partial w_k} \quad (17)$$

$$= \frac{1}{N} \sum_{t=1}^{N} \varepsilon(t) \cdot v_k \cdot f'\big(y_k(t)\big) [(\underline{b}_k * \{i\} * \{h_k\})_t + b_{k0} \cdot i(t)]$$

3.2 Training of the Neural Network

The equations from (13) to (17) allow us to calculate the gradient of the cost function (12) with respect to all the parameters of the network. The classical gradient descent method is used as training algorithm:

$$\Gamma_{m+1} = \Gamma_m - \eta \cdot \nabla J \quad (18)$$

where Γ is the set of all network parameters, m is the iteration index and η is the learning rate. Given $N + 1$ consecutive samples of the input signal, the training proceeds exactly as for the static MLP neural networks. If the activation function of hidden neurons $f(\cdot)$ is linear, the cost function J results convex, and the convergence to the global minimum is guaranteed provided that a suitable learning rate is chosen. On the other hand, the analytical approach to find said minimum is not feasible, because the derivation with respect to the zeros z_i is approximated.

The training process is a bit more complicated if the $f(\cdot)$ is nonlinear, because the cost function J is no longer convex and the issue of the local minima has to be handled.

4 Diagnostic Approach

In this paper a Fault Detection approach is proposed, which can be applied to both linear and nonlinear systems. First a neural network like that in Fig. 1.(b) is trained off line by means of the procedure described above in order to predict one step ahead the output signal of the system. Let $N + 1$ be the number of available training samples of the signal. The number of hidden neurons and the number of state variables are determined by a trial and error procedure. When the trained network is used to predict the signal, unavoidably the output will exhibit a certain prediction error or residual. On the basis of a validation signal, different from the training signal, the statistic distribution of the residual is estimated both in the case of nominal value of all the parameters and when one or more of them have a value within the tolerance range. This allows us to determine a threshold for the detection of the fault.

As the real objective is not predicting the signal but rather detecting the occurrence of a fault, it could be difficult to establish a priori a proper goal value of the mean squared error. In fact, usually, with a high sampling rate we could obtain a small residual even with a network without dynamics, whose prediction is simply equal to the last sample. Nevertheless, such a network hasn't any information on the dynamic of the signal. Instead, our objective is to capture the dynamics of the system. To this end, the diagnosis is made performing the network as an autonomous predictor on the validation sequence. A number of neural networks with rising number of hidden neurons and state variables are first trained and then tested on the validation set. The network which correspond the best approximation is used to perform the diagnosis.

For each system parameter, four residuals are calculated, corresponding to the prediction of one or more output variables in correspondence of the nominal value, an extreme value within the tolerance range and a variation of ±10% with respect to the nominal value. The average and the variance of each residual are used as feature for the diagnosis. A greater number of experiments can aid to better describe the frontier decision for the fault detection, whereas in general further test points improve the detectability of faults.

5   Results

In order to show the suitability of the method, even for the diagnosis of dynamic systems, where the behavior in faulty state often cannot be predicted, a chaotic system has been chosen to validate the proposed method. In particular the Rössler's hyperchaotic system [15] has been used, whose paradigm was initially defined to describe chemical processes, but one demonstrated that it is suitable to model a wide class of chaotic physical systems. In [16] an electronic circuit implementing system defined in [15] is described. The difficulty to interpret the residual signal is due to the fact that the trajectory in the state space critically depends on the initial state of the system, therefore one cannot define a fault free behavior as reference. The equations that describe the system are:

$$\begin{cases} \dot{x}_1 = -x_2 - x_3 \\ \dot{x}_2 = x_1 + a \cdot x_2 + x_4 \\ \dot{x}_3 = x_1 \cdot x_3 + b \\ \dot{x}_4 = -c \cdot x_3 + d \cdot x_4 \end{cases} \tag{19}$$

where a=0.25; b=3; c=0.5; d=0.05. A hyperchaotic behavior can be observed starting from the following initial state condition: (-15;11;0.2;23).

In order to show the suitability of the method, we tried to detect a variation of parameter a of the model. A 1-3-1 network structure has been adopted, with 2 state variables in the hidden neurons and hyperbolic tangent activation function. Such network is trained over 200 samples in order to iteratively predict one step ahead the state x_1 on the basis of the past samples, when all the system parameters are at their nominal value. In Fig. 2, the trend of mean squared error on the training set is reported.

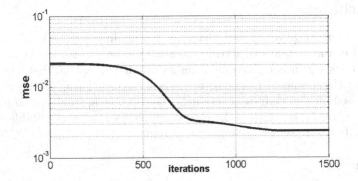

Fig. 2. Trend of mean squared error on the training set

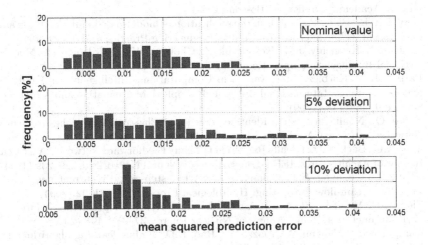

Fig. 3. Error prediction Probability Density Function

After the training phase, the reference value of the average and the variance of the residual are evaluated by performing the signal prediction being all the parameters at their nominal value but the initial state different from that one of the training signal. A number of test has been performed, by predicting the variable x_1 with all the parameters at their nominal value and with a variation of the parameter a of the 5% in the first series and of the 10% in the second one. In Fig. 3 the statistical results of this trial are summarized. As it can be observed, the statistical distribution of the prediction error is strongly affected by the value of the parameter which has been changed. This allows one to detect the occurrence of the deviation of a parameter, without the need of a preliminary classification of all the possible behaviors in presence of faults.

6   Conclusions

In this paper a neural based fault detection for nonlinear systems is proposed. The model of the system under test is realized by means of a locally recurrent neural network which is trained by using an original algorithm that allows one to represent the gradient of the cost function in explicit form. The detection of faults is based on the analysis of residuals. The approach, validated on the case of parameter deviation of the Rössler system, shows that it is suitable for the parametric faults detection in nonlinear systems.

References

1. Isermann, R.: Fault diagnosis of machines via parameter estimation and knowledge processing. Automatica 29, 815–835 (1993)
2. Chen, J., Patton, R.J.: Robust model-based fault diagnosis for dynamic systems. Kluwer Academic Publishers, Boston (1999)
3. Isermann, R., Balle, P.: Trends in the application of model-based fault detection and diagnosis of technical processes. Control Engineering Practice 5, 709–719 (1997)
4. Korbicz, J., Koscielny, J.M., Kowalczuk, Z., Cholewa, W.: Fault Diagnosis. Models, Artificial Intelligence, Applications. Springer, Berlin (2004)
5. Patton, R.J., Korbicz, J.: Advances in Computational Intelligence for Fault Diagnosis Systems. International Journal of Applied Mathematics and Computer Science 9, 468–735 (1999)
6. Nelles, O.: Nonlinear System Identification. From Classical Approaches to Neural Networks and Fuzzy Models. Springer, Berlin (2001)
7. Tsoi, A.C., Back, A.D.: Locally Recurrent Globally Feedforward Networks: A Critical Review of Architectures. IEEE Transactions on Neural Networks 5, 229–239 (1994)
8. Chang, L.C., Chang, F.J., Chiang, Y.M.: A two-step-ahead recurrent neural network for stream-flow forecasting. Hydrological Processes 18, 81–92 (2004)
9. Zio, E., Di Maio, F., Stasi, M.: A data-driven approach for predicting failure scenarios in nuclear systems. Annals of Nuclear Energy 37, 482–491 (2010)
10. Campolucci, P., Uncini, A., Piazza, F., Rao, B.D.: Online learning algorithms for locally recurrent neural networks. IEEE Transactions on Neural Networks 10, 253–271 (1999)
11. Boi, P., Montisci, A.: A Neural Based Approach and Probability Density Approximation for Fault Detection and Isolation in Nonlinear Systems. In: Iliadis, L., Jayne, C. (eds.) EANN/AIAI 2011, Part I. IFIP AICT, vol. 363, pp. 296–305. Springer, Heidelberg (2011)
12. Cannas, B., Cincotti, S., Marchesi, M., Pilo, F.: Learning of Chua's circuit attractors by locally recurrent neural networks. Chaos, Solitons and Fractals 12, 2109–2115 (2001)
13. Peng, C.C., Magoulas, G.D.: Nonmonotone BFGS-trained recurrent neural networks for temporal sequence processing. Applied Mathematics and Computation 217, 5421–5441 (2011)
14. Patan, K.: Stability Analysis and the Stabilization of a Class of Discrete-Time Dynamic Neural Networks. IEEE Transactions on Neural Networks 18, 660–673 (2007)
15. Rössler, O.E.: An equation for hyperchaos. Physics Letters A 71, 155–157 (1979)
16. Wang, G.Y., Zhang, X., Zheng, Y., Li, Y.X.: A new modified hyperchaotic Lü system. Physica A 371, 260–272 (2006)

Measurement Correction for Multiple Sensors Using Modified Autoassociative Neural Networks

Javier Reyes Sanchez, Marley Vellasco, and Ricardo Tanscheit

Departmente of Electrical Engineering
Pontifical Catholic University of Rio de Janeiro
Rua Marquês de São Vicente, 225
22451-900 Rio de Janeiro, RJ, Brazil
javier.rys@gmail.com, {marley,ricardo}@ele.puc-rio.br

Abstract. In industrial plants, the analysis of signals provided by monitoring sensors is a difficult task due to the high dimensionality of the data. This work proposes the use of Autoassociative Neural Networks trained with a Modified Robust Method in an online monitoring system for fault detection and self-correction of measurements generated by a large number of sensors. Unlike the existing models, the proposed system aims at using only one neural network to reconstruct faulty sensor signals. The model is evaluated with the use of a database containing measurements collected by industrial sensors that control and monitor an internal combustion engine. Results show that the proposed model is able to map and correct faulty sensor signals and achieve low error rates.

Keywords: sensors, calibration, fault detection, autoassociative neural networks, signal monitoring system.

1 Introduction

Instrument selection, distribution, installation and control are considered to play an important and key role in a company's engineering operations. The existing fieldbus protocols, also known as industrial network protocols (Profibus, Modbus, Hart, ASI), allow for better communication and interaction between operators and engineers and the field equipment. As a result, reliable measurements are obtained as well as information regarding possible system failures [1]. However, it is not unusual for a sensor, and sometimes the most critical one, to have a degradation that is overlooked by the operator and for this reason be the cause of an undesired shutdown of the production process [2].

The last few decades have witnessed the development of technologies for monitoring industrial process conditions during plant operations [3]. To this end, industries have been seeking to replace periodic maintenance by condition-based maintenance strategies as a means by which to obtain a potentially more efficient online method.

Computational Intelligence techniques have been used in the development of fault detection methods [4][5][6]. These methodologies, in which computational models are designed for the purpose of predicting real system outputs, is unsuitable for performing sensor diagnosis because it is based on correct input data (measurements

C. Jayne, S. Yue, and L. Iliadis (Eds.): EANN 2012, CCIS 311, pp. 135–144, 2012.
© Springer-Verlag Berlin Heidelberg 2012

via sensors) and assumes that the inputs into the real system and into the model are fault-free. When there is a noticeable difference between the output of the real system and the output of the model, one assumes that there is a problem in the real system. In the case of sensor monitoring and diagnosis, the goal is to find the malfunctioning sensors (system inputs) that cause such problems [7].

The motivation of this work was to develop a model that would be able to perform online monitoring and self-correction of multiple sensor measurements in order to reduce maintenance costs, minimize the risk of using uncalibrated or faulty sensors, increase instrument reliability and consequently reduce equipment inactivity. The model is based on Autoassociative Neural Networks [8] with a modified training procedure (M-AANN).

2   Autoassociative Neural Network Architecture

Autoassociative neural networks are inspired by the NLPCA (nonlinear principal component analysis) methodology [9][10]. In simplified terms, in this method the system input data are mapped to the output data by means of a nonlinear function G. The reconstruction of the original data is carried out by a "demapping" function expressed by a nonlinear function H. Functions G and H are selected so as to minimize information loss during the mapping-demapping process. The mapping-demapping procedure is implemented by an autoassociative neural network (AANN), whose architecture consists of two serially connected neural networks responsible for the implementation of the mapping function, G, and of the demapping function, H, as shown in Fig. 1.

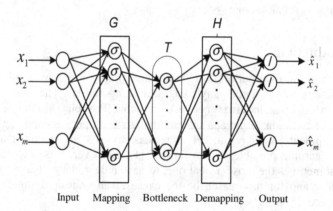

Fig. 1. Autoassociative neural network

Autoassociative neural networks have already been used for correcting signals provided by a single sensor [8][11]. The model proposed in this work employs AANNs with a modified training procedure for correcting signals from multiple sensors.

3   Self-correction Model

The objective is to correct the measurements from a set of sensors when their performance is degraded due to improper sensor installation, aging or deformation as a result of exposure to unsuitable ambient conditions. Sensor measurements are the inputs to the self-correction model, which calculates the best estimate of the input vector. The main element of the self-correction model is the Autoassociative Neural Network (AANN), which is used to reconstruct the signals. The AANN plays the role of an identity matrix when the measurements are fault-free, and of a nonlinear function that is able to reconstruct signals when any degradation is detected in them. The development of the self-correction model comprises the following steps.

Step 1 - Sensor Correlation: the degree of correlation between the variables is a significant aspect when an AANN neural network is to be used as a monitoring system [8][12]. In situations where the correlation is high, drift, offset or noise in one of the sensors do not have significant effects on the response of the AANN because its output is related to all the inputs by means of a large number of weights and patterns. It was observed in this study that although all the sensors were monitoring the same plant, they did not always show a high degree of correlation. Thus, when there is a large number of sensors and correlations are not uniform, it is proposed that the sensor set be subdivided into smaller groups with similar correlations such that better signal approximations can be obtained.

Step 2 – Data Preprocessing: the database must be cleaned by removing inconsistent or incomplete data before training the AANN. Then, data normalization is performed. In this work, where the activation function of the hidden layer neurons is of the *tansig* (tangent sigmoid) type, measurements were mapped to a range of [-1, 1] in order to simplify the training process.

Step 3- Estimating the number of neurons: the complexity of an AANN is defined by the number of neurons in the bottleneck layer [8][13] (see Fig. 1). Although the number of neurons in the mapping and demapping layers affects the neural network performance, the number of neurons in the bottleneck layer has a more significant effect on the quality of the response. Distinct topologies are trained and tested in order to evaluate the number of neurons required to provide the AANN with the ability to correct a sensor's measurements. The network must be trained with a database containing disturbed data. The number of neurons in the mapping and demapping layers is then selected with the aim that the desired response can be obtained under noise-free conditions. The network will then be able to reconstruct a sensor signal with the help of the existing correlation with the signals from other sensors.

Step 4- Modified Robust Training Method (M-AANN): the aim of this step is to provide the network with the ability to reconstruct signals that present abrupt errors at the same time. The robust training method proposed by [11] can reconstruct measurements resulting from faulty sensors, provided the faults are not simultaneous, using one or more AANNs [14][15]. The robust AANN is able to reconstruct sensor measurements when a single sensor in the group has failed, but in the reconstruction of the faulty signal there are deviations in the outputs of the fault-free sensors.

With the objective of improving the response of the AANN network in the presence of multiple faults and of developing a method that can be employed in real systems, this work proposes a modification of the robust training method. Let us consider a training set $X= [X_{1 \, x \, n}, X_{2 \, x \, n}, ... X_{m \, x \, n}]$, where $X_{i \, x \, n}$ represents the vector of n samples from sensor i. By inserting noise only in the data of sensor i, for example, it is possible to observe at the output of the neural network how the reconstruction of the signals from all the sensors is carried out. Assuming that the measurements from two other sensors ($i+2$ and $i+4$, for example) have not been reconstructed satisfactorily, random noise is added, at the input of the network, to the values measured by those two sensors. This procedure is repeated for two other sensors until the data of all sensors, always in groups of two, have been presented to the network in a corrupted form. The name given to the network trained in this way is M-AANN (Modified Autoassociative Neural Network).

4   Case Study

The case study was chosen in order to show that the M-AANN is able to perform online reconstruction of sensor measurements affected by drifts, offsets and noise. The model was evaluated with the use of a database containing measurements from sensors that control and monitor an internal combustion engine coupled with an alternator that generates the power required to feed two electric motors responsible for the rear wheel traction of a mining truck. The database provided contains values that had been previously calibrated and filtered by the ECM (Engine Control Module) during the signal acquisition stage. Among the 40 measurement variables provided, 32 were selected, corresponding to the measurements taken by distinct pressure and temperature sensors mounted externally to the engine. The variables corresponding to calculations performed by the ECM, to PWM (Pulse Width Modulation) measurements and to sensors that delivered ON-OFF type outputs were disregarded.

Differently from other studies, in this work the signals produced by the sensors are not ideal, that is, they vary according to the engine's effort during the transport of the material. Of the 2000 samples, 800 were used in the training phase, 200 for validation and 1000 for testing. In order to avoid overfitting, the early-stopping technique was used in the training phase of all the neural networks tested.

Based on the methodology described in Section 3, which takes into account the correlations between sensors measurements, these were divided into three groups and a modified autoassociative neural network (M-AANN) was associated to each one of these groups. Fig. 2 presents a block diagram of the self-correction model. The tests performed with each of the three groups are detailed below.

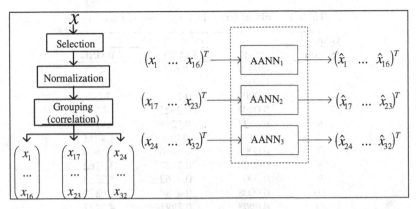

Fig. 2. Self-correction Model

Group 1 – the 16 sensors selected for this group measure the temperature in the internal combustion chamber of the eight cylinders of the engine and showed a correlation above 0.87. Distinct M-AANN topologies were tested by varying the number of neurons in the hidden layers and inserting noise into the sensor measurements. Based on the errors obtained in the validation phase, the network having the highest generalization capability was selected via cross-validation. In this case 6 neurons were selected for the bottleneck layer.

The next step was to train the M-AANN to reconstruct measurements when noise, drifts or offsets were present in the input vectors. The modified robust method (Section 3) was then used, with a random noise of 10% of the measurement range. The number of neurons in the mapping and demapping layers was increased until a response with a maximum MAPE error of 2% was obtained.

The network was then tested for its ability to correct failures in three or more faulty sensors. Data corresponding to the measurements from three randomly selected faulty sensors were presented to the M-AANN$_1$ (the drift rate was identical to the one in the previous test). The results for all sensors have been summarized in Table 1, where sensor 1, 5 and 11 are considered as faulty. Fig. 3 presents the response of the M-AANN$_1$ for sensors 1 and 2, with regard to drifts in sensors 1, 5 and 11. It may be noticed that the M-AANN$_1$ is able to perform self-correction for the measurements affected by sensor drifts, with a MAPE value of less than 1%.

Group 2 – Sensors measurements in this group were found to have a correlation of approximately 0.6. In keeping with the proposed methodology, distinct topologies were tested. It was considered that there would be 9 to 16 neurons in the mapping and demapping layers and 1 to 6 neurons in the bottleneck layer. Next, the modified training method was used, in which errors equivalent to 20% of the sensor measurement range were inserted in the input vectors. The number of neurons in the mapping and demapping layers was increased until a response with a maximum

Table 1. Drift of 100 units for sensors 1, 5 and 11

Group 1	16-18-6-18-16		
Sensor	MSE	MAPE (%)	RMSE
1	*0.0019*	*0.655*	*6.4436*
2	0.0002	0.255	3.7725
3	0.0002	0.2366	3.245
4	0.0004	0.2622	4.01
5	*0.0021*	*0.7299*	*7.9174*
6	0	0.088	1.4911
7	0.0006	0.3436	4.4874
8	0.0003	0.2727	3.7162
9	0.0005	0.3862	5.2263
10	0.0008	0.4989	6.6872
11	*0.0003*	*0.2991*	*4.5121*
12	0.0006	0.4446	6.0273
13	0.0003	0.2389	3.6136
14	0.0004	0.4016	5.688
15	0.001	0.5091	6.5595
16	0.0006	0.4262	6.1295

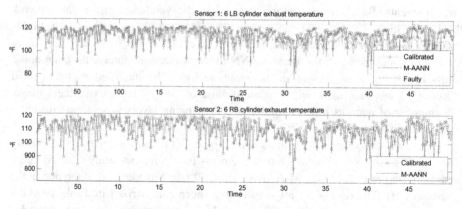

Fig. 3. Sensors response to a low drift of 0.7% per minute in sensors 1, 5 and 11

MAPE value of 2% was obtained. At the end of the training process, the topology obtained presented 15 neurons in the mapping and demapping layer and 3 neurons in the bottleneck layer.

The M-AANN$_2$ was then tested for its ability to deal with multiple faults of the offset type. The measurements from sensors 2, 4, 5, 6 and 7 (selected randomly) were corrupted by the addition of an offset of 4 units at the same time. Table 2 summarizes the errors (MAPE) obtained for situations with and without disturbance. The mean of the difference between these two situations increases from 0.51% to 0.85% (MAPE). Therefore, the errors are within the limits established for the sensor set analyzed herein.

Table 2. Response of M-AANN$_2$ to faults in sensors 2, 4, 5, 6 and 7 (offset=4 units)

Group 2	7-15-3-15-7	
Sensor	MAPE (%) Undisturbed	MAPE (%) Disturbed
1	0.5881	1.7353
2	0.5901	*1.2726*
3	0.4154	0.3847
4	0.3233	*0.4628*
5	0.6845	*0.8208*
6	0.4406	*0.6704*
7	0.5917	*0.6408*

Group 3 - the nine sensors in this group present a mean of 0.33 in the correlations of their measurements. Since there are measurements with correlations of less than 0.2, this group constitutes a hard test with respect to the reconstruction of signals in the presence of abrupt sensor faults. The same method was used again to train the M-AANN$_3$. This network was tested for simultaneous faults in sensors 1, 3, 7 and 9, with an offset of 5 units at the same time. The results for these faults are summarized in Table 3 and Fig. 4 presents the responses. It can be observed that the M-AANN$_3$ was able to reconstruct the sensor signals with a resulting MAPE of less than 2%. In addition, the mean of the difference between these latter cases and those in which the sensor signals were not disturbed was 0.34%. Therefore, the errors are within the limits established for the sensor set analyzed herein.

A fundamental conclusion, which also asserts the validity of the model proposed in this work, is that the three networks proved to have excellent generalization capabilities. When trained using disturbed measurements for only two sensors, for example, during the testing phase they were able to reconstruct the signals generated by all the sensors, with minimum error.

Table 3. Responses of M-AANN$_3$ to faults in sensors 1, 3, 7 and 9 (offset=5 units)

Group 3		9-15-7-15-9	
Sensor	MSE (e-3)	MAPE (%)	RMSE
1	*0.0063*	*0.0118*	*0.0357*
2	0.0058	0.052	0.0055
3	*0.1267*	*0.1596*	*0.1647*
4	0.2235	0.2087	0.2087
5	0.016	0.0504	0.0039
6	0.0188	0.0553	0.1489
7	*0.0217*	*0.0277*	*0.0699*
8	0.2588	0.2309	0.2011
9	*0.0165*	*0.065*	*0.141*

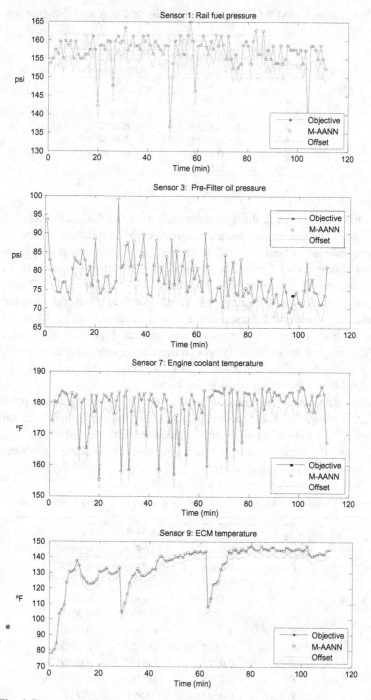

Fig. 4. Responses of M-AANN$_3$ to faults in sensors 1, 3, 7 and 9 (offset=5 units)

5   Conclusion

This work has presented in detail a proposal for a model for online monitoring of industrial sensors based on the use of autoassociative neural networks. The model proposes to organize sensors with similar degrees of correlation into different groups, assign a neural network to each of these groups and have the network perform self-correction of the faulty sensor measurements. The modified robust method proposed in this work enabled the model to generalize and reconstruct measurements when the monitoring system detected simultaneous faults associated with different sensors in a same group. The model was evaluated with the use of measurements from sensors installed in a real engine. The three M-AANN neural networks corresponding to the three groups of sensors identified in the case study proved to be able to perform self-corrections with resulting MAPE values of less than 2%, even when there were errors of up to 100 units in the sensor measurement.

References

1. Tian, G.Y., Zhao, Z.X., Baines, R.W.: A Fieldbus-based Intelligent Sensor. Mechatronics 10, 835–849 (1999)
2. Afonso, P., Ferreira, J., Castro, J.: Sensor Fault Detection and Identification in a Pilot Plant under Process Control. Chemical Eng. Research and Design 76, 490–498 (1998)
3. Monsef, W.A., Fayez, A.: Design of a Neural - PLC Controller for Industrial Plant. In: Int. Conf. on Mach. Learning - Models, Technologies & Applications, Las Vegas, USA (2007)
4. Garcia-Alvarez, D., Fuente, M.J., Vega, P., Sainz, G.: Fault Detection and Diagnosis using Multivariate Statistical Techniques in a Wastewater Treatment Plant. In: 7th IFAC International Symposium on Advanced Control of Chemical Processes, Turkey (2009)
5. Koscielny, J., Syfert, M.: Fuzzy Diagnostic Reasoning that takes into Account the Uncertainty of the Relation between Faults and Symptoms. Int. J. Appl. Math. Comput. Sci. 16, 27–35 (2006)
6. Theilliol, D., Noura, H., Ponsart, J.: Fault Diagnosis and Accommodation of a Three-Tank System based on Analytical Redundancy. The Instrumentation, Systems, and Automation Society 41(3), 365–382 (2002)
7. Najafi, M., Culp, C., Langari, R.: Enhanced Auto-Associative Neural Networks for Sensor Diagnosis (E_AANN). In: Int. J. Conf. on Neural Networks & IEEE Int. Conf. on Fuzzy Systems, Hungary (2004)
8. Kramer, M.A.: Nonlinear Principal Component Analysis using Autoassociative Neural Networks. A.I.Ch.E. Journal 37(2), 233–243 (1991)
9. Coura, R., Seixas, J., Soares, W.: Classificação de Sinais de Sonar Passivo Utilizando Componentes Principais Não-lineares. Learning and Nonlinear Models 2, 60–72 (2004)
10. Cuenca, W., Seixas, J., Levy, A.: Análise de Componentes Principais para Identificar Descargas Parciais em Transformadores de Potência. In: Brazilain Symposium on Neural Nets, Rio de Janeiro, Brazil (2004)
11. Kramer, M.A.: Autoassociative Neural Networks. Computers in Chemical Engineering 16(4), 313–328 (1992)
12. Wrest, D., Hines, W., Uhrig, R.: Instrument Surveillance and Calibration Verification through Plant Wide Monitoring Using Autoassociative Neural Networks. University of Tenn-Knoxville, USA (1996)

13. Hines, J., Garvey, D.: Process and Equipment Monitoring Methodologies applied to Sensor Calibration Monitoring. Wiley InterScience Quality and Reliability Engineering International 23, 123–135 (2007)
14. Marseguerra, M., Zoia, A.: The Autoassociative Neural Networks in Signal Analysis II: Application to on-line monitoring of a simulated BWR component. Annals of Nuclear Energy 32, 1207–1223 (2005)
15. Hines, J., Grinok, A., Attieh, I., Urigh, R.: Improved Methods for On-line Sensor Calibration Verification. In: 8th Int. Conf. on Nuclear Engineering, Baltimore, USA (2000)

Visual Based Contour Detection
by Using the Improved Short Path Finding

Jiawei Xu and Shigang Yue

Department of Computer Science, University of Lincoln, Brayford Pool, Lincoln,
LN6 7TS, United Kingdom
{jxu,syue}@lincoln.ac.uk

Abstract. Contour detection is an important characteristic of human vision perception. Humans can easily find the objects contour in a complex visual scene; however, traditional computer vision cannot do well. This paper primarily concerned with how to track the objects contour using a human-like vision. In this article, we propose a biologically motivated computational model to track and detect the objects contour. Even the previous research has proposed some models by using the Dijkstra algorithm [1], our work is to mimic the human eye movement and imitate saccades in our humans. We use natural images with associated ground truth contour maps to assess the performance of the proposed operator regarding the detection of contours while suppressing texture edges. The results show that our method enhances contour detection in cluttered visual scenes more effectively than classical edge detectors proposed by other methods.

Keywords: Contour detection Multi-direction searching Short path finding.

1   Introduction

Contour detection in real images is a fundamental problem in many computer vision tasks. Contours are distinguished from edges as follows. Edges are variations in intensity level in a gray level image whereas contours are salient coarse edges that belong to objects and region boundaries in the image. By salient is meant that the contour map drawn by human observers include these edges as they are considered to be salient. However, the contours produced by different humans for a given image are not identical when the images are of complex, natural scenes. In such images, multiple cues are available for the human visual system (HVS) [2] - low level cues such as coherence of brightness, texture or continuity of edges, intermediate level cues such as symmetry and convexity, as well as high level cues based on recognition of familiar objects.

Recent research has shown that the perception of contour derived from animated human actions is located in the right posterior superior temporal sulcus (STS), although some have reported bilateral activations on the STS anterior to Visual area

C. Jayne, S. Yue, and L. Iliadis (Eds.): EANN 2012, CCIS 311, pp. 145–151, 2012.

MT [5]. In many articles, it was showed that common eye contour detection areas were activated by both types of stimuli, indicating that the ventral and dorsal visual streams are activated by the recognition of biological motion stimuli. Taken together, this information suggest that contour detection is promoted by specific brain regions that are distinct from those involved in the perception of other types of motion.

2   Methodology and Proposed Scheme

The flowchart of our proposed idea is as follows, the input images are converted into binary image, then we start tracing the half contour clockwise whilst other half contour anti-clockwise, after shortest path is found, the we draw the contour line by using the red curve and the starting points by the blue block, as we illustrated in the figure 1.

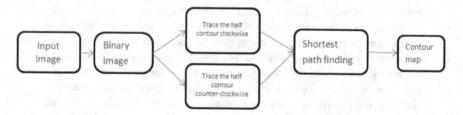

Fig. 1. Contour detection scheme

2.1   Image Binary

Our binarization techniques used in contour detection tasks are aimed at simplifying and unifying the image data at hand. The simplification is performed to benefit the oncoming processing characteristics, such as computational load, algorithm complexity and real-time requirements in industrial-like environments. One of the key reasons why we are implementing binary images is to enhance the real-time ability on contour detection and simplify our model to the maximum extent.

2.2   Pseudo Code of Dijkstra Algorithm

Dijkstra algorithm is the typical shortest path finding scheme for computing one node to any other nodes on the virtual network. In many modern courses such as data structure, graphics and operational research, Dijkstra algorithm is one of the most representative and functional methods. As we illustrated in the table1, the original flowchart is such as the following.

Table 1. Dijkstra pseudo code

```
function Dijkstra(Graph, source):
        for each vertex v in Graph:             // Initializations
                dist[v] := infinity ;            // Unknown distance function from
source to v
                previous[v] := undefined ;       // Previous node in optimal path
from source
        end for ;
        dist[source] := 0 ;                      // Distance from source to source
        Q := the set of all nodes in Graph ;     // All nodes in the graph are un-
optimized - thus are in Q
        while Q is not empty:                    // The main loop
            u := vertex in Q with smallest distance in dist[] ;
            if dist[u] = infinity:
                    break ;                      // all remaining vertices are
inaccessible from source
            end if ;
            remove u from Q ;
            for each neighbor v of u:            // where v has not yet been removed
from Q.
            alt := dist[u] + dist_between(u, v) ;
            if alt < dist[v]:                    // Relax (u,v,a)
                dist[v] := alt ;8                     previous[v] := u ;
                decrease-key v in Q;             // Reorder v in the Queue
            end if ;1              end for ;
        end while ;
        return dist[] ;
    end
```

3   Experimental Results

To evaluate the accuracy of contour detection, we have tested the performance on one image dataset. We have implemented the proposed algorithm using MATLAB2010b on a platform with Intel Core2 3.0 GHZ CPU and 3G memories.

3.1   Experiment Setup and Image Resource

The experiment is based on the mix programming of C++ and MATLAB, Dijkstra algorithm is realized by using Visual Studio 6.0 and then, converted into MEX file, which can be thought as the DLL file, alternatively. Then we can use this function easily without considering the functions written by C/C++.

Most of the methods for the evaluation of edge and contour detectors use natural images with associated desired output that is subjectively specified by the observer

We tested the performance of the proposed scheme on 15 natural images from a database designed to evaluate the performance of contour detection. For each test image, an associated desired output binary contour map that was drawn by human is given.

The average running time is about 0.52s of the whole images and we can say this algorithm has a real-time ability.

3.2   Results

Fig. 2. Results of contour detection on test images. The original images are displayed on the left column and our results are illustrated on the right column.

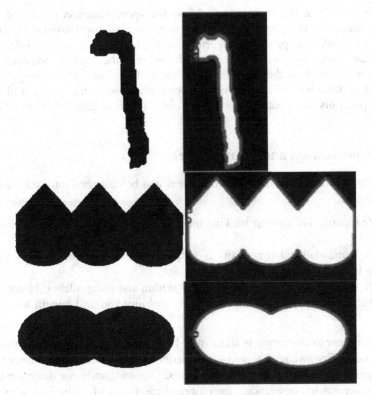

Fig. 2. *(Continued)*

4   Discussions

In practice, contour detection is an intermediate level operation in computer vision with its output often used as input for further stages performing higher level

Fig. 3. comparison with the previous approaches, there exists some break points on the face contour, while the applications provided by OpenCV(Open Source Computer Vision)[4] are mostly detected on the regular perimeters, our algorithm can detect irregular traces as illustrated in Figure 2.

processing. It is hence of interest to know the appropriateness of its use given a specific high level task. In other high-level tasks such as shape-based recognition and image retrieval, the proposed algorithm can play a very useful role in their performance improvement. As illustrated in Figure 3, the tradition contour detection is shown below and they are deviated from the ground truth images. In our experiment described in Figure 2, objects contour is close to the ground truth images, they depict only the contour of objects, while omitting edges that are caused by texture.

5   Conclusions and Future Work

Generally, the main innovations of this paper can be described into three aspects as follows:

1. We separate our contour tracking into multi-direction which is more close to the reality.

2. The computational complexity is reduced by a lot since our contour detection methods is not a full search on the whole image.

3. We designed an improved Dijkstra algorithm and encapsulate it in our toolbox, this enhances the robustness and real-time capability and will benefit to our future work.

The experiment to this extent is successful. However, there is still a lot of work to do in the future. First, we need to add the cluttered background onto the images to prove this algorithm still works under different noise. Second, when our detection target is in motion or not so conspicuous, the proposed method needs to be improved to filter visual cues from the motion vector fields and set up a new robust model to extract our vision attention and detect the attention contours. This work will be discussed in our further work.

Acknowledgements. Thanks to all of the collaborators whose modeling work is reviewed here, and to the members of school of computer science, at the University of Lincoln, for discussion and feedback on this research. This work was supported by the grants of EU FP7-IRSES Project EYE2E (269118).

References

1. Joshi, G.D., Sivaswamy, J.: A simple scheme for contour detection. In: International Conference on Computer Vision Theory and Applications (VISAPP), pp. 236–242 (2006)
2. Koivisto, M., Mantyla, T., Silvanto, J.: The role of early visual cortex(V1/V2) in conscious and unconscious visual perception. NeuroImage 51, 828–834 (2010)
3. Baumann, R., van der Zwan, R., Peterhans, E.: Figure-ground segregation at contours: a neural mechanism in the visual cortex of the alert monkey. European Journal of Neuroscience (1997)
4. OpenCV C interface,
   http://opencv.willowgarage.com/documentation/c/index.html

5. Benoit, A., Caplier, A., Durette, B., Herault, J.: Using human visual system modeling for bio-inspired low level image processing. Computer Vision and Image Understanding 114, 758–773 (2010)
6. Larsson, J., Heeger, D.J., Landy, M.S.: Orientation selectivity of motion-boundary responses in human visual cortex. Journal of Neurophysiology 104, 2940–2950 (2010)
7. Montaser-Kouhsari, L., Landy, M.S., Heeger, D.J., Larsson, J.: Orientation-selective adaptation to illusory contours in human visual cortex. Journal of Neuroscience 27, 2186–2195 (2007)
8. Cavanaugh, J., Bair, W., Movshon, J.: Nature and interaction of signals from the receptive field center and surround in macaque v1 neurons. Journal of Neurophysiology (2002)
9. Dobbins, A., Zucker, S.W., Cynader, M.S.: Endstopped neurons in the visual cortex as a substrate for calculating curvature. Nature (1987)
10. Dubuc, B., Zucker, S.: Complexity, confusion and perceptual grouping. part ii: mapping complexity. International Journal on Computer Vision (2001)
11. Grigorescu, C., Petkov, N., Westenberg, M.: Contour detection based on nonclassical receptive field inhibition. IEEE Transactions on Image Processing (2003)
12. Canny, J.: A computational approach to edge detection. IEEE Transactions on Pattern Analysis and Machine Intelligence (1986)

Analysis of Electricity Consumption Profiles by Means of Dimensionality Reduction Techniques

Antonio Morán[1,*], Juan J. Fuertes[1], Miguel A. Prada[1], Serafín Alonso[1],
Pablo Barrientos[1], and Ignacio Díaz[2]

[1] SUPPRESS research group, Esc. de Ing. Industrial e Informática, Campus de Vegazana s/n,
24071, León, Spain
{a.moran, jj.fuertes, ma.prada, saloc, pbarf}@unileon.es
http://suppress.unileon.es

[2] Dept. de Ing. Elétrica, Electrónica, de Computadores y Sistemas, Universidad de Oviedo,
Campus de Viesques s/n, Ed. Departamental 2, 33204, Gijón, Spain
idiaz@isa.uniovi.es

Abstract. The analysis of the daily electricity consumption profile of a building
and its correlation with environmental factors make it possible to estimate its
electricity demand. As an alternative to the traditional correlation analysis, a new
approach is proposed to provide a detailed and visual analysis of the correlations
between consumption and environmental variables. Since consumption profiles
are normally characterized by many electrical variables, i.e., a high dimensional
space, it is necessary to apply dimensionality reduction techniques that enable
a projection of these data onto an easily interpretable 2D space. In this paper,
several dimensionality reduction techniques are compared in order to determine
the most appropriate one for the stated purpose. Later, the proposed approach
uses the chosen algorithm to analyze the profiles of two public buildings located
at the University of León.

Keywords: Dimensionality reduction, information visualization, electricity consumption profiles.

1 Introduction

The search of the energy efficiency in buildings, particularly in public buildings, is a
mandatory step to ensure a sustainable development. It is absolutely necessary to make
an effort in order to define strategies that reduce electricity consumption while maintaining the quality of energy services. In this sense, the first step should be to carry
out an analysis of the daily electricity consumption profiles. The analysis of electricity consumption in a building is quite important to determine the time dependence of
the consumption of that building [1], i.e., whether the consumption pattern is repeated
daily, reaching the consumption peaks at the same time, or otherwise the consumption

* Supported by a grant from the Consejería de Educación de la Junta de Castilla y León and the
European Social Fund. This work was supported in part by the Spanish *Ministerio de Ciencia
e Innovación* (MICINN) and the European FEDER funds under project CICYT DPI2009-
13398-C02-02.

C. Jayne, S. Yue, and L. Iliadis (Eds.): EANN 2012, CCIS 311, pp. 152–161, 2012.
© Springer-Verlag Berlin Heidelberg 2012

is not clearly influenced by the hour of the day [2]. It is also useful to study the correlation between the consumption patterns and certain environmental variables such as temperature, solar radiation, humidity, etc. with the aim of estimating the most likely consumption profile with regard to the weather forecast. Although statistical methods such as correlation analysis [3] are useful to analyze dependences among variables, it would also be useful to visualize the correlations jointly for every profile. This way, an untrained user could intuitively determine consumption patterns and estimate future behaviors.

In order to understand the consumption of a building, it is useful to analyze its daily consumption curve, i.e., the 24-hour curve which constitutes the consumption profile. Since the profile is composed of a large number of points, the input space is high dimensional and therefore its visualization is difficult. For that reason, the application of a dimensionality reduction technique is proposed to generate a 2D visualization space which preserves most of the structure and patterns of the input space. Many algorithms have been exposed for this purpose. A review of the most important ones is presented in [4], which classifies them with regard to the aim, either the preservation of spatial/geodesic distances or the preservation of the topology with a fixed or adaptive lattice.

Among these techniques, the MDS (*Multi-Dimensional Scaling*) [5] tries to achieve a set of points in the low-dimensional space whose mutual distances are as similar as possible to the ones in the input space. The Isomap algorithm [6] is a variant of the multi-dimensional scaling that uses geodesic distances and then, linear MDS is applied on the minimum distance matrix to obtain global optimization. The algorithm can deal with a great amount of data, but the manifold learning is hindered by the scarcity of training data, the strong dependence on the neighborhood and its high computational complexity. Sammon's *Nonlinear Mapping* (NLM) [7] emphasizes the preservation of the local distances. The main advantage of this approach is its computational simplicity. The CCA (*Curvilinear Component Analysis*) was proposed by [8] as an improvement to the self-organizing maps [9].Its computational complexity is lower than other nonlinear algorithms and performs better on strongly folded data structures. CCA allows also to interpolate and extrapolate new data. A noteworthy modification of CCA is the *Curvilinear Distance Analysis* (CDA) [10], which uses geodesic distances to establish better connections between neighbor units.

Locally Linear Embedding (LLE) [11] uses *conformal mapping*, which replaces each point in the data set by a linear combination of its neighbors. This way, the geometry of the manifold can be represented in terms of linear coefficients, thus simplifying the algorithm. Even though the LLE depends on the local information of the manifold, it is able to use the information about the interconnection among the points, guaranteeing an optimal convergence of the system. The LE (*Laplacian Eigenmaps*) [12] can be seen as a similar technique to LLE, but considering only the k nearest neighbors to preserve locally the topology of data. LE works similarly to clustering and requires few parameters. On the other hand, the SNE (*Stochastic Neighbor Embedding*) [13] is a probabilistic technique to project data, or rather their dissimilarities. In this case, the aim is not to preserve distances between data samples but the probabilities that those samples are neighbors. An improvement, called t-SNE [14], was proposed to alleviate the crowding problem. This is a common problem to most techniques. It provokes the points at a moderate distance to appear too far away on the projected space when the algorithm tries to model small distances accurately. The t-SNE approach uses a symmetric cost function [15] and the Student-t distribution instead of the Gaussian one to compute similarity in the low-dimensional space. The SNE technique tends to preserve better the local groups. However, the t-SNE is not well suited to project onto spaces with a high dimension, so it is generally used for visualization.

Each of the algorithms presented above has different advantages and disadvantages. For that reason, it is necessary to select the dimensionality reduction technique that is more suitable for the purpose of this work, which is to project consumption profiles in a low-dimensional representation that allows to visualize the dependencies of electricity on buildings. This approach will be tested on buildings at the University of León (Spain), which are subject to different usage patterns and therefore, it can be assumed distinct consumption profiles and correlations. The proposed approach to define profiles, select the dimensionality reduction algorithm and visualize dependencies is presented in section 2. In section 3, the experiments are defined and the results are presented. Section 4 discusses the conclusions.

2   Proposed Methodology

Consumption profiles are useful to determine how the building is used during a particular day. Profile analysis enables the study of the correlation between the electricity consumption and the hour of the day. This is useful for detecting deviations in consumption, anomalous patterns and, also, estimating electricity consumption in a simple way. The consumption profiles use the active power as an indicator. The input data set to our approach comprises a set of vectors which are formed by:

$$\mathbf{x} = \left\{ P_0, P_1, P_2, \ldots, P_{23} \right\}, \tag{1}$$

where P_h is the mean active power during the hour h of that day. This way, there are N vectors, each one representing the consumption curve for a certain day. Each consumption profile vector can be associated to an additional vector with its corresponding environmental variables (temperature, solar radiation, humidity,...).

The method proposed to visualize the consumption profile data is shown in Figure 1. For that purpose, consumption profiles are obtained for each day as explained above. On the other hand, vectors with the daily average values of the environmental variables are also built. When computing the projection of the consumption profile, the information provided by the environmental vector is not considered. Nevertheless, once the projection is obtained, the correlations between the projected profiles and the environmental variables are studied. For each point in the output space, there will be an environmental vector associated so the projection can display information of environmental variables with either color, shape or radius of the point projection.

As exposed in section 1, there is a number of alternative techniques to project data onto a low-dimensional space. Although each technique has its own advantages and disadvantages, its performance depends strongly on the characteristics of the data set it is applied to. Few methods and heuristics are available to choose the most suitable

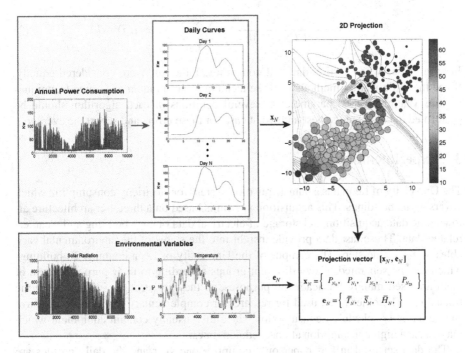

Fig. 1. Methodology for visualizing the daily consumption curves

technique with regard to a certain data set. Indeed, the usual procedure is to visualize the projected data and let the users select the preferred one according to their own experience and criteria [16].

Nevertheless, the selection method used in this work relies on two informed and quantitative metrics: trustworthiness and continuity. This method was introduced in [17] and it is based on the assumption that proximities in a good visualization should hold as well in the original data, whereas the projection should keep all the proximities of the original data. A metric similar to trustworthiness is also used in [18] to select the optimal value of a parameter (perplexity) of the t-SNE projection.

A projection is considered *trustworthy* when the k nearest neighbors of a point in the output space are also neighbors in the input space. Let N be the number of samples, $r(i,j)$ be the order of sample j according to its distance to i in the input space and $U_k(i)$ be the set of samples that belong to the k-sized neighborhood of sample i in the output space but not in the input space. The value for trustworthiness is given by:

$$F(k) = 1 - \frac{2}{Nk(2N - 3k - 1)} \sum_{i=1}^{N} \sum_{j \in U_k(i)} (r(i,j) - k). \qquad (2)$$

The *continuity* can be computed analogously to the trustworthiness. If $V_k(i)$ is the set of samples in the neighborhood of i in the input space but not in the visualization space, and $\hat{r}(i,j)$ is the order of sample j in the list arranged according to the distances from i in the output space, continuity is defined as:

$$C\left(k\right) = 1 - \frac{2}{Nk\left(2N - 3k - 1\right)} \sum_{i=1}^{N} \sum_{j \in V_k(i)} \left(\hat{r}\left(i, j\right) - k\right). \tag{3}$$

In case of ties when sorting data, all compatible rank orders are considered equally. Since the initialization methods of these techniques are random and some techniques require a neighborhood parameter k, several iterations of each algorithm should be performed to calculate the best and worst case for these errors and report the average.

3   Experimental Results

The University of León has a data acquisition system for electricity consumption which covers all its buildings. This acquisition system is based on a three-tier architecture allowing he data acquisition and storage of electrical data of each building and weather-related data. These last data provide insight into the status of the environmental variables at each time step. The campus of the University of León contains 20 buildings which can be separated in two different groups according to their purpose. Some of them are used mainly for academic purposes, so they are strongly influenced by the timetables. The other ones, used for research or complementary services, do not have such a strongly scheduled consumption, since they usually contain equipment which may cause a high consumption at unscheduled times.

The data set used in this paper only comprise one year and the daily profiles are formed by the consumption data sampled every minute so the dimension of the input dataset is 1440×365. In order to test the proposed approach, first the visualization technique is selected by means of a visual comparison of the techniques and the discussed metrics to quantify the effectiveness of the projection. Subsequently, the selected method is applied to a building of each type in order to visualize the shape of the projection as a function of time and environmental variables correlation.

3.1   Selection of the Projection Technique

The dimensionality reduction technique will be selected among all the techniques presented in Section 1. The chose technique will be the one that provides an understanding visualization and a good result in the metrics. A set of representative data of electricity consumption of the campus will be applied to select the better technique. These data were selected considering data that show some correlation with the daily hour, since the main objective is the visualization of this correlation. A set of random samples will be collected among the buildings that have a time correlation.

Figure 2 shows the results of applying the projections techniques to this data set. In these graphs the color of the point depends on the average power consumed on the day, so red values indicate high consumption and blue values indicate low consumption. These maps show that the LLE can not find a satisfactory solution and it generates a projection in which all points are mixed.

The remaining projections manage to spread the data logically and gather the days which have a reduced power consumption, while high consumption points are far from these. All projections sort data according to two major directions or curves. In general,

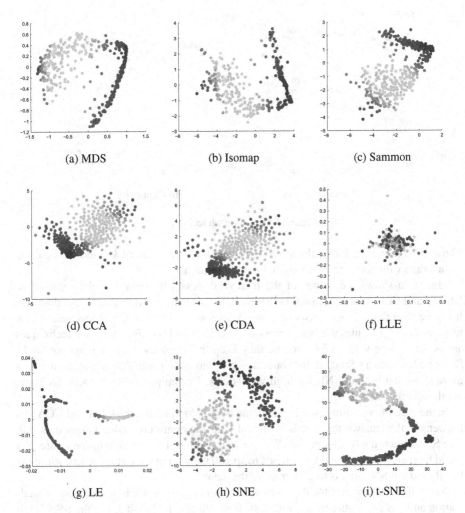

Fig. 2. Comparison of dimensionality reduction techniques applied to daily electricity consumption profiles

the first direction corresponds to the level of power consumption throughout the day. The points of low consumption are in one end while the high consumption points are in the opposite end. The second direction corresponds to the shape of the consumption pattern for days when the consumption level is similar. The shape of the profile depends on the value of the peak consumption, the number of peaks and the time at which they occur. Techniques such as t-SNE tends to group the data more densely, while the other techniques spread them more.

For these techniques, the metrics of trustworthiness and continuity are calculated to choose dimensionality reduction technique more appropriate. Since these techniques have a random initialization the result will vary with each execution of the algorithms.

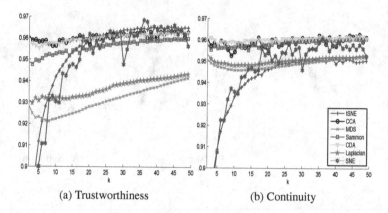

(a) Trustworthiness   (b) Continuity

Fig. 3. Metrics for each technique

Thus, these metrics are calculated several times for each value of k [1] and the result is the average value between the maximum and minimum.

Figure 3a shows the values of the trustworthiness with regard to the value of k. Isomap and LLE does not appear in the graph because the metric value is less than 0.9. It can be seen that the best trustworthiness values are given for CCA at low values of k whereas for high values the best techniques are SNE and t-SNE. These two techniques are best for large values of k because they keep the topology of large neighborhoods. Since t-SNE uses a t-Student distribution to give more weight to distant points, it works more consistently than SNE for high values of k. Techniques whose results are poor, are the LE and the MDS.

In the case of continuity (see Fig. 3b), the best techniques are the CDA and CCA. As happens in the trustworthiness, Isomap and LLE the continuity value is less than 0.9. The SNE match the results of the CDA for high values of k. Already in the figure 2, it could be seen that CDA and CCA tends to spread the data in the output space, while the t-SNE and the SNE, tends to gather more the data.

According to these results, the chosen technique is SNE since it gives a good visualization and has the higher trustworthiness. It would also be possible use the t-SNE, but the results of continuity are worse than SNE. Although CCA and CDA have comparable trustworthiness and higher continuity, it spreads the data making the visualization more difficult to interpret. Figure 2 shows that the SNE gives a visualization which is halfway betwwen CCA and t-SNE

3.2 Analysis of the Electricity Consumption Profiles

The comparison of electrical profiles allows us to observe the behavior of a building based on its consumption. Figures 4 and 5 show the results of applying SNE technique to data obtained from a teaching and a research building respectively. It can be seen that

---

[1] It must be remembered that k is the number of neighbors to consider in determining the trustworthiness and continuity of the technique and does not match the neighborhood parameter necessary for some techniques.

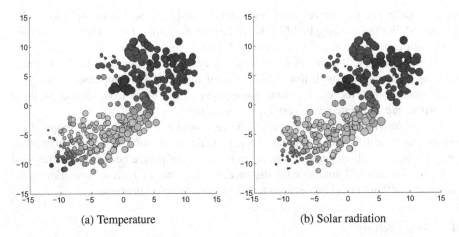

(a) Temperature   (b) Solar radiation

Fig. 4. SNE projection of the electricity consumption profiles for a teaching building

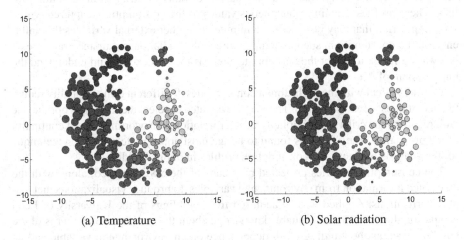

(a) Temperature   (b) Solar radiation

Fig. 5. SNE projection of the electricity consumption profiles for a research building

daily profiles are ordered according to their average consumption, which is denoted by the color of the points. The red colors reflect higher consumption than the blue ones. The dot size is related to the value of the environmental variable that is displayed. As an example, the values for daily mean temperature and average solar radiation are shown in the aforementioned figures.

The difference in the shape of the projection gives an insight of the behavior of the buildings. It can be seen that the teaching building has a high correlation with the hour (peak power profile) and therefore the main direction of the projection takes a linear shape. On the other hand, the research building has less correlation with the hour (flat power profile), so consumption stays approximately the same. Therefore, the main direction of the map takes a circular shape. This indicates that the distance between the points of high and low consumption is greater for the teaching building than for the research one. Since the size of the points represents the value of the environmental variable under analysis, correlations between them and the power consumption can also be studied. For the teaching building, it can be seen that days of low solar radiation and temperature match up with the points of high power consumption.

On the contrary, the points corresponding to high consumption in the research building do not match with low values of solar radiation and temperature, i.e., with small-sized points. Indeed, for some points, specifically those from weekends, these low values are associated with very low consumption.

Thus, through these projections it can be seen as in cases in which the shape of the profile along the different consumption is kept, making they different only in magnitude or in the hour at which peaks occur, the components of profile vector that correspond to the maximums will dominate the distance, i.e., they have a more contribution to the distance. For that reason, projections tend to show a simpler direction.

4 Conclusions

This paper presents an approach for the analysis of daily electricity consumption profiles. These profiles contain instantaneous values of the consumption, captured with a sampling period that may be less than a minute, and other external variables that influence on the consumption, such as temperature, solar radiation, occupancy rate, etc. It is always difficult to select the appropriate analysis tools to exploit and understand the huge amount of data.

In order to deal with the high-dimensional profiles, 9 different dimensionality reduction techniques were considered. They were evaluated to select the most appropriate one with regard to the criteria of simplicity of interpretation, trustworthiness and continuity of the projection. The SNE was found to be the most appropriate projection technique for the analysis of electricity consumption profiles in public buildings.

The consumption profiles projected by means of the SNE are used, along with the associated information from environmental variables, to produce visualizations that display correlations. Applied to data acquired from buildings of the University of León campus, it allows to extract valuable knowledge about the consumption patterns of the building, such as the buildings with dependence on an environmental variable and its degree. In addition, the visual character of the proposed methodology helps us to determine the time dependence of the consumption in the buildings.

References

1. Fay, D., Ringwood, J.V., Condon, M., Kelly, M.: 24-h electrical load data-a sequential or partitioned time series. Neurocomputing 55(3-4), 469–498 (2003)
2. Nizar, A.H., Dong, Z.Y., Jalaluddin, M., Raffles, M.: Load profiling method in detecting non-technical loss activities in a power utility. In: IEEE International Power and Energy Conference, pp. 82–87 (2006)
3. Kendall, M.G.: Rank Correlation Methods. Charles Griffin and Company (1948)
4. Lee, J.A., Lendasse, A., Verleysen, M.: Nonlinear projection with curvilinear distances: Isomap versus curvilinear distance analysis. Neurocomputing 57, 49–76 (2004)
5. Kruskal, J.B., Wish, M.: Multidimensional scaling. Sage university paper series on quantitative application in the social sciences, pp. 7–11 (1978)

6. Tenenbaum, J.B., de Silva, V., Langford, J.C.: A global geometric framework for nonlinear dimensionality reduction. Science 290, 2319–2323 (2000)
7. Sammon Jr., J.W.: A non-linear mapping for data structure analysis. IEEE Transactions on Computers 18, 401–409 (1969)
8. Demartines, P., Hérault, J.: Curvilinear component analysis: a self organizing neural network for non linear mapping of data sets. IEEE Transactions on Neural Networks 8, 148–154 (1997)
9. Kohonen, T.: The neural phonetic typewriter. Computer, 11–22 (Marzo 1988)
10. Lee, J.A., Lendasse, A., Donckers, N., Verleysen, M.: A robust nonlinear projection method, pp. 13–20 (2000)
11. Roweis, S.T., Saul, L.K.: Nonlinear dimensionality reduction by locally linear embedding. Science 290, 2323–2326 (2000)
12. Belkin, M., Niyogi, P.: Laplacian eigenmaps and spectral techniques for embedding and clustering. In: Advances in Neural Information Processing Systems, vol. 14, pp. 585–591. MIT Press (2001)
13. Hinton, G.E., Roweis, S.: Stochastic neighbor embedding. In: Becker, T.S., Obermayer, K. (eds.) Advances in Neural Information Processing Systems, vol. 15, pp. 833–840. MIT Press, Cambridge (2002)
14. Van der Maaten, L., Hinton, G.: Visualizing data using t-SNE. Journal of Machine Learning Research 9, 2579–2605 (2008)
15. Cook, J., Sutskever, I., Mnih, A., Hinton, G.: Visualizing similarity data with a mixture of maps. In: Proceddings of the 11th International Conference on Artificial Intelligence and Statistics, vol. 2, pp. 67–74 (2007)
16. Lee, J.A., Verleysen, M.: Nonlinear Dimensionality Reduction. Information Science and Statistics. Springer (2007)
17. Venna, J., Kaski, S.: Comparison of visualization methods for an atlas of gene expression data sets. Information Visualization 6, 139–154 (2007)
18. Van der Maaten, L.: Learning a parametric embedding by preserving local structure. In: Proceedings of the Twelfth International.Conference on Artificial Intelligence and Statistics (AI-STATS), vol. 5, pp. 384–391. JMLR W&CP (2009)

Neural Networks for the Analysis of Mine-Induced Vibrations Transmission from Ground to Building Foundation

Krystyna Kuzniar[1] and Lukasz Chudyba[2]

[1] Pedagogical University of Cracow, ul. Podchorazych 2, 30-084 Krakow, Poland
kkuzniar@up.krakow.pl
[2] Cracow University of Technology, ul. Warszawska 24, 31-155 Krakow, Poland
lchudyba@poczta.onet.pl

Abstract. Problem of the transmission of mine-induced ground vibrations to building foundation is analysed in the paper. The maximal values of horizontal vibrations velocities (horizontal vibration components and resultant vibrations) are taken into account. Application of neural networks for the prediction of building foundation vibrations on the basis of ground vibrations is proposed. Standard back-propagation neural networks as well as recurrent cascade neural network systems were used. Experimental data obtained from the measurements of ground and actual structure vibrations were applied as the neural network training, validating and testing patterns. The obtained results lead to a conclusion that the neural technique gives results accurate enough for engineering practice.

Keywords: Neural Networks, Vibrations Transmission, Mining Tremors.

1 Introduction

Soil-structure interaction plays an important role in the design process of structures subjected to ground motion and it is a very important problem from the engineering point of view. It should be noted that the soil-structure interaction problems are extensively studied with respect to earthquake excitation. However the ground motion can be induced not only by earthquakes, but also by human activity as so-called paraseismic sources.

Some of the paraseismic excitation sources as for instance traffic vibrations, underground explosions, industrial explosions and mining tremors in strip mines may be inspected and controlled. On the other hand, mining tremors resulting from underground raw mineral material exploitation are random events. Mine-related underground rockbursts excite seismic waves that reach the surface of the earth and induce the building vibrations. Although these tremors are connected with the human activity and can usually be observed only in the mining regions, they differ considerably from the other paraseismic vibrations. They are not subject to human control and they are random events with respect to the time, place and magnitude

C. Jayne, S. Yue, and L. Iliadis (Eds.): EANN 2012, CCIS 311, pp. 162–171, 2012.

likewise earthquakes. However some parameters of such ground vibrations (for instance: dominating frequencies, duration) are different from earthquake-induced ground vibrations [1].

Comparison of a huge number of records of velocities of vibrations induced by mining tremors measured at the same time on the ground and on the building foundation level leads to conclusion that they differ significantly. The analogous differences in case of acceleration and displacement response spectra are shown in the papers [2, 3]. Additionally, the evaluation of mining tremors transmissions to the building is very difficult. The influence of rockbursts parameters as mining tremor energy, epicentral distance, direction of vibrations as well as dominating frequencies of ground vibrations on the soil-structure interaction effect can be observed. However the prediction of the precise relation between ground and foundation records of velocities is not possible.

The more precise estimation of the harmfulness of mine-induced vibrations to actual buildings can be performed on the basis of building foundation vibrations. With respect to the fact that in many cases, for example in the design procedure of new structures as well as in the dynamic analysis of existing buildings, the measured ground vibrations are accessible only, the prediction of foundation vibrations is necessary. Then estimation of the way of the ground vibration transmission to building basement is indispensable.

Taking into account the difficulties in the soil-structure interaction analysis in the case of vibrations induced by mining tremors, the application of neural networks for the prediction of building foundation vibrations on the basis of ground vibrations taken from measurements is proposed in the paper. An application of neural networks for evaluation of soil-structure interaction in the case of transmission of velocities of horizontal mine-induced ground vibrations to building foundation is shown. Standard back-propagation neural networks as well as recurrent cascade neural network systems were used.

The problem is analysed with respect to apartment five-storey building founded directly on the ground using concrete strip foundations. This building is typical, representative example of objects from the wide class of medium-height buildings with similar dynamic properties. Mining tremors in the most seismically active mining region in Poland with underground copper ore exploitation – Legnica-Glogow Copperfield (LGC) (measurements come from the surface seismological measurement stations) – were the sources of ground and building vibrations. The database created of the results of long-term experimental monitoring of ground and actual structure vibrations makes it possible to use them as the patterns to design the neural network analysers for considerations of the transmission of ground vibration velocities (horizontal vibration components and resultant) induced by mining tremors to building foundation.

2   Results of Experimental Tests

Full-scale tests were performed many times in a period of more than 11 years (monitoring) [2, 3]. The analysis concerns mining tremors with energies from the

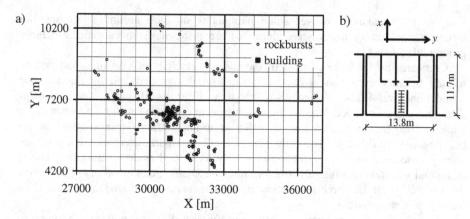

Fig. 1. (a) Schematically shown locations of analysed rockbursts and building using local seismological coordinates (X, Y); (b) Plan of the analysed building

range En = 7.4E3-2E9 J and epicentral distances from the range re = 270-5815 m. Fig. 1a schematically shows the locations of analysed rockbursts and building using local seismological coordinates (X, Y). It is visible that the ranges of mining tremors energies, epicentral distances and wave propagation directions are very wide.

Accelerations were measured. The accelerometers were placed on the ground in the front of the building (in six meters distance) and on the foundation level in the building. The main attention was devoted to measurements of the horizontal vibration components in the directions parallel to the transverse (x) and longitudinal (y) axis of the building, see Fig. 1b. The records of vibration velocities v were obtained by the acceleration records integration. Resultant velocities of ground and foundation vibrations were computed on the basis of horizontal vibration components in x and y directions.

The velocities of vibrations recorded at the same time on the ground and on the foundation level are taken into account in the analysis. Examples of the horizontal components of velocities in time domain recorded at the same time on the ground in front of the building and in the building on the foundation level as well as the resultant vibrations are presented in Fig. 2. It can be seen that the records of vibrations registered at the same time on the ground in front of the building and in the building on the foundation level can differ significantly. It has to be noted here too, that such differences were observed in the all analysed cases of vibrations [2, 3].

The comparison of maximal values (amplitudes) of velocities recorded at the same time on the ground (v_{gmax} and PGV in the case of vibrations in x or y directions and in the case of resultant vibrations, respectively) and on the foundation level (v_{fmax} and PFV, respectively) was the way of estimation of the vibrations transmission from the ground to the building. For this purpose the ratio $rv = v_{fmax}/v_{gmax}$ or the ratio $rrv = PFV/PGV$ was computed, respectively.

In total, 928 records of component velocities were considered (464 pairs ground – foundation). In the case of resultant vibrations, 226 pairs ground – foundation velocities from the same mining tremors were analysed (904 records of ground and foundation velocities in x and y directions).

a)

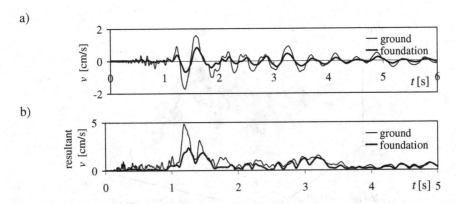

b)

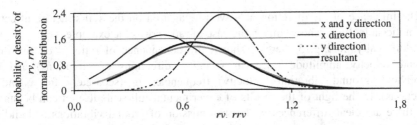

Fig. 2. Time history of component velocities in x direction from the tremor: En=1.2E8J, re=1536m (a) and resultant velocities from the tremor: En=1.7E8J, re=912m (b) in the same time on the ground and building foundation

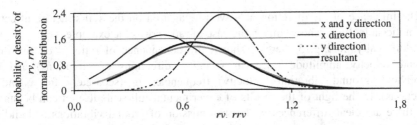

Fig. 3. Probability density of ratio rv and rrv normal distribution

The probability densities of the ratio rv normal distributions computed separately for x direction, separately for y direction, x and y directions as well as the probability density of the ratio rrv normal distribution are presented in Fig. 3.

Average values of ratio rv and rvv evaluated in several of the successive ranges of the mining tremors parameters are collected in Table 1.

Table 1. Average values of ratio rv and rvv evaluated in the successive ranges of the parameters of mining tremors

Rockbursts parameters		Average values of rv direction			Average values of rrv
		x	y	x and y	
En [J]	≤ 5E7	0.46	0.80	0.64	0.64
	> 5E7	0.62	0.91	0.76	0.74
re [m]	$re \leq 800$	0.73	0.89	0.81	0.81
	$800 < re \leq 1500$	0.42	0.76	0.59	0.56
	$re > 1500$	0.54	0.88	0.71	0.75

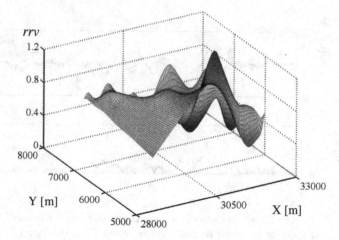

Fig. 4. Dependence of the ratio *rrv* on the seismological coordinates (X, Y)

The influence of the direction of wave propagation on the soil-structure effect is also noticeable. Fig. 4 shows this in the case of ratio *rrv* as the example.

Additionally, in Fig. 5 there is shown the dependence of ratio *rv* on maximal ground velocity amplitude v_{gmax} (in *x* and *y* directions of vibrations) and on dominating ground vibration (velocity) frequencies *fg* (separately for *x* and *y* directions). In the light of the results of experimental full-scale tests it can be stated that there are clear differences in the transmission of ground vibrations to building foundation in transverse (*x*) and longitudinal (*y*) directions. It can be also seen the influence of ground vibrations dominating frequencies on the soil-structure interaction effect.

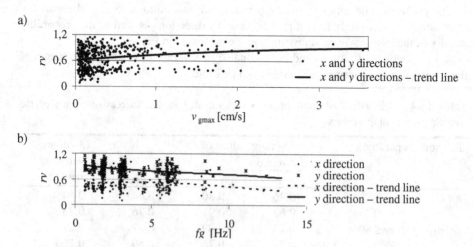

Fig. 5. Dependence of ratio *rv* on maximal ground velocity amplitude v_{gmax} (a) and dominating ground vibration (velocity) frequencies *fg* (b)

However, looking at the experimentally obtained results it is clear that the evaluation of the precise relation between the ground and the foundation records of velocities measured at the same time is not possible and the prognosis of the mine-induced ground vibration transmissions to the building foundation is very difficult.

3   Prediction of Ground Velocities Transmission to Building Foundation Using Neural Networks

3.1   Introductory Remarks

Back-propagation neural networks (BPNNs) with the Levenberg-Marquardt learning method and sigmoid activation function [4, 5, 6] were trained, validated and tested on the basis of experimental data obtained from long-term measurements performed on the ground and actual structure. The lowest validation error was the criterion used for stopping the NNs training. Recurrent cascade neural networks (RCNNs) also were used in order to increase the accuracy of neural approximations [7, 8]. Results of neural network analysis were compared with the results of experiments.

The accuracy of the network approximation was evaluated by the mean-square-error $MSE(Q)$ and the relative errors ep:

$$MSE(Q) = \frac{1}{Q} \sum_{p=1}^{Q} (z^{(p)} - y^{(p)})^2 , \tag{1}$$

$$ep = |1 - y^{(p)}/z^{(p)}| \cdot 100\% , \tag{2}$$

where: $z^{(p)}$, $y^{(p)}$ – target and neurally computed outputs for p-th pattern; $Q = L, V, T$ – number of the learning (L), validating (V) and testing (T) patterns respectively.

The numerical efficiency of the trained network also was estimated by the success ratio SR. This function enables us to compute what percentage of patterns $SR[\%]$ gives the neural prediction with the error not greater than $ep[\%]$.

3.2   Application of Standard BPNNs

The aim of the study is to apply neural networks for the prediction of ratios rv and rrv on the basis of the corresponding mining tremor and ground vibration parameters. The following neural network input parameters were taken into consideration: v_{gmax} - maximal value (amplitude) of velocities recorded on the ground in the case of vibrations in x or y directions, k - parameter related to direction of vibrations (values $k = 0.4$ and $k = 0.7$ were arbitrarily assumed for the transverse direction x and longitudinal direction y, respectively in order to differentiate the directions), PGV - maximal value (amplitude) of velocities recorded on the ground in the case of resultant vibrations, En - mining tremor energy, re - epicentral distance, X and Y - local seismological coordinates, fg - dominating ground vibration (velocity) frequency. Variant neural network input vectors were analysed taking into account the various combinations of input parameters. The corresponding values of ratio rv or rrv were expected as the output of the neural network, respectively in the case of component or resultant velocities transmission from the ground to building foundation.

From the experimental data, $P = 464$ patterns were arranged for each pair of neural network input vector-output vector in the case of the velocities of component vibrations transmission. Randomly selected 50% of the patterns were taken as the learning set. The validation and testing sets were composed of the remaining of them - 25% per validation set and 25% per testing set. Analogous splitting up of patterns was proposed for the resultant velocities: $P = 226, L = 113, V = 56, T = 57$.

Parameters of the neural network pairs of input-output vectors, the neural network architectures and the errors corresponding to the networks training, validating and testing processes for selected neural networks are given in Table 2.

Table 2. BPNNs input-output parameters, structures and training, validating and testing errors

Study cases	Input parameters	Output parameter	Neural network structure	$MSE(Q)$		
				L	V	T
NN1	v_{gmax}, En, re	rv	3-5-1	0.0092	0.0107	0.0097
NN2	v_{gmax}, En, X, Y	rv	4-3-1	0.0124	0.0127	0.0112
NN3	v_{gmax}, En, re, fg	rv	4-24-1	0.0110	0.0122	0.0118
NN4	v_{gmax}, En, re, X, Y	rv	5-12-1	0.0129	0.0131	0.0113
NN5	v_{gmax}, En, re, fg, k	rv	5-8-1	0.0059	0.0066	0.0057
NN6	v_{gmax}, En, re, X, Y, fg	rv	6-12-1	0.0095	0.0119	0.0112
NN7	v_{gmax}, En, re, X, Y, k	rv	6-11-1	0.0079	0.0072	0.0055
NN8	v_{gmax}, En, re, X, Y, fg, k	rv	7-17-1	0.0089	0.0058	0.0053
NN9	PGV, En, re	rrv	3-6-1	0.0052	0.0062	0.0040
NN10	PGV, En, X, Y	rrv	4-11-1	0.0034	0.0049	0.0032
NN11	PGV, En, re, X, Y	rrv	5-7-1	0.0032	0.0051	0.0033

In Fig. 6 there is shown the Success Ratio SR for the neural prediction of the transmission of mine-induced ground velocities to building foundation obtained using some of the neural networks presented in Table 2, as the examples. The influence of various input parameters on the neural network approximation accuracy is visible.

3.3   Recurrent Cascade Neural Network Systems

In the first ($i = 1$) step in the RCNN systems the initial neural network architectures for each one of the standard BPNNs were formulated, see Table 2. Then sequence of neural networks [7, 8] was proposed as the RCNN system ascribed to each one of the study cases collected in Table 2:

$i = 1$ (the 1st cycle of RCNN system): {initial NN input parameters} $- H - r_1$,

$i > 1$ (the i-th cycle of RCNN system): {initial NN input parameters, r_{i-1}} $- H - r_i$, where: i – the number of iteration cycle, initial NN input parameters – as in Table 2, $r_1 = rv$ or $r_1 = rrv$ – outputs of the initial NNs, H – the number of neurons in the NN hidden layer.

It should be noted that the number of neurons in the hidden layer (H) of RCNNs is selected in the first cycle and it is fixed in the next cycles. Hence practically, in the second and the next cycles, the number of testing patterns $TC = V + T$ because of the fact that it is not necessary to design the neural network architecture. The output of the previous neural network of cascade comes out for the second and the successive cycles in the input vectors of networks. Therefore the number of inputs increases.

The application of recurrent cascade neural network (RCNN) systems for the prediction of ground vibration transmission to building foundation is shown in NN8 (prediction of x and y velocity components transmission) and NN10 (prediction of resultant velocities transmission) study cases (cf. Table 2), as the examples.

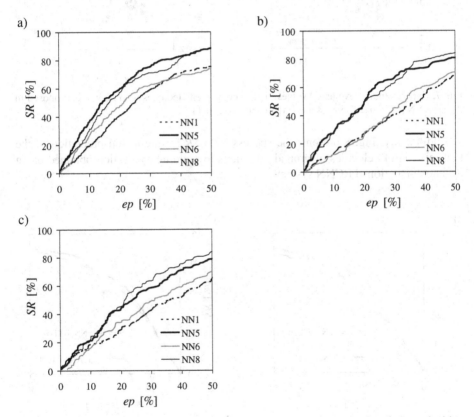

Fig. 6. Success Ratio SR versus relative error ep for neural prognosis of the velocities transmission from the ground to the building foundation in the cases of some selected networks: (a) training patterns; (b) validating patterns; (c) testing patterns

The initial neural network architectures 7-17-1 for NN8 and 4-11-1 for NN10 were formulated, see Table 2. It was the first ($i = 1$) step in the RCNN systems. The sequence of neural networks of structures 8-17-1 and 5-11-1 were applied for the next cycles, respectively.

As the example, the results of rrv obtained for $i = 1$ and $i = 11$ iteration cycles in the case NN10 are compared with the experimental ones in Fig. 7.

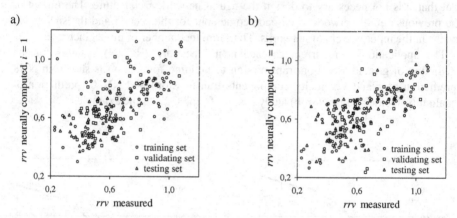

Fig. 7. Experimentally evaluated values of rrv versus predicted ones with networks proposed in the case NN10: (a) $i = 1$, NN: 4-11-1; (b) $i = 11$, NN: 5-11-1

Additionally, Fig. 8 presents the success ratio SR for the neural prediction of the transmission of velocities of ground vibrations to building foundation obtained using the above-mentioned RCNN systems.

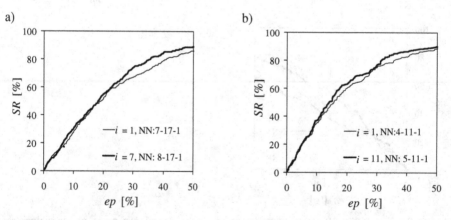

Fig. 8. Success Ratio SR for the neural prediction (all patterns) of the transmission of velocities of ground vibrations to building foundation in the cases of: (a) rv; (b) rrv

4   Conclusions

The important influence of rockburst parameters (mining tremor energy, epicentral distance, wave propagation direction) as well as ground vibrations parameters (amplitude, direction of vibrations, dominating frequency) on the values of ratios rv and rrv is clearly visible from the experimental data and from the neural network analysis. But it can be stated that it is very difficult to precisely evaluate the way of the mine-induced ground vibrations transmission to the building foundation.

Looking at the difficulties in prognosis of differences between the mine-induced ground and the building foundation vibration velocities it is visible from the results obtained that the application of neural networks enables us to predict the reduction of maximal values (amplitudes) of horizontal component and resultant vibrations (velocities) recorded at the same time on the ground and on the building foundation level with satisfactory accuracy. It was stated that certainly the more precise prognosis of ratios rv and rrv is obtained for the richer neural network input information. The analysis carried out leads to conclusion that the application of recurrent cascade neural networks system causes an increase of accuracy of obtained results. The neural technique gives results accurate enough for engineering practice.

References

1. Zembaty, Z.: Rockburst Induced Ground Motion – a Comparative Study. Soil Dynamics and Earthquake Engineering 24(1), 11–23 (2004)
2. Kuzniar, K., Maciąg, E.: Neural Network Analysis of Soil-Structure Interaction in Case of Mining Tremors. In: Doolin, D., Kammerer, A., Nogami, T., Seed, R.B., Towhata, I. (eds.) Proc. 11th International Conference on Soil Dynamics and Earthquake Engineering and the 3rd International Conference on Earthquake Geotechnical Engineering, Berkeley, USA, vol. 2, pp. 829–836 (2004)
3. Kuzniar, K., Maciąg, E., Tatara, T.: Prediction of Building Foundation Response Spectra from Mining-Induced Vibrations using Neural Networks. Mining & Environment 4(4), 50–64 (2010) (in Polish)
4. Bishop, C.M.: Pattern recognition and machine learning. Springer, New York (2006)
5. Haykin, S.: Neural networks – a Comprehensive Foundation, 2nd edn. Prentice Hall Intern. Inc., Upper Saddle River (1999)
6. Demuth, H., Beale, M., Hogan, M.: Neural Network Toolbox for use with Matlab 5. User's Guide. MathWorks Inc. (2007)
7. Klos, M., Waszczyszyn, Z.: Modal Analysis and Modified Cascade Neural Networks in Identification of Geometrical Parameters of Circular Arches. Computers and Structures 89, 581–589 (2011)
8. Piatkowski, G., Waszczyszyn, Z.: Hybrid FEM&ANN&EMP Identification of Mass Placement at a Rectangular Steel Plate. In: Ziemianski, L., Kozlowski, A., Wolinski, S. (eds.) Proc. 57th Annual Scientific Conference. Scientific Problems of Civil Engineering, pp. 286–287. Rzeszow-Krynica (2011)

Backpropagation Neural Network Applications for a Welding Process Control Problem

Adnan Aktepe[1], Süleyman Ersöz[1], and Murat Lüy[2]

[1] Kırıkkale University, Faculty of Engineering, Department of Industrial Engineering, 71450, Yahşihan, Kırıkkale, Turkey
[2] Kırıkkale University, Faculty of Engineering, Department of Electrical and Electronics Engineering, 71450, Yahşihan, Kırıkkale, Turkey
{aaktepe,sersoz71}@gmail.com, mluy@kku.edu.tr

Abstract. The aim of this study is to develop predictive Artificial Neural Network (ANN) models for welding process control of a strategic product (155 mm. artillery ammunition) in armed forces' inventories. The critical process about the production of product is the welding process. In this process, a rotating band is welded to the body of ammunition. This is a multi-input, multi-output process. In order to tackle problems in the welding process 2 different ANN models have been developed in this study. Model 1 is a Backpropagation Neural Network (BPNN) application used for classification of defective and defect-free products. Model 2 is a reverse BPNN application used for predicting input parameters given output values. In addition, with the help of models developed mean values of best values of some input parameters are found for a defect-free weld operation.

Keywords: Backpropagation neural networks, welding process control, artillery ammunition.

1   Introduction

The material and mechanic characteristics determine the quality of most welding operations. The input parameters have a significant effect on the quality of welding process. For most of the cases, it is nearly impossible to predict the effect of each input parameter on output parameters. In addition, some parameters are controllable and some of them are not. By using the prediction characteristic of Artificial Neural Networks (ANN), different ANN models are developed, implemented and results are discussed in this study.

In the literature welding processes are modeled with several approaches according to type and characteristics of the problem. Benyounis and Olabi [1] discuss statistical and numerical techniques for optimization of welding processes. In their study they give a detailed literature on the use of numerical techniques on welding problems which are factorial design, linear regression, response surface methodology, Taguchi experimental design, ANN and hybrid techniques combining two or more Artificial Intelligence (AI) techniques. Generally, the quality of a weld joint is directly

C. Jayne, S. Yue, and L. Iliadis (Eds.): EANN 2012, CCIS 311, pp. 172–182, 2012.

influenced by the welding input parameters during the welding process; therefore, welding can be considered as a multi-input multi-output process. Traditionally, it has been necessary to determine the weld input parameters for every new welded product to obtain a welded joint with the required specifications [1]. The problem studied in this paper is a multi-input multi-output process. In order to control the process, the interaction among input and output parameters must be predicted.

ANN as an AI technique is used in a variety of areas such as function approximation, classification, association, pattern recognition, time series analysis, signal processing, data compaction, non-linear system modeling, prediction, estimation, optimization and control [2]. ANN models also provide effective solution approaches to problems and have a wide application area on materials science. Sha and Edwards [3] and Chertov [4] in their studies give a detailed literature study on parameter estimation, identification, optimization and planning of parameters used in materials based science research. In the literature, there are several applications of ANN models on welding processes. Tay et al. [5], Luo et al. [6], Martin et al. [7], Kim et al.[8], Özerdem et al. [9], Martin et al. [10] and Mirapeix et al. [11] study on welding process modeling and optimization. Pal et al. [12] in their study develop a multilayer neural network model to predict the ultimate tensile stress of welded plates. Six process parameters, namely pulse voltage, back-ground voltage, pulse duration, pulse frequency, wire feed rate and the welding speed, and the two measurements, namely root mean square values of welding current and voltage, are used as input variables of the model and the UTS of the welded plate is considered as the output variable. Ateş [13] in his study presents a novel technique based on ANN for prediction of gas metal arc welding parameters. Input parameters of the model consist of gas mixtures, whereas, outputs of the ANN model include mechanical properties such as tensile strength, impact strength, elongation and weld metal hardness. As observed in the literature studies, ANN models are used for several classification and prediction applications for different kind of welding problems.

In this study, for controlling the welding process of 155 mm. artillery ammunition, which is produced in Mechanical and Chemical Industry Corporation (MKEK) Ammunition Factory, located in the city of Kırıkkale/Turkey, ANN models are developed and results are discussed. In the next section, the details of welding process and problem definition are discussed.

2   Welding Process Control Problem

In the welding process a rotating band is welded to the body of ammunition. After welding channel is created on the body in a computer numerically controlled station, it is transferred to the welding machine. Here, the first operation is preheating (Fig. 1.1). After preheating operation, the body is transferred to the welding workbench. The welding workbench is seen in Fig. 1.2. In this station, firstly the body is tied between stitch and panel. Then rotating manually, the welding channel is rubbed down, cleaned by an alcohol-soaked cloth and the body gets ready for welding operation. Meanwhile copper and brass wires are prepared. This process is shown in Fig. 1.3. The welding machine gets these wires and welding operation is realized with these metals. After cleaning process on the body, the torch distances for copper and brass wires, water discharge, gaseous fill rate, copper and brass wires' speed, torch-nozzle distance values are entered to the control panel manually. Finally welding operation starts. Welding operation carried out in the welding workbench is a gas metal arc welding. The operation is carried out creating arc between Argon gas and metal wires. When the welding operation is started, firstly an arc is formed at the torch where copper metal wire meets. And after 3-4 oscillations, as a second torch with brass metal wire an alloy is created in weld zone. When the torches make 120-140 oscillations where copper and brass metal wires meet, weld zone is filled and the operation ends automatically. After welding operation the ammunition body is picked up from welding workbench and weld zone is grinded at beginning, middle and end parts. A chemical analysis is applied to these three zones. This operation is the quality control operation shown in Fig. 1.4. In quality control operation, the zinc (ZnR), iron (FeR) and copper ratio (CuR) of weld is measured. According to quality control specifications, metal ratios in the weld must be between lower and upper limits. For a defect-free weld ZnR must be between %8-%12, FeR must be between %0,5-%4 and CuR must be between %84-%91,5.

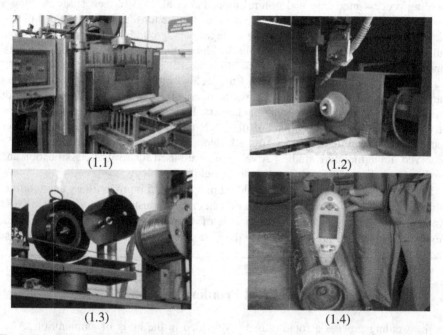

(1.1)   (1.2)

(1.3)   (1.4)

Fig. 1. This figure shows the steps of welding operation (1.1.Preheating Operation, 1.2.Welding Workbench, 1.3.Preparation of Copper and Brass Metal Wires, 1.4.Quality Control Operation)

The rotating band is the part that enables the ammunition to rotate in the barrel and it affects the velocity and quality of the ammunition. After welding treatment, the metal proportions (ZnR, FeR, CuR) in the weld region must be at the required level. For a defective product, for example if the iron ratio is lower, the rotating band may detach in the barrel. If the ratio is higher, the barrel may be ruined.

The product is highly demanded due to its strategic importance and because of the problems in welding process more than 50% of the annual demand cannot be matched. In addition, about 30% of the products are defective and for defective products reproduction and salvage costs are incurred. For 18% of the products retrieval costs are incurred, for 10% of the products scrap costs are incurred and for 2% of the products both retrieval and scrap costs are incurred.

For 155 mm. artillery ammunition discussed the main problem is: The interaction among input and output parameters of welding process cannot be predicted and therefore optimum parameter configuration cannot be determined. In order to control the welding process described interaction among input and output parameters must be resolved. By using the classification and prediction characteristics of ANNs, backpropagation ANN models are developed and implemented to solve the problem which are discussed in the application section. The next section is a brief discussion of general concept of ANN and backpropagation neural networks.

3   Method: Backpropagation Neural Networks

Neural network is an artificial intelligence model originally designed to replicate the human brain's learning process. A typical network is composed of a series of interconnected nodes and the corresponding weights. It aims at simulating the complex mapping between the input and output. The training process is carried out on a set of data including input and output parameters. The learning procedure is based on the training samples and the testing samples are used to verify the performance of the trained network. During the training, the weights in the network are adjusted iteratively till a desired error is obtained [14]. In this study the best architecture is found by a systematic trial and error approach.

One of the most commonly used supervised ANN model is backpropagation network that uses backpropagation learning algorithm. Backpropagation algorithm is one of the well-known algorithms in neural networks [15, 16, 17, 18]. In this study, feed forward multi layered perceptron neural networks are used for modeling. The reason for using this type of neural network is that it is a standard in the solution of problems related with identifying figures by applying the supervised learning and the backpropagation of errors together [19].

The training of a network by backpropagation involves three stages: the feed forward of the input training pattern, the calculation and backpropagation of the associated error and the adjustment of the weights (20).

4   Neural Network Applications for Welding Process Control

In this study 2 different models have been developed for welding process control problem defined in the Section 2. Model 1 is used for classification of defective and defect-free products and Model 2 is used to predict input parameter values given the output vales (which is a reverse BPNN application).

The input variables of the model are copper wire drawing speed (CDS), brass wire drawing speed (BDS), copper torch rate (CTR), copper torch angle (CTA), oscillation rate (OSR), center deviation of wires (CDW), water discharge (WAD), welding current (WCU) and voltage (VOL). The output variables are zinc ratio (ZnR), iron ratio (FeR) and copper ratio in percentages (CuR) of the weld.

4.1   Model 1: Classification Model

Neural networks have proven themselves as proficient classifiers and are particularly well suited for addressing non-linear problems. Given the non-linear nature of real world phenomena, neural networks are certainly a good candidate for solving the problem [20].

In this work, the first application is carried out to determine which input parameter set results with a defective product and which ones with a defect-free product. In order to solve this problem three different BPNN models (Model 1.1: Feed-forward backpropagation network, Model 1.2: Cascade-forward backpropagation network and Model 1.3: feedforward backpropagation network with feedback from output to input) are developed for classifying the products. In feed-forward backpropagation networks the first layer has weights coming from the input. Each subsequent layer has a weight coming from the previous layer. In cascade forward backpropagation networks the first layer has weights coming from the input. Each subsequent layer has weights coming from the input and all previous layers. In feed-forward backpropagation network with feedback from output to input, the first layer has weights coming from the input. Each subsequent layer has a weight coming from the previous layer and there exists a feedback from output layer to inputs. For all 3 models all layers have biases, the last layer is the network output.

ANN models developed in this study are created with MATLAB R2009a software package. ANN model architectures used in Model 1 are composed of an input layer, 1 hidden layer and an output layer. There are 9 neurons (9 input parameters) in input layer, 10 neurons in hidden layer, and 3 neurons (3 output parameters) in output layer (9x10x3). The best network architecture is found with trial and error. Several runs are conducted and the structure with minimum Mean Squared Error (MSE) is chosen. Performance of networks is discussed in "Conclusions" section. The outputs in ANN model are represented by unit vectors as:[1 1 1] = defective weld, [0 0 0] = defect-free weld. Each neuron in the output vector represents a situation whether ZnR, FeR and CuR is between quality specifications respectively or not. Therefore [1 1 1] means that output quality specifications are not met and [0 0 0] means output quality specifications are met.

The network architecture used in classification models is summarized in Fig. 2. The network is trained with Levenberg–Marquardt Algorithm (LMA). For this "trainlm" learning function is used in the MATLAB software. LMA is often the fastest backpropagation algorithm and it is highly recommended as a first-choice supervised algorithm, although it does require more memory than other algorithms.

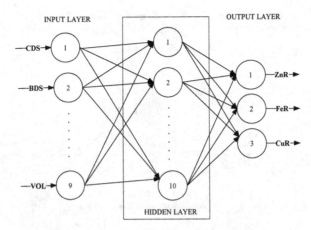

Fig. 2. This figure shows the architecture of Model 1. In this model there are 3 layers (input layer with 9 neurons, 1 hidden layer with 10 neurons and output layer with 3 neurons.

LMA provides a numerical solution to the problem of minimizing a function, generally nonlinear, over a space of parameters of the function. These minimization problems arise especially in least squares curve fitting and nonlinear programming [21]. Given a set of m empirical datum pairs of independent and dependent variables, (x_i, y_i), optimize the parameters β of the model curve $f(x,\beta)$, so that the sum of the squares of the deviations becomes minimal as in Equation (1).

$$S(\beta) = \sum_{i=1}^{m} [y_i - f(x_i, \beta)]^2 \tag{1}$$

In order to measure the network performances Mean Squared Error (MSE) is used as a performance measure. MSE is a network performance function. It measures the network's performance according to the mean of squared errors.

The MSE of an estimator λ with respect to the estimated parameter θ is defined as in Equation (2):

$$MSE(\lambda) = E([\lambda - \theta]^2) \tag{2}$$

In Fig. 3, training performance of Model 1.1, 1.2 and 1.3 are given respectively.

4.2   Model 2: Input Parameter Prediction Model

In this model, values of 3 input variables (BDS, CTR and OSR), which are 3 important classifiers found with classification algorithm, are predicted according to given output variables. For this reason a reverse BPNN model is used. Here the purpose is determining the values of most important input variables according to given output values and finding best input values for a defect-free product.

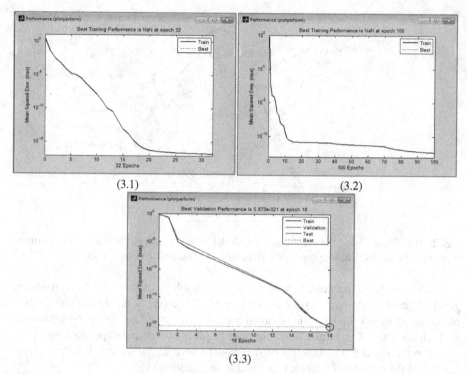

(3.1)   (3.2)

(3.3)

Fig. 3. This figure shows training performances of models 1.1 (Fig. 3.1), 1.2 (Fig. 3.2) and 1.3 (Fig. 3.3).

The architecture of Model 2 is given in Fig. 4. Here using outputs and inputs in a reverse provided the advantage of finding optimal values of input variables for a defect-free product.

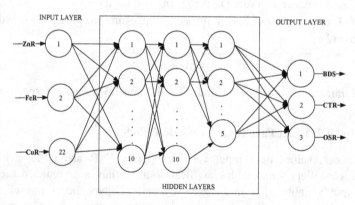

Fig. 4. In this figure architecture of Model 2 is summarized. In this model there are *5 layers* (input layer with *3 neurons*, *3 hidden layers* with *10-10-5* neurons and output layer with *3 neurons*).

For Model 2 a different structure from Model 1 is developed. The best structure is again found with trial and error under several scenarios (by changing the number of hidden layers, neurons, type of transfer functions each layer etc.). The structure giving minimum MSE in selected. The training performance of Model 2 is shown in Fig. 5.

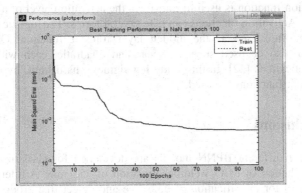

Fig. 5. This figure shows training performance of Model 2

In this model output variables are used as input neurons for predicting input values of BDS, CTR and OSR which are critical input parameters found with RandomTree classification tree algorithm with WEKA 3.6 software package (parameters that have highest standard deviation). RandomTree is used for classification for constructing a tree that considers k random features at each node. As we know the ZnR, FeR and CuR ratios for a defect-free product (which are quality specifications determined by Quality Control department), optimal BDS, CTR and OSR values are predicted.

Decision tree developed with classification tree algorithms is given in Fig. 6. According to this classification tree for a defect-free product OSR must be lower than 1155, CTR must be equal or greater than 77 and BDS must be smaller than 6,65 (which are mean values of the parameters). The BPNN model is developed to find the mean values of best input parameter combination.

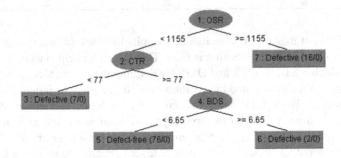

Fig. 6. This figure shows an example decision tree for classification. The variables determined with *RandomTree* algorithm (*OSR, CTR* and *BDS*) are used in application of Model 2.

For both Model 1 and Model 2 data of 101 ammunitions for 9 input variables and 3 output variables are used. 81 of 101 data were used for training of the networks. Remaining 20 data were used for testing the results. Test data are selected in a systematic way (Every 5th. instance of 101 data). MSE values are calculated according to difference between original results of 20 data and test results.

An activation function is used to transform the activation level of a neuron into an output signal. Activation functions can take several forms. The logistic and hyperbolic functions are often used as the hidden layer transfer function. Other activation functions can also be used such as linear and quadratic, each with a variety of modeling applications [22]. In this study log-sigmoid, hard limit, hyperbolic tangent sigmoid transfer functions are used.

5   Conclusions

In this work, 2 different BPNN models are developed for classification and input parameter prediction of MIG welding operation of 155 mm. Artillery ammunition. Model 1 is used for classification of defective and defect-free products. Model 2 is used for input parameter prediction.

For both of the models weld operation data of 101 ammunitions are used for classification and input parameter prediction analysis. 81 of the data were used for training of the networks. Remaining 20 instances are used for testing the network performance.

The performance of 3 different structures developed for Model 1 is shown on Table 1. Model 1 is useful for determining which instances (set of input values) produces non-defective outputs.

Table 1. Classification Results of Model 1

Model	Correctly Classified Test Instances	Incorrectly Classified Test Instances
Model 1.1	19	1
Model 1.2	20	0
Model 1.3	20	0

BPNN does not explain the classification results by rules. In order to extract rules *RandomTree* classification algorithm is used. This algorithm helped us to find the best classifiers which are BDS, CTR and OSR. By defining these variables as outputs for a non-defective combination of output values (which are used as inputs in Model 2) optimal values of BDS, CTR and OSR are found. In Model 2 there are 9 neurons (representing 9 input variables) in input layer, 10 neurons in first hidden layer, 10 neurons in second hidden layer, 5 neurons in third hidden layer and 3 neurons (representing 3 output variables) in output layer. Transfer functions are sigmoid-hyperbolic tangent-sigmoid-sigmoid-sigmoid respectively for each layer. MSE value is 0,009. The results of Model 2 show that best values for BDS, CTR and OSR are 6.3 m/min., 80 mm. and 1150 cycle/min.

The main quality indicator of a neural network is to predict accurately the output of unseen test data. In this study, we have benefited from classification and prediction success of BPNNs. The classification and prediction performance results show the advantages of backpropagation neural networks: it is rapid, noninvasive and inexpensive. Another advantage of the models is, with any input parameter combination, we can execute the model and find the output values in a few seconds.

In the future studies we plan to develop an electronic control board that can be used by workers. This panel will work as an interface for workers.

Acknowledgements. This study is supported by a grant from Industry Thesis Program of Ministry of Science, Industry and Technology of Turkey (Grant No:00748.STZ.2010-2).

References

1. Benyounis, K.Y., Olabi, A.G.: Optimization of different welding processes using statistical and numerical approaches – A reference guide. Advances in Eng. Software 39, 483–496 (2008)
2. Öztemel, E.: Yapay Sinir Ağları, İstanbul; Papatya Yayıncılık (2006)
3. Sha, W., Edwards, K.L.: The use of artificial neural networks in materials science based research. Materials and Design 28, 1747–1752 (2007)
4. Chertov, D.A.: Use of Artificial Intelligence Systems in the Metallurgical Industry (Survey). Metallurgist 47, 7–8 (2003)
5. Tay, K.M., Butler, C.: Modeling and Optimizing of a Mig Welding Process-A Case Study Using Experimental Designs and Neural Networks. Quality and Rel. Eng. Int. 13, 61–70 (1997)
6. Luo, H., Zenga, H., Hub, L., Hub, X., Zhoub, Z.: Application of artificial neural network in laser welding defect diagnosis. Journal of Materials Processing Technology 170, 403–411 (2005)
7. Oscar, M., Manuel, L., Fernando, M.: Artificial neural networks for quality control by ultrasonic testing in resistance spot welding. J. of Mat. Processing Technology 183, 226–233 (2007)
8. Kim, S., Sona, J.S., Leeb, S.H., Yarlagaddac, P.: Optimal design of neural networks for control in robotic arc welding. Robotics and Computer-Integrated Manufacturing 20, 57–63 (2004)
9. Özerdem, M.S., Sedat, K.: Artificial neural network approach to predict the mechanical properties of Cu–Sn–Pb–Zn–Ni cast alloys. Materials and Design 30, 764–769 (2009)
10. Oscar, M., Pilar De, T., Manuel, L.: Artificial neural networks for pitting potential prediction of resistance spot welding joints of AISI 304 austenitic stainless steel. Corrosion Science 52, 2397–2402 (2010)
11. Mirapeix, J., García-Allende, P.B., Cobo, A., Conde, O.M., López-Higuera, J.M.: Real-time arc-welding defect detection and classification with principal component analysis and artificial neural Networks. NDT&E International 40, 315–323 (2007)
12. Pal, S., Pal, K.S., Samantaray, K.: Artificial neural network modeling of weld joint strength prediction of a pulsed metal inert gas welding process using arc signals. Journal of Materials Processing Technology 202, 464–474 (2008)

13. Ateş, H.: Prediction of gas metal arc welding parameters based on artificial neural Networks. Materials and Design 28, 2015–2023 (2007)
14. Su, T., Jhang, J., Hou, C.: A Hybrid Artificial Neural Network and Particle Swarm Optimization for Function Approximation. International Journal of Innovative Computing 4(9) (2008)
15. Russell, S., Norvig, P.: Artificial Intelligence: A Modern Approach, 2nd edn. P. Hall, Inc. (2003)
16. Nong, Y., Vilbert, S., Chen, Q.: Computer intrusion detection through EWMA for auto correlated and uncorrelated data. IEEE Trans. Reliability 52, 75–82 (2003)
17. Li, C., Li, S., Zhang, D., Chen, G.: Cryptanalysis of a Chaotic Neural Network Based Multimedia Encryption Scheme. In: Aizawa, K., Nakamura, Y., Satoh, S. (eds.) PCM 2004, Part III. LNCS, vol. 3333, pp. 418–425. Springer, Heidelberg (2004)
18. Shihab, K.: A Backpropagation Neural Network for Computer Network Security. Journal of Computer Science 2(9), 710–715 (2006)
19. Hardalac, F., Barisci, N., Ergun, U.: Classification of aorta insufficiency and stenosis using MLP neural network and neuro-fuzzy system. Physica Medica XX(4) (2004)
20. Fausett, L.: Fundamentals of neural Networks, architectures, algorithms and applications. Prentice Hall (1994) ISBN: 0-13-334186-0
21. Marquardt, D.: An Algorithm for Least-Squares Estimation of Nonlinear Parameters. SIAM Journal on Applied Mathematics 11, 431–441 (1963)
22. Khashei, M., Bijari, M.: An Artificial Neural Network Model (p,d,q) for Timeseries Forecasting. Expert Systems with Applications 37, 479–489 (2010)

Elastic Nets for Detection of Up-Regulated Genes in Microarrays

Marcos Levano and Alejandro Mellado

Universidad Católica de Temuco, Facultad de Ingeniería, Escuela de Ingeniería Informática
Avda. Manuel Montt 056, Casilla 15-D, Temuco - Chile
{mlevano,amellado}@inf.uct.cl

Abstract. DNA analysis by microarrays is a powerful tool that allows replication of the RNA of hundreds of thousands of genes at the same time, generating a large amount of data in multidimensional space that must be analyzed using informatics tools. Various clustering techniques have been applied to analyze the microarrays, but they do not offer a systematic form of analysis. This paper proposes the use of Zinovyev's *Elastic Net* in an iterative way to find patterns of up-regulated genes. The new method proposed has been evaluated with up-regulated genes of the Escherichia Coli k12 bacterium and is compared with the Self-Organizing Maps (SOM) technique frequently used in this kind of analysis. The results show that the proposed method finds *87%* of the up-regulated genes, compared to *65%* of genes found by the SOM. A comparative analysis of Receiver Operating Characteristic with SOM shows that the proposed method is *12%* more effective.

Keywords: Elastic net, microarrays, up-regulated genes, clusters.

1 Introduction

Modern deoxiribonucleic acid microarray technologies [1] have revolutionized research in the field of molecular biology by enabling the study of hundreds of thousands of genes simultaneously in different environments [1].

By using image processing methods it is possible to obtain different levels of expression of thousands of genes simultaneously for each experiment. In this way these techniques generate thousands of data represented in multidimensional space. The process is highly contaminated with noise and subject to measurement errors, finally requiring experimental confirmation. To avoid repeating the whole process experimentally gene by gene, pattern recognition techniques are applied that make it possible to select sets of genes that fulfil given behavior patterns at their gene expression levels.

The most widely used method to determine groupings and select patterns in microarrays is the Self-Organizing Maps (SOM) technique [2-4]. One of the problems of SOM is the need to have an initial knowledge of the size of the net to project the data, and this depends on the problem that is being studied. On the other hand, since SOM is based on local optimization, it presents great deficiencies by restricting data projections only to its nodes.

C. Jayne, S. Yue, and L. Iliadis (Eds.): EANN 2012, CCIS 311, pp. 183–192, 2012.

The Elastic Net (EN) [5] method generates a controllable net described by elastic forces that are fitted to the data by minimizing an energy functional, without the need of knowing its size a priori. This generates greater flexibility to adapt the net to the data, and like the SOMs it allows a reduction in dimensionality, that improves the visualization of the data, which is very important for bioinformatics applications.

ENs have been applied to different problems in genetics, such as analysis of base sequence structures (adenine, cytosine, guanine and thymine), where base triplet groupings are discovered [6]; automatic gene identification in the genomes of the mitochondria of different microorganisms [4]. But as far as we can tell, there is no application for finding patterns in microarrays.

This paper proposes the use of ENs to divide clusters iteratively, together with the k-means method and using indices to measure the quality of the *clusters*, making it possible to select the number of groups formed in each iteration.

To evaluate the results, data from the most widely studied microorganism, the bacterium Escherichia Coli K12 (E.Coli), were used. The levels of gene expression of a set of *7,312* genes were analyzed by means of the microarrays technique. In this set there are *345* up-regulated genes that have been tested experimentally [6] and must be detected with the new method. The results are compared with those of the traditional SOM method.

2   Method

2.1   Theorical Foundation

Zinovyev defines the Elastic Net [5] as a net of nodes or neurons connected by elastic forces (springs), where $Y = \{y^i, i = 1..p\}$ is a collection of nodes, $E = \{E^{(i)}, i = 1..s\}$ is a collection of edges, and $R^{(i)} = \{E^{(i)}, E^{(k)}\}$ is the combination of pairs of adjacent edges called *ribs* denoted by $R = \{R^{(i)}, i=1..r\}$. Each edge $E^{(i)}$ starts at node $E^{(i)}(0)$ and ends at node $E^{(i)}(1)$. The *ribs* start at node $R^{(i)}(1)$ and end at node $R^{(i)}(2)$, with a central node $R^{(i)}(0)$. The data to be analyzed are $x^j=[x^j{}_1,...,x^j{}_M]^T \in R^M$, where M is the dimension of the multidimensional space and $j =1..N$ is the number of data.

The set of data closest to a node is defined as a taxon, $K^i = \{x^j : \| x^j - y^i \| \rightarrow min\}$. It is clear that there must be as many taxons as nodes. Here $\| x^j - y^i \|$ is the norm of the vector $(x^j - y^i)$, and the Euclidian norm is used. This means that the taxon K^i contains all the vectors of the x^j data whose norms with respect to node y^i are the smallest.

Energy $U^{(Y)}$ between the data and the nodes is defined by equation 1:

$$U^{(Y)} = \frac{1}{N}\sum_{i=1}^{P} \sum_{x^j \in K^i} \left\| x^j - y^i \right\|^2$$

(1)

where each node interacts only with the data of its taxon. An elastic energy between the nodes $U^{(E)}$ is added by equation 2:

$$U^{(E)} = \sum_{l=1}^{s} \lambda_i \left\| E^i(1) - E^i(0) \right\|^2$$

(2)

where λ_i are the elasticity constants that allow the net's elasticity to be controlled. Additionally, a deformation energy $U^{(R)}$ between pairs of adjacent nodes, is also added by equation 3:

$$U^{(R)} = \sum_{I=1}^{R} \mu_i \left\| R^i(1) - 2R^i(0) + R^i(2) \right\|^2 \tag{3}$$

where μ_i are the deformability constants of the net. The same values of λ and μ are chosen for all the λ_i and μ_i. The total energy is now minimized by equation 4 with respect to the number and position of the y^i nodes for different μ and λ:

$$U = U^{(Y)} + U^{(E)} + U^{(R)} \tag{4}$$

We used the VIDAEXPERT implementation, which can be found in Zinovyev et al. [5]. In addition to the flexibility offered by the ENs to fit the net to the data, the projections of the data to the net can be made over the edges and at points within the net's cells, and not only over the nodes as required by the SOMs. This leads to an approximation that has a better fit with the real distribution of the data in a smaller space. This property is very important for applications in bioinformatics, where the specialist has better feedback from the process.

2.2   Method for Searching Patterns

The algorithm used to find groups of genes that have the same behavior patterns consists of four fundamental phases: data preprocessing, EN application, pattern identification, and finally a stopping criterion and cluster selection based on the level of expression and inspection of the pattern that is being sought. Figure 1 shows the diagram of the algorithm's procedure.

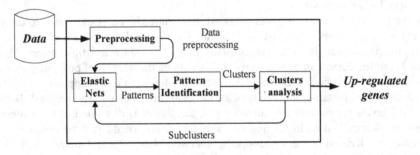

Fig. 1. Process diagram

Phase 1: Preprocessing
The set of N data to be analyzed is chosen, $x^j = [x^j_1,...,x^j_M]^T$, $j = 1...N$, where M is the dimension of the multidimensional space. For this application, N corresponds to the *7,312* genes of the E.coli bacterium and M to the *15* different experiments carried out on the genes, and x^j is the gene expression level. The data are normalized in the form $\theta^j = ln(x^j - min(x^j) + 1)$ which is used as a standard in bioinformatics [4], [6].

Phase 2: Application of the Elastic Net (EN)
The package of Zinovyev et al. [5], which uses the following procedures, is applied:

(a) The data to be analyzed are loaded.
(b) The two-dimensional net is created according to an initial number of nodes and the values of the elastic and deformability constants λ and μ.
(c) The net is fitted to the data, minimizing the energy U. For that purpose the initial values of λ and μ are reduced three times (four pairs of parameters are required to be entered by the user). The decrease of λ and μ results in a net that is increasingly deformable and less rigid, thereby simulating annealing [7], allowing the final configuration of the EN to correspond to an overall minimum of U or a value very close to it [5].
(d) The data are projected over the net on internal coordinates. In contrast with the SOM, in which *piecewise constant projecting* of the data is used (i.e., the data are projected on the nearest nodes), in this method *piecewise linear projecting* is applied, projecting the data on the nearest point of the net [5]. This kind of projection results in a more detailed representation of the data.
(e) Steps (c) and (d) are repeated for different initial values of the nodes, λ and μ, until the best resolution of the patterns found is obtained.

Phase 3: Pattern identification
The data are analyzed by projecting them on internal coordinates for the possible formation of clusters or other patterns such as accumulation of clusters in certain regions of the net or a typical dependence of the data in a cluster with respect to the dimensions of the multidimensional space. In the latter case, the average of the data for each dimension is calculated (cluster's centroid for the dimension).

For the formation of possible clusters the k-means method is used [7] together with the quality index I [8], which gives information on the best number of clusters. The centroids of each cluster are graphed and analyzed to find possible patterns.

Phase 4: Cluster analysis
Once the best number of clusters is obtained, the centroids' curves are used to detect and extract the possible patterns. In general, the centroid curve of a cluster may present the pattern sought, may be a constant, or may not show a definite trend. Also, the values of the curve can be in a range that is outside the interest of possible patterns (low levels of expression). To decide if the clusters found in a first application of the EN contain clear patterns, the behavior of the centroids' curves are analyzed. If the centroids' levels are outside the range sought, the cluster is discarded; if the patterns sought are detected, the cluster that contains the genes sought will be obtained (in both cases the division process is stopped), otherwise phases 2 and 3 are repeated with each of the *clusters* and the analysis of phase 4 is carried out again, repeating the process.

2.3   Data Collection

The data correspond to the levels of gene expression of *7,312* genes obtained by the microarray technique of E.Coli k12 [6]. These data are found in the GEO database

(Gene Expression Omnibus) of the National Center for Biotechnology Information [1]. The work of Liu et al. [6] provides the *345* up-regulated genes that were tested experimentally. Each gene is described by *15* different experiments (which correspond to the dimensions for the representation of each gene) whose gene expression response is measured [6] on glucose sources. Specifically there are *5* sources of glucose, *2* sources of glycerol, *2* sources of succinate, *2* sources of alanine, *2* sources of acetate, and *2* sources of proline. The definition of up-regulated genes according to [4], [6] is given in relation to their response to the series of sources of glucose considering two factors: that its level of expression is greater than *8.5* on a $\log_2$ scale, and that its level of expression increases at least *3* times from the first to the last experiment on the same scale. For our evaluation we considered a less restrictive definition that includes the genes that have only an increasing activity of the level of expression with the experiments; since the definition given in [6] for up-regulated genes contains very elaborate biological information for which a precise identification of the kind of gene to be detected is required.

The original data have expression level values between zero and hundreds of thousands. Such an extensive scale does not offer an adequate resolution to compare expression levels; therefore a logarithmic normalization is carried out. In this case we preferred to use the natural logarithm [7] instead of the base 2 logarithm used by Liu, because it is a more standard measure. The limiting value for the expression level was calculated using our own algorithm by determining the threshold as the value that best separates the initial clusters (θ_{min}). This expression level allows discarding groups of genes that have an average level lower than this value.

Once the normalization and the initial filtering have been made, the data are ready for the proposed method to be applied.

3   Results

First, the net's parameters were calibrated, i.e. the size of the net was set and the series of pairs of elasticity (λ) and deformability (μ) parameters were selected. The strategy chosen consisted in evaluating different net sizes and pairs of parameters λ and μ for the total data set that would allow minimizing the total energy U.

The minimum energy was obtained with a mesh of 28x28 nodes that was used throughout the whole process. Implementation of the EN [5] requires a set of at least four pairs of λ and μ parameters to carry out the process, because it adapts the mesh's deformation and elasticity in a process similar to simulated annealing [7] that allows approximation to overall minimums. The set of parameters that achieved the lowest energy values had λ with values of *{1.0; 0.1; 0.05; 0.01}* and μ with values of *{2.0; 0.5; 0.1; 0.03}*. For the process of minimizing the overall energy U, *1,000* iterations were used. Then the cluster subdivision iteration process was started.

Figure 2 shows the representation of the first division on internal coordinates and the expression levels of the centroids for the two clusters selected by the index I (for this first iteration). The expression level value equidistant from the two clusters corresponds to $\theta_{min}=5.5$.

---

[1] http://www.ncbi.nlm.nih.gov/sites/entrez?db=ncbisearch

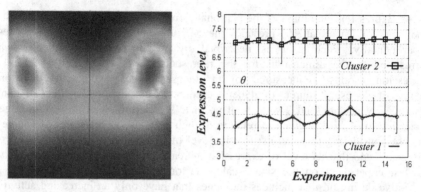

Fig. 2. First iteration of the method. Left: projections on internal coordinates. Right: centroids and choice of the minimum expression level.

The iteration process generates a tree where each node is divided into as many branches as clusters are selected by the index I. In the particular case of E.Coli k12, a tree of depth five is generated. The generation of the tree is made together with a pruning by expression level, i.e., only those clusters that present an expression level greater than $\theta_{min} \geq 5.5$ are subdivided.

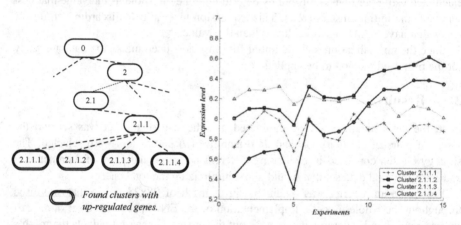

Fig. 3. Prepruned subcluster generation tree. Every leaf shown contains the set of genes with increasing activity where the up-regulated genes to be evaluated are found. The coding of each node shows the sequence of the nodes through which one must go to reach each node from the root.

Finally, to stop the subdivision process of the groups that have an expression level greater than θ_{min}, the behavior of the expression level in the experiments was examined. In this case we only looked for a simple increasing pattern of the expression level in the centroids through the *15* experiments (the strict definition of up-regulated genes given in [6] was not used). Figure 3 shows the tree of subclusters generated by the process applied to the genes of E.Coli k12.

Figure 4 shows the behavior of increasing expression level for the penultimate division of the 2.2.2 clusters.

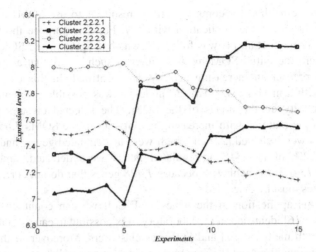

Fig. 4. Subdivision patterns of the 2.2.2 cluster. Two clusters with increasing activity (2.2.2.2 and 2.2.2.4) are shown where the division process ends, and clusters 2.2.2.3 and 2.2.2.1 which are discarded.

Table 1 shows the clusters chosen at the end of the process (tree leaves in Fig. 3), indicating the total number of genes with increasing activity detected, the number of up-regulated genes.

Table 1. Clusters that contain the genes with increasing activity chosen by the proposed method, and number of up-regulated genes in each group

Clusters	Genes with increasing activity	Up-regulated genes
2.1.2.3	172	68
2.1.1.3	238	53
2.1.1.2	276	49
2.2.2.2	153	41
2.2.2.4	197	33
2.1.2.1	313	29
2.2.1.1.1	68	13
2.2.1.1.2	104	5
2.2.1.2.2	58	8
Totals	1,579	299

The results show that the process chooses *1,579* genes, of which *299* correspond to up-regulated genes of the *345* that exist in the total data set, i.e. *86.7%* of the total number of up-regulated genes. From the practical standpoint for the biological field, only *19%* effectiveness has been achieved because there are *1,280* genes that are not up-regulated, which must be discarded using biological knowledge or by means of individual laboratory tests.

An alternative method for comparing these results is to use SOMs with the same data and conditions of the application with EN. For this purpose the methodology proposed by Tamayo et al. [4] was followed, which suggests using SOMs in a single iteration, where the initial SOM mesh is fitted in such a way that at each node the patterns that present an increasing activity are identified. In this case the process shows that with a mesh of size *5x6* (*30* nodes) it was possible to obtain patterns of increasing activity on the nodes of the SOM. The selected clusters are obtained directly from the patterns with increasing activity. With the SOMs *1,653* increasing activity genes were selected, *225* of which were up-regulated genes, and therefore in this case *65.2%* of the *345* up-regulated genes were detected, and a practical efficiency of *13.6%* was achieved, because *1,428* genes that do not correspond to up-regulated genes must be discarded.

Since in this application to the genes of E.Coli we can count on the *345* up-regulated genes [6] identified in the laboratory, it is possible to carry out an evaluation considering both methods (EN and SOM) as classifiers. Moreover, if the expression level θ is considered as a classification parameter, it is possible to make an analysis by means of Receiver Operating Characteristic (ROC), varying the expression level θ over an interval of [*4.4 – 8.9*]. Figure 5 shows the results of the ROC curves for both methods.

The optimum classification value for EN is achieved at $\theta^*=5.6$. At this point a sensitivity of *86%* and a specificity of *82%* were reached, covering an area of *0.87* under the ROC curve. When the same data, normalization values and expression level ranges were considered for SOM, an optimum classification value of $\theta^*=5.9$ is obtained, achieving a sensitivity of *65%*, a specificity of *84%*, and an area under the ROC curve of *0.78*.

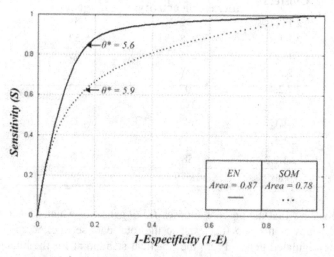

Fig. 5. ROC curves for EN and SOM

4   Discussion

When the results of the proposed method (which uses EN) are compared with those of the traditional SOM method, it is seen that the EN method detects *74* up-regulated genes more than the SOM, which correspond to *21.5%* of those genes. For practical purposes it must be considered that these genes are not recoverable in the case of the SOM because they are mixed up with the group of *5,659* undetected genes. On the other hand, the efficiency of the method that uses the EN is better, because it requires discarding *1,280* genes that are not expressed, compared to the *1,428* that must be discarded with the SOM. Since the final objective of the experiment with E.Coli k12 consists in detecting the up-regulated genes, it is possible to consider the EN and SOM methods as classifiers and carry out an analysis of the merit of the classification by means of an ROC curve.

When considering an overall analysis of the classifier using the expression level θ as a parameter, it is important to consider the area under the ROC curve. In this case the area for the proposed method is *0.87*, compared to *0.78* for the SOM, which represents an *12%* improvement. In relation to the sensitivity at the optimum decision level, the proposed method is *21%* more sensitive than the SOM.

The numerical advantages derived from the application of the proposed method for the detection of the up-regulated genes of E.Coli are clear, but there are other aspects that must be analyzed with the purpose of projecting these results to the search of genes expressed in microarrays. The ENs present several advantages that allow reinforcing the proposed method of iteration divisions. On the one hand, the ENs have a greater capacity for adapting the net to the data because they have a set of parameters that control the deformation and elasticity properties. By carrying out the minimization of the overall energy in stages (evaluating different combinations of parameters λ and μ), a process similar to annealing is induced, making it possible to approach the overall minimum and not be trapped in local minimums. The same minimization methods allow the automatic selection of parameters that are fundamental for the later development of the process, such as the minimum expression level θ_{min} and the size of the net.

The other important advantage of the ENs refers to their representation capacity, because the use of *piecewise linear projecting* makes it possible to increase the resolution of the data projected on the space having the lowest dimensions (internal coordinates). In the case of the microarray analysis this better representation becomes more important, since a common way of working in the field of microbiology and genetics is based on the direct observation of the data. On the other hand, the SOMs only allow a projection on the nodes when using *piecewise constant projecting* or the alternative U-matrix projections [2], which approximate only sets of data to the plane but do not represent directly each data.

A valid point that should be analyzed when comparing SOMs with ENs is to consider the argument that an iteration process of divisions with SOMs can improve the results of the method. But the iteration process presented is based on the automatic selection of parameters (particularly the size of the net and the minimum expression level) for its later development, which is achieved by a global optimization method like EN. The SOM does not allow the expression level to be determined automatically, and that information must come from the biological knowledge of the expression levels of particular genes. The alternatives of using the minimum error of vector quantization of SOM as an alternative the minimum energy of EN did not produce satisfactory results.

5   Conclusions

This paper proposes the use of Elastic Nets as an iteration method for discovering clusters of genes expressed in microarrays. This proposal has a general character for application in the field of bioinformatics, where the use of microarrays is gaining increasing importance day by day.

The results of the application to the discovery of up-regulated genes of E.Coli k12 show a clear advantage of the proposal over the traditional use of the SOM method.

We chose to carry out a comparison with well established methods that are used frequently in the field of bioinformatics, but it is also necessary to evaluate other more recent alternatives such as flexible EN [9].

There is also the possibility of studying improvements of the present proposal, exploring the alternative of adapting the deformation and elasticity parameters at each iteration, or using a more powerful clustering method alternative to k-means, such as Block-Entropy [10].

References

1. Molla, M., Waddell, M., Page, D., Shavlik, J.: Using machine learning to design and interpret gene-expression microarrays. Artificial Intelligence Magazine 25, 23–44 (2004)
2. Kohonen, T.: Self-organizing maps. Springer, Berlin (2001)
3. Hautaniemi, S., Yli-Harja, O., Astola, J.: Analysis and visualization of gene expression microarray data in human cancer using self-organizing maps. Machine Learning 52, 45–66 (2003)
4. Tamayo, P., Slonim, D., Mesirov, J., Zhu, Q., Kitareewan, S., Dmitrovsky, E., Lander, E., Golub, T.: Interpreting patterns of expression with self-organizing maps: Methods and application to hematopoietic differentiation. Genetics 96, 2907–2912 (1999)
5. Zinovyev, A.Y., Gorban, A., Popova, T.: Self-organizing approach for automated gene identification. Open Sys. and Information Dyn. 10, 321–333 (2003)
6. Liu, M., Durfee, T., Cabrera, T., Zhao, K., Jin, D., Blattner, F.: Global transcriptional programs reveal a carbon source foraging strategy by *Escherichia coli*. J. Biol. Chem. 280, 15921–15927 (2005)
7. Duda, R., Hart, P., Stork, D.: Pattern classification. John Wiley Sons Inc. (2001)
8. Maulik, U., Bandyodpadhyay, S.: Performance evaluation of some clustering algorithms and validity indices. IEEE PAMI 24, 1650–1654 (2002)
9. Lévano, M., Nowak, H.: New aspects elastic nets of the elastic nets algorithm clusters analysis. J. Neural Computing & Applications 20(6), 835–850 (2011)
10. Larson, J.W., Briggs, P.R., Tobis, M.: Block-Entropy Analysis of Climate Data. Procedia Computer Science 4, 1592–1601 (2011)

Detection and Classification of ECG Chaotic Components Using ANN Trained by Specially Simulated Data

Polina Kurtser, Ofer Levi, and Vladimir Gontar

Department of Industrial Engineering and Management, Ben-Gurion University of the Negev,
P.O. Box 653, Beer-Sheva 84105, Israel
kurtser@post.bgu.ac.il

Abstract. This paper presents the use of simulated ECG signals with known chaotic and random noise combination for training of an Artificial Neural Network (ANN) as a classification tool for analysis of chaotic ECG components. Preliminary results show about 85% overall accuracy in the ability to classify signals into two types of chaotic maps – logistic and Henon. Robustness to random noise is also presented. Future research in the form of raw data analysis is proposed, and further features analysis is needed.

Keywords: ECG, Deterministic chaos, Artificial Neural Networks.

1   Introduction

Various suggestions regarding the physiological origin of chaotic component in ECG were made since Guevara et al. [1] introduced nonlinear approaches into heart rhythm analysis. For example, Voss et al. [2], discussed the chaotic component in different biological systems. They formulated three possible sources of non linear behavior: adaptation of the system to its physiological needs using control feedback loops, adaption of a sub-system to changed basic conditions as a result of pathophysiological process and compensation of a failing subsystem by other subsystems. Therefore, it's reasonable to assume that a disconnection between the heart and its feedback loop may result in reduced chaotic behavior. This supports Guzzetti et al [3] findings, who reported loss of complexity in heart-transplanted patients, as a result of loss of the neural modulation of heart rate.

The importance of chaotic component of heart rate variability (HRV) was discussed by Loskutov [4] who has shown differences in value ranges of correlation dimensions for a number of cardiac disorders. In complementation to traditional HRV measures Ho et al. [5] demonstrated how nonlinear HRV indices may contribute prognostic value to predicting survival in heart failure.

Despite the promising results of those methods they are still limited by the need of pre-processing for the detection of QRS complexes. A QRS complex is a repeating pattern of an ECG signal. It represents the depolarization of ventricles of the heart. This complex is usually the most visually obvious part of the ECG wave. Thus in order to track HRV many choose to detect the QRS complex and measure the time

C. Jayne, S. Yue, and L. Iliadis (Eds.): EANN 2012, CCIS 311, pp. 193–202, 2012.

between two neighbor QRS complexes. The need to detect a QRS complex is open to errors, since the detection algorithms have certain limits of accuracy. Recently published algorithms by Abibullaev et al [6] reported accuracy of 97.2% in correctly detected QRS complexes, which still results in about 120 false QRS complex detections per hour.

Another approach, taken by several research groups ([7], [8]), is the analysis of the chaotic component of the ECG signal itself, by extracting Lypunov exponents and correlation dimension. Owis et al [7] attempted to detect and classify arrhythmia. They reported significant statistical differences in extracted features value ranges, but poor classification ability. Übeyli [8] made an attempt to classify an ECG signal into different pathological types of ECG beats using the mentioned above features along with wavelet coefficients and the power levels of power spectral density (PSD).

Even though the chaotic component analysis presented above show high potential in revealing valuable diagnostic information, some papers emphasize the challenges one should consider while using these features. Mitschke et al [9], and Govindan et al [10] clamed Lyapunov exponents, correlation dimension and entropies to be misleading in determining and analyzing chaotic components of a signal. Mitschke et al [9] raised the problem of distinction between stochastic and deterministic chaotic data. Govindan et al [10] who were looking for evidence of deterministic chaos in ECG and HRV have also addressed the problem, and therefore chose to apply surrogate and predictability analysis of normal and pathological signals, instead of applying those measures. They have shown that the dynamics underlying the cardiac signals is nonlinear, and further results indicated the possibility of deterministic chaos.

Due to the difficulties arisen from the analysis of the chaotic component it's important to emphasize the reasons we insist to do so. Therefore some classical linear methods should be mentioned.

Gothwal et al [11] used Fast Fourier transforms (FFT) to identify QRS complexes using two pass filter. Artificial Neural Network (ANN) was then applied to identify cardiac arrhythmias. The input features were maximum, minimum and average QRS width and the heart rate. They showed impressive results of 98.5% overall classification accuracy. These results show the primary advantage of the use of ANN for classification as fast and accurate tool of classification. But one should remember that the variation in heart rate and QRS size in different arrhythmias is well known and can be easily detected by cardiologists. Therefore the results do not reflect more subtle changes that might not be visually detected by a physician. In addition, the used training set was raw data signals which are exposed to errors.

Dokur et al [12] compared between the performance of discrete wavelet transform (DWT) and discrete Fourier transform (DFT) for ECG classification showing 89.4% accuracy for DFT and 97% accuracy for DWT. Bigger et al [13] have statistically analyzed measures of RR variability to screen groups of middle-aged persons to identify individuals who have substantial risk of coronary deaths or arrhythmic events. All of those methods show potential results and have been used by researchers and physicians for a long period of time.

Unfortunately, as Karrakchou et al [14] mentioned back in 1996, these methods are starting to show their limits. Under the assumption that chaotic components do exist in the signal the frequency space of an ECG will always include a uniform band, which will reduce accuracy if not handled properly. Statistical methods require long time measures to gather enough observation for decision making. Karrakchou and his team suggested wavelet and chaotic based algorithms as modern signal processing techniques.

As a result, evidences of existence of chaos in ECG have been presented, and the clinical importance has been stressed, but a more robust tool for the detection and classification of chaotic components is needed. Übeyli [8], used chaotic features for input of ANN and showed excellent results of 94% overall accuracy in classification of variety of disorders. But, since the authors used raw ECG signals for training, which contained both chaotic components and random noise, it's somehow difficult to assign diagnostic value to the chaotic component alone. We would like to analyze the ANN classification possibilities by examining ANN on simulated ECG where the combination between two chaotic components and noise are known. Therefore their results may be further explored and possibly improved by training with simulated data that will allow the use of simpler input features and in the same time to distinguish between different chaotic components embedded within simulated ECG and possibly existing within the raw ECG. The ability to not only detect chaos, but to classify it into different types of chaos might yield valuable diagnostic information.

The method proposed here intends to improve ECG classification and extract the chaotic component by defining optimal ANN architecture (number of hidden layers) and use minimum specially designed input features. ANN will be trained by using simulated ECG with embedded different types of chaotic components and random noise.

2   Background

Typical ECG cycle consists of a P wave, a QRS complex and a T wave. A few ECG generators are available for use [15], [16] but none of them gave us the desired flexibility in setting required parameter values and insertion of chaotic components within HRV therefore we developed our own fundamental ECG cycle generator. As seen in Fig.1, P wave was represented by a sinus with amplitude A_P and half period of Δt_P. T wave was represented the same way with amplitude A_T and half period Δt_T accordingly. QRS complex was defined by 2 lines connecting 3 points Q, R, S, each at voltage value of A_Q, A_R and A_S accordingly. Flat segments PQ and TQ are Δt_{PQ} and Δt_{ST} seconds long. Q and S sloped lines are Δt_Q and Δt_S seconds long accordingly. Δt_{HRV} represents the period of time between the end of T wave and the beginning of next cycle P wave, which allows adjustments in R-R intervals.

We suppose that real ECG contain different types of chaotic components responsible for different heart functions and disorders. To simulate ECG which contains different types of chaotic components, we will use two types of chaotic maps: logistic and Henon map. Our goal is to train ANN to enable it to distinguish between two mentioned above maps.

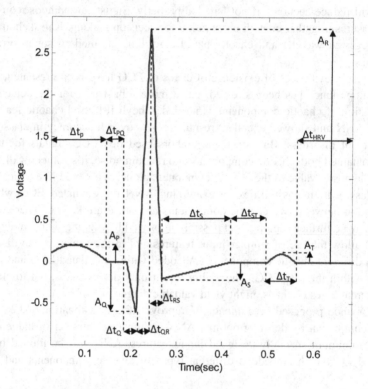

Fig. 1. Simulated fundamental ECG cycle along with adjustable parameters

Prior to implementation into the signal, each chaotic component was normalized and multiplied by a coefficient $0 < \alpha < 1$,

$$\Delta_i = \alpha \cdot \frac{\Delta_i^* - \mu}{\sigma} . \tag{1}$$

Where Δ_i^* - chaotic time series generated by logistic or Henon maps; μ - mean; σ - standard deviation.

The fundamental ECG cycle (Fig.1) was repeated nc times and sampled at the rate of Fs samples per second, forming N point long chaos and noise free discrete fundamental ECG signal. The process of applying chaos, as represented in Fig.2, includes insertion of two chaotic components. The first Δ_i^1, defined by the chosen $\alpha^{(1)}$, was added to the each sample point of the fundamental ECG. The second chaotic component Δ_i^2, defined by $\alpha^{(2)}$, was added to Δt_{HRV} at each cycle causing chaotic variations in R-R intervals.

The final signal was visually compared to signals from MIT-BIH Normal Sinus Rhythm Database [17]. Fig. 2d shows a fragment of an exemplary raw data. By comparing Fig. 2c and 2d, it can be seen that the difference between the real data and simulated data doesn't exceed 5-10%, and therefore the simulation may be used as a proper representation of real ECG signals.

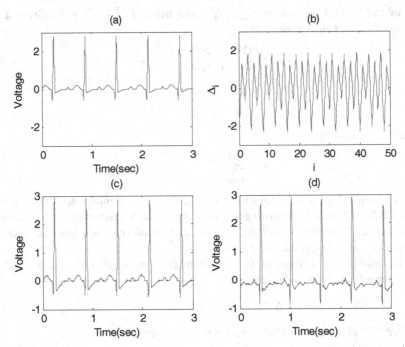

Fig. 2. (a) ECG repeating waveform; (b) Logistic chaos $\alpha=1$; (c) ECG with $\alpha^{(1)}=0.01$, $\alpha^{(2)}=0.5$ logistic chaos; (d) Extract of raw data from MIT-BIH Normal Sinus Rhythm Database, signal No 1625;

Therefore we are accomplishing our simulated ECG by adding 2 chaotic components to the simulated fundamental ECG. The proposed algorithm will further extract and analyze those components.

To analyze the simulated ECGs, ANN input features should be extracted. The choice of the right set of features is crucial for obtaining optimal classification accuracy. In this paper, after examining different possibilities, we chose to narrow our list of features into the following two: variance of variance, and ratio of ECG areas above and below y=0.

Variance of Variances. Each simulated ECG signal, N points long, is split into n equal segments $Y_1,...,Y_n$. For L=N/n the length of each segment.

$$v_i = var([Y_{i1} \dots Y_{iL}]) = \frac{\sum_{j=1}^{L}(Y_{ij} - \overline{Y_{i.}})^2}{L-1}. \tag{2}$$

The variance of variances is given by:

$$V = var[v_1 .. v_n]. \tag{3}$$

In this paper we've chosen n to be equal to 5.

Ratio of the ECG Areas above $(\sum_{i=1}^{N} Y_i')$ **and below** $(\sum_{i=1}^{N} Y_i'')$ **y=0.**. Given Y, a simulated ECG signal, N points long:

$$Y_i' = \begin{cases} Y_i & Y_i > 0 \\ 0 & Y_i \le 0 \end{cases}. \tag{4}$$

$$Y_i'' = \begin{cases} Y_i & Y_i < 0 \\ 0 & Y_i \ge 0 \end{cases}. \tag{5}$$

The ratio R is given by:

$$R = \frac{\sum_{i=1}^{N} Y_{i'}}{\sum_{i=1}^{N} Y_{i''}}. \tag{6}$$

The ANN output was the classification into logistic versus Henon map. This way the trained ANN will be able to distinguish between possible characteristic of different chaotic maps. Therefore it will be able to distinguish between possible chaotic characteristics of different cardiac disorders.

To examine the robustness of the algorithm the test set should include random noise generated by uniform distribution:

$$\Delta^{noise} \sim U(-\beta, \beta). \tag{7}$$

Where - Δ^{noise} is the noise added to each point; β – coefficient.

3   Results and Discussion

A total of 2000 training ECGs and 400 noise free test ECGs have been simulated, as shown in Table 1. Another 1600 signals at different noise levels were also generated. Each signal is simulated at heart rate of 80 bpm, nc=40 cycles, Fs=1000 Hz. For each of those signals features V and R were computed. In order to keep the input features at the same scale, V was multiplied by 10^3 and R was divided by 10 prior entering the neural network.

Currently there is no commonly agreed method for determining the number of neurons and hidden layers of an ANN. In this paper we've decided to follow a method of trial and error. The used ANN was a simple ANN with two inputs and one output. It was trained using Levenberg-Marquardt Back-Propagation algorithm. A comparison between different ANN architectures is given in Table 2 and Table 3.

Table 1. Simulated data. $\alpha^{(2)}=0.5$.

Set type	Map type	$\alpha^{(1)}$	β	No of samples
Training	Logistic	0.01	0	500
Training	Logistic	0.03	0	500
Training	Henon	0.01	0	500
Training	Henon	0.03	0	500
Test	Logistic	0.01	0	100
Test	Logistic	0.03	0	100
Test	Henon	0.01	0	100
Test	Henon	0.03	0	100
Test	Logistic	0.01	0.001	100
Test	Logistic	0.03	0.001	100
Test	Henon	0.01	0.001	100
Test	Henon	0.03	0.001	100
Test	Logistic	0.01	0.005	100
Test	Logistic	0.03	0.005	100
Test	Henon	0.01	0.005	100
Test	Henon	0.03	0.005	100
Test	Logistic	0.01	0.01	100
Test	Logistic	0.03	0.01	100
Test	Henon	0.01	0.01	100
Test	Henon	0.03	0.01	100
Test	Logistic	0.01	0.05	100
Test	Logistic	0.03	0.05	100
Test	Henon	0.01	0.05	100
Test	Henon	0.03	0.05	100

Table 2. Accuracy of ANNs for simulated test sets with 1 hidden layer and 2 hidden layers for different number of neurons per layer, in classifying a given signal into logistic and Henon chaos added to simulated data with $\alpha^{(1)}=0.03$, $\alpha^{(2)}=0.5$, $\beta=0$

Hidden Layers	Neurons per layer	Overall Accuracy	True Logistic	True Henon
1	1	82.0%	74.0%	90.0%
1	5	85.0%	77.0%	93.0%
1	10	85.0%	83.0%	87.0%
1	**15**	**85.0%**	**77.0%**	**93.0%**
2	3	86.0%	83.0%	89.0%
2	5	85.0%	77.0%	93.0%
2	7	84.5%	78.0%	91.0%
2	10	86.5%	83.0%	90.0%
2	15	85.5%	80.0%	91.0%

Table 3. Accuracy of ANNs for simulated test sets with 1 hidden layer and 2 hidden layers for different number of neurons per layer, in classifying a given signal into logistic and Henon chaos added to simulated data with $\alpha^{(1)}$=0.01, $\alpha^{(2)}$= 0.5, β=0

Hidden Layers	Neurons per layer	Overall Accuracy	True Logistic	True Henon
1	1	81.5%	76.0%	87.0%
1	5	82.5%	84.0%	81.0%
1	10	83.0%	79.0%	87.0%
1	**15**	**84.0%**	**86.0%**	**82.0%**
2	3	84.0%	83.0%	85.0%
2	5	83.0%	83.0%	83.0%
2	7	83.0%	82.0%	84.0%
2	10	83.5%	88.0%	79.0%
2	15	83.5%	84.0%	83.0%

Since the chaotic component coefficient α might carry information about illness severity, another classification tool is need based on the same input features, that will allow to distinguish between different values of $\alpha^{(1)}$ for the same type of chaos. Results found in Table 4. Table 5 represents accuracy of classification into logistic and Henon chaos for applied noise, with a chosen architecture of 1 hidden layer and 15 neurons.

The results, as presented in Table 2 and Table 3, demonstrate relatively high accuracy in test sets classification. Since it's preferable to have a simpler architecture, 1 hidden layer with 15 neurons will yield best results.

As a result it's possible to conclude that the two proposed features are sufficient for extracting and classification chaotic characteristics embedded into ECG by ANN. Also, as seen in Table 4, perfect classification was achieved based on the same features for $\alpha^{(1)}$ estimation.

As for robustness to noise, based on Table 5, the tool showed adequate results for $\beta\leq0.01$ value and especially for $\alpha^{(1)}$=0.03 but decrease in classification accuracy for $\beta>0.01$ values .This stresses the ability of the chosen ANN architecture to explore chaotic behavior and the need for additional features to improve the classification results for the presence of high levels of noise.

4   Conclusions

The diagnostic and physiological significance of the chaotic behavior of an ECG signal has been stressed, along with proves of chaotic components in ECG. But the presented tool's ability to not only detect chaos within ECG but also to distinguish between different chaotic components which might be responsible for different heart abnormalities, could serve for better ECG based diagnosis.

The suggested approach, of use of simulated ECG, enables to define ANN architecture composed by 2 neurons input layer, 15 neurons hidden layer and 1 neuron output layer. This architecture have yield promising results in classification of different chaotic components (logistic and Henon) embedded within the ECG.

Table 4. Accuracy of ANN for classifying a given test signal, $\beta=0$, $\alpha^{(2)}=0.5$ into $\alpha^{(1)}=0.01$ and $\alpha^{(1)}=0.03$

Chaos	Hidden Layers	Neurons per layer	Overall Accuracy	True $\alpha^{(1)}=0.01$	True $\alpha^{(1)}=0.03$
Logistic	1	1	100%	100%	100%
Henon	1	1	100%	100%	100%

Table 5. Robustness to noise, in classifying a given test signal into logistic and Henon chaos. 1 hidden layer 15 neurons.

$\alpha^{(1)}$	β	Overall Accuracy	True Logistic	True Henon
0.01	0.05	50%	0.0%	100.0%
0.01	0.01	59.5%	88.0%	31.0%
0.01	0.005	73.5%	75.0%	72.0%
0.01	0.001	80.0%	77.0%	83.0%
0.03	0.05	50.0%	100.0%	0.0%
0.03	0.01	83.5%	80.0%	87.0%
0.03	0.005	86.5%	82.0%	91.0%
0.03	0.001	82.0%	71.0%	93.0%

Overall accuracy of 84%-85% in test sets, in distinguishing between different chaotic maps and similar accuracy for noise levels of $\beta \leq 0.01$ have shown this tool to be affective in classification. Random noise, of $\beta \leq 0.01$, as an unavoidable component of any recorded ECG, have shown to have low to moderate impact on the results and therefore it's reasonable to assume that future work, that will include testing of raw data, will yield high classification capabilities without the need of major adjustments.

Therefore future work will include analysis of real data, for the classification of cardiac disorders, and exploration of additional features that might handle noise levels of $\beta \geq 0.01$ better.

Finally, it's important to mention that the proposed ANN architecture has universal capabilities and therefore might be used in other fields and applications, which require the analysis of the chaotic component alone on signals which include random noise as well.

References

1. Guevara, M.R., Glas, L., Shrier, A.: Phase locking, period-doubling bifurcations, and irregular dynamics in periodically stimulated cardiac cells. Science 214, 1350–1353 (1981)
2. Voss, A., Schulz, S., Schroeder, R., Baumert, M., Caminal, P.: Methods derived from nonlinear dynamics for analysing heart rate variability. Philosophical Transactions. Series A, Mathematical, Physical, and Engineering Sciences 367(1887), 277–296 (2009)
3. Guzzetti, S., Signorini, M.G., Cogliati, C., Mezzetti, S., Porta, A., Cerutti, S., Malliani, A.: Non-linear dynamics and chaotic indices in heart rate variability of normal subjects and heart-transplanted patients. Cardiovascular 6363(95), 441–446 (1996)

4. Loskutov, A.: Time series analysis of ECG: a possibility of the initial diagnostics. International Journal of Bifurcation and Chaos 17(10), 3709–3713 (2007)
5. Ho, K.K.L., Moody, G.B., Peng, C.K., Mietus, J.E., Larson, M.G., Levy, D., Goldberger, A.L.: Predicting survival in heart failure case and control subjects by use of fully automated methods for deriving nonlinear and conventional indices of heart rate dynamics. Circulation 96(3), 842–848 (1997)
6. Abibullaev, B., Seo, H.D.: A new QRS detection method using wavelets and artificial neural networks. Journal of Medical Systems 35(4), 683–691 (2011)
7. Owis, M.I., Abou-Zied, A.H., Youssef, A.-B.M., Kadah, Y.M.: Study of features based on nonlinear dynamical modeling in ECG arrhythmia detection and classification. IEEE Transactions on Biomedical Engineering 49(7), 733–736 (2002)
8. Übeyli, E.D.: Detecting variabilities of ECG signals by Lyapunov exponent. Neural Computing and Applications 18(7), 653–662 (2009)
9. Mitschke, F., Dämmig, M.: Chaos versus noise in experimental data. International Journal of Bifurcation and Chaos 3, 693–702 (1993)
10. Govindan, R.B., Narayanan, K., Gopinathan, M.S.: On the evidence of deterministic chaos in ECG: Surrogate and predictability analysis. Chaos 8(2), 495–502 (1998)
11. Gothwal, H., Kedawat, S., Kumar, R.: Cardiac arrhythmias detection in an ECG beat signal using fast fourier transform and artificial neural network. Journal of Biomedical Science and Engineering 4(4), 289–296 (2011)
12. Dokur, Z., Olmez, T.: Comparison of discrete wavelet and Fourier transforms for ECG beat classification. Electronics Letters 35(18), 1502–1504 (1999)
13. Bigger, J.T., Fleiss, J.L., Steinman, R.C., Rolnitzky, L.M., Schneider, W.J., Stein, P.K.: RR variability in healthy, middle-aged persons compared with patients with chronic coronary heart disease or recent acute myocardial infarction. Circulation 91(7), 1936–1943 (1995)
14. Karrakchou, M., Vibe-Rheymer, K., Vesin, J.M., Pruvot, E., Kunt, M.: Improving cardiovascular monitoring through modern techniques. IEEE Engineering in Medicine and Biology Magazine 15(5), 68–78 (1996)
15. Losada, R.: ECG. 1988-2002 The MathWorks, Inc.
16. McSharry, P., Clifford, G.: A dynamical model for generating synthetic electrocardiogram signals. IEEE Transactions on Biomedical Engineering 50(3), 289–294 (2003)
17. Goldberger, L.A., Amaral, L.A.N., Glass, L., Hausdorff, J.M., Ivanov, P.C., Mark, R.G., Mietus, J.E., Moody, G.B., Peng, C.K., Stanley, H.E.: Resource for Complex Physiologic Signals PhysioBank, PhysioToolkit, and PhysioNet: Components of a New Research. Circulation 101(23), e215–e220 (2000)

Automatic Landmark Location for Analysis of Cardiac MRI Images

Chrisina Jayne[1], Andreas Lanitis[2], and Chris Christodoulou[3]

[1] Coventry University,
Department of Computing,
Priory Street, Coventry, CV1 5FB, UK
[2] Department of Multimedia and Graphic Arts,
Cyprus University of Technology,
31 Archbishop Kyprianos Street, P.O. Box 50329, 3603 Lemesos, Cyprus
[3] Department of Computer Science,
University of Cyprus
75 Kallipoleos Avenue, P.O. Box 20537, 1678 Nicosia, Cyprus

Abstract. This paper addresses the problem of automatic location of landmarks used for the analysis of MRI cardiac images. Typically the landmarks of shapes in MRI images are located manually which is a time consuming process requiring human expertise and attention to detail. As an alternative a number of researchers use shape modelling and image search techniques for locating the required landmarks automatically. Usually these techniques require human expertise for initializing the search and in addition they require high quality, noise free images so that the image-based landmark location is successful. With our work we propose the use of neural network methods for learning the geometry of sets of points so that it is possible to predict the positions of all required landmarks based on the positions of a small subset of the landmarks rather than using image-data during the process of landmark-location. As part of our work the performance of neural network methods like Multilayer Perceptrons, Radial Basis Functions and Support Vector Machines is evaluated. Quantitative and visual results demonstrate the potential of using such methods for locating the required landmarks on endo-cardial and epicardial landmarks of the left ventricle of MRI cardiac images.

Keywords: MRI cardiac images, automatic landmarks location, neural networks.

1   Introduction

Cardiac magnetic resonance imaging (MRI) has attracted considerable research interest in the last fifteen years [1] [2] [3]. The cardiac MRI provides information useful for detection of abnormalities and problems with the heart function such as the ejection function (EF), left myocardium mass (MM), and stroke

C. Jayne, S. Yue, and L. Iliadis (Eds.): EANN 2012, CCIS 311, pp. 203–212, 2012.
© Springer-Verlag Berlin Heidelberg 2012

volume (SV) [4]. Calculation of these quantities depends on accurately delineating the endocardial and epicardial contours of the left ventricle (LV). Manual drawing of these contours is time consuming, prone to errors, intra- and inter- observer variability [5]. Therefore automating this process is highly desirable. Many researchers have investigated this challenging problem and suggested various approaches such as for example image processing techniques including thresholding and shape extraction [1], image-driven segmentation [6], active appearance models [7], subject-specific dynamical model [8] and dynamic programming [9]. Generating-shrinking neural network classifier has been applied to the problem of segmentation methods in short axis cardiac MR images in [10] to classify tissue points into three classes: lung, myocardium, and blood and combined with a spatiotemporal parametric modelling. Reviews on cardiac image analysis and on segmentation methods in short axis cardiac MR images are given in [2] and [5] respectively. It is noted [5] that despite the developments in solving the problem of automatic segmentation of MR cardiac images, the actual problem domain is still open for further developments, as indicated by open related challenges: Cardiac MR Left ventricle Segmentation Challenge, http://smial.sri.utoronto.ca/LV_Challenge/Home.html, 2009 and RV Segmentation Challenge in Cardiac MRI, http://www.litislab.eu/rvsc, 2012.

This paper explores the potential of applying standard neural network methods such as Multilayer Perceptrons (MLPs) [11], Radial Basis Functions (RBFs) [12] and an alternative to neural networks methods, Support Vector Machines (SVMs) [13], [14] to automatically locate the endocardial and epicardial landmarks of the left ventricle of short axis MRI cardiac images based on a small number manually located landmarks. The short-axis plane is the standard imaging plane that is perpendicular to the long (apex-base) axis [5]. In this study the publically available data set of short axis cardiac MRI images provided by the Department of Diagnostic Imaging of the Hospital for Sick Children in Toronto, Canada and the manual segmentations (http://www.cse.yorku.ca/~mridataset/), [3] are utilised for the experiments. The subjects in this data set displayed different heart abnormalities including a small number with left ventricle abnormality. More details about the data set are available in [3]. Unlike most approaches described in the literature, in our approach we do not use image data for finding the locations of the landmarks, but we capitalize on shape-constraints learned by the neural network for predicting the positions of the missing landmarks. As a result human experts need to locate only a small number of landmarks which are preferably located on high contrast areas and as a result it is trivial to locate them accurately. The positions of the rest of the points which may be located on blurred image areas can be automatically positioned. This approach has the benefits of requiring the least possible human involvement and at the same time it is not required to have high quality images in order to acheive reasonable accurate landmark localization.

The paper is structured as follows: Section 2 briefly describes the methods used, Section 3 presents the experiments, Section 4 discusses the results and Section 5 draws the conclusions.

2 Methods

In this section the basic theoretical background of each of the methods under investigation is presented.

2.1 Multilayer Perceptron Method (MLP)

The Multilayer Perceptron (MLP) [11] is the most widely used neural network architecture. Typically, it consists of three layers of neurons: input, hidden and output layers fully connected with adaptive weights. The input layer passes the input values through the hidden layer while the hidden and output layer neurons are active, i.e., they contain an activation function. The activation function of the hidden layer neurons is a smooth nonlinear function (e.g., sigmoid or hyperbolic tangent). The activation function in the output layer for regression problems is usually a linear function [15]. During training, the input patterns are propagated through the network and the weights are adjusted to minimise the sum-of-square error function using a gradient descent process. In this paper a MLP with the scaled conjugate gradient algorithm is used [16], which combines the model-trusted approach and the conjugate gradient approach. It uses a numeric approximation for the second derivatives (Hessian matrix) to reduce the computations in the line search used by the traditional conjugate gradient algorithm.

2.2 Radial Basis Functions (RBF)

The Radial Basis Functions (RBF) [12] architecture is similar to the MLP architecture consisting of input, hidden and output layers of neurons, but there are some differences. Firstly the outputs in the hidden layer are not the product of the input pattern vector and the weight vector. Each neuron in the hidden layer is a centre of a cluster in the input data space found using a clustering algorithm (e.g., k-means, [17]). Secondly the transfer function associated with each hidden neuron is known as a radial basis function typically a Gaussian curve, through which the Euclidean distance between the input vector and the centre vector is passed. In this work we use thin plate spline function as the transfer function which gives better results than when the Gaussian is used. Further optimisation was achieved by varying the number of hidden nodes. The output layer uses linear activation function and the weights between the hidden and output layer are adjusted using a gradient descent algorithm. The latter parameter could also be optimised through the use of other optimisation algorithms.

2.3 Support Vector Machine Regression(SVMR)

Support Vector Machines (SVM) developed by Vapnik [14] are an alternative of the traditional neural network approaches. Their main advantage is that their formulation incorporates minimisation of the structural risk [18] which minimises an upper bound of the expected risk. This approach provides a better ability to generalise, than using the traditional approach of minimising the empirical risk [18]. SVMs have been used for both classification and regression problems [14], [18], [19],[20]. In the context of the problem for automatic location of landmarks we have multioutput regression problem. However, the standard formulation of the SVM for regression considers only the single-output problem. Multioutput formulation of SVM for regression problems (M-SVR) is proposed in [21], [22] and [23] and applied for function approximation, nonlinear channel estimation in multiple-input multiple-output systems, and remote sensing biophysical parameter estimation. The M-SVR formulated in [22] and [23] extends the single-output SVR to multiple outputs by using the ϵ - insensitive cost function based on L_2 norm, thus considering all dimensions into a unique restriction giving a single support vector for all dimensions. This solves the problem of the large complexity if each dimension is considered independently and exploits the the dependences between variables [23]. The implementation of M-SVR [21], [22] and [23] is utilised in this work.

3   Experiments

The dataset of Cardiac MR images and their manual segmentations used for the experiments are publically available online at (http://www.cse.yorku.ca/~mridataset/ and at the web site of [3]. It consists of short axis MR image sequences from 33 subjects and a total of 7980 2D images. Each subject has image sequence of 20 frames and between 8 and 15 slices acquired along the long axis. The data set contains also 5011 manually segmented MR images, with a total of 10,022 endocardial and epicardial contours. This manual segmentation carried in[3] and made available at (http://www.cse.yorku.ca/~mridataset/ provides the ground truth for each image where both endocardial and epicardial contours of the left ventricle were visible. For each z position denoting the long axis slice number z, each temporal frame t, each contour is described by 32 landmark points given in x, y pixel coordinates. We represent each contour as 66 dimensional vector which consists of the 32 landmark x, y pixel coordinates and the last couple elements are equal to the x, y coordinates of the first point. Concatenating both 66 dimensional endocardial and epicardial contour vectors gives 132 dimensional vector describing both contours. For a specific position z we construct the data set of 660 patterns, where each pattern is the 132 dimensional vector. We use leave-one-out cross-validation procedure to evaluate the models thus utilising the patterns corresponding to all frames of 32 subjects for training and all frames of the remaining 1 subject for testing. For each method described in Section 2 we investigate 13 different test scenarios. The first three scenarios correspond to having 2, 3, 4 and 8 equally distanced landmarks visible on each contour. The remaining 9 scenarios correspond to having 10% - 90% randomly selected landmarks as visible on each contour. Figure 1 shows the test scenarios when 2, 3, 10% randomly selected and 8 landmarks are manually located on each contour. For each of the methods described in Section 2 we train models that correspond to all 13 scenarios and evaluate them using leave-one-out cross-validation procedure. The methods used are based on an one-to-many mapping scheme [24] where the input vectors consist of the visible landmark coordinates and the output vector contains the coordinates of all 66 landmarks. For each model the average percentage error of the normalised Euclidian distance between the recovered and the actual landmarks for the training and test subjects are calculated, and then the average across all models for a particular scenario and a particular method are derived. In this context the average percentage error is estimated as the mean distance between actual and recovered points divided by the largest distance between any pairs of points in the given shape. Before training the models we align the visible points with the corresponding mean points using the Procrustes shape alignment method [25] so that differences between the two sets of points due to different translation, scaling and rotation are eliminated. In order to benchmark the results we replace the coordinates of the missing points with the mean values of each missing point as derived from the training set. The replacement of the positions of the missing points with the geometrically normalised average points provides a performance benchmark during our experiments. This benchmark approach is denoted as **AVG** in the result tables and figures in the next section.

It should be noted that in the experiments described it is assumed that the correspondence of the visible points in relation with the 64 landmarks located on the endocardial and epicardial contours of the left ventricle is known. This is a reasonable assumption bearing in mind that in a future application of this method the locations of the visible points will be fixed taking into account the ease of locating those landmarks by experts.

4   Results and Discussion

Table 1 presents the quantitative results from the experiments for the different test scenarios. Figure 2 illustrates visual results for each method for one test subject corresponding to the scenarios from Figure 1. The basic AVG approach gives the worst results in the majority of the test scenarios but provides a starting point for comparison with the application of the neural network methods and the support vector machine. The performance of the MLP and RBF methods is mainly influenced by the number of hidden nodes. Increasing the number of hidden nodes leads to significant improvement of the training results but at the expense of the generalisation performance and the complexity of the model. The results for the MLP method presented in Table 1 are obtained with 50 hidden nodes, learning rate equal to 0.05 and number of training epochs equal to 300. The results for the RBF method are obtained with 100 hidden nodes, using thin-plate spline function as transformation function in each node. The training time of the MLP (in order of minutes) is significantly higher than the RBF training (in order of seconds). The performance of the M-SVR model is influenced

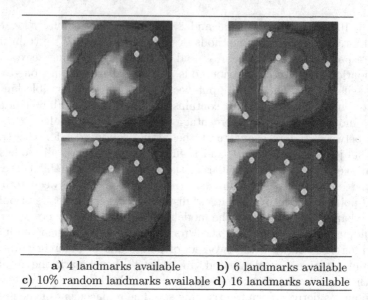

a) 4 landmarks available   **b)** 6 landmarks available
c) 10% random landmarks available **d)** 16 landmarks available

Fig. 1. Test scenarios for test subject 14, frame 12, z position 5., Manually located landmarks are shown in colour.yellow

mainly by the choice of the parameters C, ϵ and the kernel parameter. Parameter C determines the trade off between the model complexity and the degree to which deviations larger than ϵ are tolerated in optimisation formulation [26]. If C is large (infinity) then the objective is minimising the empirical risk without regard to the model complexity [26], [27]. The parameter ϵ controls the width of the ϵ - insensitive zone and the number of support vectors used to construct the regression function [28]. The results presented in Table 1 are obtained by experimenting with different values for these parameters. The kernel was set to linear, polynomial and radial basis functions. The best results are obtained with polynomial kernel of second degree, $epsilon = 0.0001$ and $C = 10000$ for the first three scenarios (3, 4 and 8 visible equally distant landmarks) and $C = 100000$ for the remaining test scenarios. The training time is in order of seconds.

5   Conclusion

A method for locating endocardial and epicardial landmarks of the left ventricle of MRI cardiac images is presented. Unlike other segmentation methods reported in the literature, the proposed method relies on the use of image data only for locating a small number of key landmarks. The positions of the remaining landmarks are predicted based on the anticipated geometry of the shapes learned through a training procedure. The training process is performed using neural network based methods. An experimental comparative evaluation using MLPs, RBFs and SVM regression is carried out in order to find the neural network

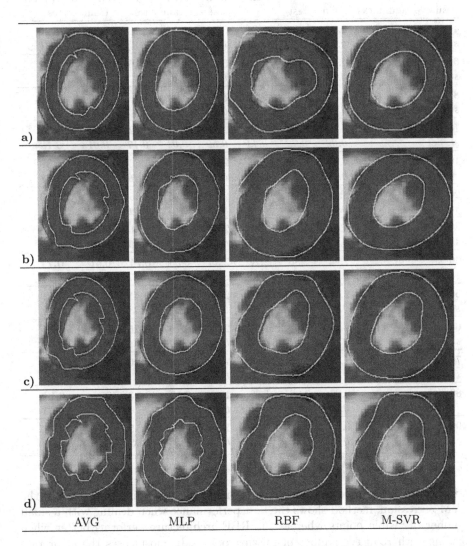

| AVG | MLP | RBF | M-SVR |

Fig. 2. Visual Results - Test subject 14, frame 12, z position 5, yellow contour based on automatically recovered landmarks, red contour based on manual landmarks. **a)** 4 landmarks available, **b)** 6 landmarks available, **c)** 10% random landmarks available, **d)** 16 landmarks available

Table 1. Mean % and standard deviation of normalised Euclidean distances between recovered and original landmarks for train and test sets,

Method/ available landmarks	AVG train	AVG test	MLP train	MLP test	RBF train	RBF test	M-SVR train	M-SVR test
4	4.93	4.99	3.64	3.76	2.66	3.78	3.06	**3.22**
st. dev	0.03	0.93	0.07	0.81	0.02	1.24	0.02	0.74
6	4.53	4.58	3.03	3.13	1.38	2.16	1.70	**1.82**
st. dev	0.03	0.85	0.25	0.62	0.09	0.73	0.01	0.39
8	4.29	4.34	2.87	3.04	0.99	1.46	1.26	**1.38**
st. dev	0.03	0.83	0.17	0.75	0.06	0.35	0.08	0.28
16	3.56	3.60	2.70	2.82	0.52	0.72	0.60	**0.67**
st. dev	0.02	0.72	0.14	0.59	0.04	0.17	0.01	0.13
10%	4.33	4.38	2.65	2.81	1.05	1.56	1.27	**1.39**
st. dev	0.03	0.82	0.10	0.59	0.04	0.40	0.08	0.28
20 %	3.95	4.01	2.49	2.62	0.95	1.35	1.10	**1.23**
st. dev	0.02	0.76	0.12	0.63	0.04	0.31	0.07	0.24
30 %	3.59	3.64	2.39	2.52	0.76	1.11	0.88	**1.03**
st. dev	0.02	0.71	0.08	0.57	0.07	0.29	0.05	0.25
40 %	2.84	2.88	2.22	2.35	0.53	0.74	0.61	**0.71**
st. dev	0.02	0.54	0.08	0.54	0.04	0.17	0.03	0.14
50 %	2.31	2.34	2.03	2.17	0.31	**0.41**	0.36	0.42
st. dev	0.02	0.45	0.07	0.48	0.04	0.08	0.02	0.07
60 %	1.89	1.91	1.91	2.04	0.19	**0.25**	0.24	0.28
st. dev	0.01	0.38	0.05	0.44	0.01	0.04	0.01	0.04
70 %	1.36	1.27	1.86	1.99	0.15	**0.20**	0.21	0.26
st. dev	0.01	0.30	0.04	0.41	0.01	0.03	0.01	0.04
80 %	0.87	0.88	1.82	1.94	0.10	**0.13**	0.15	0.18
st. dev	0.07	0.20	0.05	0.42	0.02	0.02	0.01	0.03
90 %	0.43	0.44	1.79	1.88	0.12	**0.15**	0.11	0.15
st. dev	0.03	0.10	0.05	0.37	0.03	0.03	0.01	0.02

architecture best suited to the aforementioned task. According to the results the SVM regression, method shows the best performance when dealing with a small number of visible points whereas the RBF architecture performs better when dealing with reduced number of missing points. In most cases the error rate is less than 2% when dealing with previously unseen examples, demonstrating the suitability of the proposed methods for this application. According to the quantitative and visual results, it is evident that the methods used learn both the within and intra person variability enabling the estimation of the complete endo -cardial and epicardial shape using a small number of landmarks. The importance of this work is due to the fact that based on this approach the human intervention required is limited both in terms of duration and in terms of the level of expertise required since the human operator can locate only few points located on positions that is trivial to locate even in the presence of noise. Based on the positions of easy-to-locate landmarks the overall endo-cardial and epicardial shape can be defined with reasonable accuracy.

Our future research directions involve the development of improved methods, the applications of the method to other application domains and the developments of an integrated tool that can be used for locating the endo-cardial and epicardial shape from MRI images or Ultra Sound Images. As far as the development and improvement of the method we plan to evaluate additional learning methods and different neural network architectures. Also the method will be combined with methods based on image evidence so that, where possible, the results of the shape prediction are locally refined using image-data. Also we plan to apply similar methods to other applications involving medical images or other relevant problems. For example we plan to use the proposed methodology for other problems involving missing data such as the 3D shape reconstructions using image evidence from a single view. Also in the future we plan to develop and evaluate an integrated application that can be used for extracting shape information from images using a small number of landmarks, so that the results of our research are utilized in real life applications.

References

1. Nachtomy, E., Cooperstein, R., Vaturi, M., Bosak, E., Vered, Z., Akselrod, S.: Automatic assessment of cardiac function from short-axis MRI: procedure and clinical evaluation. Magn. Reson. Imaging 16(4), 365–376 (1998)
2. Frangi, A.F., Niessen, W., Viergever, M.A.: Three-dimensional modelling for functional analysis of cardiac images: a review. IEEE Trans. Med. Imaging 20(91), 2–25 (2001)
3. Andreopoulos, A., Tsotsos, J.K.: Efficient and generalizable statistical models of shape and appearance for analysis of cardiac MRI. Med. Imag. Anal. 12, 335–357 (2008)
4. Eugene, C., Lin, M.D.: Cardiac MRI. Technical Aspects Primer (2011), http://emedicine.medscape.com/article/352250-overview
5. Petitjeana, C., Dacherb, J.-N.: A review of segmentation methods in short axis cardiac MR images. Medical Image Analysis 15, 169–184 (2011)
6. Cocosco, C.A., Niessen, W.J., Netsch, T., Vonken, E.J., Lund, G., Stork, A., Viergever, M.A.: Automatic image-driven segmentation of the ventricles in cardiac cine MRI. J. Magn. Reson. Imaging 28(2), 366–374 (2008)
7. Mitchell, S.C., Lelieveldt, B.P., van der Geest, R.J., Bosch, H.G., Reiber, J.H., Sonka, M.: Multistage hybrid active appearance model matching: segmentation of left and right ventricles in cardiac MR images. IEEE Trans. Med. Imaging 20(5), 415–423 (2001)
8. Zhu, Y., Papademetris, X., Sinusas, A.J., Duncan, J.S.: Segmentation of the Left Ventricle From Cardiac MR Images Using a Subject-Specific Dynamical Model. IEEE T. on Medical Imaging 29(3), 660–687 (2010)
9. Hong, L., Huaifei, H., Xiangyang, X., Enmin, S.: Automatic LeftVentricleSegmentation in Cardiac MRI Using Topological Stable-State Thresholding and Region Restricted Dynamic Programming. Acad. Radiol. (2012), http://dx.doi.org/10.1016/j.acra.2012.02.011

10. Stalidis, G., Maglaveras, N., Efstratiadis, S., Dimitriadis, A., Pappas, C.: Model based processing scheme for quantitative 4-D cardiac MRI analysis. IEEE Trans. Inf. Technol. Biomed. 6(1), 59–72 (2002)
11. Rumelhart, D.E., Hinton, D.E., Williams, R.J.: Learning representations by back-propagation errors. Nature 323, 533–536 (1986)
12. Powell, M.J.D.: Radial basis functions for multivariable interpolation: A review. In: IMA Conference on Algorithms for the Approximation of Functions and Data, pp. 143–167. RMCS, Shrivenham (1985)
13. Vapnik, V.: Estimation of Dependences Based on Empirical Data. Moscow, Nauka (1982) (in Russian); English Translation: Springer, New York (1979)
14. Vapnik, V.: The Nature of Statistical Learning Theory. Springer, New York (1995)
15. Bishop, C.M.: Neural Networks for Pattern Recognition. Oxford University Press, Oxford (1995)
16. Moler, M.: A scaled conjugate gradient algorithm for fast supervised learning. Neural Networks 6(4), 525–533 (1993)
17. Duda, R.O., Hart, P.E.: Pattern Classification and Scene Analysis. Wiley, New York (1973)
18. Gunn, S.R.: Support Vector Machines for Classification and Regression. Technical Report, Image Speech and Intelligent Systems Research Group, University of Southampton (1997)
19. Vapnik, V., Golowich, S., Smola, A.: Support vector method for function approximation, regression estimation, and signal processing. In: Mozer, M., Jordan, M., Petsche, T. (eds.) Neural Information Processing Systems, pp. 169–184. MIT Press, Cambridge (1997)
20. Smola, A.J., Schölkopf, B.: A tutorial on support vector regression. Stat. Comput. 14(3), 199–222 (2004)
21. Pérez-Cruz, F., Camps-Valls, G., Soria-Olivas, E., José Pérez-Ruixo, J., Figueiras-Vidal, A.R., Artés-Rodríguez, A.: Multi-dimensional Function Approximation and Regression Estimation. In: Dorronsoro, J.R. (ed.) ICANN 2002. LNCS, vol. 2415, pp. 757–762. Springer, Heidelberg (2002)
22. Sánchez-Fernández, M., de Prado-Cumplido, M., Arenas-García, J., Pérez-Cruz, F.: SVM multiregression for nonlinear channel estimation in multiple-input multiple-output systems. IEEE Trans. Signal Proc. 52(8), 2298–2307 (2004)
23. Tuia, D., Verrelst, J., Alonso, L., Pérez-Cruz, F., Camps-Valls, G.: Multioutput Support Vector Regression for Remote Sensing Biophysical Parameter Estimation. IEEE Geoscience and Remote Sensing Letters 8(4), 804–808 (2011)
24. Jayne, C., Lanitis, A., Christodoulou, C.: Neural network methods for one-to-many multi-valued problems. Neural Computing and Applications 20, 775–785 (2011)
25. Gower, J.: Generalized procrustes analysis. Psychometrika 40(1), 33–51 (1975)
26. Hastie, T., Tibshirani, R., Friedman, J.: The Elements of Statistical Learning. Data Mining, Inference and Prediction. Springer (2001)
27. Cherkassky, V., Shao, X., Mulier, F., Vapnik, V.: Model Complexity Control for Regression Using VC Generalization Bounds. IEEE T. on Neural Networks 10(5), 1075–1089 (1999)
28. Cherkassky, V., Ma, Y.: Practical selection of SVM parameters and noise estimation for SVM regression. Neural Networks 17(1), 113–126 (2004)

Learning of Spatio-temporal Dynamics in Thermal Engineering

Matthias De Lozzo[1,2], Patricia Klotz[2], and Béatrice Laurent[3]

[1] EPSILON Ingénierie
Technoparc 10, 10 rue Jean Bart, BP 97431, 31674 Labège Cedex, France
[2] ONERA, The French Aerospace Lab
BP74025, 2 avenue Edouard Belin, 31055 Toulouse Cedex 4, France
matthias.de_lozzo@onera.fr
[3] Institut de Mathématiques de Toulouse, INSA Toulouse
135 avenue de Rangueil, 31077 Toulouse Cedex 4, France

Abstract. Thermal engineering deals with the estimation of the temperature at different points and instants for a given set of boundary and initial conditions. For this, an analytic model replaces accurate but time-expensive numerical simulation models; it is independent of the boundary conditions and parameterized by the statistical learning of multidimensional temporal trajectories. This black-box model is a recursive neural network emulating the temperatures of interest over time from the only knowledge of initial conditions and exogenous variables.

The number of hidden neurons is selected by a non-asymptotic approach based upon the minimization of a penalyzed criterion. Methods like the slope heuristic and the dimension jump enable the calibration of the penalty constant in presence of a n-sample. In practice, their extrapolation to dependent data gives accurate results in the sense of the mean square error.

The surrogate model and the model selection are successfully applied to an industrial benchmark.

Keywords: recursive neural network, nonuniform time step, model selection, penalized criterion, thermal engineering.

1 Introduction

The context of this work is the estimation of the temperature at $n_{\mathbf{y}}$ points of an aeronautical equipment in steady and transient states. From the one hand, numerical simulation models are a solution for this problem based upon the spatio-temporal discretization of the nonlinear physics equations using finite elements or boundary elements methods for example. These models produce an accurate approximation of the temperatures (High Fidelity models) but they are time-expensive. From the other hand, compact thermal models [1], that is to say physical-based reduced models based upon a thermal-electrical analogy, are widely used in thermal analysis. These models quickly estimate the temperatures. Nevertheless they are less accurate (Low Fidelity models) than the first ones, in particular moving away from their associated boundary conditions [1].

C. Jayne, S. Yue, and L. Iliadis (Eds.): EANN 2012, CCIS 311, pp. 213–222, 2012.
© Springer-Verlag Berlin Heidelberg 2012

Consequently this paper deals with the construction of an accurate and boundary-conditions independent surrogate model $\hat{f}$ quickly emulating the temperatures in transient and steady state by means of the statistical learning of a dataset $\mathcal{A}$. This dataset is made up of n different numerical simulations, each of them associating to an inputs vector $\mathbf{u} \in \mathbb{R}^{n_u}$ (principally made up of boundary conditions) the corresponding temperatures vector $\mathbf{y} \in \mathbb{R}^{n_y}$ in steady state or its time-discretized evolution in transient state. The size n of $\mathcal{A}$ is small because of the restrictive cost of calls to the numerical simulation models.

Formulation of the Problem. More generally let $\mathbf{y}$ be a multidimensional quantity of interest and $\mathbf{y}^k := \mathbf{y}(t_k)$ its value at time t_k. We assume that $\mathbf{y}^k$ is function of the sets $\{\mathbf{y}(t) : t < t_k\}$ and $\{\mathbf{u}(t) : t < t_k\}$, that is to say the present state of $\mathbf{y}$ is caused by the past ones and the previous temporal values of $\mathbf{u}$. From this consideration we are interested in $\hat{\mathbf{y}}$, the estimation of the quantity of interest $\mathbf{y}$ by a surrogate model.

Moreover for the studied problems, no information is present before the initial time t_0 when are setted the initial conditions and appears the exogenous variable $\mathbf{u}$. So the estimation $\hat{\mathbf{y}}^k$ of $\mathbf{y}^k$ is function of the sets $\{\hat{\mathbf{y}}(t) : t_0 \leq t < t_k\}$ and $\{\mathbf{u}(t) : t_0 \leq t < t_k\}$.

In addition, $\mathbf{u}$ is not known continuously between t_0 and t_k but at some instants $t_0, t_1, ..., t_{k-1}$ depending on the sampling rates. Thus it is necessary to have $\hat{\mathbf{y}}^k$ function of $\{\hat{\mathbf{y}}(t) : t \in \{t_0, ..., t_{k-1}\}\}$ and $\{\mathbf{u}(t) : t \in \{t_0, ..., t_{k-1}\}\}$.

Thence, we are looking for a parametric and recursive surrogate model $\hat{f}$ parameterized by $\mathbf{w} \in \mathbb{R}^{n_w}$ of the form:

$$\hat{f} : \mathbb{R}^{n_u + n_y} \times \mathbb{R}^{n_w} \longrightarrow \mathbb{R}^{n_y}$$
$$(\mathbf{u}^{k-1}, \hat{\mathbf{y}}^{k-1}; \mathbf{w}) \longmapsto \hat{f}(\mathbf{u}^{k-1}, \hat{\mathbf{y}}^{k-1}; \mathbf{w}) =: \hat{\mathbf{y}}^k, \text{ where } \hat{\mathbf{y}}^0 := \mathbf{y}^0 \quad (1)$$

such that $\|\mathbf{y}^k - \hat{\mathbf{y}}^k\|_2^2$ is small, $\|\cdot\|_2$ being the euclidean norm on $\mathbb{R}^{n_y}$. The form (1) means that in order to estimate $\mathbf{y}$ at time t_k, the parametric function $\hat{f}$ can use as inputs no more than the exogenous variable $\mathbf{u}$ and the emulated $\hat{\mathbf{y}}$ at time t_{k-1}. Traditionally, a learning dataset $\mathcal{A} = \left(\mathbf{u}_i^{k_i}, \mathbf{y}_i^{k_i}\right)_{\substack{i \in \{1, ..., n\} \\ k_i \in \{0, 1, ..., K_i\}}}$ made up of n time-discretized trajectories of the couple $(\mathbf{u}, \mathbf{y})$ is used so as to find the vector $\mathbf{w}$ minimizing an adequacy criterion of the surrogate model $\hat{f}$. The trajectories of the dataset $\mathcal{A}$ own different lengths and time discretizations which are available for the use of $\hat{f}$.

In order to emulate a dynamic system by means of the learning dataset $\mathcal{A}$, many surrogate models based upon a recursive formulation are commonly present in the literature, e.g. time series [3], recurvise neural networks [4] and variational neural networks [2]. Some of them use uniformly time spaced data while others require more than the only knowledge of exogenous variables and outputs at the previous time. Consequently, it is not possible to directly use these tools: we have to modify them in order to satisfy the formulation (1).

Layout. In Section 2, a surrogate model of the form (1) is presented. New tools for model selection in the artificial neural network area are described in Section 3. Finally in Section 4, the surrogate model as well as the later tools are validated with an industrial benchmark.

2   A Multidimensional Temporal Surrogate Model

2.1   A Recursive Surrogate Model

The model (2) enables to fulfill the requirements (1):

$$\begin{cases} \hat{\mathbf{y}}^{k+1} = \hat{f}_N(\mathbf{u}^k, \hat{\mathbf{y}}^k, \delta_{k+1}; \mathbf{w}) \\ \hat{\mathbf{y}}^0 = \mathbf{y}^0 \end{cases} \tag{2}$$

where $\hat{f}_N$ is an artificial neural network and $\mathbf{y}_0$ the initial conditions. Among the inputs of $\hat{f}_N$, $\hat{\mathbf{y}}^k$ are the previous outputs generated by $\hat{f}_N$ and $\delta_{k+1} = t_{k+1} - t_k$ is the current time step. The latter is a way of building a model working with various time steps, which is required by the trajectories of the learning dataset $\mathcal{A}$. (2) is a kind of nonlinear time series based upon a frozen neural structure, that is to say only the last outputs and exogenous variables are used.

About Artificial Neural Networks. In presence of a nonlinear relationship of the form $\mathbf{y} = f(\mathbf{x})$ with $\mathbf{x} \in \mathbb{R}^{n_x}$, $\mathbf{y} \in \mathbb{R}^{n_y}$ and f unknown, f can be approached by a surrogate model $\hat{f}_N$, called artificial neural network [5] or multilayer perceptron (MLP), parameterized by $\mathbf{w} \in \mathbb{R}^{n_w}$:

$$\hat{f}_N(\mathbf{x}; \mathbf{w}) = \left(\hat{f}_N^{(j)}(\mathbf{x}; \mathbf{w}) \right)_{1 \leq j \leq n_y} = \left(\sum_{i=1}^{N} w_i^{(j)} h \left(\sum_{l=1}^{n_x} w_{il} x_l + w_{i0} \right) + w_0^{(j)} \right)_{1 \leq j \leq n_y} \tag{3}$$

h is the sigmoidal function defined by $\forall z \in \mathbb{R}, \ h(z) = 2/(1 + e^{-2z}) - 1$.

This metamodel has good properties including universal approximation and parsimoniousness. The second property is due to the nonlinearity of the outputs in the parameters w_{il}; thus, for a fixed accuracy of estimation and under the hypothesis that f is strongly nonlinear, a neural network requires less parameters than a polynomial, Fourier or wavelet expansion for example.

In our situation, $\mathbf{x}$ stands for $(\mathbf{u}^k, \hat{\mathbf{y}}^k, \delta_{k+1})$ whereas $\mathbf{y}$ represents $\hat{\mathbf{y}}^{k+1}$ and we have the relation $n_{\mathbf{x}} = n_{\mathbf{u}} + n_{\mathbf{y}}$.

2.2   Parametrization of the Surrogate Model

Given the dataset $\mathcal{A} = \left(\mathbf{u}_i^{k_i}, \mathbf{y}_i^{k_i} \right)_{\substack{i \in \{1, \dots, n\} \\ k_i \in \{0, 1, \dots, K_i\}}}$, the vector of parameters $\mathbf{w}$ is those minimizing $e_n = \sum_{i=1}^{n} \sum_{j=1}^{n_y} \sum_{k_i=1}^{K_i} (\hat{y}_{i,j}^{k_i} - y_{i,j}^{k_i})^2$ with $\mathbf{y}_i = (y_{i,1}, \dots, y_{i,n_y})'$; e_n is the mean square error (MSE), an adequacy criterion of the surrogate model.

The Levenberg-Marquart – LM – (see e.g. [5]) algorithm and the Resilient backPROPagation – iRPROP+ – [6] are optimization methods commonly used in the neural network community (toolbox/package **nnet** in **Matlab/R** for example) so as to find the parameters **w**. iRPROP+ implements a gradient descent based upon the signs of its components rather than on its values whereas LM is a mix between the Gauss-Newton and the gradient descent algorithms. We opt for LM to the detriment of IRPROP+ because LM gives better results in our case notably for a weak number of parameters, even if this method is more time-expensive. Both methods use the gradient of e_n according to **w**:

$$\nabla_{\mathbf{w}} e_n = \sum_{i=1}^{n} \sum_{j=1}^{n_\mathbf{y}} \sum_{k_i=1}^{K_i} \nabla_{\mathbf{w}} (\hat{y}_{i,j}^{k_i} - y_{i,j}^{k_i})^2 = 2 \sum_{i=1}^{n} \sum_{j=1}^{n_\mathbf{y}} \sum_{k_i=1}^{K_i} (\hat{y}_{i,j}^{k_i} - y_{i,j}^{k_i}) \nabla_{\mathbf{w}} \hat{y}_{i,j}^{k_i} \quad (4)$$

Because of the recursivity of the formulation (2), we use the backpropagation through time in order to obtain $\nabla_{\mathbf{w}} \hat{y}_{i,j}^{k_i+1}$ with $\frac{\partial \hat{y}_{i,j}^{k_i+1}}{\partial w}$ defined by:

$$\frac{\partial \hat{y}_{i,j}^{k_i+1}}{\partial w} = \frac{\partial \hat{f}_N^{(j)}(\mathbf{u}_i^{k_i}, \mathbf{x}, \delta_{k_i+1}; \mathbf{w})}{\partial w} \Bigg|_{\mathbf{x}=\hat{\mathbf{y}}_i^{k_i}} + \sum_{l=1}^{n_\mathbf{y}} \frac{\partial \hat{f}_N^{(j)}(\mathbf{u}_i^{k_i}, \hat{\mathbf{y}}_i^{k_i}, \delta_{k_i+1}; \mathbf{w})}{\partial \hat{y}_{i,l}^{k_i}} \frac{\partial \hat{y}_{i,l}^{k_i}}{\partial w} \quad (5)$$

The gradient $\nabla_{\mathbf{w}} \hat{y}_{i,j}^{k_i+1}$ is function of the previous gradient $\nabla_{\mathbf{w}} \hat{y}_{i,j}^{k_i}$ which implies to store this quantity (see Fig. 1).

The backpropagation through time allows simulating a time-discretized trajectory of $\mathbf{y}_i$ governed by the $\mathbf{u}_i$ one and starting at point $\mathbf{y}_i^0$. Without the second term in (5) and setting $\mathbf{x} = \mathbf{y}_i^{k_i}$ in the first one, the optimization procedure would lead to a one-step ahead predictor: the neural network $\hat{f}_N$ would need to use $\mathbf{y}_i^k$ as input instead of $\hat{\mathbf{y}}_i^k$ at each instant t_{k_i} in order to have $\hat{\mathbf{y}}_i^k \approx \mathbf{y}_i^k$. In this case, the surrogate model could not predict the trajectory of $\mathbf{y}_i$ autonomously, that is to say without the help of the vector $\mathbf{y}_i$ at the previous instants.

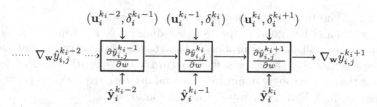

Fig. 1. Representation of the backpropagation through time (BPTT)

In this optimization algorithm in order to initialize the vector **w**, the Nguyen-Widrow method [7] is used so that the active regions of the neurons will be distributed approximately evenly over the input space.

Lastly the number of hidden neurons N in (2)-(3) is those associated to the more robust surrogate model $\hat{f}_N$, that is to say the $\hat{f}_N$ minimizing an approximation of the generalization error which is the mean of $\|\mathbf{y} - \hat{\mathbf{y}}\|_2^2$ over the exogenous inputs $\mathbf{u}$, the time-steps δ, the initial conditions $\mathbf{y}^0$ and the length K.

2.3   Intrinsic Time Step and Steady State

In some cases, the sampling rates of the learning dataset are lower than those of the physical phenomenon and the surrogate model (2) can take this into consideration. To illustrate this point, let us suppose that the phenomenon of interest has a characteristic time equal to δ_{char} where $\delta_{\text{char}} = \gamma.\delta_{\text{samp}} \leq \delta_{\text{samp}}$ with $0 < \gamma \leq 1$ and $1/\delta_{\text{samp}}$ the sampling frequency of the learning dataset. Then, a possible solution to emulate the physical phenomenon with a better fidelity is to use the model (2) with the intrinsic time step $\delta := \frac{\delta_{\text{samp}}}{\lceil \gamma^{-1} \rceil} = \sup\{\breve{\delta} \in \mathbb{R}_+ \text{ s.t. } \breve{\delta} = \delta_{\text{samp}}/k, \ k \in \mathbb{N}^* \text{ and } \breve{\delta} \leq \delta_{\text{char}}\}$ and with $\mathbf{u}^k$ as a linear interpolation between the last and next known values of $\mathbf{u}$ if $\mathbf{u}^k$ is unknown in the dataset. During the learning step, while the surrogate model $\hat{f}_N$ iterates with these δ and $\mathbf{u}^k$, its outputs computed at times present in the discretized trajectories of $\mathcal{A}$ contribute to the calculation of e_n.

Moreover in steady state, this model enables the prediction of the outputs $\mathbf{y}^\infty$ using constant exogenous variables, that is to say $\mathbf{u}^k := \mathbf{u}, \ \forall k$. In this way, for a fixed precision ε, starting with $\hat{\mathbf{y}}^0 = \mathbf{y}^0$ and repeating a finished but undeterminate number of times the step $\hat{\mathbf{y}}^{k+1} = \hat{f}_N(\mathbf{u}^k, \hat{\mathbf{y}}^k, \delta_{k+1}; \mathbf{w})$, there exists a time $t_{k'}$ when $\forall k \geq k', \|\hat{\mathbf{y}}^{k+1} - \mathbf{y}^{k+1}\|_2 \leq \varepsilon$. In other words, the outputs $\hat{\mathbf{y}}^{k+1}$ of the surrogate model (2) converge to the ones of the steady state $\mathbf{y}^\infty$ associated to the exogenous inputs $\mathbf{u}$.

Nevertheless, it is advised to learn some transient data reaching steady states so as to be able to emulate steady state temperatures in a prediction phase; otherwise, in spite of a convergence of $\hat{\mathbf{y}}^k$, it is possible that the distance between $\hat{\mathbf{y}}^k$ and $\mathbf{y}^\infty$ remains greater than the specified ε.

Finally, taking into consideration the remarks about the intrinsic time step, δ_{k+1} is considered as constant and equal to the characteristic time δ_{char} of the physical phenomenon. This time step is given by an expert or considered equal to the lowest met in the learning dataset in the absence of a such information.

In simple problems, it is easy to select an appropriate number of hidden neurons N leading to a robust surrogate model $\hat{f}_N$ by the means of the hold-out method, a validation procedure based upon the minimization of the mean square error associated to a test dataset. In the situation where the number of recourses to the numerical simulation model is limited because of the computational time, other methods of model selection have to be considered and automated.

3   A Non-asymptotic Model Selection

The augmentation of the number of hidden neurons N in an artificial neural network reduces continuously the learning error whereas the generalization error raises after a period of decrease. This phenomenon corresponds to the bias-variance tradeoff and it is important to approach the number of hidden neurons minimizing the generalization error.

Various tools like hold-out validation and K-fold cross-validation allow this goal to be reached under the hypothesis that the error associated to these methods is closed to the generalization error. Nevertheless, the first method implies putting aside some examples from the dataset, in such a way that the size of the learning dataset is reduced; for small dataset, this is not acceptable and the selection results are dependent of the test dataset. In addition, the cross-validation requires the construction of K times more neural networks, which represents a considerable limit if neural networks parameterization is time expensive.

Consequently, we need a tool using the totality of the learning dataset without constructing more surrogate models than the necessary number during the learning phase. Penalized criterions are a solution fulfilling these expectations.

3.1   Use of a Penalized Criterion

Let $\mathcal{A} = ((\mathbf{X}_1, Y_1), ..., (\mathbf{X}_n, Y_n)) \in \Xi^n$ be a n-sample where $\Xi \subset \mathbb{R}^{n_\mathbf{x}} \times \mathbb{R}$ and $(\mathbf{X}_1, Y_1), ..., (\mathbf{X}_n, Y_n)$ are i.i.d.[1]$\sim P$. Let $Y_i = s(\mathbf{X}_i) + \varepsilon_i$, $i \in \{1, ..., n\}$ be an homoscedastic regression model where $\varepsilon_1, ..., \varepsilon_n$ are i.i.d. such that $\mathbb{E}[\varepsilon_1/\mathbf{X}_1] = 0$ and $\mathbb{E}[\varepsilon_1^2/\mathbf{X}_1] = \sigma^2$. We would like to estimate s thanks to the n-sample $\mathcal{A}$ and a collection of models $(S_m)_{m \in \mathcal{M}}$ in $\mathcal{S}$ to specify, with $\mathcal{M} = \{1, 2, ..., M\}$. In the present context, S_m is the set of neural networks with N_m hidden neurons, where N_m maps monotonously $m \in \mathcal{M}$ to $N_m \in \mathbb{N}^*$ (e.g. $N_1 = 1, ..., N_5 = 5, N_6 = 10, N_7 = 15, ..., N_{10} = 30$):

$$S_m = \left\{ \hat{f}_{N_m}(\mathbf{x}; \mathbf{w}) = \sum_{i=1}^{N_m} w_i h \left(\sum_{j=1}^{n_\mathbf{x}} w_{ij} x_j + w_{i0} \right) + w_0 \text{ s.t. } \mathbf{w} \in \mathbb{R}^{n_w} \right\} \quad (6)$$

and $\mathcal{S}$ is the set of artificial neural networks. Let $\gamma(t, (\mathbf{x}, y)) = (y - t(\mathbf{x}))^2$ be the least square contrast. We look for a $s_m \in S_m$ which minimizes the generalization error $\mathcal{R}(t) = \int_\Xi (y - t(\mathbf{x}))^2 dP(\mathbf{x}, y)$. Unfortunately, $\mathcal{R}(t)$ is function of the unknown P. So we take a $\hat{s}_m \in S_m$ minimizing the empirical contrast $\gamma_n(t, \mathcal{A}) = \frac{1}{n} \sum_{i=1}^n (y_i - t(\mathbf{x}_i))^2$ over S_m. Then, in the collection of estimators $(\hat{s}_1, ..., \hat{s}_M)$, the objective is to approach the one minimizing $\mathcal{R}(t)$ which is the Oracle: it is a benchmark we cannot reach it because of its dependence to P. To do so, we look for $\hat{m}$ minimizing the penalyzed criterion $\gamma_n(\hat{s}_m, \mathcal{A}) + \text{pen}(m)$ so as to select $\hat{s}_{\hat{m}}$ from the collection of estimators. $\text{pen}(m)$ is a penalty term function of the number of observations n and of the properties of the set S_m. In Arlot $et\ al.$ [8], the ideal penalty is of the form $\kappa_{\text{id}} \text{pen}_{\text{shape}}(m)$ where κ_{id} is a numerical constant most of the time unknown. In the case of neural networks, Barron $et\ al.$ [9] propose the shape $\text{pen}_{\text{shape}}(m) = \frac{N_m(n_\mathbf{x}+1)}{n} (1 + \log(RQ(1 + n/(N_m(n_\mathbf{x}+1)))))$ where R (resp. Q) is an upper bound of $\sum_{j=1}^{n_\mathbf{x}} |w_{ij}|$ (resp. of $\sum_{i=1}^{N_m} |w_i|$). In our study we use the simplier form $\text{pen}_{\text{shape}}(m) = \frac{N_m(n_\mathbf{x}+1)}{n}$ which gives good results in practice.

To summarize, the use of a penalized criterion is made up of two steps:

1. Calculate $\hat{s}_1, ..., \hat{s}_M$ where $\hat{s}_m$ minimizes $\gamma_n(t, \mathcal{A})$ over S_m.
2. Select $\hat{s}_{\hat{m}}$ where $\hat{m}$ minimizes $\gamma_n(\hat{s}_m, \mathcal{A}) + \kappa_{\text{id}} \frac{N_m(n_\mathbf{x}+1)}{n}$.

---

[1] "$Z_1, ..., Z_n$ are i.i.d.$\sim P$" means that $Z_1, ..., Z_n$ are independent and identically distributed random variables (i.i.d.) with law P.

The dimension jump and the slope heuristic [10] allow the calibration of the constant κ_{id} necessary for Step 2. These methods are implemented in the R/Matlab toolbox/package CAPUSHE [10] and the dimension jump is the most employed. Both go by the fact that if $\hat{m}(\kappa)$ is the m minimizing $\gamma_n(\hat{s}_m, \mathcal{A}) + \kappa \frac{N_m(n_x+1)}{n}$, then:

- for small values of κ, $N_{\hat{m}(\kappa)}$ is in the highest values of N_m because the penalized criterion decreases as the complexity N_m increases;
- for values of κ greater than $\kappa_{\min}$, the penalized criterion decreases and then increases with this complexity because after having reached a certain precision, the bias is stable and the variance increases, that is to say the penalty term takes the advantage over the empirical constrast.

3.2 The Slope Heuristic and the Dimension Jump

The Dimension Jump. The first method supposes that the selected number of hidden neurons N_m minimizing $\gamma_n(\hat{s}_m, \mathcal{A}) + \kappa \text{pen}_{\text{shape}}(m)$ is very important for $\kappa < \hat{\kappa}$ and almost stable for $\kappa > \hat{\kappa}$. In others words, the application $\kappa \to N_{m(\kappa)}$ should possess an abrupt jump around $\hat{\kappa}$ and κ_{id} is estimated by $2\hat{\kappa}$. The following algorithm describes the three steps of this method:

1. Compute, for all $\kappa > 0$, $m(\kappa) \in \text{argmin}_{m \in \mathcal{M}}[\gamma_n(\hat{s}_m, \mathcal{A}) + \kappa \text{pen}_{\text{shape}}(m)]$.
2. Find $\hat{\kappa}$ such that the number of hidden neurons $N_{m(\kappa)}$ is "large" if $\kappa < \hat{\kappa}$ and has a reasonable order otherwise.
3. Select $\hat{m} = m(2\hat{\kappa})$.

In step(2), the meaning to give to "large" is related to the dimension jump: for all κ_1 and κ_2 such that $\kappa_2 = \inf\{\kappa \text{ s.t. } \kappa > \kappa_1 \text{ and } N_{m(\kappa_1)} > N_{m(\kappa)}\}$, $\hat{\kappa}$ is the κ_1 maximizing the difference $N_{m(\kappa_1)} - N_{m(\kappa_2)}$.

The Slope Heuristic. The second method relies on a robust linear regression of the couples of points $\{(\text{pen}_{\text{shape}}(m), -\gamma_n(\hat{s}_m)) \text{ s.t. } \text{pen}_{\text{shape}}(m) \geq p_{\text{thres}}\}$ where p_{thres} is a threshold beyond whose the points $\{\text{pen}_{\text{shape}}(m), -\gamma_n(\hat{s}_m)\}$ seem to be aligned. Then twice the slope of this linear part is a good estimator of κ_{id}. The estimated value, and so the selected number of hidden neurons, are sensitive to the regression method and above all to the choice of p_{thres}. These considerations are adressed in CAPUSHE [10] and the user can choose between a default and a manual configurations sometimes leading to significantly different models.

The approach of penalized criterion minimization makes the hypothesis of i.i.d. observations in $\mathcal{A}$. For toy functions in steady state such as the Rosenbrock function, this hypothesis is verified and the number of hidden neurons selected by the slope heuristic and the dimension jump correspond to the Oracle one. Unfortunately in the context of trajectories, the hypothesis of independence is unsatisfied if we consider $\mathbf{y}^k$ and $\mathbf{y}^{k+1}$ as outputs vectors associated to two different observations in a sample and not as two pieces of a same observation. Nevertheless, for a simple problem based upon the 1D heat equation in transient state, this approach gives good results and generalizes these model selection approach to more complex situations.

These methods of model selection coupled to the surrogate model (2) are used with an industrial benchmark in thermal engineering.

4   Application to the Industrial Benchmark

Description. The studied aeronautical equipment is a switchgear cubicle (see Fig. 2) made up of exterior radiators allowing the evacuation of the power generated by two autotransformators, an autotransformator power and a diode module and an exterior pierced cage for the circulation of exterior air in the PCB region. The exterior air gets in the cubicle through holes situated in the radiators, warms up and leaves the cupboard by means of an air extractor: the system is in forced convection. During extractor failures or when the extraction flow is low, the exterior air gets in through the down holes, warms up and a part or the totality leaves the cupboard by means of the up holes of a radiator: the system is in natural convection and this represents a physical behaviour completely different. Engineers are interested in the temperature evolution at $n_y = 27$ points situated on the dissipative components, the exterior faces, the air extractor and the up holes, using a model accurate for all convection situations.

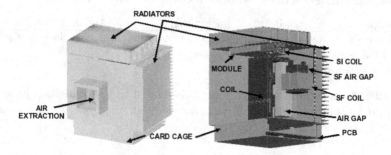

Fig. 2. Outside and inside of the switchgear cubicle

Modelization. The learning dataset $\mathcal{A}$ is made up of $n = 12$ trajectories obtained with a numerical simulation model using the software Flotherm. The time-independant exogenous inputs vector $\mathbf{u} \in \mathbb{R}^4$ is defined by u_1 – the input voltage (A), u_2 – the ambient temperature (Celsius), u_3 – the ambient pressure (mbar) – and u_4 – the flow of the extractor (g.s^{-1}) (see Fig. 3). The time step has to be taken into account because for a given trajectory it is scattered in a specific interval and takes different values; it is u_5, the fifth input. Additionally trajectories own different lengths as mentioned in Fig. 3.

The surrogate model (2) in Section 2 is used so as to build temperature predictors at the $n_y = 27$ points according to the values of the inputs vector $\mathbf{u}$ and using all the dataset $\mathcal{A}$. To do this, we use a collection $\mathcal{M}$ of $M = 13$ models S_m where the number of neurons N_m maps from $\mathcal{M}$ to $\{1, 2, 3, 4, 5, 10, 15, 20, 25, 30, 35, 40, 45, 50\}$ and for each N_m, a surrogate model minimizing e_n is created; the best number of hidden neurons in the sense of robustness and generalization is obtained by the means of the slope heuristic and jump dimension presented in Section 3.

Example	u_1	u_2	u_3	u_4	Length (s)
1	150	55	812	32	1800
2	150	20	117	5	600
3	450	70	1013	40	300
4	450	30	1013	25	300
5	750	70	1013	40	30
6	750	30	1013	25	30

Example	u_1	u_2	u_3	u_4	Length (s)
7	370	70	1013	20	300
8	370	30	1013	12.5	300
9	620	70	1013	20	30
10	620	30	1013	12.5	30
11	300	20	1013	0	60
12	450	20	1013	0	30

Fig. 3. Exogenous inputs and length of times for the learning dataset $\mathcal{A}$

Results. Globally the learning of the twelve trajectories gives accurate results with a neural network associated to a learning error $e_n^{1/2} = 5,2.10^{-3}$ for output data scaled to $[0,1]$ and a learning error of 0.6°C for unscaled output data. Furthermore 50% of the relative errors are between -0.3 and 0.3%. Fig. 4 presents the 12 trajectories from the learning dataset and their prediction for the output "SF coil": from a macroscopic point of view, the outputs and their predictions are confused whereas zooming in, some small oscillations appear which can be caused by some problem of optimization.

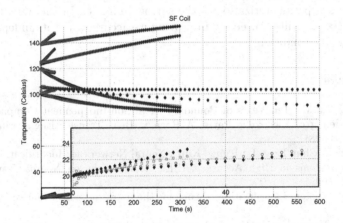

Fig. 4. Predictions (red) of the 12 trajectories (black) for the SF coil

The selected model comprises 30 hidden neurons according to the jump dimension and the slope heuristic with default configuration. After having superposed the curves from the learning dataset with their predictions for the different values of N_m, this value of 30 seems to be correct.

Finally other optimization methods could be discussed as well as the construction of a neural network by output behaviour so as to deal with weaker problems and avoid delicate situations involving a supervisory control. Indeed for a smaller number of outputs (e.g. 11 among 27), the number of parameters is smaller, previous oscillations do not occur and the slope heuristic can be less sensitive to the choice of threshold p_{thres} (see Subsection 3.2).

5   Conclusion

In a first step we presented a surrogate model based upon a recursive neural network for the prediction of temporal trajectories with multiple time steps. This model can learn data with discretization time steps greater than the characteristic one. Moreover it can estimate the quantity of interest in steady state although having a transient-state formulation. In a second step we applied an non-asymptotic penalized criterion to the neural network case, so as to select the number of hidden neurons minimizing the bias-variance tradeoff. Finally we showed that for an industrial thermal benchmark, the neural network whose number of hidden neurons has been selected by the methods associated to the penalized criterion gives accurate results for all convection situations given by the engineers, which was an industrial specification.

Beyond the current paper, this surrogate model will be a piece of a multiscale system and will interact with numerical simulation, compact thermal or surrogate models from the lower and upper levels. This is a reason for its necessary independence of the boundary conditions. Moreover in a multi-fidelity framework additional datasets with bigger sizes and lower fidelities will be available in order to facilitate the parameterization of the surrogate model. In this way, the high fidelity dataset $\mathcal{A}$ will be used for improving its accuracy and its independence of the boundary conditions.

References

1. Lasance, C.J.M.: Ten Years of Boundary-Condition-Independent Compact Thermal Modeling of Electronic Parts: A Review. Heat Transfer Engineering 29(2), 149–168 (2008)
2. Wang, Y.J., Chin-Teng, L.: Runge-Kutta Neural Network for Identification of Dynamical Systems in High Accuracy. IEEE Transactions on Neural Networks 9, 294–307 (1998)
3. Ljung, L.: System identification: theory for the user, 2nd edn. Prentice Hall PTR (1999)
4. Narendra, K.S.: Identification and control of dynamical systems using neural networks. IEEE Transactions on Neural Networks 1, 4–27 (1990)
5. Bishop, C.M.: Neural Networks for Pattern Recognition. Oxford University Press (1995)
6. Igel, C., Hüsken, M.: Improving the Rprop Learning Algorithm. In: Proceedings of the Second International Symposium on Neural Computation (2000)
7. Nguyen, D., Widrow, B.: Improving the learning speed of 2-layer neural networks by choosing initial values of the adaptive weights. In: Proceedings of the International Joint Conference on Neural Networks, vol. 3, pp. 21–26 (1990)
8. Arlot, S., Massart, P.: Data-driven Calibration of Penalties for Least-Squares Regression, vol. 10, pp. 245–279 (2009)
9. Barron, A., Birgé, L., Massart, P.: Risk bounds for model selection via penalization. Probability Theory and Related Fields 113, 301–413 (1999)
10. Baudry, J.-P., Maugis, C., Michel, B.: Slope heuristics: overview and implementation. Statistics and Computing 22(2), 455–470 (2012)

Neural Adaptive Control in Application Service Management Environment

Tomasz Sikora and George D. Magoulas

Department of Computer Science and Information Systems,
Birkbeck, University of London, Malet Street, London WC1E 7HX
{tomasz.sikora,gmagoulas}@dcs.bbk.ac.uk

Abstract. This paper presents a method and a framework for adaptive control in Application Service Management environments. The controlled system is treated as a "black-box" by observing its operation during normal work or load conditions. Run-time metrics are collected and persisted creating a Knowledge Base of actual system states. Equipped with such knowledge we define system inputs, outputs and effectively select high/low Service Level Agreements values, and good/bad control actions from the past. On the basis of gained knowledge a training set is constructed, which determines the operation of a neural controller deployed in the application run-time. Control actions are executed in the background of the current system state, which is then again monitored and stored extending the states repository, giving views on the appropriateness of the control, which is frequently evaluated.

Keywords: Application Service Management, Adaptive Controller, Service Level Agreement, Knowledge Base, Neural Networks, Performance, Metrics.

1 Introduction

Application Service Management (ASM) is a discipline which focuses on monitoring and managing performance and quality of service in complex enterprise systems. An ASM controller needs to react adaptively to changing system conditions in order to optimize a defined set of Service Level Agreements (SLA), which play the role of objective functions.

In this paper, we propose a neural network controller for ASM and explore the potential of this approach. The learning framework is based on the natural concept of comparison and learning on successes and mistakes. We develop a software framework equipped with simple statistical tools for evaluating the system. The control system takes into account current and historical system states information and generates actions for the controller. These control actions are approximated by a trained pair of Neural Networks (NN).

The software framework is able to change internal elements of runtime execution by considering control actions in background of flexible SLA definitions and

C. Jayne, S. Yue, and L. Iliadis (Eds.): EANN 2012, CCIS 311, pp. 223–233, 2012.

resources as part of the current system state. In the era of "big-data" we use an approach where all metrics are collected, e.g. actions, resources and control actions - creating a knowledge base that operates as system states repository - and reactions information after control actions. In this study the controller is only equipped with a termination actuator eliminating expensive actions (not resources tuning), and can adapt to changing conditions according to modifiable SLA definitions, without using a model of the system. The general objective is to optimize declared SLA functions values without using predictors and forecasting of service demands and resources utilization.

The next section provides an overview of previous work in ASM control. Then, Section 3 formulates the problem whilst Section 4 presents our architecture and framework. Simulation results are presented in Section 5, and the paper ends with conclusions and future work in Section 6.

2   Previous Work in ASM Control

The adaptive control of services has been the subject of substantial focus in the last decade. Perekh et al. researched the area of adaptive controllers using control theory and standard statistical models in the background of SLA [2]. More pragmatic approaches were studied in [4][5] [8], where an ASM performance control system with service differentiation used classical feedback control theory to achieve overload protection. Hellerstein and Perekh et al. introduced a comprehensive application of standard control theory to describe the dynamics of computing systems and apply system identification to the ASM field [12]. Fuzzy control and neuro-fuzzy control were proposed in [12] as promising for adaptive control in the ASM field which may also be an interesting area for further research work.

Since then many different methods of control and various techniques for actuators in ASM have been proposed, e.g. event queues for response times of complex Internet services [11], model approximation and predictors with SLA specification for different operation modes [14], observed output latency and a gain controller for adjusting the number of servers [15]. More recent works tend to focus on performance control in more distributed environments; e.g., virtualized server environments with loopback control of CPU allocation to multiple applications components in order to meet response time targets [16], cloud based solutions with QoS agreements services [18], and energy saving objectives and effective cooling in data centers [19][20].

Despite recent progress in the use of control theory to design controllers in the ASM field, the use of neural networks and their ability to approximate multidimensional functions defining system states remains under explored. Bigus nearly twenty years ago applied neural networks and techniques from control systems theory to system performance tuning [1]. Similarly to our approach he used several system performance measures such as devices utilization, queue lengths, and paging rates. Data was collected to train neural network performance models.

NNs were trained online to adjust memory allocations in order to meet declared performance objectives. Bigus and Hellerstein extended that approach in [3] and proposed a framework for adjusting resources allocation, where still no complex SLA definitions were used. NNs were used as classifiers and predictors but no extensive knowledge base was incorporated.

3   Elements of Control Theory in ASM Field

A controller working in an ASM environment running within an enterprise class system can be modeled as a dynamic system. Given a system of functions: $\mathbf{C}$: $\mathbb{R}^{n+1} \rightarrow \mathbb{R}^{n+1}, n = k + m$ in time domain, where: k is the number of actions, m is the number of system resources. All run–time characteristics of the system are defined by a set of resources and input/output actions – the current values define the *system state*.

Let us assume that $\mathbf{r}$ is the set of all system *resources*, which fully describe the system conditions. We will mainly focus on those resources which are measurable $r = \{r_1, ..., r_i, ..., r_m\}, r \subseteq \mathbf{r}$, where the i-th resource utilization is a scalar function of other measurable resources and actions in time $r_i(t) = \rho(r, a, t), r_i(t) \in [0, \infty]$. Only some measurable system resources can be controllable $r_c \subseteq r$. There are *synthetic resources* (abstract resources), which can be derived/calculated on the basis of other system parameters (resources and actions functions) $r_s \subseteq r \wedge r_s \cap r_c = \emptyset : r_s(t) = \rho_s(r(t), a(t))$, therefore cannot be controlled directly but can be used in later processing for control actions induction.

Let us assume that $\mathbf{a}$ is the set of all system *actions*, we focus on only measurable actions $a = \{a_1, ..., a_i, ..., a_k\}, a \subseteq \mathbf{a}$, where the i-th action execution is a function of resources utilized $a_i(t) = \alpha(r, a, \gamma_i(t), \mathbf{P_i}), a_i(t) \in \{0, 1\}$, triggered by the i-th event impulse with vector input parameters $\mathbf{P_i} = [P_1, ...P_X], \mathbf{P_i} \in \mathbb{R}^X$, which depends on available system resources, and consequently on other actions executions – using a shorter form $a_i^{(\mathbf{P_i})}(t) = \alpha(r, a, t)$. Similarly to resources definitions, some actions are controllable $a_c \subseteq a$. The controller can readjust execution time characteristics during the run-time of the actions, including termination, but any other change of functionality is not allowed. Many action instances of the same action type can execute concurrently, the i-th action type can have many instances executing concurrently, for different values of input parameters.

A *Service Level Agreement* (SLA) in most cases is a function of an action's execution time, but can be also built as functions of system resources $f_{SLA_i} = f_{SLA_i}(a, r, t), t \in \mathbb{R}$.

System model $\mathbf{C}$ as a vector of functions describing the system state space in discrete time domain $\mathbf{C} : [\mathbf{r}, \mathbf{a}]$, can be also presented as the following system of difference equations (state space), containing often highly non-linear functions $r_i(t) \in r, a_i(t) \in a$.

$$
\mathbf{C}: \begin{cases} r_1(t) \\ \vdots \\ r_m(t) \\ a_1(t) \\ \vdots \\ a_k(t) \end{cases} , \qquad \begin{aligned} r(t) &= \rho(r(t-1), a^{(\mathbf{P_i})}(t-1), t), \\ a(t) &= \alpha(r(t-1), a^{(\mathbf{P_i})}(t-1), t) \end{aligned} \qquad (1)
$$

4   Monitoring Environment and Controller Architecture

Figure 1 shows the software framework used in the research. All components (Monitoring agents, Collector, Loader, Evaluator, and Controller) are implemented in Java and running in isolated run–times, communicating asynchronously with the use of messaging via a JMS broker (Active MQ). This architecture provides processes separation between metrics data collection, rules generation and control execution. Processes can be deployed and run on many machines, observing system infrastructure as a whole – communication between components is asynchronous limiting time spent on waits. The collected metrics are stored in a knowledge base and are persisted in a relational database to allow flexible querying (Oracle RDBMS). Below we provide an overview of the whole environment following the data life–cycle.

4.1   Application and SLA

Control of an application has to be done in the background of defined knowledge about system functional and run–time priorities. To set such information SLA definitions are used. The SLA definitions have the form of penalty function, in \$ as units. Any of the collected metrics as part of system state can be used in SLA definitions. SLA definitions in the current implementation use SQL phrases, e.g.:

```
p_i_sla_resource_name =>
    'SLA3: 10\$ penalty for every second started of an image processing longer by average
than 10ms'
p_i_base_resource_like_filter_phrase =>
    'ExampleFilter1//HTTP/1.1://127.0.0.1(127.0.0.1):8081//dddsample/images/%.png [null]'
```

4.2   Monitoring

Metrics are acquired from monitored parts of a system by remote agents. The monitoring framework[1] uses *passive* and *active* monitoring as complementary techniques of performance tracking in distributed systems.

Active (Synthetic) monitoring operates in scheduled mode using scripts which often simulate end–user activity. The scripts continuously monitor at specified

---

[1] Allmon is a Java monitoring tool, freely available on Apache License 2.0[22].

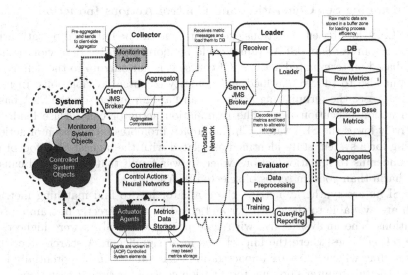

Fig. 1. Deployment diagram – a logical overview of all software components used in the complete data life–cycle for ASM control used in the research

intervals various system metrics for performance and availability, allowing access to compiled functionalities and exposing internal characteristics. Active agents are also used to gather information about resource states in the system. *Passive monitoring* enables supervision of actual (real) interactions with a system. Metrics collected can be used to determine the actual service–level quality delivered to end–users and to detect errors or potential performance problems in the system. Passive monitoring agents are deployed into the application with the use of run–time level instrumentation (often by Aspect-Oriented Programming).

4.3   Evaluator

This component selects system states stored in the database providing a view of past states and control actions. These data are processed and used to build training sets for NNs. The controller uses pairs of NNs with antagonistic rules definitions. The first NN is dedicated to detect states where potentially wrong control would be taken, the second NN is trained to confirm states where good control actions were evaluated. The proposed evaluation algorithm selects the system states where: (a) extreme SLAs values (maximum and minimum), (b) total SLAs when control was applied were lower than without. Those historical system states are merged and create training sets for the controller NNs pair. Only selected dimensions (those impacting SLA) are used to train NNs (same dimensions for both networks); aggregated normalized metrics values are passed to training sets. Trained pairs of NNs, as ready–to–use Java objects, are serialized and sent over using topic style JMS published messages to the Controller process.

4.4   Training Set Generation and Control Actions Induction

High SLA values based on actions execution time can be impacted by saturation of used resources. Some SLAs values could be more expensive than violations of SLAs based on terminated actions. Of course it is not only up to the definitions used but also to the system load and quantity of potential violations. In most practical cases the terminating actuator can have an impact on the SLA based on an action execution time. The termination actuator can release load and effectively lower the SLA value. This approach promotes termination of actions running for a system state identified as related with the highest SLA footprint, and condemns termination actions where it was found that the control caused more harm than good in terms of selected SLA levels.

As SLAs are synthetic metrics, the evaluator has to find matching metrics, which are available to the controller and could be used directly in training sets definitions. The analysis starts with time points where SLAs were highest, as these reflect times where the impact was most significant. A search algorithm helps finding which SLAs are impacted and their values can be potentially lowered in the future after applying termination control for similar situations.

Each of the controlled actions (java class.method, servlet, etc.) has its own dedicated pair of NNs instances trained to address specific for this action control in the background of declared system states conditions. NNs are trained with "bad" and "good" control actions selected from the past. The controller firstly considers "bad" control network and if the current state does not match with patterns stored in the NNs, it then begins checking "good" control.

NN are trained on normalized data, e.g. min/max boundaries – system state values which are exceeding the declared scope are not taken into account by the controller, so no control decision is taken. Of course such a situation is monitored and extends boundaries in the next evaluation iteration if necessary.

The Multi Layer Perceptron implementation with sigmoid activations of the Neuroph[2] library was applied. It was found experimentally that two hidden layers of four neurons and Momentum Backpropagation worked well in most situations with these data. Activation thresholds for termination action in control decision blocks were 0.25 and 0.75 for "bad" and "good" control respectively.

4.5   Controller and Actuator Agents

Actuator agents are weaved into an application (analogous to passive monitoring). The controller receives serialized trained NN instances from the evaluator process, which are stored in memory for easy and fast access by actuator APIs before each of the controlled actions calls. The termination actuator allows the controller to stop executing an action before its core functionality is called. The controller has access to all resources and actions metrics acquired in the plant (and it is up to monitoring configuration). The process of checking the current

---

[2] Neuroph 2.4 is an open source Java neural network framework. It contains reference to another NN library, Encog. Both of them are published under Apache 2.0 license[22].

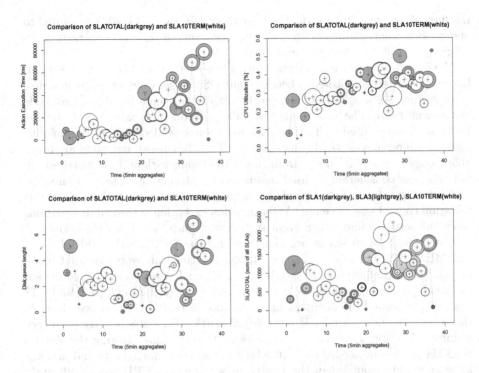

Fig. 2. Termination action and total SLA values comparison. Data collected during 3 hours simulation under two pattern loads

state and performing the control operation requires only very limited resources. The controller, as part of the model application run-time, uses a small associative table (map), which stores all decision logic as pairs of trained neural networks.

5   Simulations

The test–bed contains model application[3], load generator, monitoring and controller components described in the previous section.

Load generator: Current implementation calls a grinder[4] load script (calling web application) using an array with load distribution values and effectively transform this into run-time characteristics. During the simulation the same bimodal load pattern was used in two subsequent phases with different load–levels:

---

[3] In the research dddsample application was used as model application with a few modifications mainly changing load characteristics. It is a sample application which was developed for demonstration of Domain-Driven Design concepts[10], and is freely available under MIT License[23].

[4] Grinder is a general purpose Java load testing framework specialized in using many load injector machines for flexible load tests (jython scripts) in distributed environments/systems – it is freely available under a BSD license[24].

the first phase with 10 threads, the second with 20 threads load. We wanted to observe how the controller adaptively changes termination action characteristics adapting to different system conditions.

Monitored resources and actions: During this exercise only Operating System-OS resources were monitored. Therefore only OS metrics were present in system's state vectors. Actions were monitored by http filter agents (no direct java code instrumentation). The load generator calls only one jsp page */public/trackio* (+static assets pointed to it). The page has been extended with a block of with high IO intensive code (and consequently CPU). In case all resources are available the code execution time is from 100ms to 1100ms with uniform distribution, but this rises considerably if used resources are entering into a saturation state.

Controlled actions: Action */public/trackio* contains an object where direct termination control was woven in. A Termination exception is thrown in situations when the actuator finds the current system state in NNs pair for the action.

Evaluator: System states are generated for the CPUUSER, DISKQUEUE, and MEMORY dimensions. The evaluation described above is executed based on three SLAs definitions set: (a) SLA3 - 10$ penalty per every started second an image processing task takes longer than an average of 10ms , (b) SLA1 - 1$ penalty per second of execution of a "trackio" action if it takes over 1sec; in that case maximum penalty is 60$, (c) SLA10TERM - 20$ penalty per every terminated action. Evaluations were executed every 5mins. During the simulations the Evaluator needed from 20-90 sec to generate training sets and perform knowledge induction. When the Evaluator was running CPU was significantly utilized, what was impacting running under load application and effectively was changing the whole system characteristics. The system had to adapt to such conditions.

In our simulation, one hour sliding window was used to access system states repository data.

5.1   Data Analysis and Discussion

Simulations showed that the neural controller approach can adapt and is capable to optimize the operation of a system under certain conditions using objective functions defined as SLAs. Figure 2 shows comparison between SLAs values (circles), action execution times and main resources (CPU utilization and disk queue length) as a function of time (in 5 minutes time buckets – aggregates). All metrics were recorded during 3 hours of simulation under two load levels. The main objective is to optimize SLAs values, thus the controller tried to keep the TOTALSLA on minimal level, balancing with termination penalties (SLA10TERM), long running actions (SLA1), and potentially massive static assets execution penalties in cases of high resources consumption (SLA3). The result achieved not only manages a reasonable constant level of total SLAs, but also maintains low level of resources utilization (the controller did not allow saturation of resources).

The two load patterns used during the run are best visible on the top left chart, when action execution time rises significantly after adding twice the load to the system. At the beginning of the run no control was applied (SLA10TERM was low), because the evaluator had not established any "good" and "bad" states for potential termination control yet. Just after the 5th time bucket it took first termination actions, lowering DISKQUEUE, SLA1 and SLA3. It is worth noting the gradual decrease of cumulative SLA value during the first phase. The trend was broken after the 20th bucket when more load was added to the system. Surprisingly the first time buckets of the second phase show no termination action – that was because the new conditions were so different that the controller could not match the new system states with those represented in the trained NNs. Just after the controller begins terminating actions, massive growth of total SLA is caused (mainly penalties for termination) but reasonably low resources consumption is maintained. Around the 30th time bucket the controller changes the strict mode of operation, reducing the number of terminations applied; consequently more actions are called and the utilization of all resources increases. At the end, the controller contains state definitions, which allows the system to operate on a level of total penalty 2000$ per time bucket, with quite high execution times but reasonable resources utilized.

6   Conclusions and Future Work

In this paper we proposed a data-driven approach to design an ASM control system that allows flexible generic SLA definition for scalable distributed environments. We evaluated the use of a knowledge base containing many gigabytes of system run–time metrics. Empirical results showed that the controller equipped with a simple states evaluator is able to adapt to changing run–time conditions based on SLA definitions. We will be continuing the research in an area of black and grey–box controllers targeting more complicated cases, where more actions with different run-time characteristics and new types of actuators are present. Also, we will explore apart from actions termination, other techniques such as resources modifying, actions microscheduling, caching services tuning etc.

Huge amounts of data cause challenges for search in the systems states knowledge base; we will study more efficient mining methods for patterns search and dimensions selection (finding the most influencing dimensions for highest SLAs). Multivariate metrics search and analysis can be the basis for further predictors and classification research in the field. Furthermore, it would be potentially useful to investigate the problem in the context of Multi-Objective Optimization[21] considering the landscapes of each of the SLA functions instead of using a single aggregate objective function as it was done in this paper.

References

1. Bigus, J.P.: Applying neural networks to computer system performance tuning. In: IEEE International Conference on Neural Networks, pp. 2442–2447 (1994)
2. Parekh, S., Ghandi, N., Hellberstein, J., Tilbury, D., Jayram, T.S., Bigus, J.: Using Control Theory to Achieve Service Level Objectives In Performance Management. Real-Time Systems 23(1-2), 127–141 (2000)

3. Bigus, J.P., Hellerstein, J.L., Jayram, T.S., Squillante, M.S.: Autotune: A generic agent for automated performance tuning. In: Practical Application of Intelligent Agents and Multi Agent Technology (2000)
4. Abdelzaher, T.F., Lu, C.: Modeling and performance control of Internet servers. In: Proceedings of the 39th IEEE Conference on Decision and Control, Sydney, NSW, Australia, vol. 3, pp. 2234–2239 (2000)
5. Abdelzaher, T.F., Shin, K.G., Bhatti, N.: Performance guarantees for web server end-systems: A control-theoretical approach. IEEE Transactions on Parallel and Distributed Systems (TPDS) 13(1), 80–96 (2002)
6. Ying, L., Abdelzaher, T., Chenyang, L., Gang, T.: An adaptive control framework for QoS guarantees and its application to differentiated caching. In: Tenth IEEE International Workshop on Issue Quality of Service, pp. 23–32 (2002)
7. Zhang, R., Lu, C., Abdelzaher, T.F., Stankovic, J.A.: ControlWare: a middleware architecture for feedback control of software performance. In: Proceedings of 22nd International Conference on Distributed Computing Systems, pp. 301–310 (2002)
8. Abdelzaher, T.F., Stankovic, J., Lu, C.: Feedback performance control in software services. IEEE Control System Magazine 23, pt. 3, 74–89 (2003)
9. Lu, Y., Abdelzaher, T., Lu, C., Sha, L., Liu, X.: Feedback Control with Queueing-Theoretic Prediction for Relative Delay Guarantees. In: Ninth IEEE Real-Time and Embedded Technology and Applications Symposium (RTAS 2003) in Web Servers, Toronto, Canada (2003)
10. Evans, E.: Domain-Driven Design: Tackling Complexity in the Heart of Software, August 20. Addison Wesley (2003) ISBN: 0321125215
11. Welsh, M., Culler, D.: Adaptive overload control for busy internet servers. In: Proceedings of the 4th Conference on USENIX Symposium on Internet Technologies and Systems, USITS 2003, vol. 4. USENIX Association, Berkeley (2003)
12. Hellerstein, J.L., Diao, Y., Parekh, S., Tilbury, D.M.: Feedback control of computing systems. Wiley Interscience Publication, John Wiley & Sons (2004) ISBN 9780471266372
13. Hellerstein, J.L.: Challenges in control engineering of computing systems. In: Proceedings of the American Control Conference, Boston, MA, USA, vol. 3, pp. 1970–1979 (2004)
14. Abrahao, B., Almeida, V., Almeida, J., Zhang, A., Beyer, D., Safai, F.: Self-Adaptive SLA-Driven Capacity Management for Internet Services. In: 10th IEEE/IFIP Network Operations and Management Symposium - NOMS 2006 (2006)
15. Bodik, P., Griffith, R., Sutton, C., Fox, A., Jordan, M., Patterson, D.: Statistical Machine Learning Makes Automatic Control Practical for Internet Datacenters. In: Proceedings of the 2009 Conference on Hot Topics in Cloud Computing, HotCloud 2009. USENIX Association, Berkeley (2009)
16. Wang, Z., Chen, Y., Gmach, D., Singhal, S., Watson, B.J., Rivera, W., Zhu, X., Hyser, C.D.: AppRAISE: Application-level Performance Management in Virtualized Server Environment. IEEE Transactions on Network and Service Management 6(4), 240–254 (2009)
17. Xiong, P., Wang, Z., Jung, G., Pu, C.: Study on performance management and application behavior in virtualized environment. In: IEEE Network Operations and Management Symposium (NOMS), Osaka, pp. 841–844 (April 2010)
18. Boniface, M., Nasser, B., Papay, J., Phillips, S.C., Servin, A., Xiaoyu, Y., Zlatev, Z., Gogouvitis, S.V., Katsaros, G., Konstanteli, K., Kousiouris, G., Menychtas, A., Kyriazis, D.: Platform-as-a-Service Architecture for Real-Time Quality of Service Management in Clouds. In: Fifth International Conference on Internet and Web Applications and Services (ICIW), p. 155 (May 2010)

19. Chen, Y., Gmach, D., Hyser, C., Wang, Z., Bash, C., Hoover, C., Singhal, S.: Integrated Management of Application Performance, Power and Cooling in Data Centres. HP Laboratories (2010)
20. Bertoncini, M., Pernici, B., Salomie, I., Wesner, S.: GAMES: Green Active Management of Energy in IT Service Centres. In: Soffer, P., Proper, E. (eds.) CAiSE Forum 2010. LNBIP, vol. 72, pp. 238–252. Springer, Heidelberg (2011)
21. Coello Coello, C.A., Van Veldhuizen, D.A., Lamont, G.B.: Evolutionary algorithms for solving multi-objective problems. Genetic algorithms and evolutionary computation, vol. 5. Springer (2002) ISBN 0306467623
22. The Apache Software Foundation, Apache License, Version 2.0 (January 2004), http://www.apache.org/licenses/LICENSE-2.0
23. Open Source Initiative OSI - The MIT License, http://www.opensource.org/licenses/MIT
24. Open Source Initiative OSI - The BSD License, http://www.opensource.org/licenses/bsd-license.php

Using Varying Negative Examples to Improve Computational Predictions of Transcription Factor Binding Sites

Faisal Rezwan[1], Yi Sun[1], Neil Davey[1], Rod Adams[1],
Alistair G. Rust[2], and Mark Robinson[3]

[1] School of Computer Science, University of Hertfordshire, College Lane, Hatfield,
Hertfordshire AL10 9AB, UK
[2] Wellcome Trust Sanger Institute, Wellcome Trust Genome Campus, Hinxton,
Cambridge CB10 1SA, UK
[3] Benaroya Research Institute at Virginia Mason, 1201 9th Avenue Seattle, WA
98101, USA
{F.I.Rezwan,Y.2.Sun,N.Davey,R.G.Adams}@herts.ac.uk, ar12@sanger.ac.uk,
mrobinson@benaroyaresearch.org

Abstract. The identification of transcription factor binding sites (TF-BSs) is a non-trivial problem as the existing computational predictors produce a lot of false predictions. Though it is proven that combining these predictions with a meta-classifier, like Support Vector Machines (SVMs), can improve the overall results, this improvement is not as significant as expected. The reason for this is that the predictors are not reliable for the negative examples from non-binding sites in the promoter region. Therefore, using negative examples from different sources during training an SVM can be one of the solutions to this problem. In this study, we used different types of negative examples during training the classifier. These negative examples can be far away from the promoter regions or produced by randomisation or from the intronic region of genes. By using these negative examples during training, we observed their effect in improving predictions of TFBSs in the yeast. We also used a modified cross-validation method for this type of problem. Thus we observed substantial improvement in the classifier performance that could constitute a model for predicting TFBSs. Therefore, the major contribution of the analysis is that for the yeast genome, the position of binding sites could be predicted with high confidence using our technique and the predictions are of much higher quality than the predictions of the original prediction algorithms.

1   Introduction

Transcription Factor Binding Sites (TFBSs) are the places in DNA where regulatory proteins bind to increase or decrease the amount of mRNA that is transcribed from the genes. These proteins themselves are also encoded by other genes, which makes them a part of the regulatory interconnections of complex

C. Jayne, S. Yue, and L. Iliadis (Eds.): EANN 2012, CCIS 311, pp. 234–243, 2012.
© Springer-Verlag Berlin Heidelberg 2012

networks known as Genetic Regulatory Networks (GRNs). Therefore, identifying these sites in DNA is an important research goal. Identifying TFBSs in a genome is such a problem where the rate of false predictions is too high. Experimental methods for identifying the location of TFBSs are time consuming and costly and therefore not amenable to a genome-wide approach [1]. In this case, the computational approach is a good solution [2,3]. However, computational methods for identifying TFBSs is still a difficult task as they still produce many false predictions [1,4,5]. There are many algorithmic approaches to generate predictions for binding sites and these basic algorithms have their own particular limitations and strengths. Taken in combination, it might be expected that they provide more information about TFBSs than they do individually.

In our previous studies [6,7], we combined various algorithmic predictions and a Support Vector Machine (SVM) was used for classification after some pre-processing on the dataset. We also showed that filtering out short predictions (post-processing) produced improved predictions. It was evident from the results that the new prediction is a little better than any of the individual algorithms. One of the major problems with training a classifier on these combined predictions is that the data constituting the training set can be unreliable. In particular, the obvious approach is to take the negative examples (those nucleotides that are not part of binding sites) from the same region of the genome as the positive examples, that is from the promoter region. However, the data from base algorithms may be much more unreliable in predicting non-binding sites in promoter regions than that from the binding sites. We addressed this issue elsewhere and showed that using negative examples in the training data from sources other than promoter regions brought substantial improvement in the computational predictions of TFBSs. We used this approach of varying the negative examples on a previously used yeast dataset [6,7] with incomplete annotations and has now become outdated. In this paper, we have used the same approach on a much newer yeast dataset with improved annotations and updated algorithms/ biological evidence, in order to validate the technique. A new type of negative examples, intronic negative examples, will also be introduced and we will show that using these negative examples with the modified cross-validation method brings further improvement to the predictions.

2   Background

As mentioned before, we used the yeast (*S. cerevisae*) as an experimental organism, as it has the most completely annotated sequences. In this study we used the latest data from the resources at the *UCSC Genome Browser*, where the majority of the data is originally from *Saccharomyces Genome Database* [8] based on the assembly sequences of S288C strain. The annotations for the TFBSs were collected from *Open Regulatory Annotation Database (ORegAnno)* [9]. We chose 60 genes of 30,000 base pairs (bps) based on the highest frequency of transcription factors binding to them and to establish this criterion we used the

Table 1. Summary of the yeast data

Total number of sequences	60
Total sequence length	30,000 bps
Average sequence length	500 bps
Average number of TFBSs per sequence	3.87
Average TFBSs width	11.68 bps
Total number of TFBS	232
TFBS density in total dataset	9.03%

Table 2. The seven prediction algorithms used with the yeast dataset

Strategy	Algorithms
Scanning algorithms	Fuzznuc
	MotifLocator
	P-match
Co-regulatory algorithms	MEME
	AlignACE
Phylogenetic data	PhastCons (conserved)
	PhastCons (most conserved)

mapping of conserved regulatory sites from [10] with p-value 0.005 and moderate conservation. The details of the dataset is given in Table 1 and the seven sources of evidence used as input in this study are listed in Table 2.

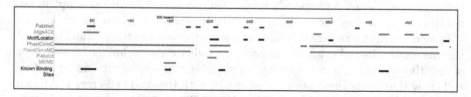

Fig. 1. Visualisation of the algorithmic predictions of the upstream region of *RPI1*. The upper seven results are from the original prediction algorithms and the final one is experimentally annotated binding sites.

Fig. 1 shows the seven algorithmic predictions together with the annotated binding sites and none of the individual algorithms can accurately predict the binding site locations. By simply concatenating the seven prediction values we constructed a 7-ary prediction vector for each position along the genome and the label of each vector is the known status of that position (that is, whether or not it is the part of a TFBS). This is illustrated in Fig. 2. Additionally we undertook pre-processing, training and testing, and post-processing described in [6,7].

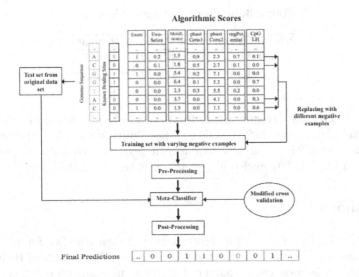

Fig. 2. The complete workflow of combining algorithmic predictions, pre-processing, classification using an SVM, and post-processing

3   Problems with the Dataset and Solutions

From Table 3, we can see that in the yeast dataset there are a number of prediction vectors that are repeated and inconsistent (same vectors with different labels). As the yeast dataset has many inconsistent data points, this suggests that this dataset is particularly unreliable for training the classifier. This happens due to very few data points have repeated many times both in the negative and positive example classes. These vectors represent points at which almost all the algorithms are predicting no binding sites, that is they are vectors constituted of zeros. So far to deal with inconsistent and repeated data, we undertook the simplest approach by removing all such vectors, but keeping one copy of the consistent vectors. As a result, nearly 70% of the yeast dataset became unusable for training.

In the yeast dataset, described so far, negative examples are taken from the promoter regions that are not annotated as TFBSs (referred to as promoter negative examples). In one of our previous studies, we introduced the concept of two further sources of negative examples namely – *distal negative examples* and *randomised negative examples*. For distal negative examples, we took intergenic regions from the yeast data that are more than 1,000 bps in length and extracted 50 bps from either side of the midpoint of the intergenic regions. There are in total 59,994 nucleotides in this negative dataset. The randomised negative example dataset was produced simply by randomising this distal negative example dataset. In this paper we have further introduced a new source of negative examples namely *intronic negative examples*. For this type of negative examples, we

Table 3. Statistics of the yeast dataset

Original	Inconsistent	Unique	Repeat
30,000	10,333	9,791	9,876
	(34.44%)	(32.64%)	(32.92%)

randomly selected 75 intronic regions from the yeast data and trimmed 10 bps of either end of a selected intron as there is possibility of some bps that are close to the end of an exon which may have a high degree of sequence conservation. There are in total 33,091 nucleotides in this negative dataset.

4   Methods

In this study, first we normalised each feature and then searched for any repetitive or inconsistent data vector in the yeast dataset. We eliminated them from the training set, as these can be potential source of misleading prediction results. We followed the same process after replacing the promoter negative examples by the distal, randomised, or intronic negative examples in training sets. After removing the repeated and inconsistent vectors from the dataset, we got a new dataset and we took two-third of it as training data. As shown from the summary of the yeast dataset that the proportion of positive and negative examples is low, we used databased sampling techniques on the training sets to make it more balanced [11] . We then randomised data rows to mix the positive and negative examples randomly. However, the test set was reconstructed from the original dataset so that it only contains biologically meaningful data. We also undertook post-processing on the final predictions to get rid of very short predictions.

4.1   The Classifier and Modified Cross-Validation

After constructing the training set, we trained an SVM with a Gaussian kernel with two hyper parameters, the *cost* and *gamma*, on the training set. It is important to find good values of these parameters, and this is normally done by a process of cross-validation. In our previous research on this problem, performance had been measured on the validation fold which was also a part of training set. However this may not be the most effective method. For training models to perform well on the test set, hence we decided to measure performance on the validation set exactly as we do on the test set. Therefore we measured the classification performance with the repeated/inconsistent vectors placed back in the validation fold and used post-processing. The step-by-step description of exactly what we have done is described in our previous work [12].

As usual we took the confusion matrix and from it defined the following measures (Equations 1 – 4) to evaluate the performance measure. Note that all four of these measures give values between 0 and 1, with the higher value being better except for *FP-rate*. As the data is imbalanced simple classification accuracy is not an adequate measure of performance. Therefore, we used *F-score* as the ideal performance measure in this study along with *FP-rate*. A higher *F-score* with a lower *FP-rate* are the most desirable performance measures in this case.

Table 4. Confusion matrix

	Predicted Negatives	Predicted Positives
Actual Negatives	True Negatives(TN)	False Positives(FP)
Actual Positives	False Negatives(FN)	True Positives(TP)

$$Recall = \frac{TP}{TP + FN} \quad (1) \quad Precision = \frac{TP}{TP + FP} \quad (2)$$

$$F\text{-}score = \frac{2 \times Recall \times Precision}{Recall + Precision} \quad (3) \quad FP\text{-}rate = \frac{FP}{FP + TN} \quad (4)$$

5   Results and Discussion

We ran four types of experiments in this study and in summary these experiments are as follows:

Experiment 1: Using negative examples sequences not annotated as TFBSs from the promoter region.

Experiment 2: Replacing promoter negative examples with distal negative examples.

Experiment 3: Replacing promoter negative examples with randomised negative examples.

Experiment 4: Replacing promoter negative examples with intronic negative examples.

Note that only the training set was modified in each experiment and the test set had never been changed.

Before presenting our experimental results, let us see how the base algorithms performed for identifying TFBSs on the same test set we used in all the experiments. We calculated the performance measures of the seven algorithms and took the best prediction result from *PhastConsMC (PhastCons most-conserved)* (see Table 5). From the results in Table 5, it is evident that even the best prediction algorithm wrongly classifies a lot of sites as binding sites which actually are not. Therefore, the *Recall* is high as well as the *FP-rate*, but the *Precision* is very low. Now we are going to discuss the results from four different experiments undertaken in this study.

Table 5. The results of base algorithms on mouse data

Base Algorithms	Recall	Precision	F-score	FP-rate
PhastConsMC	0.76	0.155	0.26	0.43

5.1 Experiment 1- Using Negative Examples Sequences Not Annotated as TFBSs

In this experiment we did not change the negative examples in the training set but varied the way in which the cross validation took place. The results of combining prediction results using an SVM with the modified cross-validation are given in Table 6. If we compare predictions from Experiment 1 with that of the best base algorithm, we can see that combining sources produces slight improvement in *F-score*. The *F-score* increased from 26% to 31% and the *FP-rate* decreased from 43% to 17%.

Table 6. The results of two-class SVM with promoter negative examples on the yeast data

	Recall	Precision	F-score	FP-rate
Yeast data+ promoter	0.48	0.23	0.31	0.17

Though this result looks promising, it is not as good as we expected. Therefore, replacing promoter negative examples by other negative examples may bring expected benefit, which we are going to discuss henceforth.

5.2 Experiment 2- Replacing Promoter Negative Examples with Distal Negative Examples

In this experiment we replaced promoter negative examples in the training set with distal negative examples with the modified cross-validation method. The prediction results, shown in Table 7, brings a huge improvement over the predictions of best base algorithm, *PhastConsMC*. The *F-score* has improved from 26% to 86% and the classifier can predict almost all the positive examples present in the test set (as the *Precision* is almost 95%). There is also a huge reduction in *FP-rate* with only a few predictions of positive examples being incorrect. This significant improvement can be due to the fact that since the distal negative examples are thousands of *bps* away from the promoter regions, there is less possibility of labelling genuine binding sites as negative; therefore it helped the classifier to characterise positive and negative examples properly.

Table 7. The results of two-class SVM with distal negative examples on the yeast data

	Recall	Precision	F-score	FP-rate
Yeast data+ distal	0.78	0.95	0.86	0.004

5.3 Experiment 3- Replacing Negative Examples with Randomised Negative Examples

In this experiment we replaced promoter negative examples in the training set with randomised negative examples and used the modified cross validation method. The results of combining prediction results using an SVM with the randomised negative examples are given in Table 8. The result from this experiment also shows a substantial improvement over the base algorithms as expected. The *F-score* has improved from 24% to 86% and there is a significant reduction in *FP-rate*. Though there is no further improvement of the *F-score* compared to that of using distal negative examples, the *FP-rate* decreases from 0.4% to 0.1%.

Table 8. The results of two-class SVM with randomised negative examples on the yeast data

	Recall	Precision	F-score	FP-rate
Yeast data+ rand	0.77	0.99	0.86	0.001

5.4 Experiment 4- Replacing Negative Examples with Intronic Negative Examples

In this experiment we replaced the negative examples in the training set with intronic negative examples and used the modified cross validation. The result in Table 9 shows further improvement over the base algorithms. The *F-score* has improved from 26% to 87% and there is a very slight improvement of F-score compared to that of using distal negative and randomised examples. The *F-score* improved due to the increase in *Recall*, however the *Precision* value decreased and as a result the *FP-rate* also increased compared to that of using randomised negative examples.

Table 9. The results of two-class SVM with randomised negative examples on the yeast data

	Recall	Precision	F-score	FP-rate
Yeast data+ intron	0.79	0.98	0.87	0.002

5.5 Visualisation of the Predictions

Here we visualised the data (like Fig. 1) to see how the predictions matched the annotations. We have taken a fraction of the yeast genome (upstream region of the gene *RPI1*) and compared the best results from different experiments along with prediction algorithms and annotation. In Fig. 3, the upper seven results are from the original prediction algorithms. The next four results are our best prediction results from four different types of experiments and the last one is the experimentally annotated binding sites from the *ORegAnno* database. Fig. 3 shows that the original prediction algorithms generated a lot of false predictions and using promoter negateive examples (Experiment 1) did not produce good prediction results as well. Whereas using distal, randomised, and intronic negative examples (Experiments 2, 3, and 4) improved the prediction considerably.

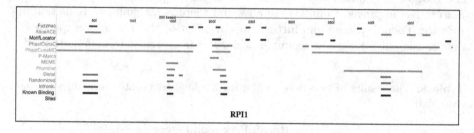

Fig. 3. Visualisation of the final predictions of the upstream region of *RPI1*. The first seven predictions are from base algorithms, the next four are from the four experiments that we undertook in this study and the last one represents annotated binding sites.

6   Conclusion

In this study, a new yeast dataset has been used with updated annotations and algorithms in order to validate the techniques we used in our previous studies. As in our earlier works, the idea of using negative examples from a source different than the promoter region together with the modified cross-validation method gave much benefit. For the yeast dataset, the *F-score* was more than 87% with almost 98% binding sites were predicted correctly. However, our original hypothesis, suggested in our previous work, that the negatively labeled promoter regions might contain many, as yet undiscovered, binding sites has proved to be incorrect. Our predictions largely coincide with the original label, hence our high *F-score* values. The results show that the algorithms collectively can identify the binding sites and non-binding sites in the promoter regions when we used with negative examples from sources other than the promoter region. Therefore, outside the promoter region the algorithms do collectively characterise these regions as containing no binding sites and this has been proven according to as the results using the distal negative examples. It was also found that in fact using randomised and intronic negative examples perform even better than distal negative examples. However, the results presented in this paper are from ongoing research and still need further validations. Therefore further work is going to be undertaken to establish the proof of concept of using negative examples from various sources to improve computational predictions of transcription factor binding sites.

References

1. Tompa, M., Li, N., Bailey, T.L., Church, G.M., De Moor, B., Eskin, E., Favorov, A.V., Frith, M.C., Fu, Y., Kent, W.J., Makeev, V.J., Mironov, A.A., Noble, W.S., Pavesi, G., Pesole, G., Régnier, M., Simonis, N., Sinha, S., Thijs, G., van Helden, J., Vandenbogaert, M., Weng, Z., Workman, C., Ye, C., Zhu, Z.: Assessing computational tools for the discovery of transcription factor binding sites. Nat. Biotechnol. 23(1), 137–144 (2005)
2. Elnitski, L., Jin, V.X., Farnham, P.J., Jones, S.J.: Locating mammalian transcription factor binding sites: a survey of computational and experimental techniques. Genome Res. 16, 1455–1464 (2006)
3. Pavesi, G., Mauri, G., Pesole, G.: In silico representation and discovery of transcription factor binding sites. Brief. Bioinformatics 5, 217–236 (2004)
4. Hu, J., Li, B., Kihara, D.: Limitations and potentials of current motif discovery algorithms. Nucleic Acids Res. 33, 4899–4913 (2005)
5. Brown, C.T.: Computational approaches to finding and analyzing cis-regulatory elements. Methods Cell Biol. 87, 337–365 (2008)
6. Sun, Y., Robinson, M., Adams, R., Rust, A.G., Davey, N.: Using Pre and Posting-processing Methods to Improve Binding Site Predictions. Pattern Recognition 42(9), 1949–1958 (2009)
7. Robinson, M., Castellano, C.G., Rezwan, F., Adams, R., Davey, N., Rust, A.G., Sun, Y.: Combining experts in order to identify binding sites in yeast and mouse genomic data. Neural Networks 21(6), 856–861 (2008)
8. Cherry, J.M., Hong, E.L., Amundsen, C., Balakrishnan, R., Binkley, G., Chan, E.T., Christie, K.R., Costanzo, M.C., Dwight, S.S., Engel, S.R., Fisk, D.G., Hirschman, J.E., Hitz, B.C., Karra, K., Krieger, C.J., Miyasato, S.R., Nash, R.S., Park, J., Skrzypek, M.S., Simison, M., Weng, S., Wong, E.D.: Saccharomyces Genome Database: the genomics resource of budding yeast. Nucleic Acids Res. 40(Database issue), D700–D705 (2012)
9. Montgomery, S.B., Griffith, O.L., Sleumer, M.C., Bergman, C.M., Bilenky, M., Pleasance, E.D., Prychyna, Y., Zhang, X., Jones, S.J.M.: ORegAnno: An open access database and curation system for literature-derived promoters, transcription factor binding sites and regulatory variation. Bioinformatics (March 2006)
10. MacIsaac, K.D., Wang, T., Gordon, D.B., Gifford, D.K., Stormo, G., Fraenkel, E.: An improved map of conserved regulatory sites for Saccharomyces cerevisiae. BMC Bioinformatics 7, 113 (2006)
11. Chawla, N.V., Bowyer, K.W., Hall, L.O., Kegelmeye, W.P.: SMOTE: Synthetic minority over-sampling Technique. Journal of Artificial Intelligence Research 16, 321–357 (2002)
12. Rezwan, F., Sun, Y., Davey, N., Adams, R., Rust, A.G., Robinson, M.: Effect of Using Varying Negative Examples in Transcription Factor Binding Site Predictions. In: Pizzuti, C., Ritchie, M.D., Giacobini, M. (eds.) EvoBIO 2011. LNCS, vol. 6623, pp. 1–12. Springer, Heidelberg (2011)

Visual Analysis of a Cold Rolling Process Using Data-Based Modeling

Daniel Pérez[1], Francisco J. García-Fernández[1], Ignacio Díaz[1],
Abel A. Cuadrado[1], Daniel G. Ordonez[1], Alberto B. Díez[1],
and Manuel Domínguez[2]

[1] Universidad de Oviedo. Área de Ingeniería de Sistemas y Automática
{dperez,fjgarcia,idiaz,dgonzalez,cuadrado,alberto}@isa.uniovi.es
[2] Universidad de León. Instituto de Automática y Fabricación
diemdg@unileon.es

Abstract. In this paper, a method to characterize the chatter phenomenon in a cold rolling process is proposed. This approach is based on obtaining a global nonlinear dynamical MISO model, relating four input variables and the exit strip thickness as the output variable. In a second stage, local linear models are obtained for all working points using sensitivity analysis on the nonlinear model to get input/output small signal models. Each local model is characterized by a high dimensional vector containing the frequency response functions (FRF) of the four SISO resulting models. Finally, the FRF's are projected on a 2D space, using the t-SNE algorithm, in order to visualize the dynamical changes of the process. Our results show a clear separation between chatter condition and other vibration states, allowing an early detection of chatter as well as being a visual analysis tool to study the chatter phenomenon.

Keywords: dimensionality reduction, cold rolling, data visualization, dynamical systems, data-based models.

1 Introduction

Steel production is usually considered an indicator of economic progress, as it is fairly related to infrastructures and development. After steel production from iron, the material needs to be treated and modified through several mechanical processes, such as the rolling process. The cold rolling of steel is a widely adopted process, in which a steel sheet is passed through a pair of rolls whereby the sheet thickness is reduced. Although this process has been studied for decades [1], many unsolved issues hold. The control of many different parameters is necessary, ranging from those related to the milling itself (force applied, torque,...) to those depending on different aspects, such as lubrication or refrigeration. Furthermore, there is an ever increasing demand for higher quality from costumers and, since it is a large and complex process that continuously evolves due to drifts, misadjustments and changes in working conditions, there is a need of continuous improvement in the efficiency of the process. Because of that, the supervision of this process is critical, in order to avoid faults that affect negatively to the material.

C. Jayne, S. Yue, and L. Iliadis (Eds.): EANN 2012, CCIS 311, pp. 244–251, 2012.

One of the most relevant faults in the cold rolling process of steel is called *chatter* [2], an unexpected powerful vibration that affects the quality of the rolled material by causing an unacceptable variation of the final thickness. The real problem of chatter is not only related to the bad quality of the manufactured product, but also to the economic losses suffered. Generally, when chatter appears, it is necessary to lower the rolling speed for a period of time, making the production rate decrease. A practical way to detect chatter is to compute the power spectral density in which this fault appears (normally 100-300Hz). However, although this procedure works well to show up the chatter condition, it fails as an early detector.

A way to predict chatter is to use a model of the rolling process [3]. However, the complexity of the whole process, with several tightly coupled phenomena (such as chemical, mechanical, and thermal) makes it difficult to build an accurate model and moreover to tune its parameters. An approach to enhance the knowledge about complex processes is visualizing their relevant information, using *dimensionality reduction* (DR) techniques [4,5]. DR techniques allow to project and study the structure of high-dimensional data into a low-dimensional space, typically a 2D/3D for visualization purposes, improving the exploratory data analysis [6].

In the DR field, several techniques have been proposed [7]. One of the first algorithms is Principal Component Analysis (PCA), described by Pearson [8]. After PCA, other DR techniques have been proposed, such as Multidimensional Scaling (MDS) methods, Independent Component Analysis (ICA)[9] or Self-Organizing Maps (SOM)[10]. In the beginning of 21st century, a new trend in DR based on nonlinear models appeared, inspiring a new collection of algorithms. These algorithms –known as *manifold learning*– involve a local optimization by defining local models of the k-nearest neighbours and an alignment in order to obtain the global coordinates of each model, usually implying a singular value decomposition (SVD). Some of the most known techniques are *Isomap* [11], *local linear embedding*(LLE) [12] and *laplacian eigenmaps* (LE) [13]. Similar to these techniques, but based on the probability distribution of data is *t-Stochastic Neighbor Embedding* (t-SNE) [14]. This technique, that has attracted attention recently [15,16], is capable of maintaining the local structure of the data while also revealing some important global structure (such as clusters at different scales), producing better visualizations than the rest.

In this paper, we propose a new approach for the study of chatter, using the DR principle for the analysis of the dynamical behavior of a model of the process. Using a novel feedforward neural network, called *extreme learning machine* (ELM) [17], the proposed approach computes a large feature vector composed of the frequency response functions (FRF) of a set of key physical variables and projects this vector into a 2D space by t-SNE algorithm. Thus, the changes in the dynamical behavior of the process are visualized. The paper is organized as follows: in section 2, a description of the method is shown; section 3 describes an experiment and the results of the method proposed and finally section 4 includes the conclusions obtained.

2   Data-Based Model Analysis through Manifold Learning Techniques

2.1   Description of the Physical Model

Classical cold rolling models try to calculate the force and the torque necessary for a given thickness reduction. As mentioned before, the complexity of an accurate model can be very high because of the assumptions taken [18]. In order to get a simple model to work with, e.g. [19], several assumptions can be done.

The classical form of a rolling force (F) model includes: the tension at the entry and exit side of a rolling stand (σ_{en} and σ_{ex}); the thickness at the entry and exit side (h_{en} and h_{ex}); the width of the strip (w); the friction coefficient (μ) and the hardness level of the material being rolled (S), see Eq. (1).

$$F = f(\sigma_{en}, \sigma_{ex}, h_{en}, h_{ex}, w, \mu, S) \tag{1}$$

A model of the rolling process makes it possible to analyze several defects arising from working operation, such as the chatter phenomenon. This phenomenon is a dynamic process, where variations in the roll force may lead to an unstable state. It is necessary to generate a model where the different factors likely to modify the force equilibrium in the rolling process are taken into account. As explained in [20], the chatter phenomenon comes from a feedback interaction with the variables entry speed, entry tension, the force of the strip on the rolls and exit thickness involved. If a dynamic model of the stand is added to this loop, a proper model to study the chatter phenomenon can be built [21,22].

2.2   Mathematical Estimation of the Model

As proposed in [23], data-based models are a practical way to develop a fault detection and prediction mechanism for complex processes. The development of data-based models provides a good feature for their application to industrial processes: fast responses to faults. According to the previous description of the rolling process, we propose a MISO model, defining the exit thickness y_k as the output of the system and force F_k, tension σ_k –used in the classical model Eq. (1)–, entry and exit speed of the strip (V_{en_k} and V_{ex_k} respectively) –due to their relevance in the chatter phenomenon–, as the inputs of the system. We also considered an autoregressive part of the output to account for internal dynamics, resulting in a NARX model.

$$y_k = f(y_{k-1}, \ldots, y_{k-n}, F_k, \sigma_k, V_{en_k}, V_{ex_k}) \tag{2}$$

A simple and fast learning algorithm for single hidden layer feedforward neural networks (SLFN's), called extreme learning machine (ELM) [17], is used to train

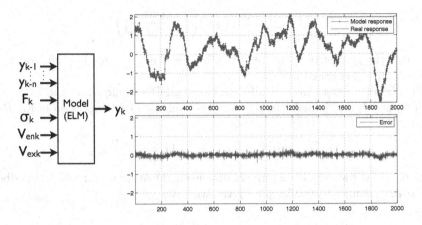

Fig. 1. Scheme of the model developed (left) and an example of its one-step-ahead prediction (right)

the NARX model. In order to obtain an optimal order of the model, we apply the Akaike Information Criterion (AIC) [24], obtaining a trade-off solution between the order and the error of the model. A scheme can be seen on the left side of Fig. 1. On the right side of Fig. 1, a comparison between the response of the real process and an example of the one-step-ahead prediction of a trained model for a testing dataset is shown.

Once the model is obtained, a local sensitivity analysis is applied to perform a system identification of the contribution of each input variable to the output. The signals are divided into M overlapped windows of size L. Let's consider $\bar{F}$, $\bar{\sigma}$, $\bar{V}_{en}$, and $\bar{V}_{ex}$, average values for the m-th window, and the delayed samples $y_{k-1}, y_{k-2}, \cdots, y_{k-n}$. The m-th local sensitivity analysis $(m = 1, \ldots, M)$ for the input F is performed adding to $\bar{F}$ a random value $\varepsilon_k \in N(0, \nu)$, being ν a small value.

$$\mathbf{u}_m^{\Delta F} = \left[y_{k-1}, y_{k-2}, \ldots, y_{k-n}, \bar{F} + \varepsilon_k, \bar{\sigma}, \bar{V}_{en}, \bar{V}_{ex} \right]^T \tag{3}$$

with an output $\mathbf{y}_m = [y_k]$, for $k = k_m, \ldots, k_m + L - 1$, being k_m the first sample of window m. Constructing $\mathbf{U}^{\Delta F} = [\mathbf{u}_m^{\Delta F}]$ and $\mathbf{Y} = [\mathbf{y}_m]$ resulting in an I/O pair $\{\mathbf{U}^{\Delta F}, \mathbf{Y}\}$. Similar to Eq. (3), we apply the same method to the other inputs, obtaining $\{\mathbf{U}^{\Delta \sigma}, \mathbf{Y}\}$, $\{\mathbf{U}^{\Delta V_{en}}, \mathbf{Y}\}$ and $\{\mathbf{U}^{\Delta V_{ex}}, \mathbf{Y}\}$ respectively. Each I/O pair defines a local SISO small-signal model of the process.

In order to estimate the dynamical behavior of each small-signal model, we compute FRF of the model. Let $P_y(m, j)$ and $P_{u_i}(m, j)$ be the power densities of the j-th frequency in the m-th small-signal model of the output and the i-th input, respectively. The SISO FRF of input i for all windows can be computed as $\mathbf{G}_i$ where

$$\mathbf{G}_i(m, j) = 10 \cdot log_{10} \left(\left| \frac{P_y(m, j)}{P_{u_i}(m, j)} \right| \right) \tag{4}$$

describes the gain of the j-th frequency bin for the m-th small-signal model, of the i-th input expressed in dB.

Finally, in order to project all data using a DR technique, all the FRF's of each input, $\mathbf{G}_i$, are joined into an augmented matrix $\mathbf{G}$, as expressed in Eq. (5).

$$\mathbf{G} = [\mathbf{G}_1\ \mathbf{G}_2\ \mathbf{G}_3\ \mathbf{G}_4] \qquad \mathbf{G} \in \mathbb{R}^{M \times D} \tag{5}$$

Each of the M rows of $\mathbf{G}$ is a large D-dimensional feature vector which describes the local dynamical behavior on a given window.

2.3   Dimensionality Reduction

The visualization of the dynamic behavior of the model is made by computing a dimensionality reduction using t-SNE [14]. This technique is based on similarities of the data by computing probabilities for original and projection space, defining neighborhoods with a value called *perplexity*. The computation of the two joint-probability distributions p_{ij} and q_{ij} corresponds to Gaussian and Student's t-distribution respectively, see Eq. (6) and (7).

$$p_{ij} = \frac{\exp(-\|x_i - x_j\|^2/2\sigma^2)}{\sum_{k \neq l} \exp(-\|x_k - x_l\|^2/2\sigma^2)} \tag{6}$$

$$q_{ij} = \frac{(1 + \|y_i - y_j\|^2)^{-1}}{\sum_{k \neq l}(1 + \|y_k - y_l\|^2)^{-1}} \tag{7}$$

With the aim of projecting new data points on the 2D map obtained with the training dataset, an out-of-sample extension of the t-SNE was developed. The addition of each new point was computed by ensuring that the sum of probabilities is equal to 1. The variance σ of a Gaussian centered at the new point is searched keeping the perplexity value fixed. The differences of the two probability distributions are measured by Kullback-Leibler divergences. Hence, the cost function is

$$C = KL(P\|Q) = \sum_i \sum_j p_{ij} \log \frac{p_{ij}}{q_{ij}} \tag{8}$$

The coordinates of the new projections are obtained minimizing this cost function, and maintaining previous coordinates of the training dataset fixed. This minimization of the cost function is made using a gradient-descent method for optimization. Finally, as a summary of the method, a flowchart of the different stages used is shown in Fig. 2.

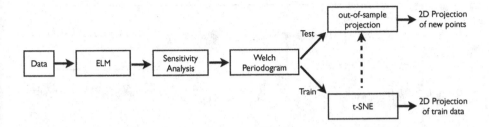

Fig. 2. Flowchart of the method

3   Experiment and Results

To validate this method, it was applied to data from a cold rolling facility. As explained in section 2.2, the variables used for modeling were: rolling force, entry tension, entry and exit speeds of the strip, and exit thickness. These variables were acquired using a data acquisition system with a sampling rate of 2000 Hz.

The training dataset was composed of signals of $N = 54300$ samples, including normal operating conditions and a chatter episode from a unique coil. In Fig. 3, the reduction of the rolling speed to avoid chatter effect can be seen, as well as the powerful variation in the other variables, especially in the thickness signal, which is closely related to the quality of the resulting product. As for testing the method, different chatter episodes of several coils were used.

To obtain the MISO model, ELM was trained using 1000 neurons in the hidden layer, with the signals normalized to zero mean and $\sigma = 1$. In Table 1, a comparison of the RMS errors of the linear regression and the EML models is shown, as well as the AIC and the Theil's Index (U) [25], for several model orders.

Table 1. Value of the RMS errors, AIC, and Theil's Index (U) for each order of the NARX model

Model Order (n)	Linear RMSE	EML RMSE	U	AIC
5	0.1856	0.1744	0.7377	$-1.5445 \cdot 10^5$
10	0.1353	0.1307	0.5529	$-1.7998 \cdot 10^5$
15	0.1252	**0.1159**	**0.4903**	$-\mathbf{1.9058 \cdot 10^5}$
20	0.1277	0.1192	0.5042	$-1.8809 \cdot 10^5$
25	**0.1251**	0.1167	0.4937	$-1.8993 \cdot 10^5$
30	0.1273	0.1194	0.5051	$-1.8787 \cdot 10^5$

According to the AIC criterion the order selected is $n = 15$. This model describes accurately the nonlinear dynamics around the working points of the process.

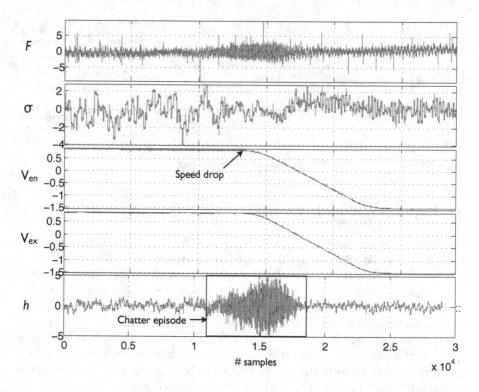

Fig. 3. Training dataset to model the process. When chatter occurs, a drop in the speed in order to reduce its effect can be seen. After achieving a proper speed level, the effect is mitigated.

For the sensitivity analysis the signals were windowed into segments of $L = 1000$ elements with an overlapping of 90%. After this analysis, the Welch Periodogram was applied to each SISO model, using a Hamming window of size 125 and an overlapping of 50%. The final size of matrix $\mathbf{G}$ is 543×516. Using this matrix, the t-SNE algorithm was applied using a value of perplexity of 30 to project data into a 2D space.

The resulting map using t-SNE technique, can be seen in Fig. 4 (a), where the entry speed is used as color and size encodings of the points. The 2D map shows two main zones of points from normal behavior of the process and a subset of points, revealing a chatter condition.

The method was applied to two subsets of data from different strips, denoted as x and +, including a chatter episode. The developed out-of-sample extension allows to project this novel test data over the trained map. These new points are placed in parts of the map that corresponds to their operating conditions (Fig. 4 (b), (c) and (d)). Their positions reveal a similar behavior to the nearest neighbors of the training points. Thus, different operating conditions are mapped in different coordinates, allowing to detect the dynamical differences in the process behavior.

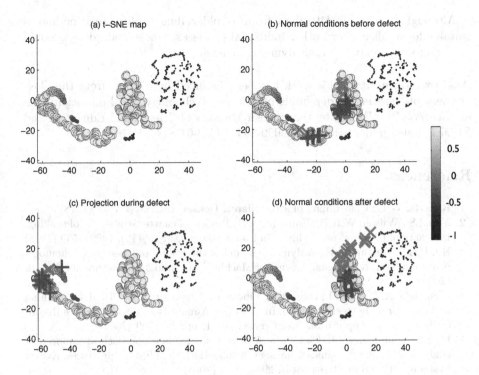

Fig. 4. (a) t-SNE projection, (b), (c), and (d) new points projected depending on operating condition

4   Conclusions

A spectral density estimation was applied to a model of a cold rolling process in order to analyze the chatter defect, which causes an unexpected vibration in the process. The analysis yields information about frequency response functions (FRF's), which can be considered as high-dimensionality data in order to apply a DR technique to visualize them. The algorithm applied (t-SNE) to a training dataset corresponding to fault situation perfectly unfolds the structure of data in the high-dimensional space, giving clear insights of this sort of faults. The method accurately defines different zones of the process, distinguishing between normal operating and chatter conditions.

The developed out-of-sample approach provides the possibility to project new data points in order to supervise the existence of problems in the process. The visualization of new projected points in the map accelerates the fault detection and helps to predict this type of faults. This procedure can be time consuming depending on the size of the dataset because once computed the map for training data the new projections are computed for each point. An alternative which reduces computational burden is to use a Nadaraya-Watson regression model [26], which estimates new projections of data based on observed values from training data.

Although it is applied to data from a cold rolling facility, this method is suitable for application to other industrial processes, whose main defects could be detected by analyzing their dynamic behavior.

Acknowledgments. This work has been financed by a grant from the Government of Asturias, under funds of Science, Technology and Innovation Plan of Asturias (PCTI), and by the spanish Ministry of Science and Education and FEDER funds under grants DPI2009-13398-C02-01.

References

1. Roberts, W.L.: Cold rolling of steel. Marcel Dekker, Inc., New York (1978)
2. Yun, I.S., Wilson, W.R.D., Ehmann, K.F.: Review of chatter studies in cold rolling. International Journal of Machine Tools and Manufacture 38(12), 1499–1530 (1998)
3. Hu, P.H., Ehmann, K.F.: A dynamic model of the rolling process. part I: homogeneous model. International Journal of Machine Tools and Manufacture 40(1), 1–19 (2000)
4. Cuadrado, A.A., Diaz, I., Diez, A.B., Obeso, F., Gonzalez, J.A.: Visual data mining and monitoring in steel processes. In: 37th IAS Annual Meeting, Conference Record of the Industry Applications Conference, vol. 1, pp. 493–500 (2002)
5. Díaz, I., Domínguez, M., Cuadrado, A., Fuertes, J.: A new approach to exploratory analysis of system dynamics using som. Applications to industrial processes. Expert Systems with Applications 34(4), 2953–2965 (2008)
6. Kourti, T., MacGregor, J.: Process analysis, monitoring and diagnosis, using multivariate projection methods. Chemometrics and Intelligent Laboratory Systems 28(1), 3–21 (1995)
7. Lee, J.A., Verleysen, M.: Nonlinear dimensionality reduction (2007)
8. Pearson, K.: LIII. On lines and planes of closest fit to systems of points in space. Philosophical Magazine Series 6 2(11), 559–572 (1901)
9. Hyvärinen, A., Karhunen, J.: Independent component analysis (2001)
10. Kohonen, T.: Self Organizing Maps. Springer (1995)
11. Tenenbaum, J.B., de Silva, V., Langford, J.C.: A global geometric framework for nonlinear dimensionality reduction. Science 290, 2319–2323 (2000)
12. Roweis, S.T., Saul, L.K.: Nonlinear dimensionality reduction by locally linear embedding. Science 290, 2323–2326 (2000)
13. Belkin, M., Niyogi, P.: Laplacian eigenmaps for dimensionality reduction and data representation. Neural Computation 15(6), 1373–1396 (2003)
14. Van Der Maaten, L., Hinton, G.: Visualizing data using t-SNE. Journal of Machine Learning Research 9, 2579–2605 (2008)
15. Bushati, N., Smith, J., Briscoe, J., Watkins, C.: An intuitive graphical visualization technique for the interrogation of transcriptome data. Nucleic Acids Research 39(17), 7380–7389 (2011)
16. Jamieson, A.R., Giger, M.L., Drukker, K., Li, H., Yuan, Y., Bhooshan, N.: Exploring nonlinear feature space dimension reduction and data representation in breast CADx with Laplacian eigenmaps and t-SNE. Medical Physics 37(1), 339–351 (2010)
17. Huang, G.B., Zhu, Q.Y., Siew, C.K.: Extreme learning machine: Theory and applications. Neurocomputing 70(1-3), 489–501 (2006)

18. Venter, R., Abd-Rabbo, A.: Modelling of the rolling process–I: Inhomogeneous deformation model. International Journal of Mechanical Sciences 22(2), 83–92 (1980)
19. Freshwater, I.: Simplified theories of flat rolling–I: the calculation of roll pressure, roll force and roll torque. International Journal of Mechanical Sciences 38(6), 633–648 (1996)
20. Paton, D.L., Critchley, S.: Tandem mill vibration: Its cause and control. In: Mechanical Working; Steel Processing XXII, Proceedings of the 26th Mechanical Working; Steel Processing Conference, pp. 247–255. Iron and Steel Soc. Inc., Chicago (1985)
21. Meehan, P.A.: Vibration instability in rolling mills: Modeling and experimental results. Journal of Vibration and Acoustics 124(2), 221–228 (2002)
22. Kimura, Y., Sodani, Y., Nishimura, N., Ikeuchi, N., Mihara, Y.: Analysis of chatter in tandem cold rolling mills. ISIJ International 43(1), 77–84 (2003)
23. Venkatasubramanian, V.: A review of process fault detection and diagnosis Part III: Process history based methods. Computers & Chemical Engineering 27(3), 327–346 (2003)
24. Akaike, H.: A new look at the statistical model identification. IEEE Transactions on Automatic Control 19(6), 716–723 (1974)
25. Theil, H.: Economics and information theory. North-Holland Pub. Co., Rand McNally, Amsterdam, Chicago (1967)
26. Simon, H.: Neural networks: a comprehensive foundation. Prentice Hall (1999)

Wind Power Forecasting to Minimize the Effects of Overproduction

Fernando Ribeiro[1], Paulo Salgado[1], and João Barreira[1,2]

[1] ECT, Universidade de Trás-os-Montes e Alto Douro,
5000-801 Vila Real, Portugal
[2] INESC TEC (formerly INESC Porto)
4200-465 Porto, Portugal

Abstract. Wind power generation increases very rapidly in the past few years. The available wind energy is random due to the intermittency and variability of the wind speed. This poses difficulty in the energy dispatched and cause costs, as the wind energy is not accurately scheduled in advance. This paper presents a short-term wind speed forecasting that uses a Kalman filter approach to predict the power production of wind farms. The prediction uses wind speed values measured over a year in a site, on the case study of Portugal. A method to group wind speeds by their similarity in clusters is developed together with a Kalman filter model that uses each cluster as an input to perform the wind power forecasting.

Keywords: Clustering, Kalman Filter, Wind Power Forecasting.

1 Introduction

Nowadays, wind power emerges as one of the most potential renewable sources of energy. This is due to its environmental characteristics and thanks to the developments that technology obtained in the past few years, which has allowed a significant increase in the energy production capacity [1]. Despite many advantages, wind power is known for its variability because of the wind speed characteristics. The increased incidence of wind power in an energy network causes an increase of the unpredictability factor of energy production. When a high increase of consumption is detected and the wind energy that is produced is slightly insignificant, it is necessary to have a rapid response of thermoelectric production (or other kinds of predictable energy). In contrast, when there is a surplus of wind power production and the power consumption is considerably low this energy may not be dispatched.

Within this context, it is important to create effective methods to predict the wind power in order to be able to make real-time correspondence between the energy consumed and the energy produced in the country. In the literature, two different methods to predict wind power have been reported: physical and statistical methods. Physical methods are based on a description of the wind farms topology, wind turbines and some information about the layout of the wind farms, and they are only

C. Jayne, S. Yue, and L. Iliadis (Eds.): EANN 2012, CCIS 311, pp. 254–263, 2012.
© Springer-Verlag Berlin Heidelberg 2012

suited for long term predictions since they used as input numerical weather predictions updated by meteorological services along with data from a large number of sources [2]. On the other hand, the time series models (or statistical) are based on historical wind data and statistical methods. They include simple linear autoregressive models (e.g. ARMA, ARIMA) [3] and recent non-linear models, such as neural networks [4], fuzzy-logic [5] and evolutionary algorithms [6]. It is difficult to compare all these methods because they depend on different situations and factors like complexity and collected data.

In this paper we present a time series model for short-term wind speed forecasting that use a bank of Kalman filter to predict the power production of a wind farm in Portugal. Wind values measured by a meteorological station over a year are the main inputs used in the speed prediction. In our approach the values of wind speeds from the previous hours are grouped into clusters accordingly with their similarity together with a bank of Kalman filter that uses each of the clusters as an input to forecast the wind power production one hour ahead.

To realize how important is forecast of wind energy production on the case study of Portugal, it is necessary to first understand which place it occupies on the national energy panorama. In section 2, it is explain the energy scenario in Portugal, the wind power production and how it is incorporate into the SEN (National Electrical System – *Sistema Eléctrico Nacional*). In section 3, the proposed approach to forecast wind power is presented and in section 4 the results from the prediction are showed using the proposed model in a wind farm. Finally, section 5 concludes the paper.

2   Wind Power Scenario in Portugal

In Portugal, REN (National Electrical Network – *Rede Elétrica Nacional*) is responsible for ensuring that the energy produced by the SEN during a day match with the energy consumed. When the demand increases the supply will also necessarily increases and when there is low demand the power production supply also become lower. Therefore, the produced energy is estimated every day by REN to ensure that most of the energy produced is consumed.

In Portugal, REN has three different types of energy production: thermoelectric, hydro, and special regime production. Thermoelectric production has a total capacity of 7 407 MW of installed power [7] and includes the coal-fired power stations, fuel oil and natural gas. Although, these types of plants are very polluting they are essential to ensure a good part of the national energy needs. Water production has a total capacity of 4 578 MW of installed power [7]. The special regime production encompasses the thermal production, cogeneration production, photovoltaic, and wind power, the last one being the most significant, with a total of 21% of the overall power switch on SEN [8].

To guarantee an uninterrupted supply of energy to the consumers, without waste, REN needs to track in real-time the energy that is produced. If the energy produced in Portugal at a certain time of the day is bigger than the consumed, it should be transported to Spain and sold to our neighbor country at fluctuating prices. Accordingly, REN has to control thermoelectric and hydroelectric power production, depending on the wind energy production and the prediction of energy consumption.

Thermoelectric production and hydroelectric power production can be regulated by the consumption needs. Thermoelectric power stations can work at maximum or minimum capacity, and hydroelectric power plants can be regulated by their consumption needs taking into account the levels of water stored.

Wind power, with a total installed power of 3 704 MW in Portugal in 2010 is the third largest source of energy in terms of total capacity [8]. Unlike other types of energy production, wind power cannot be controlled. Consequently, the production levels cannot be adjusted. As the wind energy depends directly on a renewable source which is variable, intermittent and unpredictable, it causes the most instability on the SEN. E.g. in an excellent day of winds speed the production reaches above 70 GWh, while on other days not so windy values lower below 5 GWh [11]. With this variability it becomes difficult to harness the full potential of this type of energy.

Joining the wind energy to thermoelectric power stations that are never fully switched off, at a time of low consumption we can have by itself a surplus of wind energy that is produced and not consumed. In days with excess of wind power production it is necessary to use energy storage systems to hold any excess of energy in order to use it later. Yet, it is impossible to store all energy in the form of electricity.

What happens so far is that the excess of energy produced in Portugal is mainly exported to Spain. However, the energy transported to Spain has the obstacle of being subjected to the acceptability and conditions of the only country sharing borders with Portugal. This energy is marketed at floating prices over the time of day, governed by the law of supply and demand, being in some cases marketed during the day at 0 €/MWh [9].

3   Proposed Approach

The proposed approach to forecast wind power generation is based on a wind speed forecasting. This model is divided into three phases: initially the wind speeds values of the last 12 hours are grouped into several clusters in an exclusive form. For this, it is necessary to calculate the degree of similarity between different wind curves according to their typical characteristics. In the next step, the average wind speed is forecast using a bank of Kalman filters. This forecast uses as inputs the measured wind values from the last hours associated with the cluster that most identifies with the class of winds to predict. With these values of wind speeds we can forecast the wind power production in a wind farm.

3.1   Clusters Identification

The data used in the present paper resulted from the values of wind speed measured in Alvadia (Ribeira de Pena, Vila Real, Portugal) over a period of years ranging from

2007 to 2009. These data are available online on the SNIRH (*Sistema de informação de Recursos Hídricos*) website http://snirh.pt/. However, no information regarding the measuring instruments or sampling criteria is provided.

The similarity between the different wind load-curve is evaluated by setting a metric that reflects the degree of similarity between the wind curves. This task is essential for the clustering methods [10]. In the proposed clustering approach a function of similarity S is specified and used as distance metric within it is possible to determine which cluster each element belongs (1). The method to evaluate the similarity or the proximity between the daily load curves and the cluster is obtained measuring the distance between them and the curve of the cluster center. In the approach present in this paper, the metric that gave the better results was the sum of the absolute distances between the points of the curves. However, this distance metric has a tendency contributing to the large scale attributes dominate the others. This can be avoided through standardization or through heavy distances.

$$S = \frac{1}{n_p} \times (\vec{X} - \vec{X}_m)^T \times \vec{W}. \tag{1}$$

Where n_p represents the number of points of the curve and $\vec{X}_m$ represents the average value of the attributes of all elements. For the experimental tests it was considered a horizon of 12 hours.

In order to obtain the best results, a comparative analysis was performed to determine the number of clusters used to characterize the wind speeds. Thus, we repeated the clustering process using Fuzzy C-Means and K-means algorithms. The numbers of clusters, average and maximum values of the standard deviations of the clusters used were registered. We observe that the values of the variance decreases with the increase in the number of clusters. However, for values higher than 10 the gain reduction is not very significant, so we set at 10 the number of clusters to use in the whole process of analysis and forecasting of the time series. Although both of the algorithms proved to be efficient in the grouping of wind curves based on similar form criteria, we choose to use the Fuzzy C-Mens algorithm because it presents smaller standard deviation values.

Finally, the proposed approach was applied to obtain the typical winds curves that characterize the winds regimes for a period ranging from minutes to hours, thus defining the evolution of the winds. In Fig. 1 the horizontal axis corresponds to 12 hours of the wind curved window and the vertical axis to the normalized wind speed. Each set of curves presents a very proper and distinct typology from other groups. Despite the similarity of temporal evolution of clusters mean curves, they are different concerning the amplitude variation and in the mean value. As it can be observed there are standard curves that characterize the behavior of the wind, so it is possible to find the degree of similarity that each curve has in relation to the standard curve, here determined by the clustering methods for their centers.

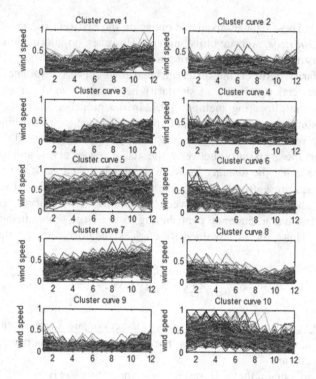

Fig. 1. Typical wind curves profiles for a horizon of 12 hours and wind curves that belong to the cluster

3.2   Kalman Filters Approach

Kalman filters in its original version assume that the system model is linear and the model inaccuracies are due to disturbances or errors of readings, which are characterized statistically by having an average value of zero and a Gaussian distribution of values [11]. However, as a result of the analysis presented in section 2, we verified that the behaviour of the wind is not always the same and in the studied location (Alvadia) the wind can be characterized in 10 different regimes. Therefore, the proposed model contains a bank of filters, one for each cluster of wind curves, so these phenomena can be modelled. Also, each of these models contains a set of 24 parameters which are linked to the behaviour of the wind at a certain time of the day. The wind speed prediction for the moment $k+1$ is provided by a linear regression of order b. The 24 parameters that make part of the proposed approach, only b parameters (related to the hours of the wind) are always used by the forecast model. As a result:

$$V_{k+1} = \sum_{i=0}^{b-1} a_{h(k-i)} V_{k-i} . \tag{2}$$

Where Vk is the wind speed recorded in the sample k, and $h(k)$ is the time of day that the sample k occurred. Function h plays an important role here since it makes the template use the parameter with an index equal to the daily time of sample k adapting the model to the correct phase of the day. The global forecast proposed model is therefore, based on the use of a bank with 10 Kalman filters corresponding each one to a different cluster (Fig. 2).

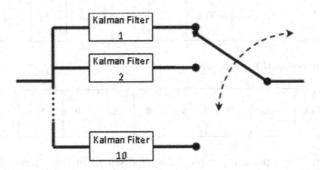

Fig. 2. Kalman filters bank

Each of the Kalman filter is tasked to adjust the parameters of the linear model (2). The observable quantity of Kalman filter is the wind speed or the wind power potential. To forecast the wind speed of the next hour it is considered the wind speed values in a past horizon of $b = 12$ hours. In this case, the structure of each of the 10 filters used in the forecasting is the same, so the model is made without reference to the number of filter being used. The model parameters to find are grouped in vector X, that is:

X=	a_0	a_1	a_2	...	a_{22}	a_{23}

Where ai is the parameter of the model that represents the i hour of the day. It is expected that the model parameters are constant from sample to sample so the random errors are added to the model:

$$X_{k+1} = X_k + W_k . \tag{3}$$

The variable factor, with a null average value and variable density of normal probability is assumed by the variable W. Thus, each of Kalman filters is tasked to estimate the parameters according to the associated model and the respective cluster. The observation of the behavior of the model is evaluated by the assertiveness of the forecast of the wind speed (or wind power) at moment k, obtained by linear regression of b past wind speeds. Such estimates go through Hk vector construction.

$H_k=$	0	...	0	V_{h-12}	...	V_{h-2}	V_{h-1}	V_h	...	0

Where $h = hour(k)$ is the time of day that succeeded sample k. Being the b value < 24 part of Hk vector elements are null values. The other elements are wind speed placed in corresponding positions to the daily time of their occurrence. Below, two examples illustrate how vector H is built:

i) is k such as $hour(k) = 12$ hours.

ii) is k such as $hour(k) = 3$ hours.

$$\mathbf{H_k=} \quad \boxed{V_{k-3}} \; \boxed{V_{k-2}} \; \boxed{V_{k-1}} \; \boxed{V_k} \; \boxed{0} \; \boxed{\cdots} \; \boxed{0} \; \boxed{V_{k-11}} \; \boxed{V_{k-10}} \; \boxed{\cdots}$$

To summarize our approach, the model is represented in the discrete form by the following equations:

$$X_{k+1} = X_k + w_k . \tag{4}$$

$$z_{k+1} = H \times X_k + v_k . \tag{5}$$

Where Xk is the vector of state (prediction model parameters) and Zk is the forecasting. The estimation of the model parameters follows the following steps:

Step 1: Calculate the Kalman gain:

$$K_k = P_k^- \times H^T \times (H \times P_k^- \times H^T + R_k)^{-1} . \tag{6}$$

Step 2: *A priori* correction of estimation in its *a posteriori* form:

$$X_k = \hat{X}_k^- + K_k \times (z_k - H \times \hat{X}_k^-) . \tag{7}$$

Step 3: Calculation of the covariance matrix of *a posteriori* error:

$$P_k = (I - K_k \times H) \times P_k^- . \tag{8}$$

Step 4: Projection of error covariance matrix for *a priori* estimation for the next interactive cycle:

$$P_{k+1}^- = P_k + Q_k . \tag{9}$$

Step 5: Calculation of the *a priori* estimation for the next iterative cycle:

$$\hat{X}_{k+1}^- = \hat{X}_k . \tag{10}$$

4   Wind Power Prediction Results

The proposed approach has been applied for wind speed prediction in a wind farm located in Serra do Alvão, near a town called Alvadia, in Portugal. The prediction of wind power production was carried out taking into account the production data from only one farm and not the entire wind production at a national level. The production forecast at national level is, despite some difficulties, more predictable and less complex. Installed in 2004, the wind farm has six Enercon wind turbines model E66 20.70 with 2 MW of power. With 12 MW of installed power, the wind farm can charge a lower power due to low winds or due to any malfunction or stoppage of wind turbines. The wind power forecast occurred as follow: in the first step, the data with the last 12 hours of wind speed information and the corresponding time were imported from the available spreadsheet to a vector v (wind speed) and to a vector t (hour and data of the day). Next, the proposed clustering approach was applied to obtain 10 different data clusters organized by the similarity between the wind curves. The clustering method was implemented in an iterative way to a maximum of 100 iterations, returning the information to which clusters belongs each wind curve. After that, the bank of Kalman filters was applied to the clustering data using the follow equation:

$$V(k) = v(K-1) \times Xaposteriori(hour(k-1) + \cdots + v(k-b) \\ \times Xaposteriori(hour(k-b) \tag{11}$$

As a result, we obtain the wind speed forecast at the moment k, based on the previous winds values ranging from k-1 to k-b. Then, the *a priori* and *a posteriori* states necessary to achieve the wind estimation are calculated like mentioned on the previous section. The accuracy of the proposed approach is evaluated in function of the actual wind speed/power that occurred. For that purpose, the mean squared error of the prediction is calculated. This means that, a lower value of error corresponds to a better forecasting. Fig. 3 shows wind speed forecasting. The prediction error for the considerate period was: 0.6254398 m/s, and the persistent reference method error was: 0.9070827 m/s.

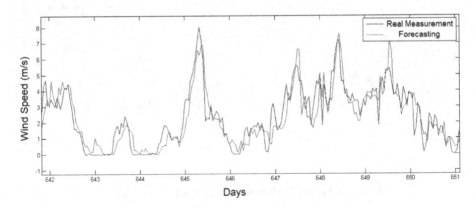

Fig. 3. Wind speed prediction

For the wind power generation forecasting the Enercom E66 20.70 wind turbine power curve is used (Fig. 4). This power curve relates the turbine's production capacity with the local wind speed, which depends on the characteristics of each wind turbine. Thus, it's possible to relate the values of wind speed with the power output generated. By multiplying this value to the number of turbines (in this case six) we obtain the prediction of power produced by each on the wind farm in relation to the wind measured. As it is well known low-amplitude wind speed (values close to zero) does not contribute to the effective production of wind power so the forecast for this range of values is not relevant.

Numerical results with the proposed approach are shown in Fig. 5. This graph shows the actual wind power (blue line), together with the forecasted wind power (red line) over a period of two years and three months. In this graph, the wind power production is forecast with an hour in advance through the values of the wind speed measured in the previous 12 hours. As shown in the figure, the forecasted wind power curve follows quite well the actual wind power curve. The prediction error for the considerate period was relatively low: 241.15 while the persistence forecast model gives a 619.65 error, which is substantially higher.

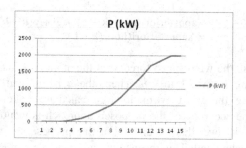

Fig. 4. Enercon E-66 20.70 power curve

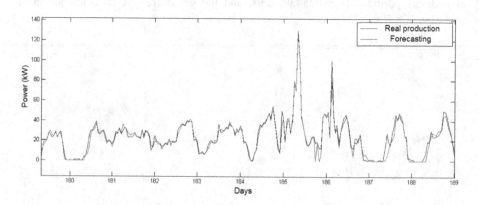

Fig. 5. Wind power prediction of the E-66 turbine (2 000 kW)

5   Conclusion

In this paper we have presented a model for short-term wind power forecasting. The proposed approach combines clustering techniques and a bank of Kalman filters: 10 different clusters characterize the local winds history and the bank of Kalman filters performs the wind speed estimation. The results from the prediction of a wind farm in Portugal show that the proposed approach provides accurate and efficient local wind power forecasting. Accordingly, this model can be used as an alternative to current methods allowing in particular, to trigger mechanisms if there is an excess of estimated wind energy production preventing this energy to be wasted, as had happened so far. As a future work, the results from the proposed model can be well extended and evaluated in other different locations.

References

1. Silva, B.: China e Índia fazem consumo mundial de energia disparar 40%. Diário Económico, 22–23 (2010)
2. Lange, M., Waldl, H.-P.: Assessing the uncetiainty of wind power predictions with regard to specific weather situations. In: Proc. of the 2001 European Wind Energy Association Conference, EWEC'UI, Copenhagen (Denmark), July 2-6, pp. 695–698 (2001)
3. Torres, J.L., García, A., De Blas, M., De Francisco, A.: Forecast of hourly average wind speed with ARMA models in Navarre (Spain). Solar Energy 79(1), 65–77 (2005)
4. Fonte, P.M., Quadrado, J.C.: ANN approach to WECS power forecast. In: 10th IEEE Conference on Emerging Technologies and Factory Automation, September 19-22, vol. 1, pp. 1069–1072 (2005)
5. Potter, C.W., Negnevitsky, M.: Very short-term wind forecasting for Tasmanian power generation. IEEE Transactions on Power Systems 21(2), 965–972 (2006)
6. Giebel, G., Brownsword, R., Kariniotakis, G., Denhard, M., Draxl, C.: The State-Of-The-Art in Short-Term Prediction of Wind Power: A Literature Overview, 2nd edn. ANEMOS.plus (2011)
7. REN - Rede Eléctrica Nacional, S.A. Caracterização da Rede Nacional de Transporte para efeitos de acesso à rede em 31 de Dezembro de 2010 (2011)
8. REN - Rede Eléctrica Nacional, S.A. A Energia Eólica em Portugal 2010 (2011)
9. OMEL Mercados A. V. Preços da tarifa de electricidade entre Espanha e Portugal, http://www.omel.es/files/flash/ResultadosMercado.swf (accessed in May 2, 2011)
10. Xu, R., Wunsch II, D.: Survey of clustering algorithms. IEEE Transactions on Neural Networks, 645–678 (2005)
11. Lee, C.R., Salcic, Z.: High-performance FPGA-based implementation of Kalman filter. Microprocessors and Microsystems 21, 257–265 (1997)

Using RISE Observer to Implement Patchy Neural Network for the Identification of "Wing Rock" Phenomenon on Slender Delta 80° Wings

Paraskevas M. Chavatzopoulos[1], Thomas Giotis[1], Manolis Christodoulou[1], and Haris Psillakis[2]

[1] Department of Electronic and Computer Engineering, Technical University of Crete
Chania, Crete, Greece
pchavatzopoulos@yahoo.gr, thgiotis@gmail.com, manolis@ece.tuc.gr
[2] Department of Electrical Engineering, Technological & Educational Institute of Crete
Heraklion, Greece
psilakish@hotmail.com

Abstract. In this paper, the "wing rock" phenomenon is described for slender delta 80° wing aircrafts on the roll axis. This phenomenon causes the aircraft to undergo a strong oscillatory movement with amplitude dependent on the angle of attack. The objective is to identify "wing rock" using the Patchy Neural Network (PNN), which is a new form of neural nets. For the update of the weights of the network, an observer called RISE (Robust Integral of Sign Error) and equations of algebraic form are used. This causes the PNN to be fast, efficient and of a low computational cost.

Keywords: wing rock phenomenon, RISE observer, Patchy Neural Network (PNN), slender delta 80° wing aircrafts.

1   Introduction

Neural networks are the future of modern computing. They are volatile, applicable to all problems solved with hardware, software or firmware, fast and most importantly they are evolving constantly. Modern aircrafts need a fast and error tolerance computing method; this makes neural networks essential for aerospace.

In this paper, we present the "wing rock" phenomenon for slender delta 80° wing aircrafts. The motion of the aircraft, during this phenomenon, in the roll axis is described by the equation given in [2]. The concern of this paper is to identify this phenomenon using the Patchy Neural Network (PNN). This is a new form of neural networks proposed in [8]. For the renewal of the network's weights, the use of an observer called RISE (Robust Integral of Sign Error) is needed and equations of algebraic form. This causes the PNN to be quicker, more efficient and to have less computational cost than neural nets used now.

The MATLAB code used for the implementation can be found in [12].

C. Jayne, S. Yue, and L. Iliadis (Eds.): EANN 2012, CCIS 311, pp. 264–271, 2012.

2   "Wing Rock" Phenomenon

Wing rock is the phenomenon during which the airplane is undergoing a strong oscillatory movement. For the purpose of this paper, the phenomenon is only approached in the roll axis.

"Wing rock" is an oscillation that occurs when the airplane increases its angle of attack(α). In this paper, we focus on this phenomenon occurring on slender tailless delta 80° wing aircraft; the model of the aircraft used to observe the phenomenon has wings swept back forming the Greek letter "Δ"(delta) and it does not have any tail wings. The aspect ratio between the wingspan and the mean value of the

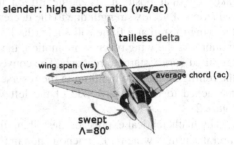

Fig. 1. Slender tailless delta 80° wing aircraft

wing's chord is high; therefore making the wings slender, the aircraft becomes aerodynamically more efficient and perfect for maneuvering in low speeds (subsonic jet speeds). The angle of the sweep Λ is the angle between the leading edge of the wing and the body of the aircraft; for this paper the aircraft has Λ=80°.

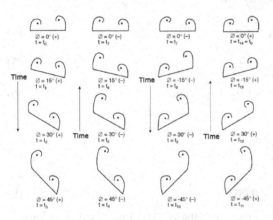

Fig. 2. Graphic representation of vortices for one cycle of wing rock

The case of the phenomenon of the "wing rock" on swept delta wings is called slender wing rock. It takes place at high angles of attack due to vortex asymmetry at the leading edge of the wing; this asymmetry can be the result of flying conditions. Furthermore, flying at high angles of attack in asymmetric flow conditions (landing in a cross-wind) or induced lateral oscillations due to unsteady flow over the wing could lead to the shedding of an asymmetric vortex.

As a result, the leading edge leeward vortex shifts outboard and the leading edge windward vortex shifts inboard causing the wing to initially roll in the positive roll direction. The sudden roll movement causes the leeward vortex on the up-going wing to compress and the windward vortex on the down-going wing to stretch, which increase the initiated rolling moment. As the roll angle increases, the kinematic-coupling between the angle of attack and the sideslip causes the effective angle of attack on the wing to decrease and the effective sideslip to increase. The increased sideslip on the wing during roll causes the windward vortex on the down-going wing to move inboard and toward the surface and the leeward vortex on the up-going wing to move outboard and lifted off. The convective time lag associated with the motion of the lifted off vortex causes the right wing to continue dipping until the lifted vortex takes its final position. Then the lift on the down-going wing together with the reduction of vortex strength due to the decrease of the effective angle of attack causes the wing to stop at a finite roll angle (the limit cycle amplitude) and then reverses its motion. As the wing reverses its motion, the effective angle of sideslip decreases and the lifted vortex starts to reattach. The convective time lag of the vortex motion helps the rolling moment to build up in the reversed direction until the vortex is completely reattached to the leeward side of the left wing. This is graphically represented in *figure 2*.

The mathematical expression describing the movement of a slender delta 80° wing aircraft during "wing rock" phenomenon in the roll axis is presented in [2].

$$\ddot{\varphi} + \omega^2\varphi = \mu\dot{\varphi} + b_1|\varphi|\dot{\varphi} + b_2|\dot{\varphi}|\dot{\varphi} + b_3\varphi^3 \tag{1}$$

where φ is the roll angle of the plane, $\dot{\varphi}$ its first derivative and $\ddot{\varphi}$ its second derivative. The terms ω^2, μ, b_1, b_2 and b_3 are given using the expressions

$$\omega^2 = -Ca_1 \tag{2}$$

$$\mu = Ca_2 - D \tag{3}$$

$$b_1 = Ca_3 \tag{4}$$

$$b_2 = Ca_4 \tag{5}$$

$$b_3 = Ca_5 \tag{6}$$

where

$$C = \frac{pU_\infty^2 Sb}{I_{xx}} = 0.354 \tag{7}$$

$$D = 0.001 \tag{8}$$

and the coefficients a_i depend on the angle of attack(α) and provided in the paper. The values of a_i are given for angle of attack 15^0 and 25^0, as well as the table with the variables used.

Table 1. Values of coefficients a_i for $\alpha=15^0$ and 25^0

α	α_1	α_2	α_3	α_4	α_5
15	-0.0102590	-0.0214337	0.05711631	-0.0619253	-0.14664512
25	-0.0525606	0.04568407	-0.17652355	0.0269855	0.06063813

Table 2. Values of variables used

Variable	Quantity	Value
ρ	air density	_1.1955 kg/m3_
S	wing area	_0.0324 m2_
I_{xx}	mass moment inertia	_0.27·10-3 kg·m2_
U_∞^2	characteristic speed	_15 m/s_

The use of MATLAB is necessary for the simulation of the movement the airplane is undergoing during "wing rock". The simulation of the movement is shown below.

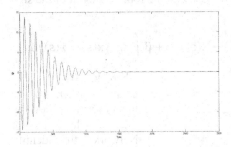   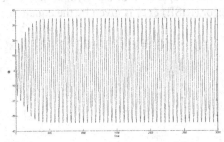

Fig. 3. Roll angle as a function of time for $\alpha=15^0$. The oscillation fades out with time.

Fig. 4. Roll angle as a function of time for $\alpha=25^0$. It starts with a relatively small magnitude and builds up in time, until it reaches a peak.

The figures above are result of the MATLAB code used to simulate the phenomenon.

Comparing these two figures, it can be deduced that a way to prevent wing rock from happening is to have an angle of attack under 20^0; wing rock can also be prevented with the use of flaps.

The simulation also outputs the phase plane of the movement in the roll axis φ to its first derivativeφ̇, which is required for the weight update of the neural network discussed later.

3   RISE Observer

When only some measurements or outputs are available for identification and control, it is usually necessary to estimate the rest of the system's states. This leads to the development of linear and non linear techniques with the use of observers or estimators. Observers give an estimation of non finite states; therefore the estimated states must be accurate in detecting possible errors and failures.

One approach of designing non linear observers is the high gain category. The design of these estimators aims in the disassociation of the linear and non linear part. Then, the gain of the estimator is selected so as the linear part to dominate over the non linear one. Choosing a very high gain can result in having a relatively small observation error, but it will cause large amplitude oscillation to the system's noise.

The noise levels and the dynamic of the system change with the passing of time and the accuracy of the methods is diminished. Thus, the use of robust observers or methods of estimation with noise tolerance is essential. One method that could be used is the RISE (Robust Integral of Sign Error) observer. This method increases the robustness of the observer and at the same time the effectiveness of error tracing in comparison to typical estimators. The observer suggested offers, beyond robust estimation in case of noise, an area of asymptotic stability for state estimation. Simulations have shown that RISE is capable to estimate states with 25% accuracy with rising noise levels.

For this paper, an observer is developed, so that it can be used by a neural network for learning and unknown state identification and RISE feedback for robustness. The full state approach of the asymptotic estimator based on the robust integral of the sign error (RISE) $\xi(t)$ is given by the equation

$$\dot{\hat{x}}(t) = \xi(t) = k\big(x(t) - \hat{x}(t)\big) + \lambda\kappa \int_0^t (x(s) - \hat{x}(s))ds + \beta \int_0^t sgn\big(x(s) - \hat{x}(s)\big)ds$$

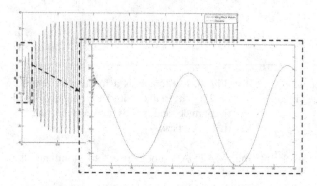

where $k, \lambda, \beta > 0$, $\dot{x}$ is $\ddot{\varphi}$ and x is $\dot{\varphi}$. To use this observer to develop the Patchy Neural Network to identify wing rock phenomenon, the variables have the following values

$$k = 2, \lambda = 1, \beta = 20.$$

These values are given after simulations, according to *Lemma 1* of reference [8]. The observer was created with the use of MATLAB which resulted in the

Fig. 5. Comparison between the simulation of the motion of the aircraft during wing rock and the estimation of the motion from RISE observer for α=25°

figures shown below. For the development of this observer, the use of the values of φ found in the simulation of the phenomenon is essential.

4   Patchy Neural Network(PNN)

For the construction of the neural network, a new category of local networks is used; it is called patchy neural network (PNN) with basis' functions that are patches of the state space. This specific network with a sufficient number of nodes can approximate a general smooth non linear function over a given compact region with the desired accuracy. PNN network is used to extract and store information obtained by the estimation of an observer, using a simple algebraic weight update law. The way PNN works is presented.

Let n-dimensioned rectangles

$$I := I_1 \times I_2 \times \cdots \times I_n$$

and δ partitions of each interval I_i mathematically expressed as

$$I_i = \bigcup_{j=1}^{N_i} A_{i,j} = \bigcup_{j=1}^{N_i} [a_{i,j-1}, a_{i,j}]$$

with $\alpha_{i,j} = a_{i,0} + j\delta$, $1 \leq i \leq n$.

Patch functions are defined on the sets $A_{1,i_1} \times \cdots \times A_{n,i_n}$, where $1 \leq i_j \leq N_i$ and $1 \leq i \leq n$

$$p_{i_1,i_2,\cdots,i_n}(x) = \begin{cases} 1, & \text{if } x \in A_{1,i_1} \times \cdots \times A_{n,i_n} \\ 0, & \text{else} \end{cases}$$

A patchy neural network is a neural network with one hidden layer and basis vector that consists of patch functions with output

$y = \sum_{i_1=1}^{N_1} \cdots \sum_{i_n=1}^{N_n} w_{i_1,\cdots,i_n} p_{i_1,\cdots,i_n}(x) = W^T P(x)$,

where

$W = [w_{1,\cdots,1}, \cdots, w_{N_1,\cdots,N_n}]^T \in \mathbb{R}^{N_1 \times \cdots \times N_n}$

and

$$P(x) = [p_{1,\cdots,1}, \cdots, p_{N_1,\cdots,N_n}]^T$$

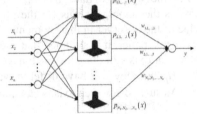

Fig. 6. Structure of Patchy Neural Network (PNN)

The network's weights are updated algebraically with the use of the equation

$$\hat{w}_{i_1,\cdots,i_n}^j(t) = \left(1 - p_{i_1,\cdots,i_n}(x(t))\right) \hat{w}_{i_1,\cdots,i_n}^j(t^-) + p_{i_1,\cdots,i_n}(x(t))\xi(t)$$

or equivalently

$$\hat{w}_{i_1,\cdots,i_n}^j(t) = \begin{cases} \hat{w}_{i_1,\cdots,i_n}^j(t^-), & \text{if } p_{i_1,\cdots,i_n}(x(t)) = 0 \\ \xi_j(t), & \text{if } p_{i_1,\cdots,i_n}(x(t)) = 1 \end{cases}$$

where $j = 1,2,\ldots,n$ with initial values $\hat{w}_{i_1,\cdots,i_n}^j(0) = 0$ for $1 \leq i_1 \leq N_1, \ldots, 1 \leq i_n \leq N_n$ and $1 \leq j \leq n$. The vector $\hat{W}^T P(x)$ with $\hat{W} := [\hat{W}_1 \cdots \hat{W}_n]$, with $\hat{W}_i = [\hat{w}_{1,\cdots,1}^i, \cdots, \hat{w}_{N_1,\cdots N_n}^i]^T$ can be used to estimate $f(x)$.

Specifically, the figure of the equation to be estimated is given. I_i corresponds to the i axis, which consists of N_i portions $A_{i,j}$. Each

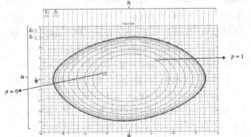

Fig. 7. Patches over the phase plane for $\alpha=25^0$. Each square box is a patch.

portion is an area $[a_{i,j-1}, a_{i,j}]$ with δ length. Each patch has such a portion and the union of these portions $\bigcup_{j=1}^{N_i} A_{i,j} = \bigcup_{j=1}^{N_i}[a_{i,j-1}, a_{i,j}]$ is I_i. If there is a point of x in that portion on the figure, then p equals 1, else is zero.

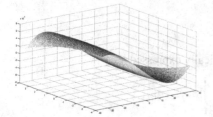 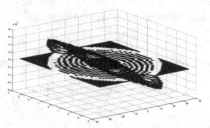

Fig. 8. Function $f(\varphi, \dot{\varphi})$ for $\alpha=25^0$. Shows the three dimensional distribution of the equation.

Fig. 9. PNN approximation for $\alpha=25^0$. It shows the networks weights of the PNN. Has the form of the phase plane shown in Figure 7.

For the weight update, the value of p is being used. If $p = 0$, then the weight remains the same as it was for the previous moment for that patch, else it takes the value of the observer in that moment.

For this paper, it is assumed that each axis will be from lowest value to the highest value that exists in the axis. Each axis will be split into the same number of patches, $N_i = N_j$, but they will have different value for δ, which depends on the length of the axis. As such, δ for each axis is calculated

$$\delta_{axis} = \frac{\max(axis) - \min(axis)}{N_{axis}}$$

For the construction of the neural network, $\dot{\varphi}$ is the input in the observer. The figure used is the phase plane given from the simulation of the movement during wing rock.

PNN is also implemented with the use of MATLAB.

5   Conclusions

For angle of attack smaller than 20^0, the oscillation occurring during wing rock fades out with time. But, in higher angles the oscillation starts with relatively small amplitude until it reaches maximum amplitude and continues "rocking" at that. The amplitude and the frequency of the oscillation in the final state do not depend on initial values, but on the angle of attack. The "wing rock" phenomenon can be eliminated with the use of the flaps of the aircraft or by reducing the angle of attack.

Creating the observer needs the use of the values of φ and $\dot{\varphi}$ obtained in the phenomenon's simulation. RISE observer is very fast in estimating the targeted function and is superior to other observers as it can estimate states with 25% accuracy with rising noise levels, increases the robustness of the observer and at the same time the effectiveness of error tracing. It also offers an area of asymptotic stability for state estimation.

Patchy Neural Network is easy and simple to its learning. PNN is capable of approximating generalized non linear function. It gives some advantages over other networks; the network is capable of learning the unknown nonlinearity in some region of the state space from a single visit of the state trajectories to the patches of the region. PNN has weight update laws of algebraic form, not given in the form of differential equations, resulting in the significant reduction of the computational cost for learning since only n ODEs are solved form the observer, in contrast to $N_1 \times \cdots \times N_n$ ODEs needed to train the weights of a neural network of $N_1 \times \cdots \times N_n$ nodes.

The MATLAB code developed in [12] makes possible the simulation of the "wing rock" phenomenon, leading to the creation of the RISE observer and the construction of the Patchy Neural Network for the identification of the phenomenon. So, this paper concludes that the identification of wing rock, with the use of PNN and the aid from RISE observer is feasible, easily implemented with the use of MATLAB code. This is a test for the PNN method on nonlinear equations. There is a linear equation tested in [8]. It proves that this method works for nonlinear, as well as linear, equations.

References

1. Guglieri, G., Quagliotti, F.B.: Analytical and experimental analysis of wing rock. Nonlinear Dynamics 24, 129–146 (2001)
2. Elzebda, J.M., Nayfeh, A.H., Mook, D.T.: Development of an analytical model of wing rock for slender delta wings. Journal of Aircraft 26(8), 737–743 (1989)
3. Guglieri, G., Quagliotti, F.: Experimental observation of the wing rock phenomenon. Aerospace Science and Tecnology, 111–123 (1997)
4. Gurney, K.: An Introduction to neural network, p. 234. ULC Press (1997)
5. Abbasi, N.: Small note on using Matlab ode45 to solve differential equations (August 2006)
6. Slender Wing Theory, http://soliton.ae.gatech.edu
7. MATLAB, The Language of Technical Computing. Getting Started with MATLAB. Version 5. The Mathworks (December 1996)
8. Psillakis, H.E., Christodoulou, M.A., Giotis, T., Boutalis, Y.: An observer approach for deterministic learning with patchy neural network. International Journal of Artificial Life Research 2(1), 1–16 (2011)
9. Wang, C., Hill, D.J.: Deterministic learning theory for identification, recognition and control, 1st edn., p. 207. CRC Press (2009)
10. Fonda, J.W., Jagannathan, S., Watkins, S.E.: Robust neural network RISE observer based fault diagnostics and prediction. In: Proc. of the IEEE Internation Conference on Neural Networks (2010)
11. Nelson, R.C., Pelletier, A.: The unsteady aerodynamics of slender wings and aircraft undergoing large amplitude maneuvers. Progress in Aerospace Sciences 39, 185–248 (2003)
12. Paraskevas-Marios, C.: Use of observer, based on neural networks, for identification of the "wing rock" phenomenon on delta 80 degrees wing aircrafts, p. 130. Technical University of Crete, Dept of Electronic and Computer Engineers, Chania (2011)

Models Based on Neural Networks and Neuro-Fuzzy Systems for Wind Power Prediction Using Wavelet Transform as Data Preprocessing Method

Ronaldo R.B. de Aquino[1], Hugo T.V. Gouveia[1], Milde M.S. Lira[1],
Aida A. Ferreira[2], Otoni Nobrega Neto[1], and Manoel A. Carvalho Jr.[1]

[1] Federal University of Pernambuco (UFPE)
Recife, Brazil
[2] Federal Institute of Education, Science and Technology of Pernambuco(IFPE),
Recife, Brazil
{rrba,milde,otoni.nobrega,macj}@ufpe.br,
{hugotvg,aidaaf}@gmail.com

Abstract. Several studies have shown that the Brazilian wind potential can contribute significantly to the electricity supply, especially in the Northeast, where winds present an important feature of being complementary in relation to the flows of the San Francisco River. This work proposes and develops models to forecast hourly average wind speeds and wind power generation based on Artificial Neural Networks, Fuzzy Logic and Wavelets. The models were adjusted for forecasting with variable steps up to twenty-four hours ahead. The gain of some of the developed models in relation to the reference model was of approximately 80% for forecasts in a period of one hour ahead. The results showed that a wavelet analysis combined with artificial intelligence tools provides more reliable forecasts than those obtained with the reference models, especially for forecasts in a period of 1 to 6 hours ahead.

Keywords: Wind Energy, Artificial Intelligence, Fuzzy Logic, Wind Forecasting, Neural Networks, Time Series Analysis, Wavelet Transforms.

1 Introduction

Among all alternative energy sources currently explored, wind energy is one of the most successful. One reason for this is the incentive policy by various countries, which ensures the purchase of electricity produced from wind power, even if it does not offer competitive prices. Germany and Denmark were the first countries to adopt policies to encourage the development of wind generation, followed by several countries, including Brazil, with the creation of the Incentive Program for Alternative Sources of Electric Energy – PROINFA.

The use of wind kinetic energy for electric power generation presents some inconveniences related to the uncertainties in the generation. Such inconveniences are basically caused by the variability of the wind speed. Therefore, the application of

C. Jayne, S. Yue, and L. Iliadis (Eds.): EANN 2012, CCIS 311, pp. 272–281, 2012.
© Springer-Verlag Berlin Heidelberg 2012

tools or techniques that are able to forecast the energy to be generated by those sources is essential to an appropriate integration of the wind source with the electric power system [1].

This paper suggests the application of Artificial Neural Networks – ANN, Adaptive Neuro-Fuzzy Inference Systems – ANFIS and multiresolution analysis by Wavelet Transform – WT toward the development of tools able to accomplish wind speed and wind power generation forecasting. The generated tools aim to subsidize planning of a hybrid generation system (hydrothermal and wind generations), mitigating some inconveniences of wind power generation [2].

The forecasting models evaluated in this study are developed from simple wind speed time series from the local wind site studied, due to a difficulty in obtaining more data from other sites. A similar kind of univariate model based on reservoir computing for wind forecasting was shown in [3].

In this paper, the created models are univariate, i.e., a single variable (wind speed) is used as an explanatory variable of the forecasting models.

The wind speed time series used to develop the models consist of hourly average values measured in two different cities in the state of Rio Grande do Norte. These cities are: MACAU and MOSSORÓ. The data were measured at a 10m height between January and December 2008 and the size of each series is equal to 8784. A detailed analysis of these and others databases was presented in [4].

2 Data Preprocessing

2.1 Normalization

The forecasting models based on ANN demand consistent treatment of the data in order to guarantee reasonably good performance and their effective application. In general, normalization is a straightforward and effective treatment for the data [1].

Normalization is carried out to assure that all variables used in the model inputs have equal importance during training; therefore the activation functions of artificial neurons are limited. In this paper, the data were normalized according to (1), also shown in [4], whose values ranged from 0 to 1.

$$\overline{X}(t) = \frac{X(t) - X_{min}}{X_{max} - X_{min}} \tag{1}$$

where: X and $\overline{X}$ are the wind speed time series non-normalized and normalized, respectively; X_{max} and X_{min} are the maximum and the minimum absolute value of the wind speed time series, respectively.

2.2 Multiresolution Analysis

The wind speed time series have a highly oscillatory behavior, so the development of models that provide satisfactory forecasts is a very complex task.

In order to mitigate the effects of wind speed variability and accomplish comparative studies among several forecasting models, a multiresolution analysis by WT is proposed in this paper. This analysis facilitates adjusting the parameters and training of the forecasting models.

The multiresolution analysis decomposes signals into different time scales resolutions. This analysis is accomplished through the pyramidal algorithm, which makes WT calculation more efficient. This algorithm makes it possible to obtain "approximation" and "details" of a given signal [1].

After preliminary tests of decomposition by wavelets using several wavelet families to identify which one could be the best to decompose the original wind speed signal, it was found that the *Daubechies* family wavelets were more robust [1].

Once defined the wavelet family to be used, an analysis was accomplished to decide on the best level to decompose the signal. In this analysis, it was found that the signals should be decomposed at level 3, because at higher decomposition levels, the detail coefficients did not present relevant information for the forecasting models [1].

2.3   Conversion of Wind Speed Forecasting to Wind Power Generation

The selected turbine was ENERCON E-70 E4, which is designed to operate at a 57m height. However, the wind speed time series were measured at a 10m height, consequently, the input and output patterns used to adjust and train the forecasting models are related to this height. Therefore, these models accomplish wind speed forecasting at a reference 10m height.

Before transforming the wind speed forecasting into wind power forecasting using the wind turbine manufacturer's power curve, it is necessary to convert the wind speed forecasting from 10m height to the turbine height of 57m.

The conversion formula is given in (2), using the coefficient of roughness $\alpha = 0.1$. In this equation, s represents the wind speed at a height of H and s_0 represents the wind speed forecasting at a reference height of H_0.

$$s = s_0 \left(\frac{H}{H_0} \right)^{\alpha}$$

(2)

3   Performance Evaluation

The three basic types of errors used in this paper to illustrate the performance of a wind speed forecasting model are: the mean absolute error – *MAE*, the mean squared error – *MSE* and the normalized mean absolute error – *NMAE* . The equations used to calculate them are defined as in (3), (4) and (5):

$$MAE_s(k) = \frac{1}{N} \sum_{t=1}^{N} |e_s(t + k \,|\, t)|, \tag{3}$$

$$MSE_s(k) = \frac{1}{N} \sum_{t=1}^{N} e_s(t + k \,|\, t)^2, \tag{4}$$

where: N - number of data used for the model evaluation; k - forecast horizon (number of time-steps); $e_s(t+k|t)$ - error corresponding to time $t+k$ for the wind speed forecast made at origin time t; $s(t+k)$ - measured wind speed at time $t+k$.

The errors defined in (3) and (4) refer to the wind speed forecasts. The error used in this paper to illustrate the performance of the power generation forecasts was the normalized mean absolute error – $NMAE$. The equation used to calculate it is defined as follows:

$$NMAE_P(k) = \frac{100}{N \cdot P_{inst}} \sum_{t=1}^{N} |e_P(t + k \,|\, t)|, \tag{5}$$

where: $e_P(t+k|t)$ - error corresponding to time $t+k$ for the power generation forecast made at origin time t; P_{inst} – installed wind power capacity.

It is worthwhile developing and implementing an advanced wind power forecasting tool if it is able to beat reference models, which are the result of simple considerations and not of modeling efforts. Probably the most common reference model used in the frame of wind power prediction or in the meteorological field is the persistence model [5]. This naive predictor tool states that the variable's future value will be the same as the last one measured, i.e.

$$y_{PERS}(t + k \,|\, t) = y(t), \tag{6}$$

where: $y_{PERS}(t+k|t)$ - wind speed (or power generation) forecast for time $t+k$ made at origin time; $y(t)$ - measured wind speed (or power generation) at time t.

4   Development of the Forecasting Models

This paper proposes six different models for the forecasts of hourly average wind speeds. Four of them use ANN and the other two use ANFIS. The power generation forecasts for each model are obtained using the forecasted speeds and the wind turbine power curve. It is important to note that each model will be applied to each hour of the day. In reference [6] was presented a comparison of various forecasting approaches, including ANN and ANFIS but without use WT. The way use of WT to build the models in subchapter 4.2 is the principal novelty of this paper.

4.1   Models: ANNLM, ANNRP *and* ANFIS

These three models have the same input and output standards, i e., four inputs and one output. The input data are the last four hourly average speeds. The output is the forecast of the hourly average speed for a given horizon. Fig. 1 schematically illustrates the models.

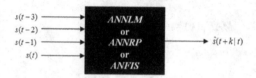

Fig. 1. Ilustration of *ANNLM, ANNRP* and *ANFIS*

The *ANNLM* and *ANNRP* models are structured with *Multilayer Feedforward* fully connected neural networks, whose training algorithms are *Levenberg-Marquardt* (LM) and *Resilient Propagation* (RP) respectively. These networks' architecture consists of an input layer, an intermediate layer (hidden layer) and an output layer.

The number of neurons in the hidden layer of neural models is determined by varying the number of neurons in this layer. The amount that provided the best performance during training was thus selected. The hidden layer neurons use the *Hyperbolic Tangent* activation function. These models' output layer consists of only one neuron, because there is only one output. This neuron's activation function is *Sigmoid Logistics*.

The *ANFIS* model also has four inputs and one output. The selection of the best inference system was performed using the *subtractive clustering* technique. The length of each cluster's influence radius was determined by varying its value. The length that provided the best performance during training was thus selected.

4.2   Models: WTANNLM, WTANNRP *and* WTANFIS

These models differ from the previous ones in the kind of inputs. Here, the inputs are the wavelet coefficients from the multiresolution analysis, using the *Daubechies* wavelet family (db10) at level 3. This causes an increase in the number of inputs from 4 to 16, since the inputs are now composed by the approximation and detail coefficients, according to the illustration shown in Fig. 2.

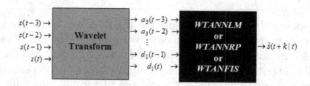

Fig. 2. Ilustration of *WTANNLM, WTANNRP* and *WTANFIS*

4.3   Databases Partition

Before performing the models' training, a matrix of input and output patterns was set up for each series of speeds, and for each forecasting horizon (1 to 24 hours ahead). For each row of these matrices standards, the first columns are the inputs and the last column is the output. For *ANNLM, ANNRP* and *ANFIS*, these arrays have five columns, while for *WTANNLM, WTANNRP* and *WTANFIS,* seventeen columns.

As stated earlier, the training standards of the models using neural networks, the matrices were normalized before performing the *ANNLM, ANNRP, WTANNLM* and *WTANNRP* training.

The database was divided into 60%, 30%, and 10% for training , validation and testing sets, respectively

4.4   Training for the Chosen Architecture

In projects involving ANN, there is a correlation among the number of patterns used in the training set, the complexity of the problem to be handled, the number of free parameters (weights and bias) and the learning process, as shown in [7]. A deterministic rule exists relating these variables but some researchers prefer certain practical rules. In this work, these variables are restricted to the number of free parameters, i.e., to the ANN architecture, because the number of training patterns is limited to the size of the database and the complexity of the problem is intrinsic to the kind of application.

In this regard, to select the ANN architecture, the rule in [4] was used. The rule states that architectures must be defined by changing the number of neurons in the hidden layer from 3 to 15, and selecting the one with the best performance (lowest *MSE*) in the validation set during the training process.

As an example, it was obtained 11, 13, 14 and 15 neurons in the hidden layer for ANNLM, ANNRP, WTANNLM and WTANNRP models, respectively, when using the Macau dataset in one hour ahead forecasting.

4.5   Choosing the Best ANN for Each Neural Model

The best ANN for each neural model was chosen by k-fold cross-validation method, where k represents the number of partitions randomly generated from the examples to train, test and validate the ANNs. In this method, the samples are divided into mutually exclusive k partitions. For each of k iterations, n different partitions are used to test the ANNs and all the others $(k - n)$ are used to train and validate [1].

It is expected that by applying this technique, the MAE value in each experiment on the test set represents the expected output of the ANN created by the k experiments, as shown in [7, 8]. Here, the k value was set equal to 10. And from the fold with lower MAE, averaged in ten initializations, it was chosen the best ANN.

4.6   Choosing the Best Inference System for Each Neuro-Fuzzy Model

The methodology used to choose the best Neuro-Fuzzy model was proposed in [4]. From the training set, the subtractive clustering technique was used in order to generate the initial Fuzzy Inference Systems – FIS with the cluster radius ranging from 0.3 to 0.7, with an increment equal to 0.1. Then, ANFIS was used to adapt the membership functions of the generated FIS. Training and validation sets were used to adapt these functions. Finally, the selected models were those that had the lowest *MAE* for the test set.

As an example, it was obtained 6 and 10 numbers of fuzzy rules for ANFIS and TWANFIS models, respectively, when using the Macau dataset in one hour ahead forecasting.

4.7   Results of the Chosen Models

The best *MAE* values for the models chosen for each of the analyzed sites are presented here.

Fig. 3 shows the results for MACAU. We can observe that for forecasts up to 7 hours ahead, models that use WT provide better forecasts than those that do not use it. From 13 hours ahead, the Neuro-Fuzzy models provide better forecasts than the others, with the *WTANFIS* model providing the best forecasts for all considered time steps.

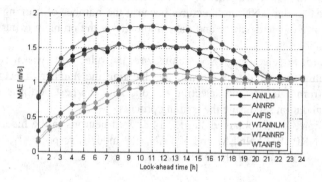

Fig. 3. Best *MAE* values for MACAU test sets

Fig. 4 shows the results for MOSSORÓ. We can see that for forecasts up to 9 hours ahead, models that use WT provide better forecasts than those that do not use it. From 19 hours ahead, the Neuro-Fuzzy models provide better forecasts than the others, with the *WTANFIS* model providing the best forecasts for all considered time steps.

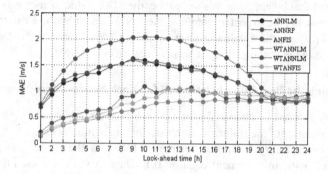

Fig. 4. Best *MAE* values for MOSSORO test sets

5   Forecasts and Comparisons between the Models

This section presents comparisons between *WTANNLM* and *WTANFIS* performances with those obtained using reference model (6)   for MACAU. *WTANNLM* and *WTANFIS* were chosen because they presented, in most cases, a smaller *MAE* for test sets.

5.1   Wind Speed Forecasts

The period chosen for the MACAU forecasts evaluation is formed by 8016 hourly average speed from 01/01/2009 to 30/11/2009.

The *MAE* values are shown in Fig. 5. We can note that the models proposed in this paper provide better forecasts than reference model for all look-ahead times. Another interesting feature of the proposed models is that up to 12 hours ahead, the average forecasts errors are significantly lower than those obtained with the reference model. From 12 to 24 hours ahead, the average errors of the proposed models hardly vary with the forecast horizon, while for the reference model these errors reduce.

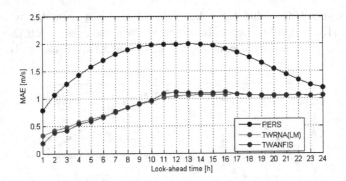

Fig. 5. *MAE* of wind speed forecasts

5.2   Power Generation Forecasts

Fig. 6 shows the power curve used to perform power generation simulations in this paper. It is a wind turbine rated at 2300 kW and the hub is 57 m high.

Manufacturers only provide a few points of the curve (red points), so to estimate the power in kW at any speed between the cut-in and cut-out, the curve was adjusted at four different intervals through the least squares method.

The *NMAE* values are shown in Fig. 7. We can observe that for the proposed models, the maximum *NMAE* is about 6.5%   of installed wind power capacity . For the persistence model, the maximum *NMAE* is about 13%   of installed wind power capacity.

According to [9], for flat terrain wind farm, typical model results for single wind forecasting are: *NMAE* around 6% of the installed capacity for the first 6 hours, rising to 9-11% for 48 hours ahead.

In [10], the *NMAE* value for six hours ahead is around 15%, while in [11], the *NMAE* is around 9% at the same horizon. In [12], *NMAE* values for forecasts from 1 to 4 hours ahead are 5.98%, 9.47%, 12.00% and 15.21%, respectively.

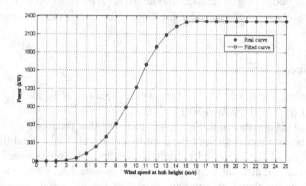

Fig. 6. Wind turbine power curve

The average *NMAE* value for 3 hours ahead forecasts with the hybrid model proposed in [13] is equal to 1.65%. In this same study, the average *NMAE* values for the persistence and new reference models are 5.24% and 5.17%, respectively.

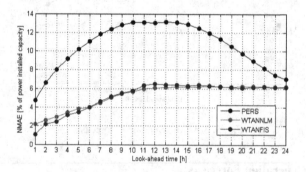

Fig. 7. *NMAE* of power generation forecasts

6   Conclusions

The models developed in this paper are based on time series analysis of wind speeds through computational intelligence. The presented forecast horizons fall within short-term forecasts for up to twenty-four hours ahead, which is an appropriate horizon to subsidize the operation planning of power systems.

All developed models provided good forecasts for all the look-ahead times considered, especially for the shorter ones (from 1 to 6 hours ahead). We found that the forecasts quality is strongly influenced by the autocorrelation series for both adopted reference models and those models that do not use WT.

Decomposition of wind speed series through WT allowed models to extract relevant information about the cyclical and seasonal behavior of wind speeds. The information presented in the wavelet transforms through its approximation and detail components was crucial to significantly improve the developed forecasts with models that use these signals as inputs.

Through the analysis, we found that the behavior of forecast errors for different horizons was very similar for all the three sites. The adopted methodology to develop the models was very satisfactory, which ensured statistically reliable forecasts, i.e., the models acquired the ability to generalize without becoming biased.

References

1. Aquino, R.R.B., et al.: Application of wavelet and neural network models for wind speed and power generation forecasting in a brazilian experimental wind park. In: Proceedings of International Joint Conference on Neural Networks, pp. 172–178. IEEE Press, Atlanta (2009)
2. Ahlstrom, M., Jones, L., Zavadil, R., Grant, W.: The future of wind forecasting and utility operations. IEEE Power & Energy Magazine 3(6), 57–64 (2005)
3. Ferreira, A.A., Ludermir, T.B., Aquino, R.R.B., Lira, M.M.S., Neto, O.N.: Investigating the use of reservoir computing for forecasting the hourly wind speed in short–term. In: Proceedings of International Joint Conference on Neural Networks, pp. 1950–1957. IEEE Press, Hong Kong (2008)
4. Gouveia, H.T.V.: Wind forecasting and wind power generation in brazilian northeast: analyzing different sites and looking for the best modeling based on artificial intelligence. M.S. thesis, Dep. Elect. Eng., UFPE University, Recife, Brazil (2011) (in Portuguese)
5. Madsen, H., Kariniotakis, G., Nielsen, H.A., Nielsen, T.S., Pinson, P.: A protocol for standardising the performance evaluation of short-term wind power prediction models. In: Proceedings of Global Wind-Power Conference (CD-ROM), Chicago (2004)
6. Sfetsos, A.: A comparison of various forecasting techniques applied to mean hourly wind speed time series. Renewable Energy 21, 23–35 (2000)
7. Haykin, S.: Neural networks: a comprehensive foundation, 2nd edn. Prentice Hall, NJ (1998)
8. Prechelt, L.: Proben1 - a set of neural network benchmark problems and benchmarking rules, Technical Report, pp. 21–94 (1994)
9. Giebel, G., et al.: The State of the Art in Short-Term Prediction of Wind Power A Literature Overview, 2nd edn., ANEMOS.plus/SafeWind, Deliverable Report (2011)
10. Miranda, V., Bessa, R., Gama, J., Conzelmann, G., Botterud, A.: New Concepts in Wind Power Forecasting Models. In: WindPower 2009 Conference and Exhibition Centre, Chicago, Illinois (2009)
11. Sideratos, G., Hatziargyriou, N.D.: An Advanced Statistical Method for Wind Power Forecasting. IEEE Transactions on Power Systems 22(1), 258–265 (2007)
12. Kusiak, A., Zheng, H., Song, Z.: Short-Term Prediction of Wind Farm Power: A Data Mining Approach. IEEE Transactions on Energy Conversion 24(1), 125–136 (2009)
13. Catalão, J.P.S., Pousinho, H.M.I., Mendes, V.M.F.: Hybrid Wavelet-PSO-ANFIS Approach for Short-Term Wind Power Forecasting in Portugal. IEEE Transactions on Sustainable Energy 2(1), 50–59 (2011)

Neural Networks for Air Data Estimation: Test of Neural Network Simulating Real Flight Instruments

Manuela Battipede, Piero Gili, and Angelo Lerro

Politecnico di Torino, 10129 Torino, Italy
Department of Mechanical and Aerospace Engineering
{manuela.battipede,piero.gili,angelo.lerro}@polito.it

Abstract. In this paper virtual air data sensors have been modeled using neural networks in order to estimate the aircraft angles of attack and sideslip. These virtual sensors have been designed and tested using the aircraft mathematical model of the De Havilland DHC-2. The aim of the work is to evaluate the degradation of neural network performance, which is supposed to occur when real flight instruments are used instead of simulated ones. The external environment has been simulated, and special attention has been devoted to electronic noise that affects each input signals examining modern devices.. Neural networks, trained with noise free signals, demonstrate satisfactory agreement between theoretical and estimated angles of attack and sideslip.

Keywords: Neural network, turbulence, noise, air data.

1 Introduction

The flight control computer (FCC) can be considered the core of modern UAVs since several autopilot modes are implemented to guarantee stability, control and navigation of the aircraft. The autopilots need several parameters, and some of them are from external environment, better known as air data, that are necessary for FCC to work properly: the angle of attack and sideslip are two of them. Today, the angle of attack can be measured using vanes [1], multi hole probes [2], or other advanced devices [3]. The angle of sideslip is often measured differentiating the two aircraft side static pressure, or using multi hole probe. Both angle of attack and sideslip can be measured using a Pitot-boom device on the aircraft nose. Any external devices is not very suitable on modern UAVs for several reasons, such as camera angle of view or stealth feature. This issues are highly enhanced when a multiple installation of the same external sensors is requested for redundancy.

For Air Data Systems (ADS), where redundancy is requested for compliance to safety regulations, a voting and monitoring capability that cross-compares the signals from different sensors is needed: this feature is therefore used to detect and isolate ADS failures down to the single sensor, depending on the level of redundancy requested [4], [5]. The increasing need of modern UAVs to reduce cost and complexity of on-board systems has encouraged the practice to substitute, whenever

C. Jayne, S. Yue, and L. Iliadis (Eds.): EANN 2012, CCIS 311, pp. 282–294, 2012.
© Springer-Verlag Berlin Heidelberg 2012

feasible, expensive, heavy and sometimes even voluminous hardware devices with executable software code. A practical example, which is referred to as analytical redundancy in the current literature, is the process of replacing some of the actual sensors with virtual sensors, which can be used as voters in dual-redundant or simplex sensor systems, to detect inconsistencies of the hardware sensors and can be eventually employed to accommodate the sensor failures. More generally, analytical redundancy identifies with the functional redundancy of the system. The idea of using software algorithms to replace physical hardware redundancy was introduced as soon as digital computers started being used in the 1970's to perform redundancy management. Approaches developed to detect and isolate sensor failures were ultimately to become important parts of later control reconfiguration schemes. An example is the Sequential Probability Ratio Tests that were flight-tested on the F-8 Fly-by-Wire demonstrator in the late 1970's [6]. Throughout the 1970s and 1980s many papers appeared describing various algorithms for managing redundant systems and redundant sensors. Many attractive advanced algorithmic solutions have been proposed especially throughout the last two decades, mostly related to model-based techniques but capable of taking into account some robustness requirements with respect to exogenous disturbances and model uncertainties.

The present air-data virtual sensor is based on neural networks (NN) to overcome the gap between the mathematical model and the real aircraft of the model-based methods, which is the main drawback of this technique. In the aerospace field, there are several examples of NNs used as system identification devices to estimate aerodynamic coefficients [7], angle of attack [8], [9], and sideslip [10] from data derived from other sources that are not vanes, differential pressure sensors or modern multifunction probes. The majority of virtual sensors described in previous works [11], [12], [2], however, need to be fed at least by an actual value of dynamic pressure. Other few examples that do not need the dynamic pressure exist [13] but some of the input variables are quite difficult to be calculated with good accuracy, such as the exact position of the center of gravity, the engine torque on the shaft and the actual thrust at any time. Due to complexity related to this kind of system, the aim of the present research is to develop a virtual sensor that can use data derived from the attitude heading reference system (AHRS) and one source of dynamic pressure in order to keep the virtual sensor complexity low. Each of these two sensors has been characterized with its own noise level according to available data, as will be described later.

The final goal of this paper is to validate the designed neural networks when external disturbances occur, as severe air turbulence, and reconstructed flight data are used as inputs, instead of the analytical ones. One of the most critical issues related to real neural networks application, is the gap between simulated and real sensors, with their own accuracy and noise level, which must be taken into account to evaluate the actual neural network performance [14].

When real tests cannot be performed or because of the lack of real data, these data are replaced with noise corrupted signals [15], [16] to understand the neural network behavior when fed by real input signals.

2   The "Beaver" Matlab Model

The FDC toolbox for Matlab was developed starting from the De Havilland DHC-2 "Beaver" aircraft [17] of Fig. 1 and it was designed to analyze aircraft non-linear dynamics and flight control systems. The aircraft is modeled as a rigid body with constant mass value. Here, the Beaver primary control surfaces are considered whereas, for example, flap deflection effects are neglected.

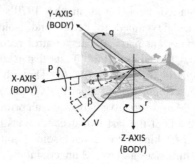

Fig. 1. De Havilland DHC-2 "Beaver"

The relationship between the wind-axis and body-axis velocities can be expressed as follows [18]

$$\begin{Bmatrix} u \\ v \\ w \end{Bmatrix} = V \begin{bmatrix} \cos\alpha\cos\beta \\ \sin\beta \\ \sin\alpha\cos\beta \end{bmatrix} \tag{1}$$

or, conversely:

$$V = \sqrt{u^2 + v^2 + w^2}$$
$$\alpha = \tan^{-1}\frac{w}{u} \tag{2}$$
$$\beta = \tan^{-1}\frac{v}{\sqrt{u^2 + w^2}}$$

Differentiating Eq.s 2, the resulting system, constituted by three ordinary differential equations, needs to be solved in order to find the analytical expressions for V, α, and β. For simplicity, attention is focused on the angle of attack, so differentiating the second of Eq. 2 it follows:

$$\dot{\alpha} = \frac{u\dot{w} - w\dot{u}}{u^2 + w^2} \tag{3}$$

Substituting u and w of Eq. 1 and rearranging the terms Eq.3 yields to:

$$\dot\alpha = \frac{\dot w \cos\alpha - \dot u \sin\alpha}{V\cos\beta} \tag{4}$$

According to [17], the relationships between axial acceleration, gravity acceleration and angular velocities in the body reference frame can be written as

$$
\begin{aligned}
\dot u &= rv - qw - g_0\sin\vartheta + F_x \\
\dot v &= -ru + pw + g_0\sin\phi\cos\vartheta + F_y \\
\dot w &= qu - pv + g_0\cos\phi\cos\vartheta + F_z
\end{aligned}
\tag{5}
$$

where the contribution of external force ($F_i = ma_i$) can also be indicated in terms of number of g acceleration $n_i = a_i/g$ ($i = x, y, z$). Introducing an expression for F_z as a function of pilot commands, or, more in general, as a function of the control surface deflections, it is possible to obtain the relationships between the normal acceleration n_z and the command actions, such as in Eq. 6, purposely written for the Beaver model [17]:

$$n_z = \frac{F_z}{m} = \frac{1}{2}\rho V^2 (C_{z0} + C_{z,\alpha}\alpha + C_{z,\alpha^3}\alpha^3 + C_{z,q}\frac{qc}{V} + C_{z,\delta_e}\delta_e + C_{z,\delta_e\beta^2}\delta_e\beta^2) \tag{6}$$

Therefore, considering Eq. 5, Eq. 4 can be rearranged as follows

$$
\dot\alpha = \frac{(qu - pv + g_0\cos\phi\cos\vartheta + n_z)\cos\alpha}{V\cos\beta} \\
- \frac{(rv - qw - g_0\sin\vartheta + n_x)\sin\alpha}{V\cos\beta} \tag{7}
$$

Then, substituting Eq. 1, we can obtain the final expression for angle of attack ODE:

$$
\dot\alpha = \frac{1}{V\cos\beta}\{[V(q\cos\alpha\cos\beta - p\sin\beta) + n_z + g_0\cos\phi\cos\vartheta]\cos\alpha + \\
-[V(r\sin\beta - q\sin\alpha\cos\beta) - g_0\sin\vartheta + n_x]\sin\alpha\} \tag{8}
$$

Eq. 8 is essential to understand the independent variables which α depends on. According to Eq.8, thus, it is possible to write the following functional relationship:

$$\alpha = f_\alpha(q_c, n_x, n_z, \beta, \vartheta, \phi, p, q, r) \tag{9}$$

where the velocity V has been substituted by dynamic pressure because it is the source from which the velocity is derived from. Following a similar procedure the functional relationship for β can be written as follows:

$$\beta = f_\beta(q_c, n_x, n_y, n_z, \alpha, \vartheta, \phi, p, q, r) \tag{10}$$

So, substituting β in Eq. 9 with its expression written in Eq. 10, collecting all the independent variables and explicitating the dependency on pilot demands, Eq.s 9, 10 can be further expanded as:

$$\alpha = F_\alpha(q_c, n_x, n_y, n_z, \vartheta, \phi, p, q, r, \delta_e, \delta_r, \delta_a), \tag{11}$$

and

$$\beta = F_\beta(q_c, n_x, n_y, n_z, \vartheta, \phi, p, q, r, \delta_e, \delta_r, \delta_a), \tag{12}$$

From Eq.s 11, (12 , it follows that the angle of attack is a function of the dynamic pressure, body axis acceleration, Euler angles and pilot commands. The advantage of using a neural network is that the information expressed by Eq.s 11, (12 is already adequately detailed to implement a virtual sensor, whereas an exact or approximated expression for the F_α and F_β function is not strictly necessary.

3   Maneuvers for Training and Testing Stages

As stated above, the training stage requires a collection of maneuvers that should be thoroughly representative of the flight envelope, so the choice can be oriented towards a flight containing a wide selection of commands, in terms of amplitude, shape and frequency, and preferable where the same excitations reach their entire span range. An example of the maneuvers performed for data gathering is given in Fig. 1.

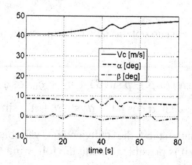

Fig. 2. Detail of aircraft dynamic of the training stage

These maneuvers have been performed exploiting the autopilot available in [17] which can follow reference signals assigned by the user before simulation using an ad-hoc written Matlab routine.

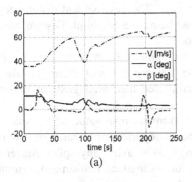 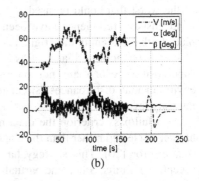

(a)   (b)

Fig. 3. Aircraft dynamics of the test stage without and with severe turbulence

In the present work, the longitudinal aircraft dynamics are controlled by assigning a well defined θ reference signal, the roll aircraft motion is obtained following ϕ reference signal, whereas the side slip angle, β, is just obtained acting on the rudder command. The simulated flight is composed by a basic set of maneuvers performed at different flight speeds (or several attitudes) in order to have a training set that covers the most part of flight envelope. Moreover, the external disturbances, turbulence and wind gusts, can be activated during the simulation when the user desires.

Usually, aircraft flight data gathering for air data calibration [19] takes place in still air in order to avoid environmental effects on measurements, and consequently corrupting the calibration. In addition, when a new aircraft prototype is developed, the first flights are flown with basic maneuvers to investigate aircraft three axis stability (Fig. 1) and its responses to pilot commands. The basic maneuvers consist of sweep, which is produced by a sine-shaped command, the doublet, which is a fast sweep, and the hold maneuver, during which one of the flight variables is hold for a certain time. Here, the training data set is supposed to be collected during the flight test data gathering of an aircraft prototype, therefore, the basic maneuver set consist of sweep, doublet and hold maneuvers performed in the longitudinal and lateral directional plane independently, since the aircraft dynamics can be uncoupled as stated in [18], under small attitude hypothesis.

The basic maneuver set is performed at constant velocity, but the whole training maneuver is composed of four basic maneuver set repeated at four different velocities, or throttle command position, from stall to maximum velocity, end lightest turbulence level activated for very little time, less than 5% of the entire training flight.

Fig. 3 shows the maneuver that has been used to test the virtual sensors based on neural networks. The test maneuver, starts from a trimmed flight condition, and then the thrust is increased in very little time at the maximum power by acting on the throttle. While the velocity is increasing, several maneuvers are performed, this time exciting both longitudinal and lateral directional dynamics at the same time. Based on the experience gained flying the "Beaver" aircraft model, this kind of maneuver is one of the worst that could be used as validation case. The environment disturbances as wind gusts or turbulence have been simulated using the Simulink Dryden block in place of that present in the original Beaver model.

It can be noticed that in this kind of test maneuvers there are many motion superposition which are not present in the training set and there are also the environmental disturbances, simulated as turbulence and wind gusts at maximum intensity level.

As a fail/pass criterion the maximum error from nominal value on test maneuver has been selected rather than the classical concept of performance based on root mean square error [20]. This strategy has been required to avoid any spike greater than acceptance limits, since the virtual sensor can be used as real-time discriminator source, while the root mean square is calculated over an entire flight.

4   Noise and Turbulence Details

If compared to simulated signal, actual inertial signals, are always affected by noise deriving from several sources, such as engine induced vibrations and on board sensor electronic noise. The electronic noise here is represented using the data specifications of commercial MEMS Attitude Heading Reference Signals (AHRS) [21]. The noise on dynamic pressure, q_c, has been modeled according to some available data sheets [22], [23].

The real aircraft vibration frequency spectra consists of a broadband background with superimposed narrow band spikes. The background spectrum results mainly from the engine combined with many lower level periodic components due to the rotating elements (engines, gearboxes, shafts, etc.) associated with propeller, and some specifications are collected in ref. [24] for several aircraft categories. The issue in using these data is that the vibration sensed by on board instruments, such as AHRS, depends on the kind of material used for aircraft structure, the method adopted to isolate vibrations, and so on. So, in this work, only the electronic noise due to real flight instruments will be considered to avoid any comment linked to a particular aircraft that cannot have a general validity.

Therefore, the data from Ref.s [21]-[23] were used to define conservative peak to peak noise values in order to develop a realistic noise model, as reported in Table 1.

The discrepancies between the mathematical model and the real aircraft can be seen as another source of noise, especially when the sensor sampling is close to the first structural mode frequencies, such as in slender bodies. As it is always quite complicated to keep separate each source of noise, therefore any single signal has been marked with its own noise level.

Even if the degradation of signals depends on the specific electronic devices and the type of aircraft, it has been decided to adopt a conservative approach based on the worst-case scenario, incorporating the combination of highest noise level detected within the references here considered.

Table 1. Noise level used to corrupt mathematical model data

Variable	Noise Level (peak to peak)
q_c	±0.03 hPa
n_x	±0.005 g
n_y	±0.005 g
n_z	±0.005 g
ϑ	±0.01 deg
ϕ	±0.01 deg
p	±0.1 deg/s
q	±0.1 deg/s
r	±0.1 deg/s

A white noise with zero mean value have been added to each signal and the peak to peak values have been selected according to Table 1.

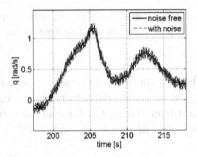

Fig. 4. Example of pitch rate with and without noise

Fig. 4 shows an example related to the pitch rate. It must be noted that for the angle of attack estimation no electronic filters have been considered.

The authors only decided to activate the highest turbulence level during the validation stage, because it is very realistic: usually, on real aircraft, first flight data gathering are scheduled when the weather is favorable: clean sky and very light wind.

5   Results

To assess the feasibility of the method it has been decided to simulate the best-case scenario, implementing a neural network trained and tested on noise-free signal: performance are evaluated on the bases of ±1 deg error specification for the angle of attack. The virtual sensor has been developed using a state-of–the-art tool, developed by third parts under the Matlab simulation environment [25]. The kind of neural network used is a multi layer perceptron (MLP) it has one hidden layer with 15 neurons characterized by hyperbolic tangent activation function and one output layer, as described in Fig. 5.

The system identification exploits the NNOE model trained with the Marquardt method [25]. The input variables have been chosen from Eq. 11 and they are directly processed by the neural network without any kind of buffer of memory, which would have been needed in the case that past inputs were also requested.

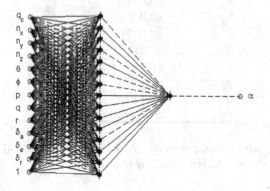

Fig. 5. Default neural network scheme used for AOA and AOS estimation: the solid lines stand for positive weights, the dashed lines stand for negative weights, the red circles are the network neurons

This kind of network has the enormous advantage to be light in terms of hardware resources: it only needs to be connected with the requested variables in input ports at current time t to give an estimation of $\alpha(t)$. Here, neural networks will be identified as NNA if designed for angle of attack, α, and NNB if designed for angle of sideslip, β.

5.1   Angle of Attack

The estimation error for the angle of attack is plotted in Fig. 6. The maximum value is less than 0.1 deg (peak to peak).

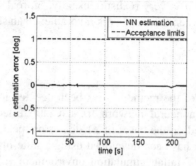

Fig. 6. Estimation error obtained using the test maneuver without turbulence and noise for neural network NNA

The advantage of using the Beaver mathematical model is that turbulence and noise contributions can be investigated one by one. The absolute maximum estimation error due to the severe level of turbulence is less than 0.4 deg, as shown in Fig. 7.

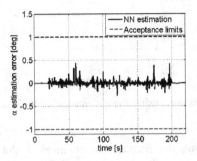

Fig. 7. Estimation error obtained using the test maneuver with turbulence and without noise for neural network NNA

Fig. 8 shows the estimation error on test maneuver considering turbulence and noisy input signals. It can be seen that the input signal noise introduces such noise on output contained within ±0.1 deg.

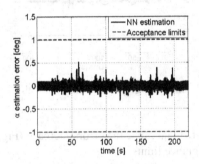

Fig. 8. Estimation error obtained using the test maneuver with turbulence and noise for neural network NNA

The neural network designed for angle of attack estimation has been tested using reconstructed actual signal with acceptable performance.

5.2   Angle of Sideslip

The estimation error for the angle of attack is plotted in Fig. 6. The maximum value is less than 0.2 deg (peak to peak), higher than for angle of attack.

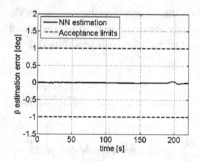

Fig. 9. Estimation error obtained using the test maneuver without turbulence and noise for neural network NNB

The advantage of using the Beaver mathematical model is that turbulence and noise contributions can be investigated one by one. The absolute maximum estimation error due to the severe level of turbulence is less than 0.4 deg, as shown in Fig. 7.

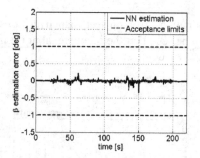

Fig. 10. Estimation error obtained using the test maneuver with turbulence and without noise for neural network NNB

Corrupting data with artificial noise the errors increase (Fig. 11) but, however, it is contained inside the acceptance limits.

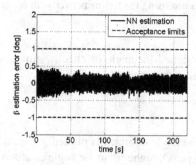

Fig. 11. Estimation error obtained using the test maneuver with turbulence and noise for neural network NNB

Comparing Fig. 10, Fig. 11, it can be noted that the noise introduces different bandwidth on angle of attack and sideslip, respectively ±0.1 and ±0.5 deg. This phenomena require a more accurate investigation that goes beyond the aim of this work.

5.3   Conclusion

The present work presented an analysis of the degradation which occurs on a neural virtual air-data sensor, when noise-corrupted data are processed.

The first part of the paper deals with the development of a virtual sensor based on noise-free signals: the functional relationship between the angle-of-attack, α, angle of sideslip, β, and the most relevant flight variables has been derived and assessed in an ideal environment. In particular, the virtual air-data sensor, implemented ad-hoc for the De Havilland DHC-2 "Beaver" analytical model, has been conceived as a function of data derived from gyros, accelerometers, pilot commands and a Pitot tube (dynamic pressure). The external disturbances have been considered using the Simulink Dryden wind turbulence model, which is able to simulate wind gusts and several turbulence level according to Dryden model. Te results have shown very good performance when the virtual sensors are coupled with real instruments. The future developments of this work, shall be to take into account the actual aircraft noise, collecting some real flight data, in order to evaluate the performance of the designed virtual sensors and try to train the neural network using these data.

The results have shown that keeping neural network dimensions within certain limits (in terms of hardware resources) is feasible: the state-of-the-art in the neural network field proposes many techniques that involve more complex neural network architectures which can be expressed with a smaller amount of parameters. The optimization of the neural technique can be regarded as an option in this first phase, but it remains an issue that must be addressed, especially in consideration of future implementation of the virtual sensor as an analytical redundancy tool for the on-board sensor management system.

Acknowledgments. The authors gratefully thank Alenia Aermacchi for supplying the Ph.D. scholarship making this work possible. In particular, the inestimable experience that Mr. Silvio Caselle provided to this activity.

References

1. Sakamoto, G.M.: Aerodynamic Characteristics of a Vane Flow Angularity Sensor System Capable of Measuring Flightpath Accelerations for the Mach Number Range from 0.40 to 2.54. NASA-TN-D-8242, p. 45 (1976)
2. Calia, A., et al.: Multi-hole probe and elaboration algorithms for the reconstruction of the air data parameters. In: IEEE International Symposium on Industrial Electronics, ISIE 2008 (2008)
3. Aerosonic Corporation:
   http://www.aerosonic.com/wp-content/uploads/2011/06/Sensors-201106.pdf

 4. Napolitano, M.R., An, Y., Seanor, B.A.: A fault tolerant flight control system for sensor and actuator failures using neural networks. Aircraft Design 3(2), 103–128 (2000)
 5. Oosterom, M., Babuska, R.: Virtual sensor for fault detection and isolation in flight control systems - fuzzy modeling approach. In: Proceedings of the 39th IEEE Conference on Decision and Control (2000)
 6. Tomayko, J.E.: Computers Take Flight: A History of NASA's Pioneering Digital Fly-By-Wire Project. NASA 2000. SP-2000-4224 (2000)
 7. Norgaard, M., Jorgensen, C.C., Ross, J.C.: Neural Network Prediction of New Aircraft Design Coefficients. NASA Technical Memorandum (112197) (1997)
 8. Oosterom, M., Babuska, R.: Virtual Sensor for the Angle-of-Attack Signal in Small Commercial Aircraft. In: 2006 IEEE International Conference on Fuzzy Systems (2006)
 9. Samara, P.A., Fouskitakis, G.N., Sakellariou, J.S., Fassois, S.D.: Aircraft Angle-Of-Attack Virtual Sensor Design via a Functional Pooling Narx Methodology. In: Proceedings of the European Control Conference (ECC), Cambridge, UK (2003)
10. Xiaoping, D., et al.: A prediction model for vehicle sideslip angle based on neural network. In: 2010 2nd IEEE International Conference on Information and Financial Engineering (ICIFE), pp. 451–455 (2010)
11. Rohloff, T.J., Whitmore, S.A., Catton, I.: Air Data Sensing from Surface Pressure Measurements Using a Neural Network Method. AIAA Journal 36(11)
12. Samy, I., Postlethwaite, I., Green, J.: Neural-Network-Based Flush Air Data Sensing System Demonstrated on a Mini Air Vehicle. Journal of Aircraft 47(1)
13. McCool, K., Haas, D.: Neural network system for estimation of aircraft flight data
14. di Fusco, C.: Ricostruzione degli Angoli Di Incidenza e di Derapata del Velivolo Mediante Elaborazioni dei Dati Aria Basate su Reti Neurali, Department of Aerospace Engineering. University of Pisa, Pisa (2006)
15. Rowley, H.A., Baluja, S., Kanade, T.: Neural network-based face detection. IEEE Transactions on Pattern Analysis and Machine Intelligence 20(1), 23–38 (1998)
16. Svobodova, J., Koudelka, V., Raida, Z.: Aircraft equipment modeling using neural networks. In: 2011 International Conference on Electromagnetics in Advanced Applications (ICEAA) (2011)
17. Rauw, M.: FDC 1.2 – A Simulink Toolbox for Flight Dynamics and Control Analysis (2001)
18. Etkin, B.: Dynamics of flight: stability and control. Wiley (1982)
19. Haering Jr., E.A.: Airdata Measurement and Calibration, N.A.a.S. Administration, Editor (1995)
20. Skapura, D.M.: Building Neural Networks. ACM Press (1996)
21. Gladiator Technologies, I.: High Performance MEMS AHRS "LN Series" (2011)
22. Sensortec, B.: BMP085 Digital Pressure Transducer (2011)
23. Honeywell Model AS25D Aerospace Pressure Transducers
24. DEFENSE, D.O.: Environmental Engineering Considerations and Laboratory Tests, MIL-STD-810E
25. Norgaard, M.: The nnsysid toolbox for use with Matlab (2003), http://www.iau.dtu.dk/research/control/nnsysid.html

Direct Zero-Norm Minimization for Neural Network Pruning and Training

S.P. Adam[1,2], George D. Magoulas[3], and M.N. Vrahatis[1]

[1] Computational Intelligence Laboratory, Dept. of Mathematics,
University of Patras, Rion - Patras, Greece
[2] Dept. of Informatics and Telecommunications,
Technological Education Institute of Epirus, Arta, Greece
[3] Dept. of Computer Science and Information Systems, Birkbeck College,
University of London, United Kingdom

Abstract. Designing a feed-forward neural network with optimal topology in terms of complexity (hidden layer nodes and connections between nodes) and training performance has been a matter of considerable concern since the very beginning of neural networks research. Typically, this issue is dealt with by pruning a fully interconnected network with "many" nodes in the hidden layers, eliminating "superfluous" connections and nodes. However the problem has not been solved yet and it seems to be even more relevant today in the context of deep learning networks. In this paper we present a method of direct zero-norm minimization for pruning while training a Multi Layer Perceptron. The method employs a cooperative scheme using two swarms of particles and its purpose is to minimize an aggregate function corresponding to the total risk functional. Our discussion highlights relevant computational and methodological issues of the approach that are not apparent and well defined in the literature.

Keywords: Neural networks, pruning, training, zero-norm minimization, Particle Swarm Optimization

1 Introduction

Despite the popularity of neural network models in engineering applications designing a neural network for a particular application remains a challenge. For example, in system identification using neural networks, finding the optimal network architecture is not straightforward [1]. In addition designing a neural network for a control system is critical for optimal performance [2]. Besides the type of the nodes' activation functions and network training parameters this design also involves selecting the right number of hidden layer nodes and their interconnection. Thus, practitioners and researchers invested in pruning techniques for defining the best available neural network architecture [1], [2].

The question on the number of hidden layers and more specifically on the number of nodes per hidden layer necessary to approximate any given function $f : \mathbb{R}^n \to \mathbb{R}^m$, has been tackled by several researchers, as noted in [3], and has

C. Jayne, S. Yue, and L. Iliadis (Eds.): EANN 2012, CCIS 311, pp. 295–304, 2012.

been studied as part of the problem of the density of Multi Layer Perceptrons (MLPs). Furthermore, work introduced in [4], [5], [6] and [7] highlighted important theoretical aspects of the density problem and derived bounds on the number of nodes of the hidden layer for one hidden layer MLPs. However, as noted in [6], "in applications, functions of hundreds of variables are approximated sufficiently well by neural networks with only moderately many hidden units". No precise rules exist regarding the necessary and sufficient architecture of the neural network to deal with some specific problem. It is common when designing a network to apply a pruning technique. This consists in designing a "fat" network with many nodes in the hidden layer and fully interconnected layers. Then proceeding with detecting those connections and/or nodes that are superfluous, with respect to the mapping function of the network, and removing them. A number of pruning techniques and related research are reported in [8].

The approach proposed in this paper is a network pruning technique. Section 2 is devoted to the problem background and a brief literature review. In section 3 the problem of zero-norm minimization is formulated. Section 4 discusses how cooperative swarms are used in this minimization problem and section 5 presents application experiments. The paper ends with some concluding remarks.

2   Background and Previous Work

The whole problem is defined as the complexity-regularization problem and falls within Tikhonov's regularization theory, [9]. From that point of view, defining the architecture of a network and training it in order to maximize generalization performance in the context of supervised learning, is equivalent to minimizing the total risk given by the following expression, as noted in [10],

$$\mathcal{R}(w) = \mathcal{E}_S(w) + \lambda \mathcal{E}_C(w) \,, \tag{1}$$

where $\mathcal{E}_S(w)$ is the standard performance measure, typically the error function of the network's output, $\mathcal{E}_C(w)$ is a penalty term, a functional that depends exclusively on the network architecture, that is the network's complexity, and λ is a real number whose value determines the importance attached to the complexity penalty term when minimizing the total risk.

The so called penalty term methods are derived from the formulation of (1). These methods tend to minimize the total risk functional by defining weight decay or weight elimination procedures which are applied during network training with back-propagation and hunt out "non-significant" weights by trying to drive their values down to zero. This means that the corresponding connections are eliminated and in consequence nodes with no connections at all are also eliminated. Typical methods in this category are proposed in [11] and [12].

Another group of methods tend to eliminate connections and/or nodes by calculating various sensitivity measures such as the sensitivity of the standard error function to the removal of units (Skeletonization, [13]), the removal of connections [14] and LeCun's Optimal Brain Damage, [15]. In general, such methods act on the network architecture and modify it once the network is trained. Among several other methods that have been proposed it is worth mentioning those that are hessian-based, [16], or variance-based pruning techniques, see [8].

A number of methods have also been defined based on Evolutionary Computation. Hancock in [17] analyzes the aspects of pruning neural networks using the Genetic Algorithm (GA). Methods reported in the literature use the GA for network architecture optimization as well as for weight training. However, as noted in [18], GA may not yield a good approach to optimizing neural network weights because of the Competing Convention Problem, also called the Permutations Problem. Some research efforts propose hybrid approaches where the network training task is carried out by backpropagation.

In particular, it is worth mentioning here that some recent pruning techniques adopt the Particle Swarm Optimization (PSO) paradigm. One approach is proposed in [19] and uses a modified version of PSO in order to determine the network architecture (nodes and connections) as well as the activation function of the nodes and the best weight values for the synapses. Results of the experiments presented give relatively complex networks having intra-layer connections, direct input-to-output connections and nodes of the same layer with different activation functions. The subsequent question deals with how easily these networks can be implemented for real life applications. Another approach is presented in [20] and uses a so-called cooperative binary-real PSO to tune the structure and the parameters of a neural network. Researchers' main assumption is that the signal flow via the connections of the nodes is controlled by ON/OFF switches. This engineering consideration is implemented using a binary swarm together with a real valued swarm which is used to train the network. Despite the results reported, our opinion is that, there are two points that remain unclear in this paper. The first deals with the exact mode of interaction between the two swarms. The second concerns the problem of state space exploration with particles traveling around in different subspaces of the state space. The latter will be discussed and clarified later in section 4. Lastly, some hybrid approaches can be found in the literature using both PSO and classical backpropagation for weight training.

The approach presented in this paper is a penalty term method for neural network regularization. The penalty term is based on the zero-norm of the vector formed by the weights of the network. In contrast to [20] our approach has a clearly defined model of interaction between the two swarms and an effective mechanism for state space exploration. Lastly, our work is underpinned by a mathematical theory for dealing with the neural network pruning problem.

3   Minimizing the Zero-Norm of Weights

Let $\bar{w}$ denote the vector formed by the weights of the network connections arranged in some order. Then the zero-norm of this vector is the number of non-zero components, also defined by the expression, $\|\bar{w}\|_0 = card\{w_j, \ w_j \neq 0\}$. Given that, pruning a network is eliminating connections, or in other words zeroing corresponding weights, then pruning a network can be considered as minimizing the zero-norm of the weight vector of the network. Use of the zero-norm has been proposed, in the machine learning literature, as a measure of the presence of some features or free parameters in learning problems when sparse learning machines are considered. The reason for using zero-norm minimization in machine learning is that, seemingly, it provides a natural way of directly addressing two objectives, feature selection and pattern classification, in just a single optimization, [21]. In fact, these two objectives are directly interconnected, especially regarding regularization which enforces sparsity of the weight vector. One may easily notice that pruning a network can, also, be considered as a feature selection problem, and, thus, the above considerations, again, justify its rephrasing as a zero-norm minimization problem.

Although zero-norm minimization has been used by many researchers in various machine learning contexts; [22], it is important to point out an important theoretical issue in the context of neural networks that is computational complexity when minimizing the zero-norm of some vector. Amaldi and Kann in [23], proved that minimizing the zero norm of some vector, let $\|\bar{w}\|_0$, subject to the linear problem $y_i \left(wx_i + b\right) \geq 1$ is a problem that is NP-complete. In practice researchers address zero-norm minimization using some approximate form such as, $\sum_i \left(1 - e^{-\alpha|w_i|}\right)$, where α is a parameter to be chosen, or $\sum_i \log_{10} \left(\varepsilon + |w_i|\right)$, where $0 < \varepsilon \ll 1$.

Driving weights to zero or deleting weight in order to eliminate connections between nodes has been fundamental in a number of network regularization approaches. Such pruning techniques are weight decay procedures which are applied combined with gradient descent-based training methods. These weight decay procedures use a differentiable form which is based on some norm of the weight vector. Typical examples are the methods proposed by Hinton, [11], and Moody and Rögnvaldsson, [24]. It is worth noting here that Rumelhart, as reported in [25], investigated a number of different penalty terms using approximate forms of some weight vector norms. In particular, the penalty term $\sum_i \left[\left(w_i^2/w_0^2\right) / \left(1 + w_i^2/w_0^2\right)\right]$ proposed by Weigend et al. [12], constitutes an approximate form of the zero-norm of the weight vector.

Our approach to network pruning adopts direct minimization of the zero-norm of the weight vector. Hence, the values of selected weights are zeroed without decay (directly) and thus corresponding node connections are cutoff abruptly. Such a consideration gives rise to the connectivity pattern of a network, i.e. a binary vector where each component denotes whether a connection is present, with 1, or 0 if it is deleted. This vector can be considered as the connectivity adjoint of the weight vector. Thus, using an indicator function the weight vector $\bar{w} = (w_1, w_2, ..., w_n)$, is replaced by the binary vector defined as $\mathbb{1}_{\bar{w}} = (\mathbb{1}_{w_1}, \mathbb{1}_{w_2}, ..., \mathbb{1}_{w_n})$, where $\mathbb{1}_{w_i} = \begin{cases} 1 & w_i \neq 0 \\ 0 & w_i = 0 \end{cases}$, $1 \leq i \leq n$. In consequence, the penalty term of Equation (1) becomes $\mathcal{E}_C(w) = \lambda \|w\|_0$ or using the connectivity vector representation $\mathcal{E}_C(w) = \lambda \sum_{i=1}^{n} \mathbb{1}_{w_i}$. Finally, if the mean squared error of the network output is used as the standard measure of the network performance the aggregate function (1) takes the form,

$$\mathcal{R}(w) = \frac{1}{p}\frac{1}{n}\sum_{k=1}^{p}\sum_{i=1}^{n}\left(d_i^k - o_i^k\right)^2 + \lambda\sum_{i=1}^{n}\mathbb{1}_{w_i}. \tag{2}$$

This form of the penalty term is not differentiable. So its minimization cannot be based on classical back-propagation and an evolutionary approach should be adopted. Minimizing an aggregate function of the form (2) has been addressed as a multi-objective optimization problem and several methods have been proposed in the context of evolutionary computation. Our approach is based on PSO and specifically on a cooperative scheme that uses two swarms. This scheme is relative to the approaches developed in [26] and [27].

4   Implementation of Cooperative Swarms

The cooperative scheme adopted for the minimization of this aggregate function makes use of two swarms; let them be TrnSrm (Training Swarm) and PrnSrm (Pruning Swarm). TrnSrm undertakes the minimization of the term corresponding to the standard error of the neural network output and so it performs network training. On the other hand, PrnSrm aims at minimizing the zero-norm of the network's connectivity vector, thus implementing the pruning function. It is obvious that these two terms of the risk functional and therefore the two swarms are very strongly coupled. During optimization we expect PrnSrm to derive the thinnest possible network architecture that performs well in terms of training and generalization that is permitting TrnSrm to reach some "good" local minimum for the network error function.

MLPs considered here have only one hidden layer and are fully interconnected without lateral or backward connections between the nodes. The assumption regarding the number of layers is not restrictive for the proposed method as none of the statements, herein, depends on this specific hypothesis. Let N be the number of weights of such a neural network. Each particle in TrnSrm corresponds to a network that constitutes a possible solution to the standard error term of Equation (2). Such a particle is a vector in $\mathbb{R}^N$ whose values correspond to the network weights. Particles in PrnSrm are binary. Each particle describes a possible network configuration represented by a binary vector in $\{0,1\}^{N-B}$, where B is the number of bias connections. Hence, only connections between nodes are used to form the binary particles. There is a one-to-one correspondence between particles in TrnSrm and particles in PrnSrm. Thus, any change in a particle in PrnSrm implies a change in the architecture of the corresponding particle in TrnSrm. The fitness function for TrnSrm is either the mean squared or the mean absolute error of the network output, while the fitness function for PrnSrm is $\sum_{i=1}^{N-B}\mathbb{1}_{w_i}$.

A methodological issue when defining and minimizing the aggregate form (1) is the choice of λ. If the form $\mathcal{R}(w) = \lambda_1\mathcal{E}_S(w) + \lambda_2\mathcal{E}_C(w)$, where $\lambda_1 + \lambda_2 = 1$ is used instead of (1) then the relation between λ_1 and λ_2 determines the priority given by the minimization process to each of the two terms. In this context, promoting some network architecture, detected by PrnSrm, depends on the estimation that this architecture is likely to lead or even be an optimal solution for the minimization problem. The measure of this estimation is λ_2. In our approach defining λ_1 and λ_2 relies on the following considerations.

The overall minimization task is carried out by two populations which, though tightly coupled, operate separately. However, when the pruning swarm is left to operate alone, observation shows that about 80% of the components of a binary vector are zeroed in about 10% of the total number of iterations, Fig. 1. During this number of iterations for a simple problem such as the Iris classification benchmark, using a 4-5-3 network, the training swarm operating alone achieves to reduce the network output error by about 40% of the initial value, Fig. 1. This rough observation underlines the consideration that, for every network architecture offered by PrnSrm as a solution, TrnSrm should be given the time (number of iterations) to verify that the proposed architecture can be trained to minimize the standard error function. Furthermore some estimation should be made about the reliability of this result. This suggests setting $\lambda_1 = 0.7$ and $\lambda_2 = 0.3$.

Effective implementation of the proposed approach is equivalent to having

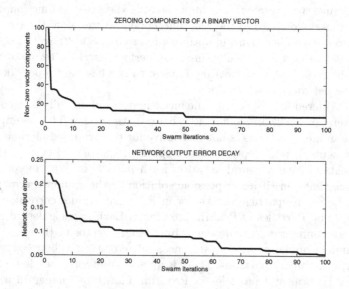

Fig. 1. Indicative minimization rate for the two swarms

a computing scheme consisting of two separate processes operating in parallel that need to be synchronized in order to achieve the optimal solution in terms of network complexity and performance. At any time the aggregate function reflects the problem's best state. The training swarm operates in order to reduce the output error of the best network architecture to some significant level. For our experiments this level is set to 70% of the output error computed for the best network architecture selected every time the training swarm is launched. If this goal is reached then the pruning swarm is activated to further reduce the number of connections of the network architecture contributing the aggregate function. Once such a new architecture is found the training swarm is launched again.

Another important issue for the exploration process concerns the order in which connections are eliminated. For instance suppose that there are K particles in each swarm. Following an iteration of PrnSrm these K particles suggest k possible network architectures, where $k \leq K$. Considering that, when a connection is eliminated the corresponding dimension in the N-dimensional weight space is not explored by the optimization process, it is easy to realize that a random cutoff of network connections disperses the particles around in different subspaces of the N-dimensional weight space. Thus the exploration process weakens in the sense that a global best may become non-significant for particles exploring different subspaces. The obvious consequence is that the optimization process may fail to converge. The solution we adopted for this problem is to apply connection cutoff in some ordered way by progressively eliminating input connections and output connections of successive nodes of the hidden layer. This is a kind of "normalized connection elimination" which ensures that at any time there exists a subspace of the weight space that is common to and explored by all particles in the swarm.

5   Experimental Evaluation

In order to validate the proposed approach we carried out a number of experiments on four well known real world problems, the Fisher Iris classification benchmark, the Servo prediction, the Solar sunspot prediction problem and the Wine classification problem. All data sets are from the UCI repository of machine learning database. Note that two alternative approaches were used for the Solar sunspot prediction problem. The first (Solar1) is the most typical one found in the literature, while the second (Solar2) is used for comparison with the approach presented in [20]. Table 1, summarizes typical network architectures found in the literature and structures used for the experiments.

The experimental setup for the proposed approach comprises two swarms as described above, each one consisting of 50 particles. The classic algorithm, [27], without enhancements is used for updating the velocities of the particles. The max number of iterations for the swarms is set to 1000 and values for the cognitive and the social attraction parameters are set to 0.5 and 1.25 respectively. For each network used, a set of 100 instances were derived and pruning experiments were executed using the proposed approach and two classical pruning methods, the Optimal Brain Damage (OBD) [15] and the Optimal Brain Surgeon (OBS), [16], as implemented in NNSYSID20, [1].

Table 1. Number of nodes and connections for networks used in experiments

	Typical network architectures		Network architectures used	
	Nodes	Connections	Nodes	Connections
Iris	4-2-3	14 (+5 bias)	4-20-3	140 (+23 bias)
			4-10-3	70 (+13 bias)
Servo	12-3-1	39 (+4 bias)	12-8-1	104 (+9 bias)
Solar1	12-5-1	65 (+6 bias)	12-15-1	195 (+16 bias)
Wine	13-6-3	96 (+9 bias)	13-15-3	240 (+18 bias)
Solar2	3-3-1	12 (+4 bias)	3-8-1	32 (+9 bias)
			3-15-1	60 (+16 bias)

Table 2, hereafter, summarizes the results obtained for the following parameters: Nodes pruned (Npr) in hidden layer, Connections pruned (Cpr) and Generalization achieved (Gen). Values reported for generalization for the prediction problems are mean value and standard deviation. Generalization performance is the mean value of the percentage of correctly classified test patterns for the Iris and the Wine problems. For the Servo and the Solar sunspot prediction problems generalization is the mean absolute error between the network output and the expected output of the test patterns.

Table 2. Results of the experiments

	Proposed Method			OBS			OBD		
	Npr	Cpr	Gen	Npr	Cpr	Gen	Npr	Cpr	Gen
Iris	12	86	95.56%	0.6	0	34%	0	0	34%
	6	45	97.78%	2	41	86.67%	6	40	68.89%
Servo	5	65	0.0515 0.0091	1	42	0.0721 0.0309	1	43	0.0716 0.0329
Solar1	9	114	0.0905 0.0092	0	1	0.9094 0.2468	0	1	0.9094 0.2468
Wine	9	139	97.78%	0	0	34%	0	0	34%
Solar2	6	23	0.0770 0.0050	13	3	0.0693 0.0085	2	10	0.2190 0.3000
	10	39	0.0765 0.0058	8	28	0.0693 0.0164	24	11	0.2190 0.3925

It is worth noting that due to the way connections are eliminated the proposed method does not provide sparse networks in terms of having few connections between dispersed nodes in the hidden layer. One may notice that the two classic methods OBS and OBD fail when faced with really "fat" networks, while our approach achieves to eliminate "fat" from the network even in difficult situations. When compared against the relative method presented in [20] (experiments under the Solar2 label) the proposed method demonstrates clearly better performance in terms of pruning, while generalization cannot be compared as the authors in [20] do not provide enough data. Finally we need to note that despite the small number of iterations (1000) for the swarms the method converged in more that 80% of the cases. This score overrides the defect reported in the literature that classic weight decay methods fail to converge, [25], and so they are not used in practice.

6 Concluding Remarks

The proposed methods introduces zero-norm minimization for pruning the weights of an MLP. Besides the definition of a solid mathematical basis the method clarifies two important implementation points which are not clear with other methods [19], [20] and greatly affect convergence of the minimization process. Experiments carried out and results obtained, in Table 2 above, clearly reveal the ability of the proposed method to eliminate connections and nodes while training the network. Networks obtained after pruning are considered optimal in the sense that the number of nodes and connections are very close to the values typically found in the literature for the benchmarks under consideration. When compared against two classic pruning methods the proposed approach performs better both in terms of pruning and training, and thus it breaks the non-convergence deficiency associated with typical weight decay methods. However, in terms of training the approach can be further improved considering that the ability of classic PSO algorithm is relatively poor regarding local search. In this paper we did not elaborate this issue which will be further investigated in future work.

References

1. Norgaard, M.: Neural Network Based System Identification Toolbox, version 2. Technical report, 00-E-891, Dept. of Automation, Technical University of Denmark (2000)
2. Stepniewski, S.W., Keane, A.J.: Topology Design of Feedforward Neural Networks by Genetic Algorithms. In: Ebeling, W., Rechenberg, I., Voigt, H.-M., Schwefel, H.-P. (eds.) PPSN 1996. LNCS, vol. 1141, pp. 771–780. Springer, Heidelberg (1996)
3. Pinkus, A.: Approximation theory of the MLP model in neural model. Acta Numerica, 143–195 (1999)
4. Jones, L.K.: A simple lemma on greedy approximation in Hilbert space and convergence rates for projection pursuit regression and neural network training. The Annals of Statistics 20, 601–613 (1992)
5. Barron, A.R.: Universal approximation bounds for superposition of a sigmoidal function. IEEE Trans. Inform. Theory 39, 930–945 (1993)
6. Kůrková, V., Kainen, P.C., Kreinovich, V.: Estimates of the number of hidden units and variation with respect to half-spaces. Neural Networks 10, 1061–1068 (1997)
7. Hornik, K.: Approximation capabilities of multilayer feedforward networks. Neural Networks 4, 251–257 (1991)
8. Reed, R.: Pruning algorithms - A Survey. IEEE Trans. Neural Networks 4, 740–747 (1993)
9. Tikhonov, A.N., Arsenin, V.Y.: Solution of Ill-posed Problems. W.H. Winston, Washington, DC (1977)

10. Haykin, S.: Neural networks: A comprehensive Foundation. Prentice-Hall, Upper Saddle River (1999)
11. Hinton, G.E.: Connectionist learning procedures. Artificial Intelligence 40, 185–234 (1989)
12. Weigend, A.S., Rumelhart, D.E., Huberman, B.A.: Generalization by weight-elimination with application to forecasting. In: Lippmann, R., Moody, J., Touretzky, D. (eds.) Advances in Neural Information Processing Systems (3), pp. 875–882. Morgan-Kaufmann, San Mateo (1991)
13. Mozer, M.C., Smolensky, P.: Skeletonization: A Technique for Trimming the Fat from a Network via Relevance Assessment. In: Touretzky, D.S. (ed.) Advances in Neural Information Processing Systems (1), pp. 40–48. Morgan Kaufmann, San Francisco (1989)
14. Karnin, E.D.: A simple procedure for pruning back-propagation trained neural networks. IEEE Trans. Neural Networks 1, 239–242 (1990)
15. LeCun, Y., Denker, J.S., Solla, S.A.: Optimal Brain Damage. In: Touretzky, D.S. (ed.) Advances in Neural Information Processing Systems (2), pp. 598–605. Morgan Kaufmann, San Francisco (1990)
16. Hassibi, B., Stork, D.G.: Second order derivatives for network pruning: Optimal Brain Surgeon. In: Hanson, S.J., Cowan, J.D., Giles, C.L. (eds.) Advances in Neural Information Processing Systems (5), pp. 164–172. Morgan-Kaufmann, San Mateo (1993)
17. Hancock, P.J.B.: Pruning neural networks by genetic algorithm. In: Aleksander, I., Taylor, J.G. (eds.) Proc. of the International Conference on Artificial Neural Networks, pp. 991–994. Elsevier, Brighton (1992)
18. Whitley, D.: Genetic Algorithms and Neural Networks. Genetic Algorithms in Engineering and Computer Science, pp. 191–201. John Wiley (1995)
19. Garro, B.A., Sossa, H., Vazquez, R.A.: Design of artificial neural networks using a modified particle swarm optimization algorithm. In: Proc. IEEE International Joint Conference on Neural Networks, Atlanta, pp. 938–945 (2009)
20. Zhao, L., Qian, F.: Tuning the structure and parameters of a neural network using cooperative binary-real particle swarm optimization. Expert Systems with Applications (2010)
21. Weston, J., Elisseeff, A., Schölkopf, B., Tipping, M.: Use of the zero-norm with linear models and kernel methods. J. Machine Learning Res. 3, 1439–1461 (2003)
22. Fung, G.M., Mangasarian, O.L., Smola, A.J.: Minimal kernel classifiers. J. Machine Learning Res. 3, 303–321 (2002)
23. Amaldi, E., Kann, V.: On the approximability of minimizing non zero variables or unsatisfied relations in linear systems. Theoretical Computer Science, 237–260 (1998)
24. Moody, J.E., Rögnvaldsson, T.: Smoothing regularizers for projective basis function networks. In: Mozer, M., Jordan, M.I., Petsche, T. (eds.) Advances in Neural Information Processing Systems (9), pp. 585–591. MIT Press, Denver (1997)
25. Hanson, S.J., Pratt, L.Y.: Comparing biases for minimal network construction with back-propagation. In: Touretzky, D.S. (ed.) Advances in Neural Information Processing Systems (1), pp. 177–185. Morgan Kaufmann, San Francisco (1989)
26. Parsopoulos, K.E., Tasoulis, D.K., Vrahatis, M.N.: Multi-objective optimization using parallel vector evaluated particle swarm optimization. In: Proc. of the IASTED International Conference on Artificial Intelligence and Applications (AIA), Innsbruck, vol. 2, pp. 823–828 (2004)
27. van de Bergh, F., Engelbrecht, A.P.: A cooperative approach to particle swarm optimization. IEEE Trans. Evolutionary Computation 8, 1–15 (2004)

3D Vision-Based Autonomous Navigation System Using ANN and Kinect Sensor

Daniel Sales, Diogo Correa, Fernando S. Osório, and Denis F. Wolf

University of São Paulo, Mobile Robotics Lab, Av. Trabalhador São-Carlense, 400,
P.O. Box 668, São Carlos, Brazil
{dsales,diogosoc,fosorio,denis}@icmc.usp.br

Abstract. In this paper, we present an autonomous navigation system based on a finite state machine (FSM) learned by an artificial neural network (ANN) in an indoor navigation task. This system uses a kinect as the only sensor. In the first step, the ANN is trained to recognize the different specific environment configurations, identifying the different environment situations (states) based on the kinect detections. Then, a specific sequence of states and actions is generated for any route defined by the user, configuring a path in a topological like map. So, the robot becomes able to autonomously navigate through this environment, reaching the destination after going through a sequence of specific environment places, each place being identified by its local properties, as for example, straight path, path turning to left, path turning to right, bifurcations and path intersections. The experiments were performed with a Pioneer P3-AT robot equipped with a kinect sensor in order to validate and evaluate this approach. The proposed method demonstrated to be a promising approach to autonomous mobile robots navigation.

Keywords: Mobile Robotics, Autonomous Navigation, Kinect, Artificial Neural Networks, Finite State Machine.

1 Introduction

AI techniques implementation on Autonomous Mobile Robots and Intelligent Vehicles has an important role on international scientific community [9][13][14]. One of the most desirable features for a mobile robot is the autonomous navigation capability. There are many known relevant works on this research field, as for example the Darpa Challenges (2004 and 2005 editions on desert and 2007 in Urban environment)[7][8] and ELROB [15][16].

Autonomous mobile robots usually perform three main tasks: localization, mapping/planning and navigation [17]. Localization task is related to estimate robot´s position in a well-known environment using its sensorial data. Mapping consists on creating an environment representation model, based on robot´s localization and sensorial data. Navigation is the capability to process environment information and act, moving safely through this environment.

C. Jayne, S. Yue, and L. Iliadis (Eds.): EANN 2012, CCIS 311, pp. 305–314, 2012.
© Springer-Verlag Berlin Heidelberg 2012

In order to an autonomously navigate into structured environments composed by streets or corridors, the robot must know its approximate position, the environment map and the path to be performed (source/destination). So, navigation in this environment consists on following a well-defined path, considering the available navigation area.

The main focus of this work is the implementation of a Topological Autonomous Navigation System able to recognize specific features in a path on indoor environments (composed by corridors and intersections). This navigation system is intended for indoor service robots in several different tasks, from the simplest ones as carrying objects until critical ones as patrolling. The implemented system for these applications must be easy to configure and use. It must be also robust allowing the robot to both navigate and detect possible abnormalities.

The proposed approach does not require a well-defined environment map, just a sketch representing the main elements, resulting in a simple path sight. Furthermore, this approach does not require an accurate robot´s pose estimation. The main goal is to make the robot navigate in an indoor structured environment deciding when and how to proceed straight, left or right, even when these three possibilities are detected simultaneously (intersections).

The topological approach uses an ANN [19] to classify sensor data and a FSM to represent the steps sequence for each chosen path. The ANN learns all possible states, and a FSM generator is responsible to convert any possible path into a sequence of states. This way, the system combines this deliberative topological behavior with a simple reactive control for a safe navigation.

The next topics of this paper are organized as follows: section 2 presents some realted works; section 3 presents the techniques and resources used for state detection; section 4 presents the experimental results; section 5 presents the conclusion and possible future works.

2   Related Works

Several approaches have been used for navigation, using many different sensors (for example laser, sonar, GPS, IMU, compass) singly or fused [9][17][18]. One of the most used approaches recently is the Vision-Based navigation [20]. This method uses video cameras as the main sensor. Cameras are very suitable for navigation and obstacle avoiding tasks due to its low weight and energy consumption [1]. Furthermore, one single image can provide many different types of information about the environment simultaneously. It is also possible to reduce costs by using cameras rather than other types of sensors [2]. The Vision-based approach implementation is already usual in navigation systems for structured or semi-structured environments [3][7][9][10][11]. These systems classify the image, with track segmentation for safe navigable area identification.

Although these works present good results, the scope for a conventional camera is restricted, and many implementations demand camera data fusion with laser sensors (Sick Lidars, IBEO, Velodyne), radars or even special vision systems such omnidirectional cameras [7][8][9][18]. This fusion becomes necessary specially when depth information is needed. This kind of information is not originally provided by a conventional camera.

Is worth noting that fusion-based approaches are usually expensive, so the use of Kinect sensor can lead to lower cost solutions. Kinect is a low-cost 3D sensor developed by Microsoft for XBOX 360 videogame which allows the player to use its own body as the controller in games. It is composed by a RGB camera associated to an infrared transmitter and receiver allowing depth estimation of the environment elements. Since its sensorial advantages were found out, many independent researches were being held in order to explore this device features in applications from health to robotics [25]. The main advantage of this sensor is the capability of depth maps construction, allowing a very accurate distance estimation for the obstacles detected in front of the robot.

In order to implement an autonomous navigation system, purely reactive approaches are not suitable. Immediate reaction to sensor captured data is not enough to ensure a correct robot control in a more complex path. A more robust system must be developed, providing sequence and context information that are absent in purely reactive models.

In Robotics, FSM-based approaches [5] are widely used. FSMs are useful because the system can be easily described as a sequence of states (context changes). Inputs (sensors data) are considered to allow state changes and to define the adequate reaction for each state (motor action). The proposed system in this work is based on this idea, so the path is described as a FSM and the current state observed through captured sensor data processing.

The use of a Machine Learning technique such ANN is a very interesting way of process sensorial data [6]. ANNs are robust to noise and imprecision on input data. They are also able to detect states and transitions between these states and very efficient to generalize knowledge. This way, this method is very useful for features detection e state recognizing.

The association of ANNs to FSMs has been researched since the 90s [21][22][23][24], when the ANN models were developed and improved, occuping an important place on AI and Machine Learning Researches. Recently, some works were developed using the association of these techniques to robotics problems, from car parking [18] until robots and vehicles navigation in indoor and outdoor environments [4]. These applications have two main problems: the high cost of its main sensor (laser) and the low amount of environment information (bidimensional detection only, with a depth information of a planar cut).

Thus, the proposed solution with kinect sensor shows extremely lower costs in addition to a more complete and accurate environment information, since tridimensional detection is performed simultaneously to conventional image capture.

3   System Overview

The proposed system is composed by three main steps. The first one is the ANN training in order to recognize all possible states (features) using previously collected environment data. The second one is the FSM Generation for any chosen path.

The third step is the autonomous navigation combining this deliberative control (topological path planning) with a reative control to keep the robot aligned. Figure 1 shows the system setup and navigation overview.

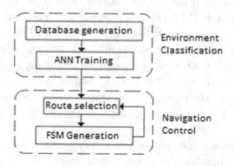

Fig. 1. System overview

3.1   ANN Training

This step consists on training the ANN to efficiently recognize all possible situations in a specific environment. ANN inputs are obtained from sensorial data, and the output is the detected class (state). For the environment used in this work, eight classes were defined (they are better described on next section). The dataset is created saving a "log" of collected input data moving the robot through the environment. The adopted learning algorithm is supervised, so a specialist must "manually" classify this data before ANN training, forming the input/output pairs. Several topologies were tested, and the best results were obtained from a feed-forward MLP, with 632 neurons on input layer, 316 neurons on hidden layer and 8 neurons on output layer. Figure 2 shows a simple graphic representation of this MLP.

Kinect captures are stored on a 640x480 matrix (one element per pixel), and each value is the distance between the sensor and the object represented in that pixel. For navigation purposes, there is no need to use all this 307200 points (this could also make ANN training impracticable due to the high ammount of input neurons). This way, an interest zone on Kinect´s capture is selected, in which the information is enough for state detection.

Kinect sensor has a very interesting property: each line of the depth matrix can be compared to a standard laser scan. This way, it is possible to say that a single kinect data capture provides the same data amount than 480 consecutive laser scans.This is another great advantage of using Kinect, high amount of available data for each single scan. An interesting way to minimize the effects of noisy inputs is to rely on data redundancy. This can be done processing various lines of the depth matrix for state detection. As mentioned earlier, it is impracticable to use all the 480 lines of information as inputs for the ANN, so an "interest area" must be defined.

For this work, 80 lines of depth matrix were selected, the correspondent pixels are represented at Figure 3. The mean of each column is calculated, so the result is one line only with the mean information, similar to a conventional laser scan. This line is used as the ANN input.

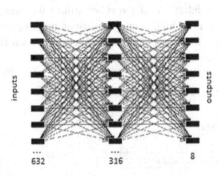

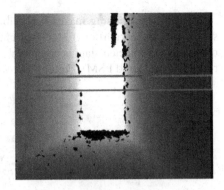

inputs

outputs

632   316   8

Fig. 2. ANN Topology   **Fig. 3.** Interest area on captured kinect´s frame

Each output neuron is associated to one of the 8 implemented states. The ANN is trained once, and must be able to work for any chosen path.

3.2 FSM Generation

Once the topological map for the environment is known, it is easy to establish a route between two points of this map, manually or with an algorithm.

Every route can be seen as a sequence of steps (states), so it is trivial to generate a FSM to represent this path. A single algorithm converts any possible path into a sequence of states and expected actions (also considering that one state can have more than one associated action). This way, after path selection the FSM is stored on system to be used by control unit.

3.3 Navigation Control

The hybrid control combines the deliberative control obtained from FSM-based topological navigation with a reactive control which must keep the robot into its expected route, avoiding collisions.

The Topological Navigation allows the robot to follow its planned path and also know its approximate position, but doesn´t controls the robot "into" every situation (state). When the robot is in a corridor ("straight state"), a reactive control is activated to keep the robot centered and aligned in this corridor. This is the main benefit of this hybrid control: take advantage of deliberative model for path control and simultaneously guarantee a safe navigation with reactive control.

For this implementation, it is assumed that robot´s initial position is always known, as the topological map also. The current position is estimated based on current state detection.

In this approach it isn´t necessary to estimate the robot´s exact position, it "knows" its approximate position based on current state and there is a reactive control responsible for keeping it safe.

Input data processing makes it possible to determine if the robot still at the same state (part of the path) or if a context change is needed. State transitions must occur only with the detected state is compatible with next expected state (this information is related to the stored FSM). Figure 4 shows the navigation control flowchart.

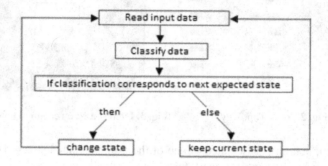

Fig. 4. Navigation control flowchart

4   Experiments and Results

Experiments were carried out in a real environment using a Pioneer P3-AT robot equipped with a dual-core computer for processing and Kinect as the only sensor.

The indoor environment used in the tests was represented with 8 states, illustrated at Figure 5. The created states are: "straight path", "left turn", "right turn", "left and right turns", "left turn and straight", "right turn and straight", "blocked" and "intersection".

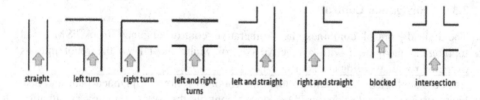

Fig. 5. Possible situaions on environment

The implemented ANN was designed and trained with Stuttgart Neural Network Simulator (SNNS), and then converted to C language with SNNS2C tool to be integrated with the robot´s control unit.

The ANN training database was collected after controlling the robot manually through the environment in several angles and positions, and then sliced into the 8 classes. At this step, about 180 examples were collected for each class, resulting in a database with 1421 input/output pairs, used for supervised learning.

The training algorithm used was Resilient Propagation (R-Prop). This algorithm is achieving great results for feed-forward networks in many applications due to its good training time and convergence. Training parameters were set up as follows: $\delta 0 = 0.1$, $\delta max = 50$, $\alpha = 5.0$ and number of epochs = 1000.

Five different topologies were tested, with different number of hidden layer neurons and number of hidden layers. The tests were held with 16, 80 and 316 neurons on hidden layer. These amounts were considered after empirical tests. The input layer is composed by 632 neurons (vector with the mean of the 80 lines of depth matrix), and output layer is composed by 8 neurons, 1 neuron for each possible class.

A great variation on training times was observed. With 1000 epochs, the training time for 632-16-8 net was 20 minutes, 2 hours for 632-80-8 net, and 8 hours for 632-316-8 approximately.

ANN validation was done with stratified 5-fold cross-validation method. This way, 5 train and test sets were generated, with 80-20 proportion on data (80% used for training and 20% for test, with same proportion of elements from the 8 class on the datasets). The networks with best results were 632-316-8 and 632-80-80-8, with 92,2% and 92% accuracy respectively, as can be observed at Table 1. The main difference between these networks is the training time: 8 hours for the first one versus 2 hours for the second one.

Table 1. ANN Accuracy after 1000 training cicles

ANN	Partition 1		Partition 2		Partition 3		Partition 4		Partition 5		Average	
	Train	Test	Train	Test	Train	Test	Train	Test	Train	Test	Train	Test
632-16-8	89	88	95	87	92	81	92	86	91	85	91,8	85,4
632-80-8	97	89	96	91	97	90	97	92	96	89	96,6	90,2
632-316-8	99	92	99	93	98	92	98	91	98	93	98,4	92,2
632-16-16-8	90	80	93	85	94	85	90	86	94	86	92,2	84,4
632-80-80-8	98	92	99	91	98	91	99	92	99	94	98,6	92

The confusion matrix for partition 2 in 632-316-8 net is shown next, on Figure 6. It is possible to note that error per class is close to zero, which means that very few classification errors occur.

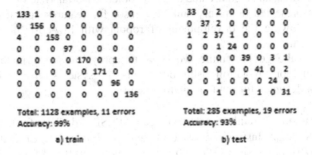

Fig. 6. Train confusion matrix (a) test confusion matrix (b)

Despite the excellent results considering the accuracy mean, a 100% safe navigation is guaranteed only if no errors occur in state detection. As a 100% accuracy is not achieved with the ANN, something must be done to guarantee that no unexpected state changes occur due to a wrong classification. This way, a "filter" was implemented, removing possible oscillations resulting from classification errors. This means that a state transition will occur only after some consecutive detections of the expected state, indicating confidence on transition detection.

After testing and validating the ANN, it was implemented on the real robot, recognizing features for a specific indoor environment. Figure 7 shows a successful classification of the "left turn and straight" state. The frame on the left is a graphic representation of the depth matrix, and the frame on the right is a "real" frame captured from kinect.

Fig. 7. State "left turn and straight" classification

5   Conclusion

Excellent results were achieved, with high accuracy level for the ANN individually, and 100% accuracy on navigation task after filter implementation on all experiments carried out. This shows that the association of ANN and FSM is a very suitable approach for autonomous robotic navigation.

This system is very flexible, as it can be re-trained to recognize new situations, settings and features, and also use and combine other sensorial systems.

Kinect was presented as a very suitable sensor for features detection on indoor environmets, allowing the development of robots with low-cost 3D vision-based navigation systems.

The main challenge for future works is to apply this same methodology for autonomous outdoor navigation, using other sensorial systems (also fusing sensors), as kinect is not designed for outdoor environments.

Acknowledgments. The authors acknowledge FAPESP and CNPq for their support to INCT-SEC (National Institute of Science and Technology - Critical Embedded Systems - Brazil), processes 573963/2008-9 and 08/57870-9 and financial support to authors (master´s grant).

References

1. Zingg, S., Scaramuzza, D., Weiss, S., Siegwart, R.: MAV Navigation through Indoor Corridors Using Optical Flow. In: IEEE International Conference on Robotics and Automation (ICRA 2010), Anchorage, Alaska (May 2010)
2. Scaramuzza, D., Siegwart, R.: Appearance Guided Monocular Omnidirectional Visual Odometry for Outdoor Ground Vehicles. IEEE Transactions on Robotics 24(5) (October 2008)
3. Shinzato, P.Y., Wolf, D.F.: Features Image Analysis for Road Following Algorithm Using Neural Networks (September 2010)
4. Sales, D., Osório, F., Wolf, D.: Topological Autonomous Navigation for Mobile Robots in Indoor Environments using ANN and FSM. In: Proceedings of the I Brazilian Conference on Critical Embedded Systems (CBSEC), São Carlos, Brazil (2011)
5. Hopcroft, J.E., Ullman, J.D.: Introduction to Automata Theory, Languages and Computation. Addison - Wesley (1979)
6. Marino, A., Parker, L., Antonelli, G., Caccavale, F.: Behavioral Control for Multi-Robot Perimeter Patrol: A Finite State Automata approach. In: ICRA (2009)
7. Thrun, S., et al.: Stanley: The Robot that Won the DARPA Grand Challenge. Journal of Field Robotics 23(9), 661–692 (2006)
8. Urmson, C., et al.: Autonomous driving in urban environments: Boss and the Urban Challenge. Journal of Field Robotics 25(8), 425–466 (August 2008); Special Issue on the 2007 DARPA Urban Challenge, Part I
9. Buehler, M., Iagnemma, K., Singh, S. (eds.): The 2005 DARPA Grand Challenge: The Great Robot Race (Springer Tracts in Advanced Robotics), 1st edn. Springer (October 2007)
10. Nefian, A.V., Bradski, G.R.: Detection of Drivable Corridors for Off-Road Autonomous Navigation. In: ICIP 2006: Proceedings of the IEEE International Conference on Image Processing, pp. 3025–3028 (2006)
11. Álvarez, J.M., López, A.M., Baldrich, R.: Illuminant Invariant Model-Based Road Segmentation. In: IEEE Intelligent Vehicles Symposium, Eindhoven, Netherlands (June 2008),
    http://www.cvc.uab.es/adas/index.php?section=publications
12. Bishop, C.M.: Neural Networks for Pattern Recognition. Oxford University Press, Oxford (1995)
13. Bishop, R.: Intelligent vehicle applications worldwide. IEEE Intelligent Systems and their Applications 15(1), 78–81 (2000)
14. Bishop, R.: A Survey of Intelligent Vehicle Applications Worldwide. In: Proceedings of the IEEE Intelligent Vehicles Symposium 2000, pp. 25–30 (2000)
15. Kuhnert, K.-D.: Software architecture of the Autonomous Mobile Outdoor Robot AMOR. In: Proceedings of the IEEE Intelligent Vehicles Symposium 2008, pp. 889–894 (2008)
16. Schilling, K.: Assistance Systems for the Control of Rovers. In: SICE Annual Conference, Tokyo (October 2008)
17. Wolf, D.F., Osório, F.S., Simões, E., Trindade Jr., O.: Robótica Inteligente: Da Simulação às Aplicações no Mundo Real [Tutorial]. In: de Leon F. de Carvalho, A.P., Kowaltowski, T. (org.) JAI: Jornada de Atualização em Informática da SBC, vol. 1, pp. 279–330. SBC - Editora da PUC. RJ, Rio de Janeiro (2009)
18. Goebl, M., Althoff, M., Buss, M., Farber, G., Hecker, F., Heissing, B., Kraus, S., Nagel, R., Leon, F.P., Rattei, F., Russ, M., Schweitzer, M., Thuy, M., Wang, C., Wuensche, H.J.: Design and capabilities of the Munich Cognitive Automobile. In: IEEE Intelligent Vehicles Symposium 2008, pp. 1101–1107 (2008)

19. Haykin, S.: Neural Networks: A Comprehensive Foundation. Prentice Hall PTR, Upper Saddle River (1998)
20. Sales, D., Shinzato, P., Pessin, G., Wolf, D., Osório, F.: Vision-based Autonomous Navigation System using ANN and FSM Control. In: Proceedings of the IEEE Latin American Robotics Symposium (LARS), São Bernardo do Campo, Brazil (2010)
21. Giles, C.L., Horne, B.G., Lin, T.: Learning a class of large finite state machines with a recurrent neural network. Neural Networks 8(9), 1359–1365 (1995)
22. Omlin, C.W., Giles, C.L.: Constructing Deterministic Finite-State Automata in Recurrent Neural Networks. Journal of the ACM 43(6), 937–972 (1996)
23. Frasconi, P., Gori, M., Maggini, M., Soda, G.: Representation of finite state automata in Recurrent Radial Basis Function networks. Machine Learning 23(1), 5–32 (1996)
24. Cleeremans, A., Servan-Schreiber, D., Mcclelland, J.L.: Finite State Automata and Simple Recurrent Networks. Neural Computation 1(3), 372–381 (1989)
25. KinectHacks Supporting the Kinect Hacking news and community, http://www.kinecthacks.com/ (accessed in: February 10, 2012)

Hybrid Computational Model for Producing English Past Tense Verbs

Maitrei Kohli[1], George D. Magoulas[1], and Michael Thomas[2]

[1] Department of Computer Science and Information Systems,
[2] Department of Psychological Sciences,
Birkbeck College, University of London, UK
{maitrei,gmagoulas}@dcs.bbk.ac.uk,
m.thomas@psychology.bbk.ac.uk

Abstract. In this work, we explore the use of artificial neural networks (ANN) as computational models for producing English past tense verbs by combining them with the genetic algorithms (GA). The principal focus was to model the population variability exhibited by children in learning the past tense. This variability stems from genetic and environmental origins. We simulated the effects of genetic influences via variations in the neuro computational parameters of the ANNs, and the effects of environmental influences via a filter applied to the training set, implementing variation in the information available to the child produced by, for example, differences in socio-economic status. In the model, GA served two main purposes - to create the population of artificial neural networks and to encode the neuro computational parameters of the ANN into the genome. English past tense provides an interesting training domain in that it comprises a set of quasi-regular mappings. English verbs are of two types, regular verbs and irregular verbs. However, a similarity gradient also exists between these two classes. We consider the performance of the combination of ANN and GA under a range of metrics. Our tests produced encouraging results as to the utility of this method, and a foundation for future work in using a computational framework to capture population-level variability.

Keywords: Feed forward neural networks, imbalanced datasets, hamming distance, nearest neighbour, genetic algorithms, English past tense verbs, quasi regular mappings.

1 Introduction

Artificial neural networks (ANNs) are computational abstractions of the biological information processing system. In this work, we combine ANNs with genetic algorithms (GA) to develop a new computational model for learning English past tense verbs.

The English past tense has been widely studied as a testing ground for theories of language development. This is because it is quasi-regular mapping, comprising both a productive rule (add –ed to the verb stem to produce the past tense) and a set of

C. Jayne, S. Yue, and L. Iliadis (Eds.): EANN 2012, CCIS 311, pp. 315–324, 2012.

exceptions to this rule. This raises the question of the processing structures necessary to acquire the domain. Until now substantial amount of work has used ANNs as cognitive models of the acquisition process (see [1] for review). No work to date, however, has considered how to capture the wide range of variability that children exhibit in acquiring this aspect of language. Since ANNs constitute parameterised learning systems, they provide a promising framework within which to study this phenomenon [2].

Factors affecting language development are attributed to genetic and environmental influences. To model genetic influences, we use GA as a means to encode variation in the neuro computational parameters of the ANNs, thereby modulating their learning efficiency. These parameters are responsible for how the network is built (e.g., number of hidden units), its processing dynamics (steepness of the activation function), how it is maintained (weight decay), and how it adapts (learning rate, momentum). To model environmental influences, we apply a filter to the training set to alter the quality of the information available to the learning system. One candidate causal factor in producing environmental variation is socio-economic-status (SES). A body of research suggests that children in lower SES families experience substantially less language input and also a narrower variety of words and sentence structures [3]. The filter creates a unique subsample of the training set for each simulated individual, based on their SES.

This paper is organised as follows. First we provide information about the problem domain. Then the methodology adopted in our approach is described. Next we discuss the datasets used. In section 4 the experimental setups and performance assessment techniques are described. Finally we present the experimental results and discuss the findings.

2   The English Past Tense Problem

The English past tense is an example of a quasi-regular domain. This problem domain has dual nature – the majority of verbs form their past tense by following a rule for stem suffixation, also referred to as + ed rule. This rule allows for three possible suffixes - /d/ e.g. – tame – tamed; /t/ e.g. – bend – bent and /ed/ - e.g. – talk – talked. However, a significant number of verbs form their past tenses by exceptions to that rule (example: go – went, hide - hid) [4]. The verbs adhering to the former rule based approach are called regular verbs, while the verbs belonging to the second category are called irregular verbs. Also some of the irregular verbs share the characteristics of the regular verbs. For instance, many irregular verbs have regular endings, /d/ or /t/ but with either a reduction of the vowel, example: say – said, do - did or a deletion of the stem consonant, example: has – had, make – made [5]. This overlap between regular and irregular verbs is also a challenge for the model.

Our base model, prior to implementing sources of variation, was inspired by that proposed by Plunkett and Marchman [4], though see [6] for more recent, larger scale models. Plunkett and Marchman suggested that both the regular and the exception verbs could be acquired by an otherwise undifferentiated three-layer backpropagation network, trained to associate representations of phonological form of each verb stem to a similar representation of its past tense.

3   Methodology

A synergy of ANN and GA is applied to model the system for acquisition of English past tense verbs. The GA component is used to create a population of ANNs and to encode the neuro computational parameters of the ANN into the genome.

The methodology can be summarized as follows:

1.   The first step is to design ANNs incorporating a set of computational parameters that would constrain their learning abilities. In our case, we select 8 parameters. These parameters correspond to how the network is built (number of hidden units, architecture), its activation dynamics (the slope of the logistic activation function), how it is maintained (weight decay), how it adapts (learning rate, momentum, learning algorithm), and how it generates behavioral outputs (nearest neighbour threshold).

2.   The next step concerns the calibration of the range of variation of each of these parameters. Encoding the parameters within a fixed range allows variation in the genome between members of population, which then produces variations in computational properties. The range of variation of the parameter values serves as the upper and the lower bound used for converting the genotype (encoded values) into its corresponding phenotype (real values).

3.   The third step consists of encoding the range of parameter variation in the artificial genome using a binary representation. We are using 10 bits per parameter; overall the genome has 80 bits. The parameters used and their range of variation are given in Table 1.

Table 1. The Genome describing the neuro - computational parameters and their range

Parameter	Range of variation
No of Hidden Units	6 - 500
Learning Rate	0.005 – 0.5
Momentum	0 – 0.75
Unit threshold function (steepness of logistic function)	0.0625 - 4
Weight Decay	0.2 – 0.6
Nearest Neighbour Threshold	0.05 – 0.5

The remaining two parameters are the learning algorithm and the architecture, where Backpropagation training and a 3-layer feed forward network are adopted respectively. In our initial implementation, these parameters were not varied. Then, the methodology continues as follows:

4.   The fourth step concerns breeding the population of 100 ANNs using this genome.

5. The fifth step focuses on implementing the variation in the quality of environment, accounted for by SES, by means of filtered training sets. An individual's SES is modeled by a number selected at random from the range 0.6-1.0. This gives a probability that any given verb in the full training set would be included in that individual's training set. This filter is applied a single time to create the unique training set for that individual. The range 0.6-1.0 defines the range of variation of SES, and ensures that all individuals are exposed to more than half of the past tense domain.

6. The last step is about training and evaluating training performance and generalisation.

4   The English Past Tense Dataset

The English past tense domain is modeled by an artificial language created to capture many of the important aspects of the English language, while retaining greater experimental control over the similarity structure of the domain [4]. Artificial verbs are monosyllabic and constructed from English phonemes. There are 508 verbs in the dataset. Each verb has three phonemes – initial, middle and final. The phonemes are represented over 19 binary features. The network thus has $3*19 = 57$ input units and $3*19 + 5 = 62$ units at the output. The extra five units in the output layer are used for representing the affix for regular verbs in binary format.

In the training dataset there are 410 regular and 98 irregular verbs. As this is a radically imbalanced dataset generating a classifier is challenging as the classifier tends to map/label every pattern with the majority class. The mapping of the training set is given a frequency structure, called the token frequency, representing the frequency with which the individual encountered each verb. Some verbs are considered of high frequency whilst others of low frequency. The token frequency is implemented by multiplying the token frequency bit with the weight change generated by the difference between the actual output and the target output. In our experiments, the weight change of high frequency verbs was multiplied by 0.3 and of the low frequency verbs by 0.1.

A second dataset is created to assess the generalisation performance of the model. The main intent is to measure the degree to which a network can reproduce in the output layer properly inflected novel items presented in the input. The generalisation set comprises 508 novel verbs, each of which share at least two phonemes with one of the verbs in the training set, for example *wug – wugged* [7].

5   Experimental Settings and Performance Assessment

A population of 100 ANNs, whose parameters are generated by the GAs, was trained in two different setups. In the first setup, the population of ANN was trained using the full training set, i.e. it contains all the past tense verbs, along with their accepted past tense forms (henceforth referred as the Non Family setup). In the second setup, we used the filtered training sets, by taking samples from the perfect training set to create subsets, for each member of the population (henceforth referred as the Family setup). This arrangement ensures that each member of the population has a different environment or training set, and thus simulated the effect of SES. Though the networks are trained according to their filtered or Family training sets, the performance is always assessed against the full training set. A comparison of Non Family and Family setups demonstrates the impact of variability in the environment, independent of the learning properties of the ANNs.

We report below results from training 100 feed forward nets, using the batch version of RPROP algorithm. The stopping condition was an error goal of 10^{-5} within 1000 epochs. The performance was assessed using two modes - the MSE with weight decay and the recognition accuracy using nearest neighbours based criteria. The first criterion employed the Hamming distance while the second one was threshold based and used the Root Mean Square (RMS) error.

5.1   Nearest Neighbour Technique Based on Hamming Distance

In the training set, there were 508 monosyllabic verbs, constructed using consonant-vowel templates and the phoneme set of English. Phonemes were represented over 19 binary articulatory features.

The nearest neighbour accuracy was measured between the actual and the target patterns on a phoneme–by–phoneme basis using the Hamming distance. In information theory, the Hamming distance between two strings of equal length is the number of positions at which the corresponding symbols are different. In other words, it measures the minimum number of substitutions required to change one string into the other, or the number of errors that transformed one string into the other. This method provides an efficient way of calculating the nearest neighbours. The algorithm for calculating the Hamming distance is listed below.

1. Take the first pattern from the actual output and the desired output.
2. Calculate the Hamming distance between these two patterns, individually, for all three phonemes. This implies that phoneme 1 of actual output is matched with phoneme 1 of the desired output. Similarly, phoneme 2 and phoneme 3 are matched with corresponding phonemes in desired output.
3. **IF** the Hamming distance between all three phonemes is less than 2, then calculate the Hamming distance between last five bits of both patterns.
4. **IF** this distance is equal to zero, then pattern is counted as correct classification, **ELSE** it is counted as an error or misclassification.
5. In the case of misclassification, the last five bits of both the actual output and the desired output are compared with the allomorph (which consists of all possible classes with their binary representations), to find out the actual assigned class and the desired class.
6. **IF** the last five bits of the actual pattern do not match with any pattern of allomorph, then that pattern is classified as 'random'.
7. In case the **IF** condition specified in **step 3** does not hold, then the same pattern of actual output is matched with the next pattern from the desired output set. This process continues till either the IF condition (of step 3) is

satisfied OR till all patterns in the desired output set have been scanned through. In the latter case, if no match is found, then that pattern of actual output set is considered as 'Not Classified'. Repeat the process with next patterns of the actual output set.

This method gives us the total number of correct classifications, total number of errors and the types of errors for each network. The allomorph used in this algorithm is as follows: [0 0 0 0 0] represents Irregular (Ir) verbs; [0 0 1 0 1] denotes a Regular verb with +d rule (R^d); [0 1 1 0 0] stands for a Regular verb with +t rule (R^t); [0 1 0 1 0] denotes a Regular verb with +ed rule (R^ed).

5.2   Nearest Neighbour Technique Based on RMS Error Threshold

We tested the performance of the networks with an alternative technique, nearest neighbour threshold. The process is as follows:

1. Take the first pattern from the actual output set.
2. For each actual output pattern, starting from the first target pattern, consider all available target patterns.
3. Calculate the root mean square error between the actual and target pattern on a phoneme–by–phoneme basis. This implies, calculating the RMS between phoneme 1 of the actual output pattern and phoneme 1 of target set pattern, and then the RMS values for phoneme 2 and phoneme 3 as well. This results in a 508 row by 508 column array of RMS values, where each array element contains 3 values corresponding to the RMS error between the three phonemes.
4. Based on the range of the nearest neighbour values, as specified in the genome of the artificial neural networks, apply threshold on the RMS values of three phonemes taken together.
5. Select only those neighbours whose RMS error values are lower than the corresponding threshold values.
6. Compare the last five bits of the actual pattern and the selected nearest neighbours with the allomorph to determine their respective classes.
7. Select the 'majority' class from amongst the neighbours and compare it with the class assigned to the actual output pattern. If these matches, then count success else count miss classification. Repeat this process for all the patterns of the actual output set.

6   Results

We report on the classification accuracy and generalisation performance, with respect to three measures – the MSE, the Hamming distance and lastly the nearest neighbour threshold.

In terms of measuring performance based on Hamming distance for all types of verbs, i.e. irregulars, R^d, R^ed and R^t, Table 2 lists the mean values for classification success. Table 3 contains the types of miss classifications the networks made and the mean values of those errors and finally Tables 4 and 5 list the improved results after applying some post processing techniques, discussed below.

Table 2. Mean classifications success per category

Type of verb	Mean Classification on Training Set		Mean Classification on Generalisation Set	
	Non Family Networks	Family Networks	Non Family Networks	Family Networks
R^d	252.24	242.60	265.28	251.69
R^t	71.80	57.56	71.57	57.06
R^ed	10.70	10.89	7.21	7.46
Irreg	5.29	7.22	N.A.	N.A.

Table 3. Types of miss classification errors and their mean values

Assigned Category	Desired Category	Training Set		Generalisation Set	
		Non Family	Family	Non Family	Family
Irreg	R^d	3.68	4.35	3.12	3.97
Random	R^d	8.58	12.05	9.70	12.12
R^d	R^ed	5.06	6.44	8.00	10.08
Random	R^ed	34.70	31.68	41.40	38.93
R^t	R^d	9.15	7.36	9.69	8.00
R^t	R^ed	2.71	1.77	4.39	3.60
Random	R^t	14.47	23.89	17.80	25.99
R^d	R^t	8.81	9.13	7.28	7.66
Irreg	R^ed	4.74	7.61	4.99	6.24
R^d	Irreg	2.80	3.17	0	0
Irreg	R^t	1.26	3.38	1.30	4.54
Random	Irreg	11.69	10.21	0	0
R^ed	Irreg	1.39	1.58	0	0
R^ed	R^t	0.89	0.92	0.85	0.90
R^t	Irreg	1.30	0.87	0	0

Table 3 lists the types of misclassifications made by the population of networks. The most frequent misclassification was classifying a regular verb as a regular but in the wrong category, that is, the incorrect allomorph, e.g. instead of talk – talked (+ed), networks convert it as talk – talkd (+d or +t). The second most frequent mistake was classifying regular or irregular verbs as *random*. In most cases, this happens due just to the difference of one bit between the actual output affix (last five bits) and the target verb affix.

We do not consider the aforementioned two misclassifications as errors on the following grounds.

- In the first case, it is evident that the network(s) applied the production rule for forming past tense. This implies that the methodology used for converting verb to its past tense is correct.
- In the latter case, the network(s) produces all three phonemes correctly (the phonemes of the actual and target patterns match). The difference of one bit occurs in the last five bits (past tense affix). This indicates that the mechanism followed is correct, especially since the network does not categorise the verb in an incorrect category.

Therefore, we applied post-processing techniques in these two cases, which improved the accuracy of the model.

Table 4. Average performance and improvements on training set

	Non Family Networks		Family Networks	
	Correct in %	Error in %	Correct in %	Error in %
Actual Results	66.9	21.9	72.7	21.6
Improved Results	84.4	4.4	82.8	11.1

Table 5. Average performance and improvements on the generalisation set

	Non Family Networks		Family Networks	
	Correct in %	Error in %	Correct in %	Error in %
Actual Results	67.7	21.3	62.2	24.0
Improved Results	80.0	9.9	75.2	11.1

Improving Performance on training set: We employed three different techniques in order to improve the performance.

1. Considering misclassification amongst regular verbs as okay.

2. Regular verb patterns classified as random due to difference in just 1 bit were considered okay.

3. Irregular verb patterns classified as random due to difference in just 1 bit were considered okay.

Improving Performance on the generalisation set: The following technique was applied to improve generalisation performance.

1. Verb patterns classified as random due to difference in just 1 bit were considered okay.

Our model achieved 84.4 % and 80.0% accuracy on training datasets when used in the Non Family mode and an accuracy of 82.8% and 75.2% on generalisation dataset when tested in Non Family and Family modes, respectively. The initial results indicate that classification accuracies are not significantly different in the two modes.

For example, in Table 4, the population has the average/mean accuracy of 84.4% when exposed to full training set (Non Family mode) and of 82.8% when exposed to filtered training set (Family mode). This held for generalisation performance as well, as described in Table 5. These results indicate that for the ranges of genetic and environmental variation considered, genetic variation has more influence in determining performance while acquiring past tense.

6.1   Performance Based on Mean Square Error

As the second measure of performance assessment, the MSE was used to predict the accuracy. The minimum, maximum, mean and the standard deviation of time taken, performance and epochs is reported in Table 6.

Table 6. Performance based on MSE

	Non Family				Family			
	Min	Max	Mean	STD	Min	Max	Mean	STD
Time (seconds)	170	164,653	1,930	387	170	164,668	1,366	976
Performance	0.0502	42.5200	0.1229	0.3765	0.0360	42.4800	0.1100	0.3600
Epochs	425	1000	484	280	542	1000	491	284

6.2   Performance Based on Nearest Neighbour Threshold

The performance is reported in terms of number of correct classifications and the number of miss classifications made in Table 7.

Table 7. Performance based on nearest neighbour threshold using RMSE

	Non Family				Family			
	Min	Max	Mean	STD	Min	Max	Mean	STD
Correct Classifications	365	414	387	17	364	427	393	25
Misclassifications	94	143	121	17	81	144	115	25

The results show that the average correct classification performance is 76.2% and 77.4% in the discussed modes.

7   Conclusion

In this paper, we explored the use of artificial neural networks as computational models for producing English past tense verbs, and proposed a synergistic approach to capture population variability by (a) combining ANN with genetic algorithms and (b) applying a filter to the training set to simulate environmental influences such as socioeconomic status. The performance of the model was assessed using three different measures and in two different setups. Our tests produced encouraging results as to the utility of this method, and a foundation for future work in using a computational framework to capture population-level variability. These results indicate that for the ranges of genetic and environmental variation considered, genetic variation has more influence in determining performance while acquiring past tense. Our next steps are to consider the impact of different respective ranges of genetic and environmental variation, along with exploring different neural architectures.

References

1. Thomas, M.S.C., McClelland, J.L.: Connectionist models of cognition. In: Sun, R. (ed.) Cambridge Handbook of Computational Cognitive Modelling, pp. 23–58. Cambridge University Press, Cambridge (2008)
2. Thomas, M.S.C., Karmiloff-Smith, A.: Connectionist models of development, developmental disorders and individual differences. In: Sternberg, R.J., Lautrey, J., Lubart, T. (eds.) Models of Intelligence: International Perspectives, pp. 133–150. American Psychological Association (2003)
3. Thomas, M.S.C., Ronald, A., Forrester, N.A.: Modelling socio-economic status effects on language development (2012) (manuscript submitted for publication)
4. Plunkett, K., Marchman, V.: U-shaped learning and frequency effects in a multilayered perceptron: Implications for child language acquisition. Cognition 38 (1991)
5. Lupyan, G., McClelland, J.L.: Did, Made, Had, Said: Capturing quasi – regularity in exceptions. In: 25th Annual Meeting of the Cognitive Science Society (2003)
6. Karaminis, T., Thomas, M.S.C.: A cross-linguistic model of the acquisition of inflectional morphology in English and modern Greek. In: Proceedings of the 32nd Annual Meeting of the Cognitive Science Society, August 11-14 (2010)
7. Thomas, M.S.C., Ronald, A., Forrester, N.A.: Modelling the mechanisms underlying population variability across development: Simulating genetic and environmental effects on cognition. DNL Tech report 2009-1 (2009)

Characterizing Mobile Network Daily Traffic Patterns by 1-Dimensional SOM and Clustering

Pekka Kumpulainen[1] and Kimmo Hätönen[2]

[1] Tampere University of Technology, Department of Automation Science and Engineering,
Korkeakoulunkatu 3, 33720 Tampere, Finland
pekka.kumpulainen@tut.fi
[2] Nokia Siemens Networks, T&S Research, Espoo, Finland
kimmo.hatonen@nsn.com

Abstract. Mobile network traffic produces daily patterns. In this paper we show how exploratory data analysis can be used to inspect the origin of the daily patterns. We use a 1-dimensional self-organizing map to characterize the patterns. 1-dimensional map enables compact visualization that is especially suitable for data where the variables are not independent but form a pattern. We introduce a stability index for analyzing the variation of the daily patterns of network elements along the days of the week. We use clustering to construct profiles for the network elements to study the stability of the traffic patterns within each element. We found out that the day of the week is the main explanation for the traffic patterns on weekends. On weekdays the traffic patterns are mostly specific to groups of networks elements, not the day of the week.

Keywords: 1-D SOM, clustering, mobile network, daily pattern, exploratory data analysis.

1 Introduction

Mobile networks are large and complex systems. Huge amounts of data are collected from various parts of networks to support the operators in maintaining them. Efficient network management requires summarizing the essential information from the vast amounts of data.

Traffic delivered through the network elements (NE) is one of the most significant indicators of the networks performance. The daily traffic in each NE depends on its location and behavior of the subscribers that are connected to it, the rhythm of life of surrounding society [1]. In addition to the total volume of the traffic, finer details of how the traffic is distributed during a day, daily patterns, are of great importance. The daily traffic patterns can be utilized for example in data compression [2]. The daily traffic has often been observed to be behaving according to the day of the week. On the other hand, some NEs have similar patterns from day to day. In this paper we propose explorative data analysis procedures for determining to what extent the daily patterns are originated from the day of the week or if they are features of the NE itself.

C. Jayne, S. Yue, and L. Iliadis (Eds.): EANN 2012, CCIS 311, pp. 325–333, 2012.

In the following section we introduce the daily pattern data with examples. In section 3 we use a 1-dimensional self-organizing map (SOM) to visualize the variety of daily patterns. We analyze the significance of the day of the week on the daily patterns by a novel index that measures the stability of the data projected on the SOM. In section 4 we perform two additional levels of clustering on the SOM to construct behavioral patterns for each NE. This enables analyzing the relevance of the NEs on the traffic pattern types. Finally we summarize the results in the last section

2   Mobile Network Traffic Patterns

The traffic volume time series of each NE are split into daily patterns covering a 24 hour period of time. The patterns are concatenated to form a data matrix that has 24 columns, one for every hour of the day. The number of rows in the data matrix equals the product of the number of days covered by the measurement period and the number of network elements, $N = N_{days}*N_{NE}$. The traffic data are collected to matrix $X = [x_1^T\ x_2^T\ ...\ x_N^T]^T$. Each row $x_i = [x_{i0}\ x_{i1}\ ...\ x_{i23}]$, called a *daily pattern*, contains hourly traffic of one cell for one day.

The data set in this study consists of 100 NEs and 6 weeks measurement period, thus $N_{NE} = 100$ and $N_{days} = 42$. Some values are missing from the database leaving a total of 3865 daily patterns in the data set. Due to large number of missing values at midnight we include only the hourly traffics from 1 to 23 o'clock.

One week of traffic of one NE is exemplified at the top of Fig.1.

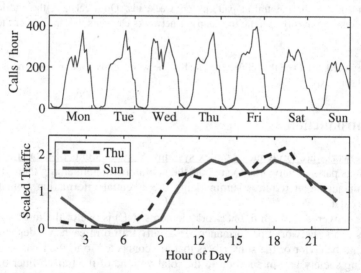

Fig. 1. One week of hourly traffic in one network element (*top*). Two scaled daily patterns presenting different behavior (*bottom*).

The maximum daily values in the data set vary from 23 to over 1600 calls in an hour. In this application we are only interested in the shapes of the daily patterns. Therefore we scale each pattern by dividing each daily pattern by its mean value. The scaled patterns from Monday and Saturday are presented at the bottom of Fig. 1.

3   Monitoring Traffic Patterns by SOM

Self-Organizing Map (SOM) is an unsupervised neural network which approximates multi-dimensional data [3]. SOM consists of a regular grid of map units, each of which contains a code vector of the same dimension as the input data, representing a point in the data space. During the training phase, the code vectors are computed to optimally describe the distribution of the data, producing vector quantization. SOM also preserves topology, i.e. the map units that are close to each other on the SOM grid will have code vectors that are close to each other in the data space.

After training the data are projected to the map, i.e. each data point is associated with the map unit whose code vector describes the point best. Minimum Euclidean distance is typically used to select those units, called best matching units (BMU).

BMUs are used to calculate topographic error of the map. It is defined as the proportion of observations for which the two best matching map units are not adjacent units on the map [4] and is formulated in equation (1).

$$\varepsilon_t = \frac{1}{N} \sum_{k=1}^{N} u(x_k),$$ (1)

where $u(x_k) = \begin{cases} 1, \text{best - and second best matching units not adjacent} \\ 0, \text{otherwise} \end{cases}$

Visualization is one of the main advantages of SOM, which is typically constructed as a 2-dimensional grid. The most common visualizations of SOM include component planes, hit histogram and U-matrix [5]. The component planes present the values of the code vectors of the map units. For a 2-d map that requires a separate 2-d plane for each variable. Hit histogram value for each map unit is constructed by counting the data points that have the map unit as a BMU, i.e. how many data points hit each map unit. U-matrix presents the distances between the code vectors of the adjacent map units on the map grid [6]. Both, hit histogram and U-matrix of 2-d SOM can be presented either color coded on 2-d planes or 3-dimensional surfaces. Hit histogram can also be presented as markers sized by the hit count.

3.1   Visualizing the Daily Patterns with 1-Dimensional SOM

We trained a 1-dimensional SOM with the data set of daily patterns. We trained several maps varying the number of map units from 50 to 100 with a step of 5 and further from 100 to 200 with a step of 10. We selected the map with 55 units, which minimized the topographic error given in equation (1). For training we used the batch algorithm provided in the SOM toolbox for MATLAB [7] (available at: http://www.cis.hut.fi/projects/somtoolbox). All the maps were initialized according to the first principal component (eigenvector of the autocorrelation matrix that has the largest eigenvalue). This is the best linear approximation of the data thus reducing the need for rough training and allows the map to converge faster [3]. This initialization is deterministic for a given data set and the map converges to a very similar basic structure regardless of the number on the map units.

A 1-dimensional SOM allows more compact visualization than a 2-dimensional map. All 2-dimensional planes for visualization can be replaced by line plots. The 23 dimensional code vectors that represent the daily patterns can be combined into one map of typical patterns. This is a huge advantage especially in this application where the variables are not independent but form a pattern. The topology preservation feature of SOM ensures that the code vectors of the adjacent units are similar, thus resulting in a smooth map of prototype patterns, depicted at the top of Fig. 2. Darker grayscales represent higher traffic.

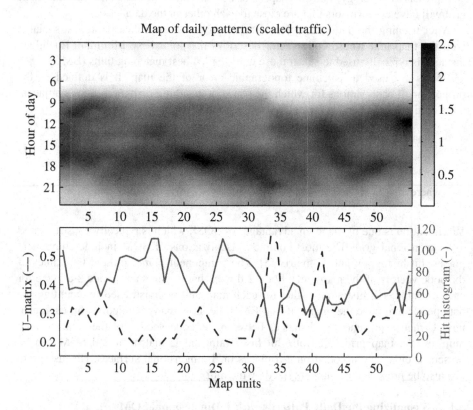

Fig. 2. Daily pattern prototypes identified by SOM (*top*). Darkness refers to the value of the scaled traffic as mapped in the color bar (*top right*). U-matrix and hit histogram (*bottom*) reduce to lines in 1-dimensional SOM.

U-matrix and hit histogram can be presented by lines as depicted at the bottom of Fig. 2. U-matrix values are simply Euclidean distances between the code vectors of the adjacent map units. The hit histogram line shows the number of data points that hit each map unit. Map units surrounded by high U-matrix values (as between units 33, 34 and 35) denote that their code vectors are further apart in the data space, representing a sparse part of the space. Such units have typically relatively small number of hits as depicted in Fig. 2.

Traffic patterns during weekends can be expected to be different from those on weekdays. We calculated hit histogram separately for each day of the week. On a 1-dimensional SOM they can be visualized on one compact hit map in Fig.3.

The patterns of weekends are clearly separated from the weekdays and concentrated to map units from 35 to 55. There are only few hits from weekdays on that area of the map. Saturday and Sunday overlap partly but most of their hits are separated, Saturday concentrating on the very end of the map.

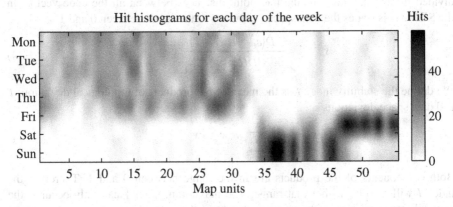

Fig. 3. Hit histograms of subsets of data on 1-D SOM can be combined to a surface. Here hit histograms for each day of the week are presented. Daily patterns during weekend differ from those on weekdays and are concentrated to the map units from 35 onward.

3.2   Stability Index

In this section we study how much the daily patterns depend on the day of the week. Thus we study the stability of the patterns in each weekday. We introduce a stability index of projection (SIP) on the map that can be calculated for a subset S of data that consists of N observations.

The data set consists of 6 weeks, providing $N = 6$ observations for each day of the week. Therefore, each network element NE has an associated set of BMUs (m_i) and corresponding code vectors (c_i) for each day of the week W (from Monday to Sunday):

$$S^{NE,W} = \left\{ m_i^{NE,W}, c_i^{NE,W} \right\}_{i=1}^{N}. \tag{2}$$

The m_i are the locations of the BMUs on the SOM topology. For 1-D SOM that is the index of the BMU, in our case an integer from 1 to L, the number of map units. In our example $L = 55$.

The stability index consists of a combination of the deviations on the map topology and in the original data space.

We define the topological deviations t_i as absolute deviations from the mode, most common BMU in the subset (the one that has the highest frequency in the set). We denote the mode as m_M, and the corresponding code vector as c_M. The deviations of the BMU indices are divided by L-1 so that the maximum possible topological deviation equals 1. The superscript ^{NE,W} denotes that the deviations are calculated within the set $S^{NE,W}$, defined in equation (2).

$$t_i^{NE,W} = \frac{\left| m_i^{NE,W} - m_M^{NE,W} \right|}{L-1} .$$ (3)

We define the data space deviations s_i as the Euclidean distances between the code vectors c_i and the one of the mode unit c_M, denoted by $D(c_i,c_M)$. The distances are divided by the maximum of the interpoint distances between all the code vectors in the SOM. This scales the data space deviations to the range between 0 and 1.

$$s_i^{NE,W} = \frac{D\left(c_i^{NE,W}, c_M^{NE,W}\right)}{\max\left\{D(c_j, c_k)\right\}_{j,k=1}^{L}} .$$ (4)

We define the stability index I as the mean of the products of topological deviations t_i and data space deviations s_i.

$$I^{NE,W} = \frac{1}{N} \sum_{i=1}^{N} t_i^{NE,W} s_i^{NE,W} .$$ (5)

Both components of the products are in the range between 0 and 1. Therefore the index I will also be within that range. The maximum value 1 can only occur if the data subset contains two observations that have their BMUs at the opposite ends of the map and the code vectors of those map units are separated by the maximum distance within the code vectors of the map.

Distribution of the stability indices calculated for each subset $S^{NE,W}$ is presented as a histogram in Fig. 4.

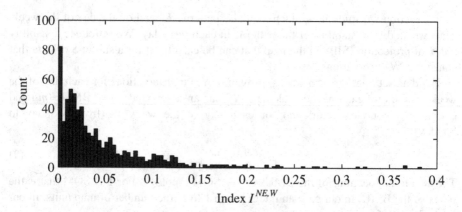

Fig. 4. Histogram of the stability indices

There is a mode at the index value of 0.01. We considered all the subsets (satisfying pairs: {NE, day of the week}) with $I<0.01$ as relatively stable. We then analyzed how these stable days, total of 142, are distributed along the NEs and days of the week.

The numbers of stable days for each NE are presented in Table 1. Top row contains the number of stable days of the week with maximum of 7, all days from Monday to Sunday. Two NEs indicate very stable behavior having all 7 days of the week considered stable and three NEs have 6 stable days. Almost half of the stable days, 48.4% are in NEs that have only one stable day of the week.

Table 1. Number and percentage of the NEs with stable days of the week. Two NEs have stable patterns in all 7 days of the week and 31 NEs (48.4%) have stable patterns in only one day of the week.

Stable days of the week	1	2	3	4	5	6	7
Number of NEs	31	12	11	3	2	3	2
%	48.4	18.8	17.2	4.7	3.1	4.7	3.1

Table 2 shows how the stable days are distributed to the days of the week. Weekend contains the most stable behavior. Both Saturday and Sunday cover about 22% of the stable days. The last row presents the deviation of the days that have no variation in their behavior, the index $I = 0$. Those 17 days are spread uniformly along all days of the week, except for Sunday, which has 7 of them.

Table 2. Distribution of the stable days along the days of the week. Significant proportion of the stable behavior (44.3%) occurs during weekends.

	Mon	Tue	Wed	Thu	Fri	Sat	Sun
Number of stable days, $I < 0.01$	13	14	17	15	20	32	31
% of stable days	9.2	9.9	12.0	10.5	14.1	22.5	21.8
Zero deviation, $I = 0$	1	3	2	2	1	1	7

4   Behavioral Profiles

In this section we introduce how the daily behavior can be summarized at the level of individual NEs. We construct behavioral profiles which summarize the daily traffic patterns each NE produces. Similar profiles have been used to characterize communication devices by the type of their internet traffic [8]

First we clustered the code vectors, which is a common procedure to further summarize the information identified in the trained SOM [9, 10]. We clustered the SOM code vectors of the 1-d SOM using hierarchical clustering [11] with Ward linkage [12]. The clusters are depicted in Fig. 5. Clusters 5 and 6 contain the patterns produced mostly on Sundays and cluster 7 on Saturdays (see Fig. 3).

For each NE we calculated the number of days that were assigned to each of the clusters. We divided those counts by the total number of observations of the NE, thus forming a 7 dimensional feature vector of proportions the NE is assigned in each cluster.

We clustered the feature vectors by K-means clustering. The number of clusters K=7 was a subjective selection. We call the centroids of these clusters behavior profiles (BP). The profiles are depicted in Fig. 6.

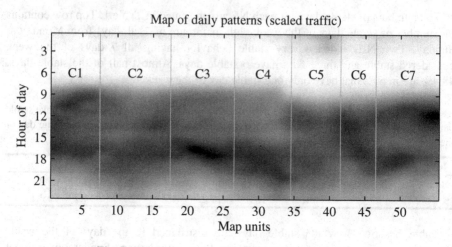

Fig. 5. Code vectors of the SOM divided into 7 clusters

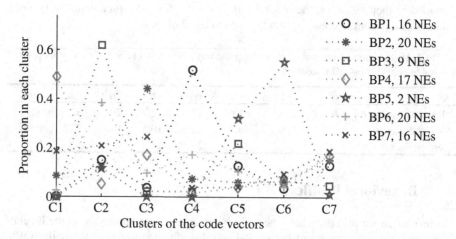

Fig. 6. Behavior profiles (BP) of the network elements (NE)

Most of the behavior profiles are concentrated in one single code vector cluster. BP7 which contains 16 NEs is an exception, spreading across several types of daily patterns. Respectively, most code vector clusters have a BP that is mostly concentrated in them. The weekend clusters 5, 6 and 7 are exceptions, they cover the patterns produced on weekends, regardless of the NE. However, the two NEs in BP5 present distinct profile. They are concentrated only on clusters 5 and 6, which are typically produced on Saturdays (see figures 5 and 3).

5   Conclusion

In this paper we demonstrated how a 1-dimensional SOM can be used to characterize and analyze daily traffic patterns. The typical visualization methods of SOM are more compact for the 1-dimensional map. This is a great advantage, especially for data where the variables in the feature vectors are not independent but form meaningful patterns.

In addition to the daily patterns, the traffic has weekly periodic behavior. We introduced a stability index of projection (SIP) on the SOM that we used to analyze the stability of the daily pattern on distinct days of the week. We found network elements that have very stable behavior, where each day of the week constantly produces similar traffic patterns. The most stable weekly behavior concentrates to the weekend.

We analyzed the daily behavior on network element (NE) level by constructing behavioral profiles which summarize the proportions of distinct daily traffic patterns of each NE. The behavioral profiles revealed that the daily patterns of individual NEs concentrated heavily to certain types of patterns.

Altogether, the day of the week is the main explanation for daily behavior mostly on Saturdays and Sundays. During the weekdays the type of patterns is determined by the NE itself, not by the day of the week. Groups of NEs producing similar patterns can be detected.

The traffic variable in this example case was the number of voice calls in hour. However, the suggested analysis procedures are applicable to any other type of traffic or other variables that produce repeating periodic patterns.

References

1. Khedher, H., Valois, F., Tabbane, S.: Traffic characterization for mobile networks. In: 56th IEEE Vehicular Technology Conference, vol. 3, pp. 1485–1489. IEEE (2002)
2. Kumpulainen, P., Hätönen, K.: Compression of Cyclic Time Series Data. In: 12th IMEKO TC1 & TC7 Joint Symposium on Man Science & Measurement, pp. 413–419 (2008)
3. Kohonen, T.: Self-Organizing Map, 2nd edn. Springer, Berlin (1995)
4. Kiviluoto, K.: Topology Preservation in Self-Organizing Maps. In: International Conference on Neural Networks (ICNN), pp. 294–299 (1996)
5. Vesanto, J.: SOM-based data visualization methods. Intelligent Data Analysis 3, 111–126 (1999)
6. Ultsch, A., Siemon, H.P.: Kohonen's Self-Organizing Feature Maps for Exploratory Data Analysis. In: International Neural Network Conference, Dordrecht, Netherlands, pp. 305–308 (1990)
7. Vesanto, J., Himberg, J., Alhoniemi, E., Parhankangas, J.: Self-organizing map in Matlab: the SOM toolbox. In: Proceedings of the Matlab DSP Conference 1999, Espoo, Finland, pp. 35–40 (1999)
8. Kumpulainen, P., Hätönen, K., Knuuti, O., Alapaholuoma, T.: Internet traffic clustering using packet header information. In: 14th Joint International IMEKO TC1+TC7+TC13 Symposium, Jena, Germany (2011)
9. Vesanto, J., Alhoniemi, E.: Clustering of the self-organizing map. IEEE Transactions on Neural Networks 11(3), 586–600 (2000)
10. Laiho, J., Raivio, K., Lehtimaki, P., Hätönen, K., Simula, O.: Advanced analysis methods for 3G cellular networks. IEEE Transactions on Wireless Communications 4(3), 930–942 (2005)
11. Everitt, B., Landau, S., Leese, M.: Cluster analysis, 4th edn., Arnold (2001)
12. Ward Jr., J.H.: Hierarchical grouping to optimize an objective function. Journal of the American Statistical Association 58(301), 236–244 (1963)

Dipolar Designing Layers of Formal Neurons

Leon Bobrowski[1,2]

[1] Faculty of Computer Science, Białystok Technical University
[2] Institute of Biocybernetics and Biomedical Engineering, PAS, Warsaw, Poland
leon@ibib.waw.pl

Abstract. A layer of formal neurons can perform separable data aggregation. The term "separable data aggregation" means that a number of input vectors belonging to one category (class) are merged by the layer in one output vector with an additional condition that input vectors belonging to different categories are not aggregated. Dipolar principles of separable layers designing are examined in the paper. Hierarchical networks can be designed from separable layers and used for aggregation of all input vectors belonging to one category in an output vector.

Keywords: separable learning sets, separable aggregation, formal neurons, dipolar layers.

1   Introduction

Formal neuron was a central part of the famous *Perceptron* model. *Perceptron* was introduced in 1952 by F. Rosenblatt as the model of neurons ability to learn [1], [2].

The layer of formal neurons transforms the input feature vectors on the output vectors with the binary components [3]. The transformation depends on parameters of particular neurons of the layer. If many input vectors are transformed at the same output vector, the number of feature vectors is decreased and, as a result, the layer aggregates feature vectors.

Separable aggregation of feature vectors is important in the context of hierarchical network designing [4]. Separable data aggregation hapens if some input vectors belonging to one category are aggregated by the layer in one output vector and input vectors belonging to different categories are not aggregated.

The principles of separable aggregation in layers of formal neurons are analyzed in this paper. The analysis focuses on the problem of the separable layer designing on the basis of learning sets representing different categories. The proposed approach to the separable layers designing from formal neurons is based on the concept of mixed dipoles division by using minimization of the convex and piecewise linear (*CPL*) criterion functions [5]. The perceptron criterion function belongs to the considered *CPL* family [3].

2   Formal Neurons

The formal neuron $FN(\mathbf{w}[n],\theta)$ with the *weight vector* $\mathbf{w}[n] = [w_1,\ldots,w_n]^T \in R^n$ and the *threshold* θ ($\theta \in R^1$) can be defined by the below decision rule $r(\mathbf{x}[n])$:

C. Jayne, S. Yue, and L. Iliadis (Eds.): EANN 2012, CCIS 311, pp. 334–342, 2012.
© Springer-Verlag Berlin Heidelberg 2012

$$r(\mathbf{x}[n]) = r(\mathbf{w}[n],\theta; \mathbf{x}[n]) = \begin{cases} 1 & \text{if } \mathbf{w}[n]^{\mathrm{T}}\mathbf{x}[n] \geq \theta \\ 0 & \text{if } \mathbf{w}[n]^{\mathrm{T}}\mathbf{x}[n] < \theta \end{cases} \tag{1}$$

The formal neuron $FN(\mathbf{w}[n],\theta)$ is activated ($r(\mathbf{x}[n]) = 1$) if and only if the weighed sum $w_1 x_1 + ... + w_n x_n$ of n inputs x_i ($x_i \in R$) is greater than the threshold θ. The decision rule $r(\mathbf{x}[n])$ of the formal neuron $FN(\mathbf{w}[n], \theta)$ depends on $n + 1$ parameters w_i ($i = 1,...,n$) and θ.

The activation field $A_{FN}(\mathbf{w}[n], \theta)$ of the formal neuron $FN(\mathbf{w}[n], \theta)$ is defined as the below half space

$$A_{FN}(\mathbf{w}[n],\theta) = \{\mathbf{x}: r_1(\mathbf{w}[n], \theta; \mathbf{x}[n]) = 1\} = \{\mathbf{x}[n]: \mathbf{w}[n]^{\mathrm{T}}\mathbf{x}[n] \geq \theta\} \tag{2}$$

The activation field $A_{FN}(\mathbf{w}[n],\theta)$ is the half space situated on the positive side of the below hyperplane $H(\mathbf{w}[n],\theta)$ in the feature space $F[n]$:

$$H(\mathbf{w}[n],\theta) = \{\mathbf{x}[n]: \mathbf{w}[n]^{\mathrm{T}}\mathbf{x}[n] = \theta\}\} \tag{3}$$

3   Separable Learning Sets

Let us assume that m objects O_j ($j = 1,....,m$) are represented as the so-called feature vectors $\mathbf{x}_j[n] = [x_{j1},...,x_{jn}]^{\mathrm{T}}$, or as points in the n-dimensional feature space $F[n]$ ($\mathbf{x}_j[n] \in F[n]$). Components x_{ji} of the feature vector $\mathbf{x}_j[n]$ represent numerical results of different measurements (*features*) x_i of a given object O_j ($x_{ji} \in \{0,1\}$ or $x_{ji} \in R^1$).

The learning set C_k contains m_k feature vectors (*prototypes*) $\mathbf{x}_j[n]$ assigned to the k-th *class* (*category*) ω_k ($k = 1,....,K$):

$$C_k = \{\mathbf{x}_j[n]: j \in J_k\} \tag{4}$$

where J_k is the set of indices j of the feature vectors $\mathbf{x}_j[n]$ assigned to the class ω_k.

Definition 1: The learning sets C_k (4) are *separable* in the feature space $F[n]$, if they are disjoined in this space ($C_k \cap C_{k'} = \varnothing$, if $k \neq k'$). This means that the feature vectors $\mathbf{x}_j[n]$ ($j \in J_k$) and $\mathbf{x}_{j'}[n]$ ($j' \in J_{k'}$) belonging to different learning sets C_k and $C_{k'}$ cannot be equal:

$$(k \neq k') \Rightarrow (\forall j \in J_k) \text{ and } (\forall j' \in J_{k'}) \ \mathbf{x}_j[n] \neq \mathbf{x}_{j'}[n] \tag{5}$$

We are also considering the separation of the learning sets C_k (1) by the hyperplanes $H(\mathbf{w}_k[n],\theta_k)$ (3) in the feature space $F[n]$.

Definition 2: The learning sets C_k (4) are *linearly separable* in the n-dimensional feature space $F[n]$ if each of the sets C_k can be fully separated from the sum of the remaining sets C_i by some hyperplane $H(\mathbf{w}_k[n],\theta_k)$ (3):

$$(\forall k \in \{1,...,K\}) \; (\exists \; \mathbf{w}_k[n],\theta_k) \; (\forall \mathbf{x}_j[n] \in C_k) \;\; \mathbf{w}_k[n]^T \mathbf{x}_j[n] > \theta_k.$$

$$\textbf{and} \quad (\forall \mathbf{x}_j[n] \in C_i, \; i \neq k) \;\; \mathbf{w}_k[n]^T \mathbf{x}_j[n] < \theta_k \tag{6}$$

In accordance with the relation (6), all the feature vectors $\mathbf{x}_j[n]$ from the set C_k (4) are situated on the positive side of the hyperplane $H(\mathbf{w}_k[n],\theta_k)$ (3) and all the vectors $\mathbf{x}_j[n]$ from the remaining sets C_i are situated on the negative side of this hyperplane.

4   Separable Data Aggregation by Layer of Formal Neurons

The layer of L formal neurons $FN(\mathbf{w}_i[n],\theta_i)$ (1) transforms feature vectors $\mathbf{x}_j[n]$ (4) into the L-dimensional output vectors $\mathbf{r}_j[L]$ with binary components:

$$(\forall k \in \{1,...,K\}) \; (\forall \mathbf{x}_j[n] \in C_k)$$

$$\mathbf{r}_j[L] = \mathbf{r}(\mathbf{W};\mathbf{x}_j[n]) = [r(\mathbf{w}_1[n],\theta_1;\mathbf{x}_j[n]),..., r(\mathbf{w}_L[n],\theta_L;\mathbf{x}_j[n])]^T \tag{7}$$

where $\mathbf{W} = [\mathbf{w}_1^T,..., \mathbf{w}_L^T]^T$ is the vector of the layer parameters.

The learning sets C_k (4) are transformed into the sets R_k of the L-dimensional output vectors $\mathbf{r}_j[L]$ with binary components:

$$R_k = \{\mathbf{r}_j[L] = \mathbf{r}(\mathbf{W};\mathbf{x}_j[n]): j \in J_k\}. \tag{8}$$

The transformed sets R_k (8) are separable if the below implication holds:

$$(k \neq k') \Rightarrow (\forall j \in J_k) \; and \; (\forall j' \in J_{k'}) \;\; \mathbf{r}_j[L] \neq \mathbf{r}_{j'}[L] \tag{9}$$

The primary goal of designing layers of formal neurons $FN(\mathbf{w}_i[n],\theta_i)$ (1) could be separable data aggregation. The separable (2) learning sets C_k (3) should be transformed into the separable sets R_k (8),

Let us change the indexing of the output vectors $\mathbf{r}_j[L]$ (7) in such a way that the different indices k and k' ($k \neq k'$) are related to different output vectors $\mathbf{r}_k[L]$ and $\mathbf{r}_{k'}[L]$.

$$(k \neq k') \Rightarrow \mathbf{r}_k[L] \neq \mathbf{r}_{k'}[L] \tag{10}$$

The transformation (7) may aggregate a number of the feature vectors $\mathbf{x}_j[n]$ (4) into one output vector $\mathbf{r}_k[L]$. Data aggregation means here decreasing the number of different feature vectors $\mathbf{r}_k[L]$ (10). Let us introduce the concept of the *active fields* S_k in order to characterise the data aggregation in the layer of L formal neurons $FN(\mathbf{w}_i[n],\theta_i)$ (1) [5].

Definition 3: The active field S_k of the layer of L formal neurons $FN(\mathbf{w}_i[n],\theta_i)$ (1) is created by those feature vectors $\mathbf{x}[n]$ which are integrated in the k-th output vector $\mathbf{r}_k[L]$ ($\mathbf{r}_k[L] \neq \mathbf{0}$).

$$S_k = S(\mathbf{W}; \mathbf{r}_k[L]) = \{\mathbf{x}[n]: \mathbf{r}(\mathbf{W}; \mathbf{x}[n]) = \mathbf{r}_k[L]\} \tag{11}$$

S_k is a convex polyhedron with the walls defined by the hyperplanes $H(\mathbf{w}_k[n],\theta_k)$ (3).

Definition 4: The *clear active field* S_k (11) contains labelled feature vectors $x_j[n]$ ($j \in J_k$) belonging to only one learning set C_k (4). The *mixed active field* S_k contains labelled feature vectors $x_j[n]$ belonging to more than one learning sets $C_{k'}$ (4).

Lemma 1: The transformed sets R_k (8) are separable (9) if and only if each active field S_k (11) of the layer of L formal neurons $FN(w_i[n], \theta_i)$ (1) is clear and each element $x_j[n]$ of the learning sets $C_{k'}$ (4) belongs to a certain active field S_k (11).

Proof: All feature vector $x_j[n]$ (4) belonging to the active field S_k (11) are aggregated into a single output vector $r_k[L]$. If the active field S_k (11) is clear, then the separability condition (9) is preserved. If the transformed sets R_k (8) are separable (9), then each active field S_k (11) has to be clear. $\square$

It can be shown that an increase of the number of layers in a hierarchical network may result only in a summation (aggregation) of some active fields S_k (11) from lower layers, but not in their division. A decrease of the number of different vectors $r_k[L']$ (10) at the output of the hierarchical network can be expected as a result of an increase of the number of the layers [4].

The layers of formal neurons $FN(w_i[n], \theta_i)$ (1) can be used for classification or decision support. The quality of the layers is evaluated in this context. Improving the quality of the layer is linked to increasing its *generalization capability* Layer should be of high quality not only on the elements $x_j[n]$ of the learning sets C_k (4), but also on previously unseen feature vectors $x[n]$ ($x[n] \notin C_k$). The generalization capability of the layer is linked to the size of the active fields S_k (11). We can expect a higher generalization capability in the layer with the larger active fields S_k (11).

Designing postulate I: The layer of formal neurons $FN(w_i[n], \theta_i)$ (1) should have the clear fields S_k (11) of a large size.   (12)

5   Dipolar Layers

The layers of formal neurons $FN(w_i[n], \theta_i)$ (1) can be designed by using the concept of dipoles [5]. The *clear dipoles* and the *mixed dipoles* are defined by using elements $x_j[n]$ ($j \in J_k$) of the learning sets C_k (4).

Definition 5: The pair of feature vectors ($x_j[n]$, $x_{j'}[n]$) constitutes the *mixed dipole* if and only if these vectors belong to different learning sets C_k and $C_{k'}$ ($j \in J_k$, $j' \in J_{k'}$, and $k \neq k'$). The pair of feature vectors ($x_j[n]$, $x_{j'}[n]$) constitutes a *clear dipole* if and only if these vectors belong to the same learning set C_k ($j \in J_k$ and $j' \in J_k$) (4).

Definition 6: The dipole ($x_j[n]$, $x_{j'}[n]$) is divided by the formal neurons $FN(w_i[n], \theta_i)$ (1) if and only if *only one* vector from this dipole activates the neuron ($r_{i'} = 1$).

Lemma 2: The necessary condition for preserving the separability of the learning set C_k (1) by the layer of L formal neurons $FN(w_i[n], \theta_i)$ (1) is the division of all mixed dipoles ($x_j[n]$, $x_{j'}[n]$) by elements of this layer.

Proof: If there exists a mixed dipole $(\mathbf{x}_j[n], \mathbf{x}_{j'}[n])$ which is not divided, then the output vector $\mathbf{r}_k[L]$ with the mixed active field S_k (11) will appear. The appearance of a mixed field excludes the separability (9) of the transformed sets R_k (8). On the other hand, if all the mixed dipoles $(\mathbf{x}_j[n], \mathbf{x}_{j'}[n])$ are divided by at least one element $FN(\mathbf{w}_i[n],\theta_i)$ (1) of this layer, there is no mixed active field S_k (11). In accordance with the *Lemma* 1, the transformed sets R_k (8) are separable (9) in such a case.

Definition 7: The layer of L formal neurons $FN(\mathbf{w}_i[n],\theta_i)$ (1) is *dipolar*, if and only if each mixed dipole $(\mathbf{x}_j[n], \mathbf{x}_{j'}[n])$ created from the elements of different learning sets C_k (4) is divided and each feature vector $\mathbf{x}_j[n]$ (1) activates ($r_i = 1$) at least one neuron of the layer.

The division of the all mixed dipoles $(\mathbf{x}_j[n], \mathbf{x}_{j'}[n])$ guarantees the implication of the transformed sets R_k (8) separability (9) from the learning sets C_k (4) separability (5). In order to achieve greater *generalization* power of the dipolar layer the below designing postulate has been formulated [5]:

Designing postulate II: Each formal neurons $FN(\mathbf{w}_i[n],\theta_i)$ (1) of the layer should divide a large number of the mixed dipoles $(\mathbf{x}_j[n], \mathbf{x}_{j'}[n])$ and a low   (13) number of the clear dipoles.

6   Convex and Piecewise Linear (*CPL*) Criterion Functions

The decision rule $r(\mathbf{w}_k[n],\theta_k; \mathbf{x}[n])$ (1) of the formal neuron $FN(\mathbf{w}_k[n],\theta_k)$ is linked to the decision hyperplane $H(\mathbf{w}_k[n],\theta_k)$ (3). In accordance with the designing postulate (13) the dipolar hyperplane $H(\mathbf{w}_k[n],\theta_k)$ (3) should divide a large number of the mixed dipoles $(\mathbf{x}_j[n], \mathbf{x}_{j'}[n])$ and a low number of the clear dipoles. In certain situations, this postulate can be achieved by minimizing the convex and piecewise linear (*CPL*) criterion functions [4]. Among others, the *perceptron criterion function* $\Phi_k(\mathbf{w}[n],\theta)$ belongs to the *CPL* family [3]. The perceptron criterion function $\Phi_k(\mathbf{w}[n],\theta)$ can be defined on the basis of the positive subset G_k^+ and the negative subset G_k^- of the labelled feature vectors $\mathbf{x}_j[n]$ (4):

$$G_k^+ = \{\mathbf{x}_j[n]: \mathbf{x}_j[n]\in C_k \text{ and } j\in J_k^+\} \text{ and}$$
$$G_k^- = \{\mathbf{x}_j[n]: \mathbf{x}_j[n]\in \bigcup_{k'\neq k} C_{k'} \text{ and } j\in J_k^-\} \tag{14}$$

where $(\forall k \in \{1,...,K\})\, J_k^+ \subset J_k$ and $J_k^- \subset (\bigcup_{k'\neq k} J_{k'})$ (4).

The positive penalty functions $\varphi_j^+(\mathbf{w}[n],\theta)$ are defined on elements $\mathbf{x}_j[n]$ of the set G_k^+

$(\forall \mathbf{x}_j[n]\in G_k^+)$

$$\varphi_j^+(\mathbf{w}[n],\theta) = \begin{cases} 1 - \mathbf{w}[n]^T\mathbf{x}_j[n] + \theta & \text{if } \mathbf{w}[n]^T\mathbf{x}_j[n] - \theta \leq 1 \\ 0 & \text{if } \mathbf{w}[n]^T\mathbf{x}_j[n] - \theta > 1 \end{cases} \tag{15}$$

Each element $\mathbf{x}_j[n]$ of the set G_k^- defines the negative penalty function $\varphi_j^-(\mathbf{w}[n],\theta)$:

$(\forall \mathbf{x}_j[n] \in G_k^-)$

$$\varphi_j^-(\mathbf{w}[n],\theta) = \begin{cases} 1 + \mathbf{w}[n]^T\mathbf{x}_j[n] - \theta & \text{if } \mathbf{w}[n]^T\mathbf{x}_j[n] - \theta \geq -1 \\ 0 & \text{if } \mathbf{w}[n]^T\mathbf{x}_j[n] - \theta < -1 \end{cases} \tag{16}$$

The *CPL* criterion functions $\Phi_k(\mathbf{w}[n],\theta)$ is the sum of the penalty functions $\varphi_j^+(\mathbf{w}[n],\theta)$ and $\varphi_j^-(\mathbf{w}[n],\theta)$

$$\Phi_k(\mathbf{w}[n],\theta) = \sum_{j\in J_k^+} \alpha_j^+ \; \varphi_j^+(\mathbf{w}[n],\theta) + \sum_{j\in J_k^-} \alpha^- \; \varphi_j^-(\mathbf{w}[n],\theta) \tag{17}$$

where the positive parameters α_j^+ and α_j^- determine the *prices* of particular feature vectors $\mathbf{x}_j[n]$. Standard values of the parameters α_j^+ and α_j^- are given below [4]:

$$\alpha_j^+ = 1 / (2m^+) \; and \; \alpha_j^- = 1 / (2m^-) \tag{18}$$

The basis exchange algorithm which is similar to the linear programming allows to find efficiently the minimal value $\Phi_k(\mathbf{w}_k^*[n],\theta_k^*)$ of the function $\Phi_k(\mathbf{w}[n],\theta)$ (17) [6].

$$min \; \Phi_k(\mathbf{w}[n],\theta) = \Phi_k(\mathbf{w}_k^*[n],\theta_k^*) = \Phi_k^* \tag{19}$$

Lemma 3: The minimal value $\Phi_k(\mathbf{w}_k^*[n],\theta_k^*)$ (19) of the criterion function $\Phi_k(\mathbf{w}[n],\theta)$ (17) is equal to zero if and only if the positive set G_k^+ and the negative set G_k^- (16) are linearly separable (6) in the feature space $F[n]$. .Moreover, the optimal hyperplane $H(\mathbf{w}_k^*[n],\theta_k^*)$ (3) divides all the mixed dipoles $(\mathbf{x}_j[n], \mathbf{x}_{j'}[n])$ and does not divide any clear dipole created from the sets G_k^+ and G_k^- (14) in this case.

It can also be proved, that the minimal value Φ_k^* (19) of the function $\Phi_k(\mathbf{w}[n],\theta)$ (17) can be decreased to zero ($\Phi_k^* = 0$) by reducing feature vectors $\mathbf{x}_j[n]$ from the subsets G'_k^+ or G'_k^- (14). As follows from the *Lemma* 3, the linear separability (6) of two learning sets C_1 and C_2 allows for fulfillment of the *Designing postulate* II (13) to the greatest extent possible.

7 Sequential Procedures of Dipolar Layer Designing

Let us assume at the beginning that the positive subset G_k^+ (14) is formed from all the elements $\mathbf{x}_j[n]$ of the learning set C_k (4) and the negative subset G_k^- is formed from all the elements $\mathbf{x}_j[n]$ of the remaining sets $C_{k'}$:

$$G_k^+ = C_k \; and$$
$$G_k^- = \bigcup_{k' \neq k} C_{k'} \tag{20}$$

The minimal value $\Phi_k(\mathbf{w}_k^*[n],\theta_k^*)$ (19) of the criterion function $\Phi_k(\mathbf{w}[n],\theta)$ (17) is equal to zero if the learning sets G_k^+ and G_k^- (20) are linearly separable (6). In this case, all elements $\mathbf{x}_j[n]$ of the set G_k^+ are located properly on the positive side $(\mathbf{w}_k^*[n]^T\mathbf{x}_j[n] - \theta_k^* > 0)$ of the optimal hyperplane $H(\mathbf{w}_k^*[n],\theta_k^*)$ and all elements $\mathbf{x}_j[n]$ of the set G_k^- are located properly on the negative side of $(\mathbf{w}_k^*[n]^T\mathbf{x}_j[n] - \theta_k^* < 0)$ of this hyperplane. The optimal hyperplane $H(\mathbf{w}_k^*[n],\theta_k^*)$ (3) divides all the mixed dipoles $(\mathbf{x}_j[n], \mathbf{x}_{j'}[n])$ and does not divide any clear dipole.

A single neuron is not a sufficient tool to divide all the mixed dipoles $(\mathbf{x}_j[n], \mathbf{x}_{j'}[n])$ if the sets G_k^+ and G_k^- (20) are not linearly separable. All the mixed dipoles $(\mathbf{x}_j[n], \mathbf{x}_{j'}[n])$ can be divided by the dipolar layer (*Definition* 10) of L formal neurons $FN(\mathbf{w}_k[n], \theta_k)$ (5). Designing dipolar layer can be carried out sequentially, by increasing the number of the neurons $FN(\mathbf{w}_k[n], \theta_k)$ (5) in the layer. The k-th formal neuron $FN(\mathbf{w}_k^*[n], \theta_k^*)$ (1) of the layer can be designed through minimization of the criterion function $\Phi_k(\mathbf{w}[n], \theta)$ (17) defined on elements $\mathbf{x}_j[n]$ of the positive subset G_k^+ and the negative subset G_k^- (14).

7.1   Ranked Strategy

Let us consider the below sequence of the learning subsets $G_{k(i)}^+$ and $G_{k(i)}^-$ (14) based on the sequence of the *admissible* subsets $R_{k(i)} = \{\mathbf{x}_j[n]\}$, where $R_{k(i)} \subset: G_{k(i)}^+$ or $R_{k(i)} \subset: G_{k(i)}^-$:

$$G_{k(1)}^+ = G_k^+ (20) \; and \; G_{k(1)}^- = G_k^- \quad (i = 1) \tag{21}$$

And

$$\textit{if } (R_{k(i)} \subset G_{k(i)}^+), \textit{ then } G_{k(i+1)}^+ = G_{k(i)}^+ / R_{k(i)} \textit{ and } G_{k(i+1)}^- = G_{k(i)}^- \tag{22}$$
$$\textit{if } (R_{k(i)} \subset G_{k(i)}^-), \textit{ then } G_{k(i+1)}^+ = G_{k(i)}^+ \textit{ and } G_{k(i+1)}^- = G_{k(i)}^- / R_{k(i)}$$

Definition 5. The formal neuron $FN((\mathbf{w}_{k(i)}^*[n], \theta_{k(i)}^*)$ (5) is *admissible* in respect to the sets $G_{k(i)}^+$ and $G_{k(i)}^-$.(22) if and only if is activated $(r_{k(i)}(\mathbf{x}_j[n]) = 1)$ by the feature vectors $\mathbf{x}_j[n]$ from only one subset $G_{k(i)}^+$ or $G_{k(i)}^-$.

The strategy of the ranked layers designing is based on a sequence of *admissible cuts* (22) of the set $G_{k(i)}^+$ or the set $G_{k(i)}^-$ by formal neurons $FN((\mathbf{w}_{k(i)}^*[n], \theta_{k(i)}^*)$ (5). The optimal parameters $\mathbf{w}_{k(i)}^*[n]$ and $\theta_{k(i)}^*$ are obtained in result of the minimization (19) of the criterion function $\Phi_{k(i)}(\mathbf{w}[n], \theta)$ (17) defined on elements $\mathbf{x}_j[n]$ of the subsets $G_{k(i)}^+$ and $G_{k(i)}^-$.(22.)

It has been shown, that the minimization (19) of the function $\Phi_{k(i)}(\mathbf{w}[n], \theta)$ (17) defined on elements $\mathbf{x}_j[n]$ of the sets $G_{k(i)}^+$ and $G_{k(i)}^-$.(22) allows to find an *admissible cut* (22) of the set $G_{k(i)}^+$ or the set $G_{k(i)}^-$ through temporary reduction of one of these sets [7].

The ranked strategy allows to design separable layers of formal neurons $FN(\mathbf{w}_k[n], \theta_k)$ (1). Moreover, the separable learning sets C_k (5) can be always transformed by the ranked layer into such sets R_k (8) which are linearly separable (6). The proof of this property can be found in the paper [7].

7.2   Dipolar Strategy

The dipolar strategy is also based on the sequence of the minimization (19) of the criterion functions $\Phi_{k(i)}(\mathbf{w}[n], \theta)$ (17) defined on elements $\mathbf{x}_j[n]$ of the subsets $G_{k(i)}^+$ and $G_{k(i)}^-$. But the subsets $G_{k(i)}^+$ and $G_{k(i)}^-$.are defined in a different way than by the relation (22). The initial sets $G_{k(1)}^+$ and $G_{k(1)}^-$ are defined by the relation (21). The minimization (19) of the criterion function $\Phi_{k(1)}(\mathbf{w}[n],\theta)$ (17) allows to define the first neuron $FN((\mathbf{w}_{k(1)}{}^*[n],\theta_{k(1)}{}^*)$ (5) of the designed layer.

Let us consider the layer of L neurons $FN(\mathbf{w}_{k(i)}[n]^*,\theta_{k(i)}{}^*)$ (5). This layer determines a number of pure and mixed active fields $S_k[L]$ (11). Particular active fields $S_k[L]$ (11) can be evaluated using some measure of *impurity* [8]. The impurity of clear active fields $S_k[L]$ (11) should be equal zero and the impurity of mixed fields $S_k[L]$ should be greater than zero. Measures of impurity used in the design of decision trees can be based, among others, on the entropy. Active fields $S_k{}^*[L]$ (11) which are characterized by a large level of *impurity* can be selected for generation of the learning sets $G_{k(L+1)}{}^+$ and $G_{k(L+1)}{}^-$ (14) used during the $(L+1)$ stage:

$$G_{k(L+1)}{}^+ = \{\mathbf{x}_j[n]: \mathbf{x}_j[n]\in C_k \ and \ \mathbf{x}_j[n] \in S_k{}^*[L]\} \ and$$
$$G_{k(L+1)}{}^- = \{\mathbf{x}_j[n]: \mathbf{x}_j[n]\in \underset{k'\neq k}{\cup} C_{k'} \ and \ \mathbf{x}_j[n] \in S_k{}^*[L]\} \tag{23}$$

What is important, more than one active field $S_k{}^*[L]$ (11) can be used to generate one pair of the subsets $G_{k(L+1)}{}^+$ and $G_{k(L+1)}{}^-$ (23).

The minimization (19) of the criterion function $\Phi_{k(L+1)}(\mathbf{w}[n],\theta)$ (17) defined on the sets $G_{k(L+1)}{}^+$ and $G_{k(L+1)}{}^-$ (23) allows to define the $(L+1)$-st neuron $FN(\mathbf{w}_{k(L+1)}{}^*[n],\theta_{k(L+1)}{}^*)$ (5) of the designed layer.

The convergence in the finite number of stages can be proved both for the designing procedures based on the ranked strategy as well as for those based on the dipolar strategy. It means, that after finite number L of stages all such mixed dipoles $(\mathbf{x}_j[n], \mathbf{x}_{j'}[n])$ based on the learning sets $G_k{}^+$ and $G_k{}^-$ (20) are divided. In other words, each mixed $(\mathbf{x}_j[n],\mathbf{x}_{j'}[n])$ with one element $\mathbf{x}_j[n]$ belonging to the learning set C_k (4) is divided by at least on formal neurons $FN(\mathbf{w}_{k(il)}{}^*[n],\theta_{k(il)}{}^*)$ (5) of the designed layer.

Let us introduce the term *epoch* in order to describe the designing process of such layer of formal neurons $FN(\mathbf{w}_{k(i)}{}^*[n],\theta_{k(i)}{}^*)$ (1), which divide all mixed dipoles $(\mathbf{x}_j[n],\mathbf{x}_{j'}[n])$ based on K ($K > 2$) learning sets C_k (4). During the first epoch, such sublayer of $L(1)$ of formal neurons $FN(\mathbf{w}^*_{k(i)}[n],\theta^*_{k(i)})$ (1) is designed which divides all mixed dipoles $(\mathbf{x}_j[n],\mathbf{x}_{j'}[n])$ with one of elements $\mathbf{x}_j[n]$ belonging to the learning set C_1 (4). Next, the learning set C_1 is removed from the family (4). During the second *epoch* the sublayer of $L(2)$ formal neurons $FN(\mathbf{w}^*_{k(i)}[n],\theta^*_{k(i)})$ (1) that divides all mixed dipoles $(\mathbf{x}_j[n],\mathbf{x}_{j'}[n])$ in which one of elements $\mathbf{x}_j n]$ belongs to the learning set C_2 (4) is designed and so on. Finally, all $(K - 1)$ sublayers are merged into a single separable layer.

In order to secure the condition, that each feature vector $\mathbf{x}_j[n]$ (4) activates ($r_i = 1$) at least one element of the dipolar layer (*Definition* 7), an additional formal neuron $FN(\mathbf{w}_k[n],\theta_k)$ (1) can be included in the last layer. The parameters $\mathbf{w}_k[n]$ and θ_k of the additional neuron can be chosen in such a manner that each feature vector $\mathbf{x}_j[n]$ (4) activates ($r_i = 1$) this neuron.

8   Concluding Remarks

Designing separable layers and hierarchical networks of formal neurons $FN(\mathbf{w}_k[n],\theta_k)$ (1) is still a challenging problem with a potentially large number of interesting applications. The data driven approach to this problem based on the concept of mixed dipoles is discussed in the paper. An assumption has been made here that data is given in the form of the standardized learning sets C_k (4) composed of multivariate feature vectors $x_j[n]$.

The discussed principle of designing layer of L formal neurons $FN(\mathbf{w}_k[n],\theta_k)$ (1) is aimed at the concept of the separable data aggregation. The proposed approach to separable data aggregation is based on the division of all mixed dipoles $(\mathbf{x}_j[n],\mathbf{x}_{j'}[n])$ by at least one element $FN(\mathbf{w}_k[n],\theta_k)$ (1) of the layer (*Lemma* 2). The division of mixed dipoles $(\mathbf{x}_j[n],\mathbf{x}_{j'}[n])$ by formal neurons $FN(\mathbf{w}_k[n],\theta_k)$ (1) can be realized efficiently through minimization of the convex and piecewise linear (*CPL*) criterion functions $\Phi_k(\mathbf{w}[n],\theta)$ (17). Tthe optimization of the separable layers aimed at increase of their *generalization capability* is still an open problem.

Acknowledgments. This work was supported by the project S/WI/2/2012 from the Białystok University of Technology and partially financed by the by the NCBiR project N R13 0014 04.

References

1. Rosenblatt, F.: Principles of neurodynamics. Spartan Books, Washington (1962)
2. Minsky, M.L., Papert, S.A.: Perceptrons. MIT Press, Cambridge (1969)
3. Duda, O.R., Hart, P.E., Stork, D.G.: Pattern classification. J. Wiley, New York (2001)
4. Bobrowski, L.: Eksploracja danych oparta na wypukłych i odcinkowo-liniowych funkcjach kryterialnych (Data mining based on convex and piecewise linear (CPL) criterion functions), Technical University Białystok (2005) (in Polish)
5. Bobrowski, L.: Piecewise-Linear Classifiers, Formal Neurons and Separability of the Learning Sets. In: Proceedings of ICPR 1996, Vienna, pp. 224–228 (1996)
6. Bobrowski, L.: Design of piecewise linear classifiers from formal neurons by some basis exchange technique. Pattern Recognition 24(9), 863–870 (1991)
7. Bobrowski, L.: Induction of Linear Separability through the Ranked Layers of Binary Classifiers. In: Iliadis, L., Jayne, C. (eds.) EANN/AIAI 2011, Part I. IFIP AICT, vol. 363, pp. 69–77. Springer, Heidelberg (2011)
8. Quinlan, J.R.: C4.5: Programs for Machine Learning. Morgan Kaufmann Publishers (1993)

Evaluating the Impact of Categorical Data Encoding and Scaling on Neural Network Classification Performance: The Case of Repeat Consumption of Identical Cultural Goods

Elena Fitkov-Norris, Samireh Vahid, and Chris Hand

Kingston University, Kingston Hill, Kingston-upon-Thames, KT2 7LB, UK
{E.Fitkov-Norris,S.Vahid,C.Hand}@kingston.ac.uk

Abstract. This article investigated the impact of categorical input encoding and scaling approaches on neural network sensitivity and overall classification performance in the context of predicting the repeat viewing propensity of movie goers. The results show that neural network out of sample minimum sensitivity and overall classification performance are indifferent to the scaling of the categorical inputs. However, the encoding of inputs had a significant impact on classification accuracy and utilising ordinal or thermometer encoding approaches for categorical inputs significantly increases the out of sample accuracy of the neural network classifier. These findings confirm that the impact of categorical encoding is problem specific for an ordinal approach, and support thermometer encoding as most suitable for categorical inputs. The classification performance of neural networks was compared against a logistic regression model and the results show that in this instance, the non-parametric approach does not offer any advantage over standard statistical models.

Keywords: neural networks, logistic regression, categorical input, encoding, scaling.

1 Introduction

Neural networks are a well established tool for tackling classification type problems as evidenced by the large number of publications comparing the performances of classification methods in wide range of disciplines [1]. These studies indicate that neural networks provide a feasible alternative to traditional statistical approaches performing to a similar, and in many cases better accuracy [1]. This is due in part to the ability of neural networks to handle the analysis of many different types of data simultaneously, including combinations of quantitative and qualitative variables [2].

The methodology for the application of neural networks to such classification type problems generally involves a data pre-processing phase which aims to improve the performance of the neural network classifier and may include steps such as feature selection [4], re-sampling [5] and feature discretisation of continuous attributes [3], [6].

C. Jayne, S. Yue, and L. Iliadis (Eds.): EANN 2012, CCIS 311, pp. 343–352, 2012.

While the impact of these data pre-processing approaches on neural network classification accuracy has been analysed in some detail [3], one area that has received little attention is the question of the impact of nominal or indicator data encoding on neural network performance. Existing research indicates that the networks classification performance is contingent on the coding scheme used and thus identifying the most appropriate nominal data coding scheme may have a significant impact on performance [3]. Despite these findings, often little or no justification is given for the use of different data encoding techniques when handling categorical, nominal or binary data with neural networks. This article sets out address this gap between theory and practice by analysing the impact of different nominal variable coding schemes on neural network classification performance in the context of predicting if a customer is likely to see a given movie in the cinema twice (investigated in [7]). Factors influencing box office performance include the movie rating, stars in the cast, genre, technical effects, number of screens and whether the movie is a sequel or not [8], [9]. Repeat viewing has received relatively little attention, [7] attempted to close that gap by analysing the profile of repeat viewers versus non-repeat viewers and building a reasonably accurate logistic regression model to predict if a subject is likely to be a repeat viewer.

In addition this article will compare the classification accuracy of neural networks against that of logistic regression modeling as tools for predicting the repeat viewing and contribute further to the large body of literature which compares the effectiveness of non-parametric versus statistical type classifiers [1], [10].

The reminder of the paper is organised as follows: Section 2 presents a literature review outlining the variety of coding approaches used in practice for coding categorical, nominal and binary data inputs in neural networks; Section 3 describes the data set, methodology and case study set up used to compare different coding techniques; Section 4 present results from the performance of neural networks under different encoding schemes and comparative analysis of the classification accuracy of the best neural network model against logistic regression modeling for predicting repeat viewing; Section 5 provides conclusions and further recommendations.

2 Categorical Data Encoding Schemes

A systematic literature review carried out by [3] identified very limited discussion and research into the impact of nominal variable encoding and scaling on machine learning algorithms including neural networks. This paper sets out to study this omission since research into the impact of different scaling and normalisation algorithms for continuous input variables is well established and shows that the predictive performance of neural networks is very sensitive to the scaling algorithm used for normalising continuous data [11].

There are five main types of categorical variable encoding used in practice: ordinal, 1-out-of-N, 1-out-of-N–1 and two types of thermometer. Each of these approaches is illustrated in Table 1 and discussed briefly below.

Table 1. Schemes for encoding categorical inputs variables

Ordinal raw value	Ordinal	N			N – 1		Thermometer (N inputs)			Thermometer (N – 1 inputs)	
	x_1	x_1	x_2	x_3	x_1	x_2	x_1	x_2	x_3	x_1	x_2
Low	1	1	0	0	1	0	1	0	0	0	0
Medium	2	0	1	0	0	1	1	1	0	1	0
High	3	0	0	1	0	0	1	1	1	1	1

Ordinal encoding allocates a unique numeric code to each category [12], [13] and its advantages are that since it each category is represented as one input, it does not increase the dimentionality of the problem space. In addition, this type of encoding lends itself to easy interpretation. However, comparative testing suggest that this encoding could have a significant negative impact on neural network's classification performance [3].

The thermometer encoding scheme codes a variable with N possible values by N inputs, with the digits adding up to the category they are coding for [14]. The units could be coded using binary inputs such as (0, 0), (0, 1) and (1, 1), however, [14] argue that the extra input connections from the additional input node to the nodes in the hidden layer increase the number of connection weights in the network and allow the network higher capacity to fit the data. This coding is well suited to categorical variables which have a meaningful order, but is not particularly well suited to nominal variables.

1-out-of-N variable encoding (also known as N encoding) is an alternative coding scheme that also creates N input variables for a variable with N categories. An observation is given a value of 1 for the category which it represents, 0 otherwise [2], [8], [15]. This coding is widely used since it allows for direct comparison of classification performance of parametric an non-parametric techniques [16]. However, it increases the dimentionality of the problem space and the difficulty in interpreting the semantics of the original variable [15]. Some researchers argue that this type of encoding will hinder the performance of a neural network classifier by introducing multicollinearity and leading to ill-conditioning [17].

1-out-of-N – 1 encoding (also known as N – 1 encoding) is similar to 1-out-of-N variable encoding, however, an attribute with N categories is substituted with N – 1 class categories. All but one observations are given a value of 1 for the category they represent. One category is encoded with all 0s, enabling it to act as a reference level for interpretational purposes and to avoid multicollinearity [17]. The 1-out-of-N – 1 encoding can also be applied to data coded using the thermometer scheme by removing the input which is coded with all 1s [3].

In addition to the coding scheme, some researchers have suggested that the scaling of categorical variable inputs may need to be changed to ensure optimal neural network classification performance. [18] argue that using 0 and 1 scaling may lead to saturation of the weights and prevent the network from converging to an optimal solution and suggest that coding of 0.1 and 0.9 should be used instead. These finding are supported by [9], [16] who show that using normalised instead of scaling values close to the boundaries of the activation function leads to better neural network classification performance. [19] propose that coding categorical binary variables using −1 and 1 instead of 0 and 1 would improve the chances of a neural network finding a good solution. To our knowledge only one comparative study has evaluated the impact of 0 to1 coding versus −1 to 1 coding on both categorical and continuous variables and the performance of the neural network was not significantly different [3].

3   Experimental Design - Repeat Movie Going Case Study

3.1   Data Set Background and Description

The data set consists of the 2002 iteration of the Cinema And Video Industry Audience Research (CAVIAR) survey which identifies the demographic characteristics of cinema-goers and if they had seen a film in the cinema more than once [7]. This annual survey (now called FAME) is undertaken by the Cinema Advertising Association (CAA), the Trade Association of cinema advertising contractors operating in the UK, which monitors and maintains standards of advertising in the UK. To remove bias from the data set, duplicate cases were removed [20]. The remaining 786 observations depict whether an individual visited the cinema to see the same movie twice or not, their age category, social class, and preference for visiting the cinema. 33% of the entries in the in the data set were repeat viewers. Details of the data set can be found in [7].

3.2   Neural Networks Overview

Neural networks were developed to simulate the function of the human brain as currently understood by neuroscientists, and in particular its ability to handle complex, non-linear pattern recognition tasks very fast. Neural networks are built from simple processing units or neurons which enable the network to learn sets of input-output mappings and thus solve classification problems. Each processing unit or neuron consist of three elements: a set of synapses or connecting links which take the input signals, an adder for summing the input signals and an activation function which limits the level of a neuron's output. In addition, each input is allocated a weight of its own, which is adjusted during training and represents the relative contribution of that input (positive or negative) to the overall neuron output. The output function of neuron k can be depicted in mathematical terms as:

$$y_k = \phi\left(\sum_{j=0}^{m} w_{kj} x_j\right). \tag{1}$$

Where - y_k is the output of neuron k, x_j denotes inputs (from 0 to m), w_{kj} denotes the synaptic weight for input j on neuron k and $\varphi(\circ)$ is the neuron activation function. The input for neuron 0 is always +1 and it acts as an overall bias, increasing or decreasing the net output of the activation function.

Multilayer feed-forward neural networks are a subtype of neural network distinguished by the presence of hidden layers of neurons, and are particularly well suited to solving complex problems by enabling the network to extract and model non-linear relationships between the input and output layers. Typically the outputs from each layer in the network act as input signals into each subsequent layer so the final output layer presents the final output from the network to different input patterns. The most common approach for building pattern recognition systems is to use feed-forward neural networks with supervised learning to enable accurate feature extraction and classification [21]. Therefore, a feed-forward neural network with either one or two hidden layers will be used for this experiment. Back propagation, essentially a gradient-descent technique and one of the most widely used algorithms will be used to train the network and minimise the error between the target and actual classifications [13], [21]. The network will be simulated using the built in Neural Network tool in PASW Statistics 18.0, which simulates networks with up to 2 hidden layers and the optimal number of neurons in each layer is determined by the software.

3.3   Logistic Regression Overview

Logistic regression uses maximum likelihood estimation to estimate a vector of coefficients, which can be used to determine the probability of a particular observation falling in the category that is being modelled. The probability that a person repeat views a film is expressed as:

$$P(Y = 1) = 1/\left(1 + e^{-(\beta_i x_i)}\right), i = 1 \ldots N. \tag{2}$$

Where β is a vector of estimated coefficients and x is a matrix of input variables.

3.4   Experimental Set up

To evaluate the performance of the neural network and logistic regression models across a range of different inputs, including new objects that the network or the logistic regression model has not seen before, it is common practice to use a holdout sample. The data requirements of neural networks and logistic regression are slightly different so each approach will be explained in turn.

In the case of neural networks the data set is split into three subsets: a training sample, a testing sample and a holdout sample [1]. The network learns pattern mappings by minimising the errors on the training set. The testing set is used as a benchmark to prevent over-fitting, while the holdout sample is used as an independent means to test the classification accuracy of the network on a sample of data that it has nor seen before (out of sample accuracy) [13]. Choosing the holdout sample randomly could lead to a bias in the accuracy estimation as the due to random sample fluctuations [10]. K-fold cross validation provides an alternative for testing the ability of a neural network to generalise by splitting an overall data sample (S) into k disjointed subsamples (folds) $\{S_1, S_2, \ldots S_k\}$ and using different combinations of folds for neural network training and testing. At each time period $t \in \{1, 2, \ldots k\}$ the network is trained on all but one fold (S/S_t) while S_t is used as a holdout sample.

This ensures that by the end of the analysis, the ability of the neural network to generalise has been tested on the entire sample [12] and that the overall classification accuracy is much more robust across different sampling variations. The 10 subsets are derived to ensure that each subset is representative of the overall data set, contain roughly the same proportion of repeaters (33%) and are approximately equal in size. Neural network classification accuracy is tested for 5 repetitions of each simulation with different random seeds. At each step eight of the subsamples are used for training, one subsample is used for testing the ability of the network to generalise (and early stopping) and one subsample is held back and used to evaluate the neural network's out of sample accuracy on that subsample. The neural network's classification performance is calculated as the average out of sample accuracy across the 10-folds.

Logistic regression does not require the use of a testing sample, therefore the 10-fold cross validation technique is adapted as follows: nine of the subsamples are used for model/coefficient estimation and the remaining subsample is used to test the out of sample accuracy of the logistic regression model. The logistic regression's classification performance is calculated as the average out of sample accuracy across the 10-folds.

Table 2. Experimental combinations for categorical input scaling and encoding

Scaling \ Encoding	N	N-1	Thermometer (N-1) inputs	Ordinal
1 hidden layer	0 or 1	exp 1	exp 4	exp 7	N/A
	-1 or 1	exp 2	exp 5	exp 8	N/A
	0.1 or 0.9	exp 3	exp 6	exp 9	N/A
	Ordinal	N/A	N/A	N/A	exp 10
2 hidden layers	0 or 1	exp 11	exp 14	exp 17	N/A
	-1 or 1	exp 12	exp 15	exp 18	N/A
	0.1 or 0.9	exp 13	exp 16	exp 19	N/A
	Ordinal	N/A	N/A	N/A	exp 20

Table 2 shows the encoding and scaling combinations, which are used to test the impact of different categorical variable encoding and scaling approaches on neural network classification performance. The N input thermometer coding described in Table 1 could not be tested as PASW Statistics 18.0 Neural Networks add on does not allow constant inputs (such as input x_1). The accuracy of the different combinations will be compared for statistically significant differences across different scaling and encoding techniques as well as to the classification accuracy of a logistic regression model.

3.5   Performance Measures

One of the most common metrics for measuring classification accuracy for categorical classifiers such as neural network and logistic regression is the confusion matrix. Various measures, derived from the confusion matrix, including overall classification accuracy, sensitivity and specificity are widely used in practice to assess classifier performance [1]. This study will use the minimum sensitivity and overall classification accuracy as the two measures of neural network and logistic regression performance.

Overall classification accuracy for a particular categorical classifier is defined as the rate of all correct predictions by the classifier. A common criticism of the overall classification accuracy measure is that it does not take into account the relative proportion of observations in different categories/classes and as a result it could lead to misleading results since the impact of underrepresented groups of observations is relatively small [16].

Minimum sensitivity is an alternative measure, which overcomes the problem of imbalanced representation. It is the minimum of the sensitivities from all different classes in the problem, in effect the worst possible performance for the classifier defined as:

$$MS = \min\{P(i); i = 1, \ldots, J\} \quad where \ P(i) = n_{ii} \bigg/ \sum_{j=1}^{J} n_{ij} . \tag{3}$$

These measures were chosen as their simultaneous use accounts for the classification performance of classifiers across of imbalanced data sets without the need for over/under sampling [22].

4   Results and Analysis

The overall classification accuracy and minimum sensitivity attained using neural network classifiers across the different combinations of scaling and encoding approaches are summarized in Table 3. It also includes accuracy and sensitivity results from a logistic regression for comparison.

Table 3. Overall classification accuracy (%) and minimum sensitivity (%) for neutral network using different combinations of scaling and encoding

Encoding \ Scaling		N OA	N MS	N-1 OA	N-1 MS	Thermometer (N-1) OA	Thermometer (N-1) MS	Ordinal OA	Ordinal MS
1 hidden layer	0 or 1	67.4	16.4*	67.9	14.2**	68.7	18.2	N/A	
	-1 or 1	67.7	15.6*	68.3	16.2*	68.7	18.2	N/A	
	0.1 or 0.9	67.9	15.3*	67.8	14.8**	68.7	18.2	N/A	
	Ordinal	N/A		N/A		N/A		69.4	17.9
2 hidden layers	0 or 1	67.3	13.5	68.3	13.7	68.8	16.2	N/A	
	-1 or 1	67.3	13.5	67.9	14.7	68.8	16.2	N/A	
	0.1 or 0.9	67.3	13.5	67.9	14.7	68.8	16.2	N/A	
	Ordinal	N/A		N/A		N/A		68.9	16.5
Logistic regression	0 or 1	N/A		67.5	21.5	N/A		N/A	

* Results significant at 5% level ** Results significant at 1% level.

4.1   Impact of Scaling, Encoding and Number of Hidden Layers on Neural Network Classification Accuracy

Estimating the impact of scaling, encoding and number of hidden layers together with combinations of these factors on classification performance shows that only encoding and number of hidden layers have an impact on neural network overall and sensitivity classification accuracy. Both effects are significant at 1% level although they are relatively small. Scaling was not found to have an impact on classification accuracy, in line with earlier findings by [3] and the frequently made assumption that scaling categorical inputs in the interval [-1, 1] is likely to improve neural network performance [19] is not supported by this research.

The impact of number of hidden layers is very small, which suggests that the effect may be constrained to just one of the dependent variables. This is confirmed by examining the between subject effects which show that the impact of hidden layers is only significant for the minimum sensitivity accuracy of a neural network. Increasing the number of hidden layers from one to two leads to a small but significant decrease in the neural network's ability to classify observations from an under-represented class (from 16.7% to 15.1%). This finding highlights that increasing neural network complexity could lead to over-fitting [19].

The effect of encoding on the minimum sensitivity and overall classification accuracy is small, but consistent across both accuracy measures. For the minimum sensitivity measure the encoding schemes split onto two sets on the basis of their sensitivity performance, with ordinal and thermometer coding significantly outperforming N and N-1 encoding.

For the overall classification accuracy, the encoding approaches split into three homogeneous sets. The mean classification accuracy when using the N encoding (67.5%), is significantly lower than the classification accuracy when using the N-1 encoding (68.0%). This in turn is significantly lower that the overall classification accuracies achieved using ordinals and thermometer encoding (69.1% and 68.7% respectively). Again, we find there is no statistically significant difference between overall network performance using the ordinal and the temperature encoding schemes.

These results suggest that the best encoding techniques for categorical data are either the ordinal or thermometer encoding schemes. These findings are inline with other researcher's findings with regards to the thermometer scheme which has been found to lead to good classification performance [3], [14]. However, the superior performance of the ordinal encoding scheme on out of sample performance in this data set is surprising. [3] concluded that neural network classification performance is very sensitive to ordinal encoding, but the different findings from this research suggest that the choice of optimal encoding may be data and problem specific. This is further confirmed by the performance of N and N-1 encoding schemes which lead to a significant deterioration is neural network sensitivity and overall classification accuracy for this problem.

Table 3 also shows the out of sample classification performance of a logistic regression model across the 10 subsamples. There is no significant difference in overall classification performance between the logistic regression and neural network with 1 hidden layer using N-1, thermometer or ordinal encoding. However, using N or N-1 encoding leads to a significant reduction in the sensitivity of neural networks as compared to the sensitivity of logistic regression. These results suggest that trying to predict repeat viewing behaviour does not benefit from the adoption of a non-parametric approach.

5   Conclusions

The results showed that neural network out of sample minimum sensitivity and overall classification performance are indifferent to the scaling of categorical inputs. However, the encoding of inputs had a significant impact on classification accuracy and utilising either ordinal or thermometer coding approaches significantly increases the out of sample sensitivity and overall accuracy of the neural network classifier. These findings confirm that the impact of categorical encoding is problem specific in the case of ordinal approach, and lend further support to thermometer encoding as an appropriate approach for categorical input encoding. Finally, the article compared the classification performance of neural networks against a logistic regression model and showed that in the case of predicting repeat movie going, the non-parametric approach does not offer any advantage over standard statistical models. This research highlights the difficulties encountered when applying neural networks to categorical type data with respect to choosing optimal input coding. In addition, the limitations of the PASW Neural Network software suggest that this problem may benefit from using alternative, more flexible software in the future.

References

1. Paliwal, M., Kumar, U.A.: Neural networks and statistical techniques: A review of applications. ESWA 36(1), 2–17 (2009)
2. Brouwer, R.: A feed-forward network for input that is both categorical and quantitative. NN (2002)
3. Crone, S., Lessmann, S., Stahlbock, R.: The impact of preprocessing on data mining: An evaluation of classifier sensitivity in direct marketing. EJOR 9(16), 781–800 (2006)
4. Niu, D., Wang, Y., Wu, D.D.: Power load forecasting using support vector machine and ant colony optimization. ESWA 37(3), 2531–2539 (2010)
5. Chawla, N.V., Bowyer, K.W., Hall, L.O., Kegelmeyer, W.P.: SMOTE: synthetic minority over-sampling technique. JAIR 16, 321–357 (2002)
6. Kim, K., Han, I.: Genetic algorithms approach to feature discretization in artificial neural networks for the prediction of stock price index. ESWA 19(2), 125–132 (2000)
7. Collins, A., Hand, C., Linnell, M.: Analyzing repeat consumption of identical cultural goods: some exploratory evidence from moviegoing. J. Cult. Econ. 32(3), 187–199 (2008)
8. Sharda, R., Delen, D.: Predicting box-office success of motion pictures with neural networks. ESWA 30(2), 243–254 (2006)
9. Zhang, L., Luo, J., Yang, S.: Forecasting box office revenue of movies with BP neural network. ESWA 36(2), 6580–6587 (2009)

10. Kim, S.: Prediction of hotel bankruptcy using support vector machine, artificial neural network, logistic regression, and multivariate discriminant analysis. Serv. Ind. J. 31(3), 441–468 (2011)
11. Mazzatorta, P., Benfenati, E., Neagu, D., Gini, G.: The importance of scaling in data mining for toxicity prediction. JCICS 42(5), 1250–1255 (2002)
12. Viaene, S., Dedene, G., Derrig, R.: Auto claim fraud detection using Bayesian learning neural networks. ESWA 29(3), 653–666 (2005)
13. Sahoo, G., Ray, C., Mehnert, E., Keefer, D.: Application of artificial neural networks to assess pesticide contamination in shallow groundwater. SCTEN 367(1), 234–251 (2006)
14. Setiono, R., Thong, J., Yap, C.: Symbolic rule extraction from neural networks: An application to identifying organizations adopting IT. Inform. & Manage. 34(2), 91–101 (1998)
15. Hsu, C.: Generalizing self-organizing map for categorical data. NN (2006)
16. Sakai, S., Kobayashi, K., Toyabe, S.I., Mandai, N., Kanda, T., Akazawa, T.: Comparison of the Levels of Accuracy of an Artificial Neural Network Model and a Logistic Regression Model for the Diagnosis of Acute Appendicitis. J. Med. Syst. 31(5), 357–364 (2007)
17. Lai, K.K., Yu, L., Wang, S., Zhou, L.: Neural Network Metalearning for Credit Scoring. In: Huang, D.-S., Li, K., Irwin, G.W. (eds.) ICIC 2006, Part I. LNCS, vol. 4113, pp. 403–408. Springer, Heidelberg (2006)
18. Basheer, I., Hajmeer, M.: Artificial neural networks: fundamentals, computing, design, and application. J. Microbiol. Methods 43(1), 3–31 (2000)
19. Kaastra, I., Boyd, M.: Designing a neural network for forecasting financial and economic time series. NC 10(3), 215–236 (1996)
20. Carter, R.J., Dubchak, I., Holbrook, S.R.: A computational approach to identify genes for functional RNAs in genomic sequences. NAR 29(19), 3928–3938 (2001)
21. Haykin, S.: Neural Netwoks and Learning Machines, 3rd edn. Pearson Intenational Edition (2009)
22. Fernández-Navarro, F., Hervás-Martínez, C., García-Alonso, C., Torres-Jimenez, M.: Determination of relative agrarian technical efficiency by a dynamic over-sampling procedure guided by minimum sensitivity. ESWA 38(10), 12483–12490 (2011)

A Hybrid Neural Emotion Recogniser
for Human-Robotic Agent Interaction

Alexandru Traista and Mark Elshaw

Department of Computing, Faculty of Computing and Engineering,
Coventry University, Coventry, UK
Mark.Elshaw@coventry.ac.uk

Abstract. This paper presents a hybrid neural approach to emotion recognition from speech, which combines feature selection using principal component analysis (PCA) with unsupervised neural clustering through self-organising map (SOM). Given the importance that is associated with emotions in humans, it is unlikely that robots will be accepted as anything more that machines if they do not express and recognise emotions. In this paper, we describe the performance of an unsupervised approach to emotion recognition that achieves similar performance to current supervised intelligent approaches. Performance, however, reduces when the system is tested using samples from a male volunteer not in the training set using a low cost microphone. Through the use of an unsupervised neural approach, it is possible to go beyond the basic binary classification of emotions to consider the similarity between emotions and whether speech can express multiple emotions at the same time.

Keywords: Emotion recognition, social robot interaction, unsupervised neural learning.

1 Introduction

This paper describes a hybrid neural approach to speech emotion recognition from the vocal waveform, which combines feature extraction and selection with a self-organising map (SOM) [12] for clustering and classification. If social robots are to be accepted by human users as companions, they must appear to offer the appropriate response related to the emotional state of the human user. The first step to achieving this is the recognition of the actual emotion expressed by the user. Even if the way emotions influence people's behaviour is not completely known, their importance cannot be underestimated. The actual meaning of an utterance can be changed greatly by the emotion that is expressed within it. For example, the utterance 'I love you' can have a very different meaning when it is said with a happy or bored emotion. It would, therefore, be beneficial in a social robot system to incorporate automatic emotion recognition as well as automatic speech recognition (ASR) to improve the user's experience.

This paper concentrates on emotion recognition from the auditory stream, since it contains relevant information that can be extracted by intelligent computer

C. Jayne, S. Yue, and L. Iliadis (Eds.): EANN 2012, CCIS 311, pp. 353–362, 2012.

algorithms. In comparison, vision based emotion recognition is much more developed and has been focused on more than auditory processing. However, visual processing demands from the user lighting conditions and geometrical constraints with respect to the person's pose and the camera field of view that is rarely the case in people's homes. Hence the paper concentrates on the auditory stream.

1.1   Emotion Recognition Applications

As robots have a more important role in society by assisting an aging population, they must offer interactions with the untrained users in a much more human way. In the case of a personal assistant robot that helps the elderly or the chronically ill to live independently, the effectiveness depends on how easily the human user can control and communicate with the robot. Such an assistant would be able to offer a more natural communication experience by reacting to the person's emotions in an anticipated manner [3].

Although the development of a robot assistant that can communicate in a manner close to human is some way off, research that has proved successful involves taking inspiration from new-borns and how they develop communication skills [18]. Two emotional autonomous robots that have received attention due to their ability to interact in a somewhat believable manner are Kismet (http://www.ai.mit.edu/projects/humanoid-robotics-group/kismet/)   and   Leonardo (http://robotic.media.mit.edu/projects/robots/leonardo/overview/overview.html).

A further application that could benefit from emotion recognition is the automation of call-centers. The robot interacting with the human senses the emotion in responses and reacts accordingly. Ultimately, if the client continues to be unhappy with the service, the robot passes the call to a human employee. In spite of several challenges, mobile device designers are also incorporating more often speech interaction in their platforms. Be it Apple's Siri or Speaktoit for Android, they both try to simulate natural dialogue but disregard completely how the words are said. The ability to identify emotions would make these applications more personal and pleasant to use.

1.2   Current State of the Art

Emotion recognition from speech has typically focused on supervised learning. An example of such an approach is EmoVoice [20] which consists of a speech corpus database, feature extraction and classifier modules. The framework's process is divided into three major steps: audio segmentation, which means finding appropriate units to be classified, feature extraction, to find those characteristics of the signal that best describe emotions and represent them as numeric vectors, and lastly, classification, determining the emotion class for a feature vector. For segmentation, EmoVoice relies on voice activity detection, which in spontaneous speech has good results as the pauses are long enough to be easily detected but might prove challenging for discourses or monologues. In the latter case, an automated speech recognition engine can help fragment the speech into words and use them as independent utterance unit. Classification is limited to one of the two classifiers integrated in the framework: a naive Bayes and a support vector machine classifier with reported accuracies of up to 70% for on-line recognition.

In terms of neural network there has been much use made of supervised networks. For instance Huang et al. (2006) [11], when developing a robot pet, combine wavelet and back propagation neural networks in a layered structure to classify emotions. Further, Zang et al. (2008) [21] trains and tests a back propagation neural network using 600 utterances containing the emotions angry, calm, happy, sad, and surprise.

While such supervised algorithms are more predictable and their behaviour easier to track, unsupervised learning offers more flexibility and is suitable to less definite problems.

Although a large amount of research is performed into emotion recognition, there is no real indication of what features extracted from the speech waveform give optimum performance. Although the basic emotions are identified in terms of features gained from pitch, intensity, energy and temporal and spectral parameters [18], there is little understanding of what these features should be. This can typically mean that intelligent techniques receive a large number of features with little understanding of what they contribute to the overall recognition process or fails to receive the critical features. In this paper we will address this issue through the description of a feature selection approach.

2   Methodology

This section of the paper will outline the overall methodology that is used to create the hybrid neural emotional speech recogniser.

2.1   Database

In this paper the Berlin Emotional Database [4] is used to train and test the unsupervised algorithm employed. It includes 535 files containing German utterances related to six emotions (anger, disgust, fear, joy, sadness, surprise) and neutral utterances that have been acted by five males and five females. To further consider the performance of the system, speech samples were recorded from an English speaking person who is not in the training set, using a low cost microphone and in a non-controlled recording environment.

2.2   Feature Generation

The human voice is a sound wave produced by a human being's vocal cords and typically within 300 and 3500 Hz. The approach for producing the sounds can be divided into three segments: diaphragm and lungs, vocal cords and articulators, all controlled by muscles. [19] Emotions affect the way these muscles are contracted and specific patterns can be distinguished.

In this research OpenSMILE [7] is used to generate 8551 features from the utterance from the Berlin Emotional Database. OpenSMILE has been configured to detect when there is voice activity and to generate a feature set instance after each utterance. These are placed in a Java queue and read in the main application. Another modification allows the library to be configured as to filter the unnecessary features.

OpenSMILE is a flexible feature extractor for signal processing with a focus on audio signals. Being fully written in C++ it allows for fast, platform specific compilation and supports multithreading to take advantage of multiple cores and this makes it suitable for real-time feature generation. The following paragraphs describe the process of generating the initial feature set.

In the first stage the sound is captured by the sound card and sampled at 48000Hz on 16 bits using a single channel, which is similar in quality to an audio CD. The auditory samples for training and testing are converted into a feature representation using the logarithmic mel-spectrum [6, 10]. In this approach the auditory waveform sample is divided by using a moving window which is 25ms in size that moves along the sample 10ms at a time.

The hammed signal $x'(n)$ is then transform to the frequency domain by the use of a discrete Fourier transform:

$$X(t) = \sum_{n=0}^{N-1} x'(n)e^{-j\pi kn/N}, \quad k = 0,..., K-1,$$ (1)

where K is the length of the discrete Fourier transform.

The magnitude spectrum is then multiplied with a triangular mel weighted filter bank and the result is summed to give the logarithmic mel-spectrum $S(m)$:

$$S(m) = \ln\left[\sum_{k=0}^{k-1} |X(k)|^2 H_m(k)\right],$$ (2)

$|X(k)|^2$ is the periodgram, $H_m(k)$ related to the mth filter, and m is the size of the filterbank.

The MFCC is a mapping of the physical frequencies such that pitches at equal distance on the scale are perceived by humans as being at equal distance from one another and is computed using:

$$mel = 1127 \log_e(1 + \frac{f_{Hz}}{700})$$ (3)

where: mel is the MFCC coefficient and f_{Hz} the frequency in Hertz.

As the instant values are not necessarily relevant by themselves, a number of functional transformations are made which generate time derivative, quartiles, moments, regression and other statistical measures.

2.3   Feature Selection

The performance of a learning algorithm can only be as high as the quality and relevance of the input data permit it. As the number of features extracted is large some have overlapping information and some are irrelevant. The ideal feature set should contain only features highly correlated with the emotion class and uncorrelated with each other. Furthermore, learning high dimensional data poses multiple challenges, including overfitting and exponential increase in the number of parameters required for constructing models [2].

Several algorithms have been developed to reduce feature space, among these are Correlation Feature Selection (CFS) [8] and Principal Component Analysis (PCA) [1]. CFS is based on the inter-feature correlation coefficient with Pearson's correlation coefficient most widely used. PCA on the other hand finds linear combinations of the variables (principal components), matching orthogonal directions that amplify variation in the data. Due to the better optimisation possible with PCA, this method was chosen to reduce the feature set from over 8000 features to 34.

2.4 Self-Organising Map (SOM)

The intelligent approach used to perform the clustering and classification of emotions from speech is the SOM (Figure 1). SOMs consist of an input and an output layer, with every input neuron linked to all the neurons in the output layer [12]. The output layer is typically a lattice of neurons that creates a topological representation of the critical characteristics of the input by creating a pattern of active and inactive units [9]. A feature of the SOM is that each sample is typically represented by a single winning unit and a surrounding neighbourhood of units on the output layer [12]. Figure 1 shows a representation of a SOM where the inputs are connected to all the units in the output layer, with the pattern of activation having a central winning unit with the highest activation marked and the surrounding units having reduced activation the further away they are from the winning unit. The output layer can be seen as a sheet of neurons in an array whose units are tuned to input signals in an orderly manner [5].

The activation of the units is calculated by multiplying the input from each input unit by its related synaptic weight and summing all the inputs for a specific unit. Learning to associate in SOMs is performed by updating the links between the input layer and the output layer via a form of Hebbian learning. SOMs have an input vector represented as $\vec{i}$. The input vector is presented to every unit of the output layer; the weights between the links in the network are provided by $\vec{w}$. The activation vector of the SOM is represented as $\vec{o}$.

Although there are various ways to determine the activation on the SOM output layer the approach used in our grounding architectures is Euclidean distance. The representation o_k of unit k is established from the Euclidean distance of the weight vector to its input, given by: $o_k = \parallel \vec{w}_k - \vec{i} \parallel$. The weights are initialised randomly and hence one unit of the network will react more strongly than others to a specific input representation. The winning unit is the unit k' where the distance $o_{k'}$ is the smallest.

The weight vector of the winning unit k' as well as the neighbouring units are updated using Equation 4 which causes the weight vector to resemble the data:

$$\Delta W_{kj} = \alpha T_{kk'} \cdot (i_j - w_{kj}) \tag{4}$$

The shape and units in the neighbourhood depends on the neighbourhood function used. The number of units in the neighbourhood usually drops gradually over time. The neighbour function in our architectures is a Gaussian: $T_{kk'} = e^{-(d_{kk'}^2/2\sigma^2)}$, where $d_{kk'}$ is the distance between unit k and the winning unit k' on the SOM output layer.

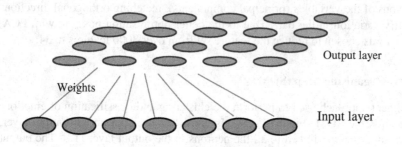

Fig. 1. Self organising map

The data received after selection has been used to train a 120x120 map, converging after approximately 1,000 iterations. A learning rate coefficient of 0.08 has been used, linearly decreasing with time. A hexagonal grid of neurons that reduces in size overtime has been found to be more effective in creating relevant neighbourhoods.

3   Results

This section of the paper will consider the performance of the hybrid neural emotion recognition system on speech samples from the Berlin Emotional Database and on a small set of recordings we produced from a male volunteer. For the Berlin Emotional Database after the initial training and convergence, a supervised class mapping is superimposed and 10-fold cross-validation evaluation carried out. This system's performance is measured based on the number and percentage of correctly recognised emotions and the recall and precision rates achieved on each emotion. Recall and precision are defined using the approaches outlined below. Where fp is false positive, tp is true positive and fn is false negative.

$$precision = \frac{fp}{tp + fp} \tag{5}$$

$$recall = \frac{fp}{tp + fn} \tag{6}$$

The number of instances correctly classified for the Berlin Emotional Database is 437 out of 535 yielding an average accuracy of 81%. Considering the confusion matrix for samples from the Berlin Emotional Database in Table 2 accuracies vary significantly with the emotion, some being easier to identify due to their higher correlation with physical characteristics of the voice while others are more subtle and context dependent. For example, boredom and angriness are being expressed more in the physical voice qualities than anxiety or happiness. When looking at the recall and precision scores for the Berlin Emotional Database in Table 1 in most cases the scores are high which indicates there are not many situations when samples are wrongly classified as a specific class or the samples are wrongly classified as another class. However, for happiness the score for recall is much lower than many of the other emotions which indicates a large number of false negatives.

Table 1. Recall and precision values for the Berlin Emotional Database and speech samples created for this study by a male volunteer

Emotion	Berlin Emotional Database		Samples form study	
	Precision	Recall	Precision	Recall
Happy	0.708	0.648	0.444	0.421
Neutral	0.779	0.848	0.75	0.375
Anger	0.835	0.835	0.593	0.667
Sadness	0.811	0.968	0.632	1
Anxiety	0.781	0.725	0.643	0.643
Boredom	0.924	0.901	0.625	0.625
Disgust	0.875	0.761	0.6	0.3

Table 2. Confusion matrix for the Berlin Emotional Database using 10-fold cross-validation (Grey square the correct emotion)

Happy	Neutral	Anger	Sadness	Anxiety	Boredom	Disgust
46	7	15	0	1	1	1
2	67	1	3	2	3	1
12	1	106	1	5	1	1
0	1	0	60	1	0	0
3	6	5	3	50	0	2
0	4	0	2	2	73	0
2	0	0	5	3	1	35

Turning to the small preliminary study related to performance of the system on speech from a male volunteer not in the training set using a low cost microphone, the performance is much worse at 59% [Table 3]. It can be seen from the recall and precision scores in Table 1 that for most of the emotions there is many false positives and false negatives. When looking at sadness the system is able to classify sad speech correctly but also has a tendency to classify disgust and boredom as sad also. This worse performance poses a challenge for emotion recognition systems that act in different environments, like the personal assistant, to accommodate invariance of background noise and whether there is a need to train with the actual user.

As well as the system offering a binary decision on the emotion recognised, it also offers the probability that the test samples belongs to different emotion classes or clusters. As can be seen in Figure 2 for this example test utterance, the system creates a different level of membership of the utterance to the classes disgust, sadness and boredom. By the use of an unsupervised approach it is possible to offer more information related to the emotions of the utterance than simply a binary decision.

Table 3. Confusion matrix for the speech samples created by the male volunteer for this study using 10-fold cross-validation (Grey square correct emotion)

Happy	Neutral	Anger	Sadness	Anxiety	Boredom	Disgust
8	0	7	0	4	0	0
0	3	0	0	0	5	0
7	0	16	0	0	0	1
0	0	0	12	0	0	0
2	0	2	0	9	1	0
0	0	0	5	0	10	1
1	1	2	2	1	0	3

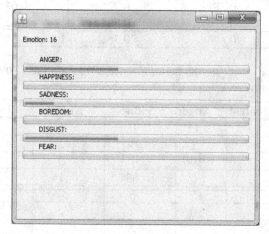

Fig. 2. An example of the emotional co-concurrency represented by the emotional recognition system

4   Discussion and Future Work

The paper presented a real-time system for emotion recognition whose accuracy is higher compared to previous work using the same Berlin Emotional Database [14, 15]. The system provides feature generation, feature selection through PCA and a self-organising Kohonen map for clustering and classification purposes and so offers new opportunities for emotion recognition. By making use of PCA, the system provides a logical approach to feature selection that overcomes some of problems associated with current approaches that simply use an educated guess to feature selection. The self-organising map projects the 34-dimensional feature space and is able to create clusters associated with emotions in 2-dimensional space. The system offers a real indication that intelligent autonomous emotion recognition could be included into assistant robots to aid interaction with human users.

The system's capacity to give indications of the probability that a sample belongs to different emotional clusters is extremely interesting. Although it might simply be seen as indicating the degree of error it is creating when producing a distinct emotional decision, the probability values might also be seen differently. It is also possible that firstly the recogniser is indicating the similarity of different emotions and secondly that humans have the capacity to represent and recognise more than one emotion at a time. The former fits in closely with the concept of the PAD emotional state model outlined by Mehrabian (1996) [13]. In this approach Mehrabian (1996) uses a three dimensional scale to represent emotions based on their levels of dominance, pleasure and arousal. This represents different emotions and their relationship to each other. The different probabilities for emotions created by the emotion recogniser might indicate that for instance sadness and disgust are close on one or more of Mehrabian axes. The latter indicates that humans face a much richer context than binary emotional classes and so social robots must decode such a situation if it is to communicate in a human-like manner [17].

Part of the reduction in performance on the speech from the male volunteer collected as part of this study might be attributed to the training set being in German and this data being recorded in English. However, it is felt that this is not the main reason as both are Germanic languages and from "western European" cultures and as such transmit non-verbal information in the same manner. Given the closeness of the origin of these two languages, it is anticipated that the hybrid emotion recognition system would in this case be language independent. This poor performance is more likely to indicate there is need to further research into how the hybrid neural recogniser can be adapted for a less than perfect environment, low quality microphones and the capacity to recognise emotion from people it is not trained on. It is possible that such performance might be improved by the use of a hierarchical model that is able to combine feature at the lower level and recognise emotions at the upper level of the hierarchy. These problems might also be overcome by the incorporation of more of the temporal nature of speech into the model. Elshaw et al. (2010) [6] presents a recurrent memory approach to speech recognition, simulating infant language acquisition to included memory of previous sections of the speech. Given the evolution of emotions, a system with temporal memory might be more feasible for individual use longer term. There is also the opportunity to combine visual features and the actual words spoken as top-down attention features with the bottom-up features already extracted from the speech waveform. However, including the visual features can increase the computational overhead greatly when performing real-time activities.

References

1. Abdi, H., Williams, L.J.: Principal Component Analysis. Wiley Interdisciplinary Reviews: Computational Statistics 2, 433–459 (2010)
2. Attias, H.: Learning in High Dimensions: Modular Mixture Models. Microsoft Research, USA (2001)

3. Breazeal, C.: The Role of Expression in Robots that Learn from People. Phil. Trans. R. Soc. B 364(1535), 3527–3538 (2009)
4. Burkhardt, F., Paeschke, A., Rolfe, M., Sendlmeier, W., Weis, B.: A Database of German Emotional Speech. In: Interspeech, Lisbon (2005)
5. Doya, K.: What are the Computations of the Cerebellum, the Basal Ganglia and the cerebral cortex? Neural Networks 12(7-8), 961–974 (1999)
6. Elshaw, M., Moore, R.K., Klein, M.: An Attention-gating Recurrent Working Memory Architecture for Emergent Speech Representation. Connection Science 22(2), 157–175 (2010)
7. Eyben, F., Woellmer, M., Schuller, B.: openSMILE - The Munich Versatile and Fast Open-Source Audio Feature Extractor. ACM Multimedia, 1459–1462 (2010)
8. Hall, M.: Correlation-based Feature Selection for Machine Learning (1999)
9. Haykin, S.: Neural Networks: A Comprehensive Foundation, Toronto, Canada. Macmillian College Publishing Company (1994)
10. Holmes, J., Holmes, W.: Speech Synthesis and Recognition. Taylor and Francis, London (2001)
11. Huang, Y., Zhang, G., Xu, X.: Speech Emotion Recognition Research Based on the Stacked Generalization Ensemble Neural Network for Robot Pet. In: Pattern Recognition, CCPR, pp. 1–5 (2009)
12. Kohonen, T.: Self-Organization of Topologically Correct Feature Maps. Biological Cybernetics 43, 59–69 (1982)
13. Mehrabian, A.: Pleasure-Arousal-Dominance: A General Framework for Describing and Measuring Individual Differences in Temperament. Current Psychology 14(4), 261–292 (1996)
14. Pan, Y., Shen, P., Shen, L.: Speech Emotion Recognition Using Support Vector Machine. International Journal of Smart Home 6(2), 101–107 (2012)
15. Shami, M., Verhelst, W.: An evaluation of the robustness of existing supervised machine learning approaches to the classification of emotions in speech. Speech Communication 49(3) (2007)
16. Slavova, V., Verhelst, W., Sahli, H.: A Cognitive Science Reasoning in Recognition of Emotions in Audio-Visual Speech. International Journal Information Technologies and Knowledge 2, 324–334 (2008)
17. Sobin, C., Alpert, M.: Emotion in Speech: The Acoustic Attributes of Fear, Anger, Sadness, and Joy. Journal of Psycholinguistic Research 28(4), 347–365 (1999)
18. ten Bosch, L., Van Hamme, H., Boves, L., Moore, R.K.: A computational model of language acquisition: the emergence of words. Fundamenta Informaticae 90, 229–249 (2009)
19. Traunmüller, H., Eriksson, A.: The Frequency Range of the Voice Fundamental in the Speech of Male and Female Adults. Department of Linguistics, University of Stockholm, Stockholm (1994)
20. Vogt, T., André, E., Bee, N.: EmoVoice — A Framework for Online Recognition of Emotions from Voice. In: André, E., Dybkjær, L., Minker, W., Neumann, H., Pieraccini, R., Weber, M. (eds.) PIT 2008. LNCS (LNAI), vol. 5078, pp. 188–199. Springer, Heidelberg (2008)
21. Zhang, G., Song, Q., Fei, S.: Speech Emotion Recognition System Based on BP Neural Network in Matlab Environment. In: Sun, F., Zhang, J., Tan, Y., Cao, J., Yu, W. (eds.) ISNN 2008, Part II. LNCS, vol. 5264, pp. 801–808. Springer, Heidelberg (2008)

Ambient Intelligent Monitoring of Dementia Suffers Using Unsupervised Neural Networks and Weighted Rule Based Summarisation

Faiyaz Doctor, Chrisina Jayne, and Rahat Iqbal

Coventry University,Applied Computing Reserch Centre,
Department of Computing, Priory Street, Coventry, CV1 5FB, UK
{faiyaz.doctor,Chrisina.Jayne2,r.iqbal}@coventry.ac.uk

Abstract. This paper investigates the development of a system for monitoring of dementia suffers living in their own homes. The system uses unobtrusive pervasive sensor and actuator devices that can be deployed within a patient's home grouped and accessed via standardized platforms. For each sensor group our system uses unsupervised neural networks to identify the patient's habitual behaviours based on their activities in the environment. Rule-based summarisation is used to provide descriptive rules representing the intra and inter activity variations within the discovered behaviours. We propose a model comparison mechanism to facilitate tracking of behaviour changes, which could be due to the effects of cognitive decline. We demonstrate using user data acquired from a real pervasive computing environment, how our system is able to identify the user's prominent behaviours enabling assessment and future tracking.

Keywords: Ambient Intelligence, Dementia Care, Unsupervised Neural Networks, Rule-based Summarisation.

1 Introduction

Dementia describes various different brain disorders that have in common a loss of brain function that is usually a progressive and degenerative condition[1]. Due to the majority of dementia sufferers being supported by carers or relatives and living in their own homes, long-term remote management of individuals is essential in order to track the stages of the disease, adjust the course of therapy and understand the changing care needs of the individual. Observing an individual's activities over time can be used to track and assess stages of the disease. For example, early stages of cognitive impairment may cause forgetting certain steps in familiar tasks or difficulty in performing new and complex tasks. Whereas in moderate stages the individual may exhibit a lack of awareness in the time of day leading to pronounced changes in behaviour from day to day, or impaired communication [2]. Ambient Intelligence (AmI) is an information paradigm, which combines pervasive computing with intelligent data representation and learning systems [3]. The health and social care domain is an important area where pervasive computing and ambient intelligence research is being increasingly applied

C. Jayne, S. Yue, and L. Iliadis (Eds.): EANN 2012, CCIS 311, pp. 363–374, 2012.
© Springer-Verlag Berlin Heidelberg 2012

to provide eHealth related services for our ageing population [4]. An increase in life expectancy has produced an older population that needs continuous monitoring and assistance, especially in case of progressive age-related illnesses such as dementia.

In this paper, we propose an AmI based system which could be deployed in a home-based environment to monitor and track the disease progression of dementia suffers, see Fig. 1. Simple unobtrusive sensors and actuator devices such as temperature, light level, pressure sensors, switches for lights, cupboards doors and household appliances would be retrofitted to positions and items at locations around the home for monitoring the individual's daily activities. The sensors and actuators are grouped into Local Sensor Groups (LSGs) for monitoring key aspects of a patient's cognitive behaviour and activities in their home. The placement of sensors would be determined by a variety of person-related factors alongside the advice of care staff and health professionals. The sensors would be accessed using open middleware component based software architecture such as Open Services Gateway initiative (OSGi) [5]. We propose using an intelligent learning and tracking approach that is based on unsupervised Snap-Drift Neural Network (SDNN) [6][7]and Weighted Rule-Based Summerisation (WRBS) [8] [9]. The SDNN would be used to identify clusters from the monitored data related to the individual's activities associated to a specific LSG. Each cluster represents a labelled set of encoded data points that capture LSG sensor and actuator states associated with prominent behaviour patterns of the individual's daily activities. WRBS is employed to extract the pattern variations and interrelationships of the multi-dimensional data points that make up each of the generated clusters. WRBS provides a means of automatically summarising the labelled data into human interpretable if-then rules. The rules provide a descriptive representation of relationships between the monitored sensor / actuator states and the labelled behaviour clusters. A model comparison mechanism is proposed to allow the SDNN behaviour model to be periodically regenerated to track subtle changes in the individuals behaviour which would be attributed to cognitive degeneration. Using data acquired from monitoring a user in a real pervasive computing environment, we demonstrated the capability of the SDNN and WRBS approaches in identifying prominent activity based behaviours of a user.

In section 2 we present a literature review discussing pervasive sensor technologies, unsupervised learning and rule base summarisation, along with referencing previous work on the development of assistive technologies for monitoring dementia suffers. In section 3 we describe the proposed AmI dementia monitoring system. In section 4, we discuss the experiments and the results. Finally conclusions and future work are presented in section 5.

2   Literature Review

It can be difficult for relatives and carers to continually observe an individual and recognise the subtle changes in an individual's behaviour and daily activities, which may signal progressively worsening stages of the disease. It can also be difficult for health professionals to accurately track the patterns of an individual's cognitive impairment and the impact this is having on daily living and wellbeing. The prevalence of relatively low cost off-the-shelf pervasive wireless sensor devices can be used for sensing motion, light level, temperature, occupancy, chair / bed pressure or appliance hazards such as checking if the cooker has been left on [10]. Simple switch sensors can be used to determine if drawers and cupboards have been opened/closed, and whether specific electronic devices have been switched on or off. For the purpose of this paper we will divide these sensors into passive input sensors such as temperature, light level or occupancy, and active output actuation based sensors such as appliance states, sensors for light switches or doors and cupboards states. The vast majority of wireless sensor devices will support one of several commercial wireless sensor networking standards such as Zigbee, ANT, Z-Wave, ONE-NET to name a few. The development of open middleware architectures such as OSGi [5] can be used to dynamically integrate heterogeneous commercial proprietary sensor devices and network technologies together for easy deployment in a residential environment. This can allow AmI systems to seamlessly record data simultaneously from various sensor devices in context of the resident's interactions with the environment [11].

Intelligent learning and adaptation systems can be embedded in the user's environment to realize AmI systems, which can recognize the individual's interactions and learn their behaviours. Soft computing approaches can be used to identify and model behaviours based on the patterns of association between sensed environmental conditions (light level, occupancy, time of day etc) and specific actions being performed by the individual (switching on the lights, changing channels on the TV or moving an object to a specified location). In particular Neural Networks (NN) clustering is based on unsupervised type of learning used to separate a finite unlabelled data into a finite and discrete set of "natural" hidden clusters found in the data [12], where these clusters imply some meaningful groups of similar patterns. Some of the most prominent NN clustering algorithms are the Self-Organising Map (SOM) [13] and the Adaptive Resonance Theory (ART) [14]. In this paper the unsupervised Snap-Drift Neural network (SDNN) [6], [7] inspired by the SOM and ART is utilized. The SDNN is used for discovering prominent behaviour groupings based on the individual's activities in the environment.

The clusters generated from the unsupervised SDNN learning approach are based on groups of data points projected in a high dimensional space centred at a generic centroid which provides a vector representation of the multivariable characteristics of each cluster. This however can be difficult to interpret by a lay person and also does not show the distribution of the inter and intra-cluster data variations found for each cluster. Rule based linguistic summarisation [9] can be used to summarise label the data points against each discovered cluster into IF-then rules. The generated If-then rules capture a linguistic interpretation of the inter and intra-cluster variability of the data points. Rule quality measures can be further used to determine the generality and reliability of the rules [9] in their ability to successfully classify the data.

The importance of pervasive continuous and reliable long-term monitoring of dementia is emphasized in [15]. Results have been obtained in the development of smart environments for monitoring various activities and behaviours as in [16] [17][18] [19] [20] [21] [22] [23]. These approaches however tend to focus on analysing very specific activities using targeted and tailored sensor devices to monitor particular indicators of cognitive decline. Our approach is based on using a holistic array of simple sensors and actuators with unsupervised learning and classification approaches being used to identify common activity and behaviour characteristics of an individual. This unlike other approaches mentioned allows us to build a linguistically interpretable behaviour model of an individual without making any prior assumptions based on specific activities and input / output classifications of the sensors and actuators. We can then continuously adapt the behaviour model to track subtle changes in the individual's behaviour which could be attributed to cognitive decline.

3   Proposed AmI Dementia Monitoring System

The placement of the sensors and actuators selected for each LSG in the individual's home would be based on factors such as the physical layout of the residence, the aspects of a individual's cognitive behaviors that need to be monitored, how advanced their dementia already is, and the kind of unassisted activities the individual is able to currently perform.

For each LSG our proposed AmI system consists of four phases for learning the individual's behaviours in the environment and tracking behaviour changes overtime, as shown in Fig. 1 and explained below.

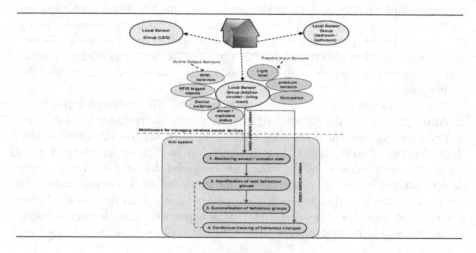

Fig. 1. Flow diagram showing the proposed AmI system for a given LSG

3.1   Monitoring Sensor/Actuator Data

The system passively monitors the user's states and actions for each LSG within the environment. Either periodically or when the user performs an action in the environment, the system will record a 'snapshot' of the current passive sensor states and active sensor states adjusted or previously set by the individual. These 'snapshots' are accumulated over a period of time (possibly over several days) so that the system observes as much of the individual's interactions within the environment as possible. The monitoring period may be designated by a carer or healthcare professional. The data collected from monitoring the individual is initially encoded by a process of mapping the sensor and actuator attribute values it to a defined set of quantifiers. In this paper these quantifiers are represented by singleton or crisp interval sets. For Boolean or categorical data attributes such as 'On' and 'Off' or 'Evening', we use singleton sets. For predefined crisp numerical ranges such as light level ranges: '20-40%', we use crisp interval sets. Both These types of quantifiers follow classical set theory where the membership of an element that belongs to a set is assessed according to a binary condition: either it belongs or does not belong to the set [24]. Each attribute's encoded data values are labelled using a numerical category representing singleton or crisp interval set encoding used. This is to allow the SDNN to process the data more effectively. The numerically labelled data will be used by the WRBC to generate the weighted inter and intra cluster rules from using the same encoded clustered data.

3.2   Identification of Users Behaviour Groupings

The SDNN is used to cluster the encoded data related to individual's identified behaviours into relatively small number of groups. The SDNN is an unsupervised modal learning algorithm able to adapt rapidly by taking advantage of the complementary nature of each mode of learning (snap and drift) [6], [7]. The learning comprises of switching between two different modes of learning snap based on Fuzzy AND and drift based on learning vector quantization (LVQ) [13]. The snap mode enables the NN model to learn the features common to all the patterns in the group, giving a high probability of rapid convergence while the drift mode ensures that average generalized features are learnt and included in the final groupings [25]. The SDNN architecture has an input, hidden and output layers. The output winning neurons from the hidden layer act as input data to the output or so called selection layer. This layer is also subject to the snap-drift learning and has an activation threshold to ensure if a good match to an existing neuron is not found a new neuron is recruited, so the progress of learning determines the number of output groups. Detailed description of the snap-drift algorithm could be found in [6] and [7].

The SDNN is trained on the encoded data acquired from monitoring the individual in their environment. During training, on presentation of each input pattern, the SDNN learns to group the input patterns according to their common features. The SDNN discovers clusters representing group profiles pertaining to specific behavior and activity characteristics based on individuals interactions with a particular LSG in the environment. Each cluster represents a set of points in a multi dimensional input feature space of the monitored characteristics exhibiting common activities in context of performed actuations and environmental states. We label the raw data based on the identified cluster. We use WRBS based on the approach described in [8] on the data to extract the pattern variations and interrelationships of the multi-dimensional feature points that make up each of the previously generated clusters.

3.3   Summarisation of Behaviour Groups

Rule based summarisation approaches [9] provide a way of automatically summarising the encoded cluster labelled data to output human interpretable If-then rules. For a given c-class pattern classification problem with n attributes (or features) a rule for a given data instance would be defined as follows:

$$IF\ x_1^{(t)}\ is\ A_1^q\ and...and\ x_n^{(t)}\ is\ A_n^q,\ THEN\ B^c. \tag{1}$$

Where $x^{(t)} = (x_1,...,x_n)^{(t)}$ represent the n-dimensional pattern vector t, A_1^q is the singleton or crisp interval set representing the numerical category for the antecedent pattern t, B^c is a consequent class (which would be one of the possible c labelled clusters), N is the number of $if - then$ rules in the rule base. This process generates $if - then$ rule for each data instance, which will consist of duplicate rules. Rule compression is performed on the data instance based rules in order to summarise the data into unique rules. This process involves a modified calculation of two rule quality measures from which we then derive the scaled weight of each unique summarisation rule. The quality measures are based on *generality* which measures how many data instances support each rule [9]and *reliability* that measures the confidence level in the data supporting each rule [9]. In our approach the rule *generality* is measured using *support* and the *reliability* of the rule is based on calculating its confidence. The support of a rule refers to coverage of data patterns that map to it [8]. The rule's *support* can be used to identify the unique rules with the most frequent occurrences of data patterns associated with them. The support of each rule is scaled based on the total data patterns for each output set so that the frequencies are scaled in proportion to the number data patterns found in each consequent set. The calculation of the scaled support for a give uniquely occurring rule and is based on the calculation described in [8]. This is used to identify and eliminate duplicate instance based rules to compress the rule base into a set of M unique and contradictory rules modelling the data. The confidence of a rule is a measure of a rule's validity describing how tightly data patterns are associated to a specific output set. The confidence value is between 0 and 1. A confidence of 1 means that the pattern described in the rule is completely unique to a single output set. A confidence of less than 1 means that the pattern that is described in the rule occurs with more than one output set, and would then be associated with the output set with the highest confidence. The rule confidence therefore quantifies the inter cluster variability in the degree of 'belongingness' of specific patterns to each cluster. The rule confidence calculation is based on the calculation described in [8]. The product of the scaled support and confidence of a rule is used to calculate the rule's *scaled weight* which is assigned to each of the M generated rules taking the following form:

$$IF\ x_1\ is\ A_1^l\ and...and\ x_n\ is\ A_n^l,\ THEN\ is\ B_1^l[scW_i]. \qquad (2)$$

The scaled weight measures the quality of each rule in modelling the data. It can be used to rank the top rules associated to each output class and choose a single winner rule among compatible rules based on methods for rule weight specification described in [8]. The generated rules provide a means by which each clusters core behaviour characteristics can be viewed in context of the distribution and variation of data points belonging to it. The rules also provide a transparent and interpretable representation of the individuals current behaviour states associated with each specific LSG in their home.

3.4   Continuous Tracking of Behaviour Changes

The AmI system has to be able to regularly update and compare its generated LSG behaviour models based on monitoring the individual's activities over time in order to track subtle changes in the activities attributing to a drift in their behaviours which could be caused from the effects of cognitive decline. The initial captured data on the individuals behaviour will produce a model that will identify behaviour clusters (groups) based on monitoring the habitual behaviours of individuals in context of the specific LSGs placed in their home. To produce a system that is able to track changes in the individuals behaviours over time, the SDNN data model has to be regenerated based on the previously accumulated data and the new data generated from re-monitoring the individuals activities. When a new model is generated it will therefore produce a new set of behaviour clusters that are based on the currently monitored and previously accumulated data up to a certain number of past iterations. The new model will be used to generate an updated set of If-then classification rules by the WRBS. Although the new behaviour clusters should incorporate behaviour characteristics encountered from the combined data, they may not include the same behaviour groups identified by the previous model. Our proposed system would therefore compare both models to check how behaviour groups have drifted and their associated rules have change between both models. This process of perpetually regenerating the data model will allow the system to identify trends in behaviour groups interpretable through the movement of the clusters and the intra and inter cluster variation in their associated If-then rules over time. These variations may signify the degeneration of mental abilities attributed to disease progression, and provide a means to mitigate these effects with targeted forms of care support. The model regeneration and comparison would be done at regular intervals up to decision point determined by a health care professional at which behaviour changes suggested a significant deterioration of the individuals executive and mental functions. At this point the individuals care plan would be updated based on any identified changes in their care needs. From this point on a new initial behaviour model would be generated and used to track ongoing disease progression.

Because our approach decomposes the pervasive environment into units of LSGs, it allows each behaviour model to be associated with a specific sensing and actuation area in the home. This allows subsets of related activities to be more accurately monitored and tracked. Sensor and actuator data can therefore be modelled without leading to curse of dimensionality issues associated the modelling and analysis of high dimensional data. Each behaviour model is also more compact due to smaller more manageable input sizes and hence tractable in terms of the size of the generated rule base. This also means that the behaviour rules are more easily interpretable to end users of the system.

4   Experiments

Experiments have been conducted on data acquired from a real pervasive computing environment test bed demonstrating the capability of our AmI system at learning the behaviour groups and associated If-then rules from monitoring a user in the environment over a period of time. The test bed used to collect the data was a room-based student bedroom environment comprising of a large number of embedded sensors, concealed actuators (e.g. buried in walls and ceiling) with the intention that the user would be completely unaware of the intelligent infrastructure of the room. The sensor and actuator devices were based around three networks, Lonworks, 1-wire (TINI) and IP which were accessed using a purpose-built Java middleware platform allowing the agent homogeneous access to the devices in the room. The student bedroom environment consisted of a variety of sensors and actuators. The sensors in the room sensed: internal light level; external light level; internal temperature; external temperature; bed pressure; chair pressure; occupancy; time measured as a continuous input on an hourly scale. The actuators in the room consisted of: 4 variable intensity spot lights located around the ceiling; desk lamp; bedside lamp; motorised window blinds; fan heater; fan cooler; and two PC based applications comprising of a media playing and word processing software application. The test bed environment demonstrates a bedroomstudy LSG that would be typically configured for our proposed system. The captured data was based on monitoring single user living in the environment for a period of three consecutive days. The raw data acquired from monitoring the user comprised of 405 instances or patterns. This was encoded into categories using a set of predefined crisp interval and singleton sets as described in section 4.1. The encoded data patterns were processed by the SDNN to find 14 clustered groups. For each cluster a centroid describes the characteristic of the cluster. From the 14 clusters found by the SDNN we have selected 4 clusters shown to define prominent user behaviours discovered in the data, which were also verified from diary studies conducted by the user during their stay in the environment. Figure 2 shows each of these selected behaviour clusters with a description of the behaviour they represent.

Selected original encoded clusters generated from NN model with behaviour descriptions																	
IN_LightLevel	Ex_LightLevel	IN_Temp	Ex_Temp	ChairPres	BedPres	Hour	Light_1	Light_2	Light_3	Light_4	Blind	BedLight	DeskLight	Heater	MSWord	MediaPlayer	Group
Medium/High/High	VHigh	*	High/VHigh	Off/On	Off	*	VLow	VLow	VLow	VLow	Open	Off	Off	Off	On	Off	19
Day-Time working behaviour, which involves the user periodically sitting at the desk as well as using the word application																	
VLow	VLow	*	Low/Medium	Off	Off/On	Night/Evening	High/VHigh	High/VHigh	*	*	Closed	On	Off	Off	Off	On	8
Evening Time Relax behaviour as person is periodically on the bed, internal lights and bedside lamp are on and the media is also on																	
VLow	VLow	*	*	On	Off	Evening/Night	VLow/Low	Low/Medium	High/VHigh	Medium	Closed	Off	On	Off	Off	On	6
Evening Time Non-Work behaviour as person is on the chair, internal lights are on and the media player is running and desk lamp is on																	
VLow	VLow	Low/Medium	Low	Off	On	Night	Low/High	Medium/High	Low/Medium	VLow	Closed	On	Off	On	Off	On	28
Night Time Feeling Cold behaviour as person is on the bed, internal and external temperatures are low, lights are on low, bedside lamp is on and the music player and heater are also on																	

Key

1 = VLow	1 = VLow	1 = VLow	1 = VLow	0 = On	0 = On	1 = Morning	1 = VLow	1 = VLow	1 = VLow	1 = VLow	0 = Closed	0 = On	0 = On	0 = On	0 = On
2 = Low	2 = Low	2 = Low	2 = Low	1 = Off	1 = Off	2 = Mid-Morning	2 = Low	2 = Low	2 = Low	2 = Low	1 = Open	1 = Off	1 = Off	1 = Off	1 = Off
3 = Medium	3 = Medium	3 = Medium	3 = Medium			3 = Afternoon	3 = Medium	3 = Medium	3 = Medium	3 = Medium					
4 = High	4 = High	4 = High	4 = High			4 = Evening	4 = High	4 = High	4 = High	4 = High					
5 = VHigh	5 = VHigh	5 = VHigh	5 = VHigh			5 = Night	5 = VHigh	5 = VHigh	5 = VHigh	5 = VHigh					
* = Any Category	* = Any Category	* = Any Category	* = Any Category			* = Any Category	* = Any Category	* = Any Category	* = Any Category	* = Any Category					

Fig. 2. Selected encoded clusters generated from SDNN representing prominent user behaviours in the environment

The key also shown describes the encoded numerical categories for each data attributes, where the ∗ notation indicates that for the given cluster and data attribute, its values can be any one of the categories pertaining to it. The encoded data patterns were then labelled according the discovered cluster they belong to and processed by the WRBS. The WRBS generated 200 rules from the 405 encoded data pattern, which is a data compression of over 50%. Figure 3 shows the top five strongest profile rules describing each of the four selected behavior clustered identified by the SDNN. The rules are shown grouped according to each labelled output cluster and ranked in order of the calculated scaled weight of each rule. Figure 3 also shows the support, scaled support and confidence for each rule. The strongest rule for each output cluster is also highlighted representing the pattern of behavior activity characteristics most strongly associated with each labelled cluster.

β-class summarisation rules describing intra and inter cluster variability																					
IN_LightLevel	Ex_LightLevel	IN_Temp	Ex_Temp	ChairPres	BedPres	Hour	Light_1	Light_2	Light_3	Light_4	Blind	BedLight	DeskLight	Heater	MSWord	MediaPlayer	Group	Support	Scaled Support	Confidence	Scaled Weight
High	VHigh	High	VHigh	On	Off	Afternoon	VLow	VLow	VLow	VLow	Open	Off	Off	Off	On	Off	19	10	0.09	1.00	0.29
Medium	VHigh	VHigh	VHigh	On	Off	Afternoon	VLow	VLow	VLow	VLow	Open	Off	Off	Off	On	Off	19	8	0.08	1.00	0.08
High	VHigh	Medium	High	On	Off	Mid-Morning	VLow	VLow	VLow	VLow	Open	Off	Off	Off	On	Off	19	8	0.08	1.00	0.08
Medium	VHigh	High	VHigh	On	Off	Mid-Morning	VLow	VLow	VLow	VLow	Open	Off	Off	Off	On	Off	19	7	0.07	1.00	0.07
Medium	VHigh	VHigh	VHigh	Off	Off	Afternoon	VLow	VLow	VLow	VLow	Open	Off	Off	Off	On	Off	19	6	0.06	1.00	0.06
VLow	VLow	High	Medium	Off	On	Night	High	High	Medium	VLow	Closed	On	Off	Off	Off	On	8	11	0.17	1.00	0.17
VLow	VLow	High	Low	Off	On	Night	Low	High	Medium	Low	Closed	On	Off	Off	Off	On	8	7	0.10	1.00	0.10
VLow	VLow	Medium	Low	Off	On	Night	High	High	Medium	VLow	Closed	On	Off	Off	Off	On	8	5	0.07	1.00	0.07
VLow	VLow	High	Medium	Off	Off	Night	VHigh	VHigh	VHigh	VHigh	Closed	On	Off	Off	Off	On	8	4	0.06	1.00	0.06
VLow	VLow	Medium	Low	Off	Off	Evening	VHigh	VHigh	VHigh	VHigh	Closed	On	Off	Off	Off	On	8	1	0.04	1.00	0.04
VLow	VLow	High	Medium	On	Off	Night	VLow	Medium	VHigh	Medium	Closed	Off	On	Off	Off	On	6	4	0.12	1.00	0.10
VLow	VLow	High	Low	On	Off	Night	VLow	Low	High	Low	Closed	Off	On	Off	Off	On	6	3	0.08	1.00	0.08
VLow	VLow	High	Medium	On	Off	Night	VLow	Medium	High	Medium	Closed	Off	On	Off	Off	On	6	3	0.08	1.00	0.08
VLow	VLow	VHigh	Medium	On	Off	Evening	Low	Low	High	Medium	Closed	Off	On	Off	Off	On	6	3	0.08	1.00	0.08
VLow	VLow	VHigh	High	On	Off	Evening	VLow	Medium	VHigh	Medium	Closed	Off	On	Off	Off	On	6	3	0.08	1.00	0.08
VLow	VLow	Low	Low	Off	On	Night	Low	Medium	Low	VLow	Closed	On	Off	On	Off	On	28	5	0.29	1.00	0.29
VLow	VLow	Medium	Low	Off	On	Night	Low	Medium	Low	VLow	Closed	On	Off	On	Off	On	28	5	0.29	1.00	0.29
VLow	VLow	Low	Low	Off	On	Night	High	High	Medium	VLow	Closed	On	Off	On	Off	On	28	3	0.18	1.00	0.18
VLow	VLow	Low	VLow	Off	On	Night	High	High	Medium	VLow	Closed	On	Off	On	Off	On	28	2	0.12	1.00	0.12
VLow	VLow	Low	Medium	Off	On	Evening	High	High	Medium	VLow	Closed	On	Off	On	Off	On	28	1	0.06	1.00	0.06

Fig. 3. WRBS generated rules for the selected clusters derived from the labeled encoded data

The top rules for each cluster provide a linguistic interpretation of the strongest pattern of behavior characteristics, which are associated with each cluster. The rules are shown to correspond to the original cluster centroids generated from the SDNN and shown in Figure 3. The rules are able to capture the intra cluster variation about the centroid which result from the SDNN model defining certain cluster attributes as having more than a single category or an unspecified category. Inter cluster variability are also captured through the confidence measure for each rule, which indicates how tightly the rule is associated to a specific cluster. In the case of the rules shown in Figure 3, due to the high number of attribute category combinations, the confidence of all these rules is equal to 1.0. This means that there are no inter cluster variability as all the rule based patterns belonging to each cluster are unique to it. If the number of attributes and category combinations were reduced, one would notice more inter cluster sharing of rules with confidence values being less that 1.0. The calculated weighting for each rule further provides end users with an explicit ranking of the rule patterns that best describe each cluster, which can be used easily track rule changes due to possible behavior changes overtime.

5 Conclusions

In this paper, we propose an ambient intelligent system to monitor and track the disease progression of dementia patients. The system could be deployed in a patient's home environment where sensors and actuators are placed in grouped units pertaining to specific locations and factors related to the patient's disease severity and the opinions of care providers. We propose using an intelligent system employing an unsupervised SDNN that will be used to discover clusters in the data pertaining to the behaviours of the patient in the environment. In addition we employ WRBS, which provide a means of summarising the labelled data into human interpretable If-then rules. The rules define relationships between the monitored sensoractuator states and the labelled behaviour clusters. We propose a continuous model regeneration and comparison mechanism for periodically remonitoring and modelling the individuals behaviours to track behaviour changes due to the effects of cognitive decline which could help carers assess the patient's cognitive abilities and care needs.

We have shown using data acquired from monitoring a user in a real pervasive computing environment that the SDNN approach can be used to identify user behaviour grouping in the data, and the WRBS can be used to provide a linguistically interpretable representation of these behaviours enabling assessment and tracking of behaviour activities. Although the experiments we have presented have been conducted with non-dementia suffers they demonstrate the capability of our proposed framework to passively monitor individuals', identify behaviour changes attributed to the observable effects of cognitive degeneration in real suffers.

For our future work we intend to carry out experiments to test the robustness of our approach and eventually aim to deploy a prototype system within a residential care setting to monitor a patient with dementia to evaluate the systems ability in providing insight into long-term behaviour variation to help assess the changing care needs of the patient.

References

1. Society, About Dementia (June 30, 2011),
   http://alzheimers.org.uk/site/scripts/documents.php?categoryID=200120
2. Agrawal, R., Srikant, R.: Fast Algorithms for Mining Association Rules. In: Proceedings of the 20th International Conference on Very Large Databases, Santiago, pp. 487–499 (September 1994)
3. Ducatel, K., Bogdanowicz, M., Scapolo, F., Leijten, J., Burgelman, J.: Ambient intelligence: From vision to reality. In: Riva, G., Vatalaro, F., Davide, F., Alcaniz, M. (eds.) Ambient Intelligence: The Evolution of Technology Communication and Cognation Towards the Future of Human-Computer Interaction. Emerging Communication: Studies in New Technologies and Practices in Communication, vol. 6. IOS Press (2003)
4. Chiarugi, F., Zacharioudakis, G., Tsiknakis, M., Thestrup, J., Hansen, K.M., Antolin, P., Melgosa, J.C., Rosengren, P., Meadows, J.: Ambient Intelligence support for tomorrow's Health Care: Scenario-based requirements and architectural specifications of the EU-Domain platform. In: Proceedings of the International Special Topic Conference on Informational Technology in BioMedicine, Ioannina, Greece, October 26-28 (2006)
5. OSGi Alliance (2011), http://www.osgi.org
6. Lee, S.W., Palmer-Brown, D., Tepper, J.A., Roadknight, C.M.: Snap-drift: real-time, performance-guided learning. In: International Joint Conference on Neural Networks, Portland, OR, USA, July 20-24, pp. 1412–1416. IEEE, Piscataway (2003)
7. Lee, S.W., Palmer-Brown, D., Roadknight, C.M.: Performance guided Neural Network for Rapidly Self Organising Active Network Management. Neurocomputing 61, 5–20 (2004)
8. Ishibuchi, H., Yamamoto, T.: Rule Weight Specification in Fuzzy Rule-Based Classification Systems. IEEE Transactions on Fuzzy Systems 13(4), 428–435 (2005)
9. Wu, D., Mendel, J.M., Joo, J.: Linguistic Summarization Using If-Then Rules. In: Proceedings of the IEEE International Conference on Fuzzy Systems, Barcelona, Spain, pp. 1–8 (July 2010)
10. Woolham, J., Gibson, G., Clark, P.: Assistive Technology, Telecare, and Dementia: Some Implications of Current Policies and Guidance. Research Policy and Planning 24(3), 149–164 (2007)
11. Conde, D., Ortigosa, J.M., Javier, F., Salinas, J.R.: Open OSGi Middleware to Integrate Wireless Sensor Devices into Ambient Assisted Living Environments. In: Proceedings of AALIANCE Conference, Malaga, Spain, March 11-12 (2010)
12. Xu, R., Wunsch, D.: Survey of Clustering Algorithms. IEEE Transaction on Neural Networks 16(3), 645–678 (2005)
13. Kohonen, T.: Self-Organisation and Asssociative Memory, 3rd edn. Springer, Heilderberg (1989)
14. Carpenter, G.A., Grossberg, S.: Adaptive Resonance Theory. The Handbook of Brain Theory and Neural Networks, 2nd edn., pp. 87–90. MIT Press, Cambridge (2003)
15. Arnrich, B., Mayora, O., Bardram, J., Troster, G.: Pervasive Healthcare: Paving the Way for a Pervasive, User-Centered and Preventive Healthcare Model. Methods of Information in Medicine 49(1), 67–73 (2010)

16. Peters, C., Wachsmuth, S., Hoey, J.: Learning to Recognise Behaviours of Persons with Dementia using Multiple Cues in an HMM-Based Approach. In: Proceedings of the 2nd International Conference on Pervasive Technologies Related to Assistive Environments, Corfu, Greece, June 23-25 (2009)
17. Mihailidis, A., Carmichael, B., Boger, J.: The Use of Computer Vision in an Intelligent Environment to Support Aging-in-Place, Safety, and Independence in the Home. IEEE Transactions on Information Technology in Biomedicine 8(3), 238–247 (2004)
18. Mynatt, E.D., Melenhorst, A.S., Fisk, A.D., Rogers, W.A.: Aware Technologies for Aging in Place: Understanding User Needs and Attitudes. Pervasive Computing 3(2), 36–41 (2004)
19. Hayes, T.L., Hunt, J.M., Adami, A., Kaye, J.A.: An Electronic Pillbox for Continuous Monitoring of Medication Adherence. In: Proceedings of the 28th IEEE EMBS Annual International Conference, New York City, USA (2006)
20. Matic, A., Mehta, P., Rehg, J.M., Osmani, V., Mayora, O.: Monitoring Dressing Activity Failures through RFID and Video. Methods of Information in Medicine 47(3), 229–234 (2008)
21. Biswas, J., et al.: Agitation Monitoring of Persons with Dementia based on Acoustic Sensors, Pressure Sensors and Ultrasound Sensors: A Feasibility Study. In: Proceedings of The International Conference on Aging, Disability and Independence, St. Petersburg, Florida, February 1-5, pp. 3–15. IOS Press, Amsterdam (2006)
22. Bonroy, B., et al.: Image Acquisition System to Monitor Discomfort in Demented Elderly Patients. In: Proceedings of the 18th ProRISC Annual Workshop on Circuits, Systems and Signal Processing, Veldhoven, The Netherlands, November 29-30 (2008)
23. Neergaard, L.: Can Motion Sensors Predict Dementia?. The Associated Press (June 19, 2007)
24. Mendel, J.: Uncertain Rule-Based Fuzzy Logic Systems: Introduction and New Directions. Prentice Hall PTR (2001)
25. Palmer-Brown, D., Lee, S.W., Draganova, C., Kang, M.: Modal learning neural networks. WSEAS Transactions on Computers 8(2), 222–236 (2009)

On the Intelligent Machine Learning in Three Dimensional Space and Applications

Bipin K. Tripathi* and Prem K. Kalra**

Harcourt Butler Technological Institute, Kanpur, India
abkt.iitk@gmail.com, pkk@iitk.ac.in

Abstract. The engineering applications of high dimensional neural network are becoming very popular in almost every intelligence system design. Just to name few, computer vision, robotics, biometric identification, control, communication system and forecasting are some of the scientific fields that take advantage of artificial neural networks (ANN) to emulate intelligent behavior. In computer vision the interpretation of 3D motion, 3D transformations and 3D face or object recognition are important tasks. There have been many methodologies to solve them but these methods are time consuming and weak to noise. The advantage of using neural networks for object recognition is the feasibility of a training system to capture the complex class conditional density of patterns. It will be desirable to explore the capabilities of ANN that can directly process three dimensional information. This article discusses the machine learning from the view points of 3D vector-valued neural network and corresponding applications. The learning and generalization capacity of high dimensional ANN is confirmed through diverse simulation examples.

Keywords: 3D motion interpretation, 3D face recognition, 3D real-valued vector, orthogonal matrix.

1 Introduction

Artificial neural networks have been studied for many years in the hope of achieving human like flexibility in processing complex information. Some of the recent researches in neurocomputing concern the development of higher dimension neurons [1–3] and their applications to the problems which deal with high dimensional information. There has been rapid development in the field of 3D imaging in last few years. This is a multidisciplinary field, which encompasses various research areas which deals with information processing in higher dimensions. It is at its infancy [4–6] and requires exploring methods based on neural networks. This paper is aimed at presenting relevant theoretical and experimental framework for machine learning based on multilayer neural networks of 3D vector-valued neurons. The 3D motion interpretation and 3D feature recognition are

---

* Associated with high dimensional computational research group in H B Technological Institute, Kanpur, India.
** The Director, Indian Institute of Technology Jodhpur, India.

C. Jayne, S. Yue, and L. Iliadis (Eds.): EANN 2012, CCIS 311, pp. 375–384, 2012.
© Springer-Verlag Berlin Heidelberg 2012

essential part of many high level image analysis and found wide practical uses in computer vision systems. Although, there are many methodologies [7, 8, 4, 5] to solve them but they use extensive mathematics and are time consuming. They are also weak to noise. Therefore, it is desirable for realistic system to consider iterative methods, which can adapt system for high dimensional applications. This paper considers 3D geometric (point set) representation of objects. The method described here is fully automatic, does not require much preprocessing steps and converges rapidly to a global minimum.

The advantages with a neural network include robustness, ability to learn, generalize and separate complicated classes [9, 3]. In 3D vector-valued neural network, the input-output signals and threshold are 3D real-valued vectors, while weights associated with connections are 3D orthogonal matrices. We will present few illustrative examples to show how a 3D vector-valued neuron can be used to learn 3D motion and used in 3D face recognition. The proposed 3D motion interpretation system is trained using only few set of points lying on a line in the 3D space. The trained system is capable of interpreting 3D motion consisting of several motion components over unknown 3D objects. Face recognition is the preferred mode of identity authentication [10–12]. The facial features have several advantages over other six biometric attributes considered by Hietmeyer [13]. It is natural, robust and uninstructive. It can not be forgotten or mislaid like other document of identification. Most of the face recognition techniques have used 2D images of human faces. However, 2D face recognition techniques are known to suffer from the inherent problems of illumination and structural variation and are sensitive to factors such as background, change in human expression, pose, and aging [14]. Utilizing 3D face information was shown to improve face recognition performance, especially with respect to these variations [8, 15].

In this article, we investigate a novel machine learning technique with 3D vector-valued neural networks through various computational experiments. Section II explains the learning rule for 3D vector-valued neural networks. The generalization ability of 3D neural network in 3D motion interpretation is confirmed through diverse test patterns in section III. Section IV is devoted for 3D face recognition for biometric applications. Section V presents final conclusion and future scope of work.

2   Learning Rule

In our multilayer network, we have considered three layers, first is of inputs, second layer is only hidden layer and an output layer. A three layer network can approximate any continuous non-linear mapping. In 3D vector-valued neural network, the bias values and input-output signals are all 3D real-valued vectors, while weights are 3D orthogonal matrices. All the operations in such a neural network are scaler matrix operations. A 3D vector-valued back-propagation algorithm is considered here for training a multilayer network, which is natural extension of complex-valued back-propagation algorithm [16, 1]. It has ability to learn 3D motion as complex-BP can learn 2D motion [2].

In a three layer network (L-M-N), first layer has L inputs (I_l), where $l = 1 .. L$, second and the output layer consists M and N vector-valued neurons respectively. By convention, w_{lm} is the weight that connects l^{th} neuron to m^{th} neuron and $\alpha_m = [\alpha_{mx}, \alpha_{my}, \alpha_{mz}]$ is the bias weight of mth neuron. $\eta \in [0, 1]$ is the learning rate and f' is derivative of a non-linear function f. Let V be net internal potential and Y be the output of a neuron. Let e_n be the difference between actual and desired value at n^{th} output, where $|e_n| = \sqrt{e_n^{x2} + e_n^{y2} + e_n^{z2}}$ and $e_n = [e_n^x, e_n^y, e_n^z]^T = Y_n - Y_n^D$.

$$W_{lm} = \begin{vmatrix} w_{lm}^x & -w_{lm}^y & 0 \\ w_{lm}^y & -w_{lm}^x & 0 \\ 0 & 0 & w_{lm}^z \end{vmatrix} \qquad W_{mn} = \begin{vmatrix} w_{mn}^x & 0 & 0 \\ 0 & w_{mn}^y & -w_{mn}^z \\ 0 & w_{mn}^z & w_{mn}^y \end{vmatrix}$$

where $W_{lm}^z = \sqrt{(w_{lm}^x)^2 + (w_{lm}^y)^2}$ and $W_{mn}^x = \sqrt{(w_{mn}^y)^2 + (w_{mn}^z)^2}$

The net potential of mth neuron in hidden layer can be given as follows -

$$V_m = \sum_l w_{lm} I_l + \alpha_m \tag{1}$$

The activation function for 3D vector-valued neuron is 3D extension of real activation function and defined as follows-

$$Y_m = f(V_m) = [f(V_m^x), f(V_m^y), f(V_m^z)]^T \tag{2}$$

Similarly,

$$V_n = \sum_m w_{mn} Y_n + \alpha_n \quad and \quad Y_n = f(V_n) = [f(V_n^x), f(V_n^y), f(V_n^z)]^T \tag{3}$$

$$Y_n = f(V_n) = [f(V_n^x), f(V_n^y), f(V_n^z)]^T \tag{4}$$

The mean square error function can be defined as :

$$E = \frac{1}{N} \sum_n |e_n|^2 \tag{5}$$

In 3D vector version of back-propagation algorithm the weight update equation for any weight is obtained by gradient descent on error function :

$$\Delta w = \eta \left| -\frac{\partial E}{\partial w^x} \quad -\frac{\partial E}{\partial w^y} \quad -\frac{\partial E}{\partial w^z} \right|^T$$

then, weights and bias in output layer can be updated as follows :

$$\Delta \alpha_n = \eta \begin{vmatrix} e_n^x \cdot f'(V_n^x) \\ e_n^y \cdot f'(V_n^y) \\ e_n^z \cdot f'(V_n^z) \end{vmatrix}$$

$$
\begin{vmatrix} \Delta w_{mn}^y \\ \Delta w_{mn}^z \end{vmatrix} = \eta \begin{vmatrix} \frac{w_{mn}^y}{w_{mn}^x} Y_m^x & Y_m^y & Y_m^z \\ \frac{w_{mn}^z}{w_{mn}^x} Y_m^x & -Y_m^z & Y_m^y \end{vmatrix} \begin{vmatrix} e_n^x \cdot f'(V_n^x) \\ e_n^y \cdot f'(V_n^y) \\ e_n^z \cdot f'(V_n^z) \end{vmatrix}
$$

Similarly, weights and bias in hidden layer neuron can be updated as follows

$$
\begin{vmatrix} \Delta \alpha_m^x \\ \Delta \alpha_m^y \\ \Delta \alpha_m^z \end{vmatrix} = \frac{\eta}{N} \begin{vmatrix} f'(V_m^x) & 0 & 0 \\ 0 & f'(V_m^y) & 0 \\ 0 & 0 & f'(V_m^z) \end{vmatrix} \sum_n \begin{vmatrix} w_{mn}^x & 0 & 0 \\ 0 & w_{mn}^y & w_{mn}^z \\ 0 & -w_{mn}^z & w_{mn}^y \end{vmatrix} \begin{vmatrix} e_n^x \cdot f'(V_n^x) \\ e_n^y \cdot f'(V_n^y) \\ e_n^z \cdot f'(V_n^z) \end{vmatrix}
$$

$$
\begin{vmatrix} \Delta w_{lm}^x \\ \Delta w_{lm}^y \end{vmatrix} = \begin{vmatrix} I_l^x & I_l^y & \frac{w_{lm}^x}{w_{lm}^z} I_l^z \\ -I_l^y & I_l^x & \frac{w_{lm}^y}{w_{lm}^z} I_l^z \end{vmatrix} \begin{vmatrix} \Delta \alpha_m^x \\ \Delta \alpha_m^y \\ \Delta \alpha_m^z \end{vmatrix}
$$

3   Learning 3D Motion

This section presents different 3D motions of objects in the physical world. The motion in a 3D space may consist of a 3D *scaling, rotation and translation* and composition of these three operations. These three basic classes of transformations convey dominant geometric characteristic of mapping. Such a mapping in space preserves the angles between oriented curves and the phase of each point on the curve is also maintained during motion of points. As described in [9, 16], the complex-valued neural network enables to learn 2D motion of signals, hence generalizes conformal mapping on plane. Similarly, a 3D vector-valued neural network enables to learn 3D motion of signals and will provide generalization of mappings in space. In contrast, a neural network in a real domain administers 1-D motion of signals hence does not preserve the amplitude as well as phase in mapping. This is the main reason as to why a high dimensional ANN can learn high dimensional mapping, while equivalent real-valued neural network cannot [16].

In order to validate the proposed motion interpretation system, various simulations are carried out for learning and generalization of high dimensional mapping. It has capacity to learn 3D motion patterns using set of points lying on a line in the 3D space and generalize them for motion of an unknown object in the space. We have used a 2-6-2 structure of 3D vector-valued neural network in all experiments of this section, which transform every input point (x, y, z) into another point (x', y', z') in the 3D space. First input of input layer takes a set of points lying on the surface of an object and second input is the reference point of input object. Similarly, the first neuron of output layer gives the surface of transformed object and second output is its reference point. Empirically it is observed that considering reference point yields better testing results. The input-output values are with in the range $-1 \leq x, y, z \leq 1$. In all simulations, the training input-output patterns are the set of points lying on a straight line

with in the space of unit sphere ($0 \leq radius \leq 1$), centered at origin and all
the angles vary from 0 to $2\,\pi$. Figure 1(a) presents an example input-output
mapping of training patterns. Following few examples depict the generalization
ability of such trained network over standard geometric objects.

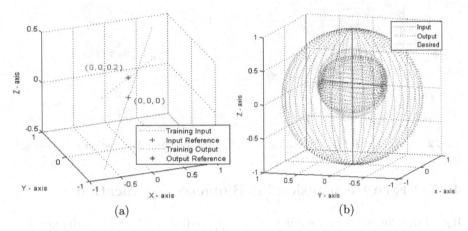

(a)   (b)

Fig. 1. (a)Training patterns : mapping shows scaling by factor 1/2, angle of rotation
$\pi/2$ and displacement by $(0, 0, 0.2)$.; (b) The tesing over a 3D object

Example 1. A neural network based on 3D vector-valued neurons has been
trained for the composition of all three transformation. Training input-output
patterns are shown in Figure 1(a). The generalization ability of such a trained
network is tested over sphere (1681 data pints). Figure 1(b) presents the generalization ability of trained network. All patterns in output are contracted by factor
1/2, rotated over $\pi/2$ radians clockwise and displaced by $(0, 0, 0.2)$. This example
demonstrate the generalization ability of considered network in interpretation of
object motion in 3D space.

Example 2. In this experiment the network is trained for input-output mapping
over a straight line for similarity transformation (scaling factor 1/2) only. The
generalization ability of such trained network is tested over cylinder containing
202 data points. The transformation result in Figures 2(a) shows the excellent
generalization with proposed methodology.

Example 3. A neural network based on 3D vector-valued neurons has been
trained with line for the composition of scaling and translation. Figure 2(b)
presents the generalization ability of this trained network. There are 451 test
data points on cylinder. All points in 3D are contracted by factor 1/2 and displaced by $(0, 0, 0.2)$. Results bring out the fact that given neural network is able
to learn and generalize 3D motion.

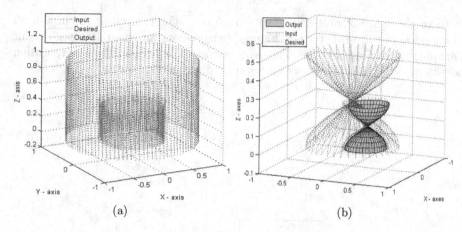

(a)   (b)

Fig. 2. (a)Similarity Transformation in 3-D; (b)Scaling and Translation in 3-D space

4   3D Face Recognition for Biometric Application

Recent developments in computer technology and call for better security applications have brought biometrics into focus. The signature, handwriting, fingerprint have a long history. More recently voice, retinal scan, iris scan and face information are considered for biometrics. When deploying a biometrics based system, we consider its accuracy, cost, ease of use, whether it allows integration with other systems and the ethical consequences of its use.

4.1   Related Work

Biometrics can be defined as the automated use of physiological (face, finger prints, periocular, iris, Oculomotor plant characteristic and DNA) and behavioral (signature and typing rhythms) characteristics for verifying the identity of living person. The physiological features are often non-alterable except severe injury, while behavioral features may fluctuate due to stress, fatigue or illness. Face recognition is one of the few biometric methods that possess the merits of both high accuracy and low intrusiveness. It is also one of the most acceptable biometrics because a human face is always bare and often used in their visual interactions. It is a potential identify of a person without document for identification.

In early methods for 3D face recognition curvatures and surface features, kept in a cylindrical co-ordinate system, were used. Moreno et al. found that curvature and line features perform better than area features [4]. Point cloud is the most primitive 3D representation for faces and Housdroff distance has been used for matching the point clouds in [7]. The base mesh is also used for alignment in [5], where features are extracted from around landmark points and nearest neighbor after that PCA is used for recognition. In [8] the analysis-by-synthesis approach that uses morphable model is detailed. The idea is to synthesize a

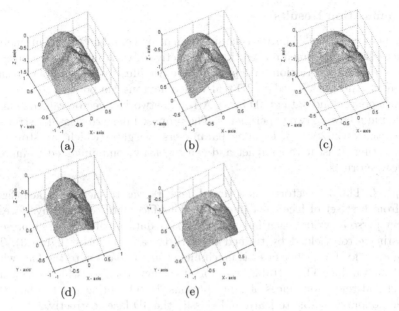

Fig. 3. Five faces of different persons considered in example 4

pose and illumination corrected image pair for recognition. Depth maps have been used in 3D imaging applications [17]. The depth map construction consists of selecting a view point and smoothing the sampled depth values. Most of the work that uses 3D face data uses a combination of representations. The enriched varieties of features, when combined with classifiers with different statistical properties, produce more accurate and robust performance. As a result of fast development in 3D imaging technology, there is strong need to address those using high dimensional neural networks.

4.2 Normalization

3D linear transformation has been considered for normalization of 3D faces. Its purpose is to align each face on a same scale and at same orientation. In order to make the standard alignment for facial features, the origin is translated to the nose tip. It is assumed that the scanned 3D face data are of front part of face and almost straight (variation $40° − 50°$ allowed) and accordingly it is translated. Logically nose tip is the peak of a face and hence can have maximum Z-coordinate value. Therefore, the Z coordinate on 3D face data is searched and their corresponding X, Y coordinates. Once the nose tip is identified, one can search in the y direction to determine the nose dip. Both nose tip and nose dip must lie on the same line. The scaling of all the faces has been done by taking distance between the nose tip and nose dip along y-axis and nose tip as the origin of the coordinate system.

4.3   Simulation Results

This paper focuses on 3D pattern classification using neural network. Our method has successfully performed recognition irrespective of variability in head pose, direction and facial expressions. We present here two illustrative examples to show how a neural network based on 3D real-valued neurons can be used to learn and recognize point cloud data of 3D faces. A 1-2-1 network of vector-valued neurons was used in following two experiments. The proposed pattern classifier structure involves the estimation of learning parameters (weights) which are stored for future testing. It is more compact and can be easily communicated to humans than learned rules.

Example 4. The 3D vector-valued neural network was trained by a face (figure 3(a)) from first set of face data (figure 3). This face data contains five faces of different persons, where each face contains 6397 data points. Table -1 presents the testing error yielded by trained network for all five faces (figure 3). The testing error for four other faces is much higher in comparison to the face which is used in training. Thus, trained network recognize the face which is taken in training and reject four faces of other persons. Results bring out the fact that this methodology is able to learn and classify the 3D faces correctly.

Example 5. In this example, the considered network was trained by first face (figure 4(a)) from the face set (figure 4). This face set contains five faces of same person with different orientation and poses. Each face contains 4663 data points. Table -2 presents the testing error yielded by trained network for all five faces (figure 4). The testing error for four other faces is also minimum and comparable to the face, which is used for training. Thus, trained network recognize all faces of same person. Thus considered methodology has successfully performed recognition irrespective of variability in head pose and orientation.

Table 1. Comparison of testing error of face set (figure 3)

MSE Training (Target error) = 5.0e-05

Test Face	3(a)	3(b)	3(c)	3(d)	3(e)
Test Error	7.79e-05	9.34e-01	2.02e-00	5.73e-02	2.61e-01

Table 2. Comparison of testing error of second face set (figure 4)

MSE Training (Target error) = 1.0e-04

Test Face	4(a)	4(b)	4(c)	4(d)	4(e)
Test Error (MSE)	3.11e-04	6.10e-03	4.90e-03	6.21e-04	7.32e-04

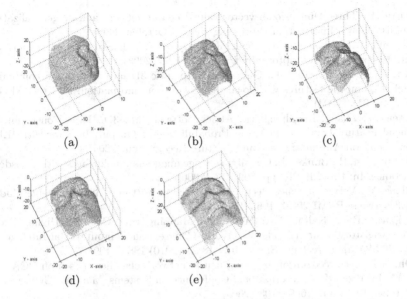

Fig. 4. Five faces of same person with different orientation and poses, Example 5

5 Inferences and Discussion

This paper presents the high dimensional engineering applications of neural networks. The generalization of 3D motion depicts that 3D vector valued network maintains the amplitude as well as phase information of each point in space during learning of 3D motion. In contrast conventional real-valued neural network administers 1D motion of signals; hence such 3D motion can not be learn and generalized by them. In the last few years more and more 2D face recognition algorithms are improved and tested. However, 3D models hold more surface information of the object that can be used for better face recognition or subject discrimination. The 3D information needs to be preprocessed after acquisition. This includes removal of artifacts, patching of holes, aligning of faces either by their center of mass or nose tip. The normalization and recognition of 3D face is presented in the paper. The work presented in this paper is basic and fundamental, but this give a direction to researchers to develop an efficient, fault tolerant, noise prone face recognition system with high dimensional neural network.

References

1. Tripathi, B.K., Kalra, P.K.: The Novel Aggregation Function Based Neuron Models in Complex Domain. Soft Computing 14(10), 1069–1081 (2010)
2. Tripathi, B.K., Kalra, P.K.: On Efficient Learning Machine with Root Power Mean Neuron in Complex Domain. IEEE Transaction on Neural Network 22(05), 727–738 (2011)

3. Nitta, T.: Three-Dimensional vector valued neural network and its generalization ability. Neural Information Processing 10(10) (October 2006)
4. Moreno, A.B., Sanchez, A., Velez, J.F., Dýaz, F.J.: Face recognition using 3D surface-extracted descriptors. In: Proc. IMVIPC (2003)
5. Xu, C., Wang, Y., Tan, T., Quan, L.: Automatic 3D face recognition combining global geometric features with local shape variation information. In: Proc. AFGR, pp. 308–313 (2004)
6. Chen, L., Zhang, L., Zhang, H., Abdel-Mottaleb, M.: 3D Shape Constraint for Facial Feature Localization Using Probabilistic-like Output. In: Proc. 6th IEEE Int. Conf. on Automatic Face and Gesture Recognition (2004)
7. Achermann, B., Bunke, H.: Classifying range images of human faces with Hausdorff distance. In: Proc. ICPR, pp. 809–813 (2000)
8. Blanz, V., Vetter, T.: Face Recognition Based on Fitting a 3D Morphable Model. IEEE Trans. PAMI 25(9), 1063–1074 (2003)
9. Tripathi, B.K., Kalra, P.K.: Functional Mapping with Complex Higher Order Compensatory Neuron Model. In: World Congress on Computational Intelligence (WCCI 2010), Barcelona, Spain, July 18-23 (2010) ISSN: 1098-7576
10. Oh, B.J.: Face recognition by using neural network classifiers based on PCA and LDA. In: Proc. IEEE International Conference on Systems, Man and Cybernetics, October 10-12, pp. 1699–1703 (2005)
11. Zhou, X., Bhanu, B.: Integrating face and gait for human recognition at a distance video. IEEE Transaction on System, Man and Cybernetics 37(5), 1119–1137 (2007)
12. Pantic, M., Patras, I.: Dynamics of facial expression: Recognition of facial actions and their temporal segments from face profile image sequences. IEEE Transaction on System, Man and Cybernetics 36(2), 433–449 (2007)
13. Hietmeyer, R.: Biometric identification promises fast and secure processing of air-line passengers. The International Civil Aviation Organization Journal 55(9), 10–11 (2000)
14. O'Tolle, A.J., Abdi, H., Jiang, F., Phillips, P.J.: Fusing face-verification algorithms and humans. IEEE Transaction on System, Man and Cybernetics 37(5), 1149–1155 (2007)
15. Abate, A.F., Nappi, M., Riccio, D., Sabatino, G.: 2D and 3D face recognition: A survey. Pattern Recognition Letters 28, 1885–1906 (2007)
16. Nitta, T.: An Extension of the Back-Propagation to Complex Numbers. Neural Networks 10(8), 1391–1415 (1997)
17. Lee, Y., Park, K., Shim, J., Yi, T.: 3D face recognition using statistical multiple features for the local depth information. In: Proc. ICVI (2003)

Knowledge Clustering Using a Neural Network in a Course on Medical-Surgical Nursing

José Luis Fernández-Alemán[1], Chrisina Jayne[2], Ana Belén Sánchez García[1], Juan M. Carrillo-de-Gea[1], and Ambrosio Toval[1]

[1] Faculty of Computer Science, Regional Campus of International Excellence "Campus Mare Nostrum", University of Murcia, Murcia, Spain
[2] Faculty of Engineering and Computing, University of Coventry, UK
chrisina.draganova@gmail.com, absanchezg@hotmail.com,
{aleman,jmcdg1,atoval}@um.es

Abstract. This paper presents a neural network-based intelligent data analysis for knowledge clustering in an undergraduate nursing course. A MCQ (Multiple Choice Question) test was performed to evaluate medical-surgical nursing knowledge in a second-year course. A total of 23 pattern groups were created from the answers of 208 students. Data collected were used to provide customized feedback which guide students towards a greater understanding of particular concepts. The pattern groupings can be integrated with an on-line (MCQ) system for training purposes.

Keywords: Neural network, clustering, nursing education.

1 Introduction

Multiple Choice Questions (MCQs) are an usual evaluation method to capture the performance of students. When students attempt such questions, invaluable data for understanding their learning process concerning a defined topic is generated. This simple depiction of their knowledge is normally lost. However, this data can be captured and automatically analyzed by a neural network. Then, lecturers are provided with groups of answers generated by the neural network, thus procuring a picture of their students' knowledge. Tutors can identify which concepts have been mastered and which have not been understood. This information can be used to create customized feedback for students. This feedback should not be confined to any particular question. In contrast, it should be obtained according to a set of common incorrect answers from the students to a set of questions on the given topic, in such a way that the learner is encouraged to think through the questions to resolve misconceptions and to gain insights independently. The time taken to create the feedback is well spent, because the feedback can be made available to any students.

This paper presents a generic method for intelligent analysis and grouping of student answers that is applicable to any area of study. It builds on previous

C. Jayne, S. Yue, and L. Iliadis (Eds.): EANN 2012, CCIS 311, pp. 385–394, 2012.

work of applying neural network based approach to the analysis of MCQs computer science tests [1]. The paper is structured as follows: after this introduction, Section 2 justifies the importance of this research by reviewing the related literature. Section 3 briefly describe the Snap-Drift Neural Network (SDNN), a neural network that supports the intelligent analysis and grouping of student answers to present diagnostic feedback. Section 4 presents the procedure followed to collect data from 208 nursing students in a medical-surgical nursing course. Section 5 discusses the results obtained. Finally, Section 6 draws some conclusions and outlines future work.

2   Literature Review

2.1   Learning Intelligent Analysis

Educational data mining is a newer sub-field of data mining which can provide insight into the behavior of lecturers, students, managers, and other educational staff and can be used to take better decisions about their educational activities [2]. Verdú et al. [3] propose a fuzzy expert system that uses a genetic algorithm in order to automatically estimate the difficulty of questions to be presented to students according to the students knowledge level. The system is successfully validated in a competitive learning system called QUESTOURnament, which is a tool integrated into the elearning platform Moodle.

Aparicio et al. [4] describes an Intelligent Information Access system which automatically detects significant concepts available within a given clinical case. This system allows students to gain understanding of the concepts by providing direct access to enriched related information from Medlineplus, Freebase, and PubMed. A trial was run with volunteer students from a second year undergraduate Medicine course. Two groups were formed: one group consisted of 26 learners who could freely seek information on the Internet, while the other group with 34 students was allowed to search for information using the developed tool. All students were provided with a clinical case history and a multiple choice test with medical questions relevant to the case. The results were slightly better in the experimental group, but were not statistically significant.

Zafra et al. [5] propose a more suitable and optimized representation based on multiple instance learning (MIL) in the context of virtual learning systems. The new representation is adapted to available information of each student and course, thus eliminating the missing values that make difficult to find efficient solutions when traditional supervised learning is used. The two learning frameworks, traditional and MIL representation, were compared using data from seven e-Learning courses with a total of 419 registered students. Significant statistically differences were found between the accuracy values obtained by different algorithms using MIL representation as compared to traditional supervised learning representation with single instance. Therefore, results showed that representation based on MIL is more effective than that of classical representation to predict student performance.

2.2   Learning Using MCQ

A number of advantages can be found in the use of MCQs [6], [7], [8], [9]: rapid feedback, automatic evaluation, perceived objectivity, easily-computed statistical analysis of test results, and the re-use of questions from databases as required, thus saving time for instructors. Moreover, MCQs are also known [10] by their ability for testing large numbers of students in a short time. MCQs have nonetheless been criticized [11], [12], [13], [14]: significant effort is required to construct MCQs, they only assess knowledge and recall, and are unable to test literacy, creativity and the synthesis and evaluation levels in the cognitive domain of Bloom's Taxonomy. However, some nurse-educators [15] [16] suggest that higher cognitive domains such as critical thinking skills can also be assessed with MCQs.

Although it can appear to be a simple task, evaluation of learning with MCQ formats are prone to construction errors [17]. Some studies [18] are found in the nursing literature claiming that MCQs can fulfill the criteria for effective assessment suggested by Quinn [19]: practicality, reliability, validity, and discrimination. However, significant effort is required in preparation to produce reliable and valid examination tools. Farley [20] estimates that it requires one hour to write a good MCQ. MCQs should be short, understandable and discriminating, with a correct grammar. They should avoid negative terms, over complicated or trick questions and ambiguity [21].

Our proposal aims at obtaining students groups (states of knowledge) by a neural network to prepare specific feedback which addresses misconceptions and guides students towards a greater understanding of particular concepts. To the best of the authors' knowledge, no other studies related to MCQs and formative evaluation have employed any similar form of intelligent analysis of the students' answers in the nursing field.

3   Clustering System

In this section, a clustering system based on SDNN to classify the students' answers and to gain insights into the students' learning needs is proposed. SDNN provides an efficient means of discovering a relatively small and therefore manageable number of groups of similar answers [1].

3.1   Snap-Drift Neural Networks (SDNN)

Neural networks-based clustering is based on an unsupervised type of learning used to separate a finite unlabeled data set into a finite and discrete set of "natural," hidden data structures [22]. Some of the most prominent neural network based clustering algorithms are the Self-Organising Map (SOM) [23] and the Adaptive Resonance Theory (ART) [24]. The Snap-Drift Neural Network (SDNN) is an unsupervised modal algorithm able to adapt rapidly by taking advantage of the complementary nature of each mode of learning (snap and drift) [25], [26]. The snap mode is based on the ART [24], while the drift mode is based on the learning vector quantization (LVQ) [23]. With the snap mode the neural network model learns the common features to all the patterns in the group, giving a high probability of rapid (in terms of epochs) convergence while the drift mode ensures that average generalised features are learnt and included in the final groupings [26].

Snap-drift is a modal learning approach [25] [26]. The SDNN switches the learning of the weights between two modes: "snap" which gives the angle of the minimum values (on all dimensions) and "drift" which gives the average angle of the patterns grouped under the neuron. Snapping ensures learning of a feature common to all the patterns in the group, while drifting tilts the weight vector towards the centroid angle of the group and ensures that an average, generalised feature is included in the final weight vector. The SDNN architecture has an input, hidden and output layers. The output winning neurons from the hidden layer act as input data to the output or so called selection layer. This layer is also subject to the snap-drift learning and has an activation threshold to ensure if a good match to an existing neuron is not found a new neuron is recruited, so the progress of learning determines the number of output groups. Detailed description of the snap-drift algorithm could be found for example in [25] [26].

3.2   Training the Neural Network

Before the neural network is endowed with capabilities for classifying, the SDNN must be trained with the students' responses to questions on a particular topic in a course. Each of the possible responses from the students is encoded into binary form, in preparation for presentation as input patterns for SDNN, as shown in Table 1. The training phase begins with the presentation of each input pattern corresponding to students' responses. The SDNN will learn to group the input patterns according to their general features. The groups are recorded, and represent different states of knowledge regarding a given topic, inasmuch as they contain the same incorrect and/or correct answers to the questions. The groups are sent to instructors in the form of templates of student responses.

The number of responses required to train the system so that it can generate the states of knowledge depends on having representative training data of the study domain. When SDNN is still creating new groups from new responses, more training data is required, because those new responses are different to previous responses. Therefore, the number of groups formed and the training of the system rely on the variation in student responses.

Table 1. Examples of input patterns for questions with five possible answers and two encoded responses

Codification	a - 00001; b - 00010 ; c - 00100; d - 01000 ; e - 10000
Response	Encoded response
[e, b, a, c]	[0,0,0,0,1,0,0,0,1,0,0,0,0,0,1,0,0,1,0,0]
[b, e, d, c, a]	[0,0,0,1,0,1,0,0,0,0,0,1,0,0,0,0,0,1,0,0,0,0,0,0,1]

4   Case Study

This section provides detailed information on the procedure followed to design and conduct an experiment to study the application of SDNN-based clustering in a nursing course.

4.1   Participants

SDNN has been used in a medical-surgical nursing course at the Catholic University of San Antonio. *Clinical Nursing I* (CN) is a second-year course which focuses on the processes to be the cause of illness, the pathophysiology of diverse health disorders and the nursing care to individuals with medical-surgical problems. Students attend 2 h/week of lectures in the first term and 14 hours of clinical skills practice. This term takes place during a 15-week period. The assignments proposed in this course cover different subjects: routes of medication administration, basic life support, wound care, drains, fluid balance, surgical instruments, central venous pressure measurement, surgical area behaviour, oxygen therapy and arterial blood gas, intravenous catheter insertion and care, urethral catheterization, enteral nutrition, breath sounds, and other related subjects.

4.2   Experiment

To investigate the effectiveness of SDNN, one experiment was designed and conducted during the first term of the academic year 2011/12. In the experiment, data was collected from 208 nursing students which performed one MCQ test at the end of the term. Most participants (82.2%) were female, with a mean age of 25.7 (SD: 7.36) years. A test consisting of ten five-choice questions related to surgical-medical nursing was prepared. The experiment was completed in a written exam session in which students were given 35 minutes time to take the test. Negative marking was used to improve the discrimination ability of the MCQ format. Table 2 shows one of the 10 questions used.

Table 2. Example of a MCQ

Diseases of the pleura:
- a. A patient with a dry pleurisy has high-grade fever
- b. Pleural effusion is a result of a chest trauma
- c. The smaller diameter tube available is used for the evacuation of a hemothorax in order to avoid causing pain to the patient
- d. Pneumothorax can be classified into two major categories: traumatic and spontaneous
- e. The onset of pneumothorax is usually insidious and slow

Facility index to inform us on the percentage of students that answer a question correctly was calculated for each question. Nine out of ten questions are between 25% and 75% which is generally considered as an acceptable range. Notice that the easier items was include at the beginning of the test. The discrimination index to attempt to differentiate the performance of one student from another

Table 3. Students' states of knowledge. FI: Facility index. DI: Discrimination index.

Question	q1	q2	q3	q4	q5	q6	q7	q8	q9	q10
FI	84%	29%	37%	70%	24%	55%	74%	71%	74%	52%
DI	0,35	0,21	0,39	0,41	0,41	0,33	0,39	0,50	0,46	0,55

Group	Size	q1	q2	q3	q4	q5	q6	q7	q8	q9	q10
1	32	c	d/n	b/e/n	c/n	*	d/n	b	d	e	*
3	7	c/d	d/n	*	c	*	d	b	n	*	*
4	6	c	n	n	c	n	n/d	b	n	n	d
5	1	n	a	b	e	n	b	b	n	e	b
6	1	d	d/n	c	b	n	e	n	n	d	d
7	1	a	d	b	d	d	b	n	n	d	d
8	68	c	e	*	c	*	d	b	d	e	d/n
9	6	c	e	*	c	*	n	b/e	*	e	n
12	14	c	*	*	b	*	*	b	d	e	*
13	3	c	*	*	b	*	*	b	*	*	*
16	1	b	n	e	c	a	d	e	d	e	d
17	6	c	e	*	*	*	n	e	d	e	d
18	2	*	e	*	d	n	*	b	d	e	*
19	3	c	e	b	n	e	n	b	e	*	d/n
20	1	c	d	c	n	b	d	b	d	c	d
22	9	c	d/n	b	c	*	*	*	d	*	*
23	2	c	*	n	*	e	*	b	*	e	*
25	13	c	e	*	*	*	d	*	d/n	e	*
26	14	c/n	d	*	*	e	d/n	*	*	e	d/n
27	5	d	e	b	c	*	d	b	d	*	n
28	2	*	e	*	*	*	d	*	*	e	n
29	9	d	e	*	c	*	d/n	b	d	e	*
30	2	c	e	*	c	*	*	e	n	*	n
31 (right)	2	c	d	b	c	c	d	b	d	e	d

was also obtained [19]. Nine out of ten questions achieved a discrimination index at 0.3 or above. Only question 3 obtained a discrimination index of 0.21, which shows that less knowledgeable students are getting more correct answers for question 2 than the more knowledgeable ones. It may indicate that the item was ambiguous.

The students' answers were used for training the SDNN. The neural network created a total of 23 pattern groups with five outlier groups (of size 1). The average group size is 9. Two groups (68 and 32) are particularly large, representing very common responses. These large groups include four or more questions which are given the same answer by the members of the group. Table 3 shows all the groups obtained by the SDNN. Let us take, for instance, group 3. All of its members chose the answer c to question 4, d to question 6, b to question 7, and *no answer* to question 8. The answers to the remaining questions vary within the group. All the students of this group answered c or d to question 1, d or *no answer* to question 2, and disparate answers to questions 3, 5, 9 and 10 (symbol *). Hence, the educator can easily spot the common mistakes in the groups of the student answers highlighted by the tool. Notice that each group is produced by the neural network on grounds of the commonality between the answers to some of the questions (to four of them in our example). In this experiment, with a 10-question test, the groups had between 3 and 10 answers in common.

The groups represent the behavior of students in terms of the states of knowledge produced by SDNN in the test carried out. For example, a student who receives feedback on state 23 (three correct answers), should progress via one

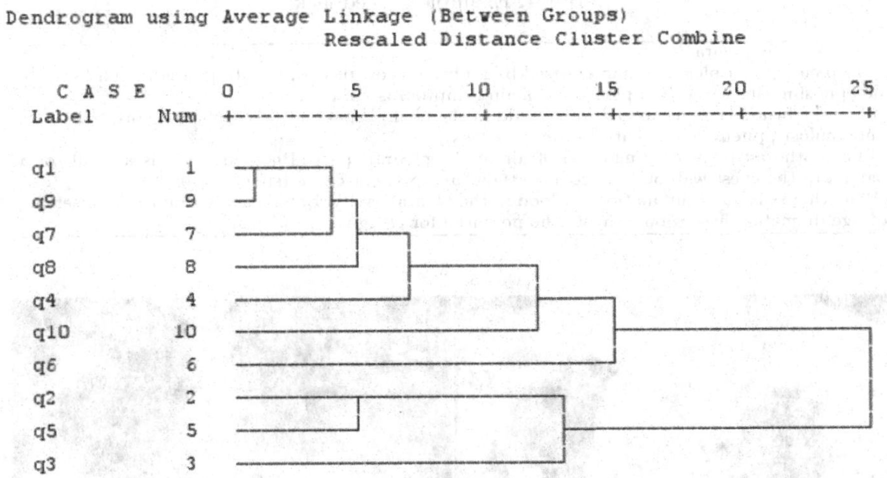

```
Dendrogram using Average Linkage (Between Groups)
                      Rescaled Distance Cluster Combine

    C A S E      0         5        10        15        20        25
    Label      Num  +---------+---------+---------+---------+---------+

    q1          1
    q9          9
    q7          7
    q8          8
    q4          4
    q10        10
    q6          6
    q2          2
    q5          5
    q3          3
```

Fig. 1. Dendrogram of a hierarchical cluster analysis

of the states in the next layer, such as state 1 (four correct answers), then state 8 (six correct answers), before reaching the "state of perfect knowledge" (state 31) which represents correct answers to all questions. In this example, students in state 1 do not understand concepts such as anamnesis or procedures such as pulmonary angiography, and are not able to raise key questions to obtain information useful in formulating a diagnosis and providing medical care to a patient with cough. Moreover, these students have difficulties in identifying breath sounds. In contrast, students in state 8 do know and understand how to perform a pulmonary angiography.

Finally, clustering was also applied to form groups of questions which were answered similarly (correctly or incorrectly) by the students. Figure 1 illustrates the similarity of the questions according to the answers of the students. For example, students who understood how to manage acute asthma (question 9) also were able to identify symptoms of dyspnea (question 1), and vice versa.

5 Discussion

The student response groups can help instructors identify misunderstood concepts, conceptual relations and understand the progress of the students. Instructors can prepare appropriate materials such as references to material that the student needs to read and feedback giving hints, which will be provided in subsequent face-to-face sessions. Then, the students have the opportunity to reflect on their answers and do some further reading. Notice that the student should not be told exactly which answer(s) is/are wrong because that would not encourage reflection and cognition. An example of feedback is provided in Table 4 and Figure 2.

Table 4. Example of feedback

Diseases of the pleura:
- Dry pleurisy is a pleurisy characterized by a fibrinous exudation, resulting in adhesion between the opposing surfaces of the pleura. Its main symptom is pain.
- Pleural effusion is an accumulation of fluid in the pleural space. It is secondary to other diseases (tuberculosis, pneumonia, heart failure, neoplasm)
- Pneumothorax is an accumulation of air in the pleural space. Pneumothorax is a result of a trauma to the chest wall or can occurs without an apparent cause (spontaneous).
- Hemothorax is an accumulation of blood in the pleural cavity. Blood can be drained by inserting a large diameter chest tube to avoid the potential for clogging.

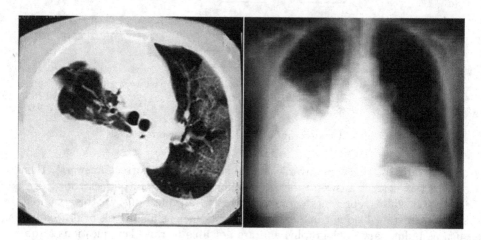

Fig. 2. Image illustrating a right sided pleural effusion as seen on an upright chest X-ray

On the other hand, the clustering of questions allows instructors to identify questions which may require mutually related knowledge. These relationships can help instructor to pinpoint collective students' misconceptions more easily and give more tailored feedback. Concept maps can be automatically created to show these relationships among concepts, to examine dependence relationships, and to organize and represent the student's knowledge. Question similarity it may also indicate that the questions measure the same or similar knowledge so they should be reformulated.

5.1   Limitations of the Study

The multiple-choice examination has been criticized for being artificial and not reflective of real life clinical situations. Authors argue, that while MCQs may have content validity, doubts can be raised about their predictive and discriminative value. However, this can be extended to any classroom assessment method to discriminate or predict performance in the clinical real life situation [27]. Therefore, MCQ-based intelligent assessment should be combined with other assessment methods to achieve an educational strategy which provide an accurate and comprehensive evaluation of student clinical practice performance. On the other hand, although negative marking may cause significant stress for students, it was used for ranking with a more normal distribution and therefore the absolute levels of marks reflects the true performance of the class better [10].

6   Conclusions and Future Work

In this paper, a knowledge clustering using a snap-drift neural network in a course on medical-surgical nursing has been described. The most innovative aspect of the proposal is the use of a intelligent analysis to discover groups of students of similar answers which represent different states of knowledge of nursing students, and groups of questions similarly answered by all of the students representing knowledge related. Feedback texts targeting the level of knowledge of individuals can be associated with each of the pattern groupings, taking into account the conceptual relations identified, to address misconceptions that may have caused the incorrect answers common to that pattern group. These data can be recorded in a database so that a tool can guide students towards a greater understanding of particular concepts, and instructors can monitor the learning progress of students. New student responses can be used to retrain the neural network. In the event that new refined groupings are created, they can be used by the educator to improve the feedback. Once designed, MCQs and feedbacks can be reused for subsequent cohorts of students.

References

1. Fernández-Alemán, J.L., Palmer-Brown, D., Jayne, C.: Effects of response-driven feedback in computer science learning. IEEE Trans. Education 54(3), 501–508 (2011)
2. Chen, C.M., Hsieh, Y.L., Hsu, S.H.: Mining learner profile utilizing association rule for web-based learning diagnosis. Expert Systems with Applications 33, 6–22 (2007)
3. Verdú, E., Verdú, M.J., Regueras, L.M., de Castro, J.P., García, R.: A genetic fuzzy expert system for automatic question classification in a competitive learning environment. Expert Systems with Applications 39, 7471–7478 (2012)
4. Aparicio, F., Buenaga, M.D., Rubio, M., Hernando, A.: An intelligent information access system assisting a case based learning methodology evaluated in higher education with medical students. Computers & Education 58, 1282–1295 (2012)
5. Zafra, A., Romero, C., Ventura, S.: Multiple instance learning for classifying students in learning management systems. Expert Systems with Applications 38, 15020–15031 (2011)
6. Epstein, M.L., Lazarus, A.D., Calvano, T.B., Matthews, K.A., Hendel, R.A., Epstein, B.B., Brosvic, G.M.: Immediate feedback assessment technique promotes learning and corrects inaccurate first responses. The Psychological Record 52, 187–201 (2002)
7. Higgins, E., Tatham, L.: Exploring the potential of multiple choice questions in assessment. Learn. and Teach. in Act. 2 (2003)
8. Kuechler, W.L., Simkin, M.G.: How well do multiple choice tests evaluate student understanding in computer programming classes? J. Inf. Syst. Educ. 14, 389–399 (2003)

9. Kreig, R.G., Uyar, B.: Student performance in business and economics statistics: Does exam structure matter? J. Econ. and Finance 25, 229–240 (2001)
10. Pamphlett, R., Farnill, D.: Effect of anxiety on performance in multiple choice examination. Med. Educ. 29, 297–302 (1995)
11. Wesolowsky, G.O.: Detecting excessive similarity in answers on multiple choice exams. J. Appl. Stat. 27, 909–921 (2000)
12. Paxton, M.: A linguistic perspective on multiple choice questioning assessment and evaluation. Assessment & Evaluation in High. Educ. 25, 109–119 (2000)
13. Nicol, D.: E-assessment by design: using multiple-choice tests to good effect. J. Further and High. Educ. 31, 53–64 (2007)
14. Masters, J.C., Hulsmeyer, B.S., Pike, M.E., Leichty, K., Miller, M.T., Verst, A.L.: Assessment of multiple-choice questions in selected test banks accompanying text books used in nursing education. J. Nurs. Educ. 40, 25–32 (2001)
15. Su, W.M., Osisek, P.J., Montgomery, C., Pellar, S.: Designing multiple-choice test items at higher cognitive levels. Nurse Educator 34, 223–227 (2009)
16. Morrison, S., Free, K.W.: Writing multiple-choice test items that promote and measure critical thinking. The Journal of Nursing Education 40, 17–24 (2001)
17. Holsgrove, G., Elzubeir, M.: Imprecise terms in UK medical multiple-choice questions: what examiners think they mean. Med. Educ. 32, 343–350 (1998)
18. Brady, A.M.: Assessment of learning with multiple-choice questions. Nurse Education in Practice 5, 238–242 (2005)
19. Quinn, F.: The Principles and Practice of Nurse Education, 4th edn. Stanley Thorne Ltd., Cheltenham (2000)
20. Farley, J.K.: The multiple-choice test: writing the questions. Nurse Educ. 14, 10–12, 39 (1989)
21. Holsgrove, G.J.: Guide to postgraduate exams: multiple-choice questions. British Journal of Hospital Medicine 48, 757–761 (1992)
22. Xu, R., Wunsch, D.: Survey of clustering algorithms. IEEE Transactions on Neural Networks 16, 645–678 (2005)
23. Kohonen, T.: Self-Organisation and Asssociative Memory, 3rd edn. Series in Information Sciences. Springer, Heilderberg (1989)
24. Carpenter, G.A., Grossberg, S.: Adaptive Resonance Theory. In: The Handbook of Brain Theory and Neural Networks, 2nd edn. MIT Press, Cambridge (2003)
25. Lee, S.W., Palmer-Brown, D., Roadknight, C.M.: Performance-guided neural network for rapidly self-organising active network management. Neurocomputing 61, 5–20 (2004)
26. Palmer-Brown, D., Draganova, C., Lee, S.W.: Snap-drift neural network for selecting student feedback. In: IJCNN, pp. 391–398 (2009)
27. Brady, A.M.: Assessment of learning with multiple-choice questions. Nurse Education in Practice 5 (2005)

A New Approach in Stability Analysis of Hopfield-Type Neural Networks: Almost Stability

Kaiming Wang*

School of Mathematics and Statistics, Xi'an Jiaotong University, Xi'an 710049, China,
School of Science, Chang'an University, Xi'an 710064, China

Abstract. In this paper,we presented a new stability concept for neural networks: almost stability. The necessary and sufficient conditions of almost stability of the Hopfield-type neural networks were proposed. Examples were also given to our conditions.

Keywords: almost stability, Hopfield neural network, almost positive definite.

1 Introduction

In applications of neural networks either as associative memories or as optimization solvers, stability of network is prerequisite. When neural networks are employed as associative memories, the equilibrium points represent the stored patterns, and, the stability of each equilibrium point means that each stored pattern can be retrieved even in the presence of noise. While when employed as an optimization solver, the equilibrium points of neural networks correspond to possible optimal solutions, and the stability of networks then ensures the convergence to optimal solutions. At the same time, stability of neural networks is fundamental for network designs, an unstable network system will be useless in practice. Due to these, stability analysis of neural networks has received extensive attentions in recent past years (see, e.g., [1–4]).

In the sense of Lyapunov's stability, an equilibrium point x^* of dynamical system is said to be stable if for any open ball centered at x^* there exists a neighborhood of x^* from which every trajectory will not escape from the ball, and moreover, x^* is said to be asymptotically stable if it is stable and attracts all points of some neighborhood of x^*. It is worth to note that, in these definitions, the "all" is required: either all trajectories initiated from some neighborhood are restricted in any given ball, or all points of some neighborhood are attracted. From the viewpoint of application, the "all" requirement is sometime unnecessary. In fact, in some practical neural networks, it is enough that there exist

---

* This work was supported by Nature Science Foundation of China under the contact no. 60970149; the Special Fund for Basic Scientific Research of Central Colleges, Chang'an University(CHD2011JC009).

C. Jayne, S. Yue, and L. Iliadis (Eds.): EANN 2012, CCIS 311, pp. 395–401, 2012.

'sufficiently many' trajectories restricted in any given ball, or/and 'sufficiently many' points are attracted. For example, when neural networks are employed as optimization solver, we only need to know which trajectories converge to the equilibrium point. However, a natural question is how to evaluate or measure the 'sufficiently many'.

Set measure, such as Lebesgue measure, may be an existing candidate that can be used to measure the 'sufficiently many'. In the sense of Lebesgue measure, A. Rantzer investigated in 2001 the almost globally asymptotical stability of nonlinear dynamical systems [5]. In his surprising research, a novel notion named density function is introduced, which can be considered as a dual part of Lyapunov function. In the later several years, many researchers applied Rantzer's 'dual theory' to optimal control to construct or seek controller (see, e.g., [6]). In 2008, Umesh Vaidya and Prashant G. Mehta [7] established a complete theory on almost stability for discrete dynamical system. In 2009, Pablo Monzón and Rafael Potrje [8] investigated the local Lyaponov's stability implication of almost global stability and gave several examples from which one can clearly separate the Lyapunov's stability from almost stability.

In this paper, we investigate the almost stability of Hopfield-type neural networks:

$$\frac{dx_i(t)}{dt} = -a_i x_i(t) + \sum_{j=1}^{n} w_{ij} g_j(x_j(t)) + I_i, \ i = 1, 2, \cdots, n \tag{1}$$

where $x_i(t)$ are the neural states, a_i the positive time constants, w_{ij} the weights between neurons i and j, g_i the transfer functions, and I_i the external inputs.

The outline of this paper is as follows. In section 2, preliminaries and notations of almost stability of the nonlinear systems are reviewed. In section 3, Applying Rantzer's dual theory to the networks, necessary conditions and sufficient conditions are obtained for network dynamic to be almost globally stable. Moreover, in section 4, some examples are presented to demonstrate the condition given in the main proposition.

2   Preliminaries and Notation

Consider the nonlinear dynamical system in $\mathbb{R}^n$:

$$z'(t) = F(z(t)), \ t \geq 0 \tag{2}$$

where F satisfies the existence and uniqueness conditions of solutions. In the following, we assume that z^* is an equilibrium point of (2), and denote $z(z_0, t)$ by the solution initiated at z_0. Moreover, we denote $B(z, r)$ by the closed ball of radius r centered at z (i.e., $B(z, r) = \{z \in \mathbb{R}^n, \|z - z_0\| \leq r, r > 0\}$), and the space of measures on $\mathbb{R}^n$ by $\mathcal{M}(\mathbb{R}^n)$. Then, following Rantzer A [5] and Umesh Vaidya and Prashant G. Mehta [7], we have definitions as follows.

Definition 1. *The equilibrium point z^* of system (2) is siad to be almost stable with respect to a measure $\mu \in \mathcal{M}(\mathbb{R}^n)$ if for each $\epsilon > 0$ there exists a $\delta > 0$ such that $\mu\{z_0 : \|z_0 - z^*\| < \delta, \|z(z_0, t) - z^*\| \geq \epsilon\} = 0$ for all $t \geq 0$.*

Definition 2. *[5, 7] The equilibrium point z^* of system(2) is said to be almost asymptotically stable with respect to measure $\mu \in \mathcal{M}(\mathbb{R}^n)$ if z^* is almost stable with respect to μ and there exists an open set $\tilde{G} \subset \mathbb{R}^n$ including z^* such that $\mu\{z_0 \in \tilde{G} : \lim_{t \to +\infty} z(z_0, t) \neq z^*\} = 0$. Furthermore, z^* is globally stable with respect to μ if $\tilde{G} = \mathbb{R}^n$.*

To the authors' knowledge, the first result on almost asymptotic stability is due to Rantzer A.

Lemma 1. *[5] Assume that $F \in C^1(\mathbb{R}^n, \mathbb{R}^n)$ and $F(0) = 0$. If there exists a non-negative function $\rho \in C^1(R^n \backslash \{0\}, R)$ such that*
   (1) the function $z \mapsto \rho(z)F(z)/|z|$ is integrable on $\{z \in \mathbb{R}^n, |z| \geq 1\}$, and
   (2) $\nabla \cdot (\rho F)(z)) > 0$ for almost all $z \in \mathbb{R}^n$ in the sense of Lebesgue's measure,
where $\nabla\cdot$ represents the diversion operator, then the equilibrium point $z^* = 0$ is almost globally stable.

In this paper we apply the above result to analysis of the almost stability of neural network (1). So, we need to transfer the networks into the form of (2). Let $x^* = (x_1^*, x_2^*, \cdots, x_n^*)^T \in \mathbb{R}^n$ be an equilibrium point of network (1), that is, x_i^* solves the following equations

$$a_i x_i^* = \sum_{j=1}^{n} w_{ij} g_j(x_j^*) + I_i, \ i = 1, 2, \cdots, n. \tag{3}$$

Make the variable transformation $z_i(t) = x_i(t) - x_i^*$ in the equations (1), and introduce the following notations:

$$z = (z_1, z_2, \cdots, z_n)^T,$$
$$f_i(z_i(t)) = g_i(z_i(t) + x_i^*) - g_i(x_i^*),$$
$$A = diag(a_1, a_2, \cdots, a_n),$$
$$W = (w_{ij})_{n \times n},$$
$$f(z) = (f_1(z_1), f_2(z_2), \cdots, f_n(z_n))^T.$$

Then, the network (1) is equivalently transformed into the compact form

$$z'(t) = -Az(t) + Wf(z(t)), \tag{4}$$

which can be further casted into the class of systems as (2) with $F(z) = -Az + Wf(z)$.

In the present investigation, we do not put additional assumption on the network parameters except the following traditional requirement on the transfer functions g_i:

(H) For each $i = 1, 2, \cdots, n$, g_i is Lipschitz continuous and monotonous increasing, that is, there is a constant ν_i such that, for all ξ_1 and ξ_2,

$$0 \leq (g_i(\xi_1) - g_i(\xi_2))(\xi_1 - \xi_2) \leq \nu_i(\xi_1 - \xi_2)^2, \tag{5}$$

or equivalently, for each $i = 1, 2, \cdots, n$,

$$0 \le sf_i(s) \le \nu_i s^2, \forall s \in \mathbb{R}. \tag{6}$$

By the qualitative theorem of ordinary differential equation, the assumption (H) can sufficiently assure the existence and uniqueness of solutions of the network system (1).

3   Almost Stability Analysis

In this section, under the existence assumption of the equilibrium of the network we considered, we focus on the analysis of its almost stability. In fact, one can find most research on the existence of equilibrium.

The next theorem proposed a method to construct a density function of such hopfield neural networks as(4). By constructing density function, the "sufficiently many" is described.

Theorem 1. *If $f(z)$ is locally Lipschitzian in $\mathbb{R}^n$, $\|f(z)\| \le N\|z\|$ (here, N is a positive constant) and there exists a positive number α such that $\left(z(t)^T P z(t)\right)^{-\alpha}$ is integrable at infinity and $\mu\{z : Q(z) \le 0\} = 0$, then the equilibrium $z = 0$ of network(4) is almost globally stable with respect to measure μ, where*

$$Q(z) = h(z)P + PA + A^T P - PWG(z) - G(z)^T W^T P,$$

$P > 0$ is a positive definite matrix, $G(z) = diag(\frac{f_1(z_1)}{z_1}, \frac{f_2(z_2)}{z_2}, ..., \frac{f_n(z_n)}{z_n})$, $z_i \ne 0 (i = 1, 2, \cdots, n)$, and $h(z) = \alpha^{-1} \sum_{i=1}^{n} (-a_i + w_{ii} f_i'(z_i))$ when f_i is differentiable at z_i.

Proof.   Let $F(z(t)) = -Az(t) + Wf(z(t))$. Since $g(z)$ is Lipschitzian by (5), $g(z)$ is differentiable almost everywhere(a.e.) in $\mathbb{R}^n$, and naturally $f(z)$ is differentiable a.e. in $\mathbb{R}^n$. We choose the density function ρ in Lemma1 as $\rho(z(t)) = \left(z(t)^T P z(t)\right)^{-\alpha}$, then we have

$$\nabla \cdot (\rho F)(z(t))$$
$$= (\nabla \rho \cdot F)(z(t)) + (\rho \nabla \cdot F)(z(t))$$
$$= \alpha \rho^{\frac{\alpha+1}{\alpha}}(z(t)) \left(\alpha^{-1} \rho^{-\frac{1}{\alpha}} divF(z(t) - \nabla \rho^{-\frac{1}{\alpha}} \cdot F(z(t))\right)$$
$$= \alpha \rho^{\frac{\alpha+1}{\alpha}}(z(t)) \left(\alpha^{-1} divF(z(t))(z(t)^T P z(t)) - 2z(t)^T P\left(-Az(t) + Wf(z(t))\right)\right)$$
$$= \alpha \rho^{\frac{\alpha+1}{\alpha}}(z(t)) z^T(t) \left(h(z)P + PA + A^T P - PWG(z) - G(z)^T W^T P\right) z(t)$$
$$= \alpha \left(z(t)^T P z(t)\right)^{-(\alpha+1)} z^T(t) Q(z(t)) z(t)$$

By Lemma1, when

$$\mu\{z : (Q(z)) \le 0\} = 0, \tag{7}$$

the equilibrium point $z = 0$ of network(4) is almost globally stable.

Remark 1. (i)$Q(z)$ in (7) is an almost positive definite matrix with size $n \times n$. This kind of matrix does exist, for example,

$$Q(z) = \begin{pmatrix} x^2 & -xy \\ xy & 1 \end{pmatrix}, z = \begin{pmatrix} x \\ y \end{pmatrix} \in \mathbb{R}^2$$

is an almost positive definite matrix as $\det(Q(z)) = 0$ if and only if $x = 0$ and $\mu\{z \in \mathbb{R}^2 : \det(Q(z)) = 0\} = 0$.

(ii)The almost positive definite is not difficult to judge. In fact, methods to judge positive definite matrix are also suitable here.

Following Monzon's work([8]), the stability information of neural network(4) can be divided into the following three conditions by the sign of $h(0)$.

Theorem 2. *If $f(z(t)) \in C^1(\mathbb{R}^n, \mathbb{R}^n)$, and $h(z), Q(z)$ are similar to Theorem 1, then the stability information of $z = 0$ in (4) can be divided into such three situations:*
(i)when $h(0) > 0$, $z = 0$ is not almost globally stable, even not locally stable.
(ii)when $h(0) < 0$,and there exists $Q(z)$ satisfying Theorem1, $z = 0$ is locally asymptotically stable. Furthermore, if $h(z) \leq 0$ and $h(z) = 0$ only if $z = 0$, then $z = 0$ is globally asymptotically stable.
(iii)when $\mu\{z, Q(z) \leq 0\} = 0$ and $\{z, Q(z) \leq 0\} \neq \emptyset$, then $h(0) = 0$ and $z = 0$ is almost globally stable but not asymptotically stable.

Proof. (i)Suppose that $h(0) > 0$, then the Jacob matrix $\frac{\partial F}{\partial z}(0)$ at least has one positive eigenvalue, so the stable manifold on $z = 0$ is of a dimension less than n and $\mu\{z_0 : \lim_{t \to \infty} z(z_0, t) = 0\} = 0$, i.e., the equilibrium cannot be almost globally stable.

(ii)when $h(0) < 0$, and there exists $Q(z)$ satisfying Theorem1, we can choose Lyapunov function as $V(z) = z^T P z$ and an open ball $U(z) = \{z \in \mathbb{R}^n : \|z\| < \delta, \delta > 0\}$, then by Lyapunov stability theorem, $z = 0$ is locally asymptotically stable and $U(z)$ is the attraction basin of the equilibrium $z = 0$.

(iii)by the Definition 2 and Theorem 1, the conclusion holds obviously.

From the condition (6), we have an important information of this neural network, that is, $0 \leq f_i'(0) \leq \nu_i, i = 1, 2, ..., n$, so the following conclusion could be easily derived.

Corollary 1. *If $f(z) \in C^1(R^n, R^n)$ and $\min_{i=1}^n w_{ii} = b > 0, \exists f_k'(0) \neq 0, 0 \leq k \leq n$ such that $bf_k'(0) > \sum_{i=1}^n a_i$, then*

$$\mu\{z_0 : lim_{t \to \infty} z(z_0, t) = 0\} = 0$$

that is, it's impossible for the equilibrium of neural network (4) to be almost globally stable.

4   Examples

In order to verify the theoretical assertions, we propose the following examples.

Example 1. Consider the following system

$$x'(t) = -x(t) + \arctan(x(t)). \tag{8}$$

Here we have $A = 1, W = 1,$ and let $P = 1, \alpha \geq 1.5, x \neq 0,$

$$Q(x) = -\frac{\alpha^{-1} x^2}{1 + x^2} + \frac{2(x - \arctan(x))}{x} > 0$$

then by Theorem1, system(8)is an almost globally stable system with $h(0) = 0$.
Furthermore, since $h(x) = -\frac{x^2}{1+x^2} \leq 0$, and $h(x) = 0$ only if $x = 0$, system(8)is globally asymptotically stable.

Example 2. Consider the networks

$$\frac{dz_i(t)}{dt} = -z_i(t) + \sum_{j=1}^{4} w_{ij} f(z_j(t)), i = 1, 2, 3, 4 \tag{9}$$

where $f(z) = \frac{1-e^{-cz}}{1+e^{-cz}} \cdot \frac{1+ke^{c'(z-h)}}{1+e^{c'(z-h)}}, c, c', h > 0, k < 0,$

$$W = \begin{pmatrix} -0.5 & -0.2 & 0.95 & 0.8 \\ 2 & 3 & 0.1 & 1.2 \\ 0.6 & 0.4 & 0.2 & 0.1 \\ -0.9 & -0.4 & 0.15 & -0.7 \end{pmatrix}. \quad F(z) = -diag(1,1,1,1)(z_1, z_2, z_3, z_4)^T +$$

$W(f(z_1), f(z_2), f(z_3), f(z_4))^T$, so by simply calculating,

$divF(z) = -4 + \sum_{i=1}^{4} w_{ii}$
$\frac{2ce^{-cz_i} + 2c(k+1)e^{(c'-c)z_i - c'h} + 2cke^{(2c'-c)z_i - 2c'h} + (k-1)c'e^{c'(z_i-h)} - (k-1)c'e^{(c'-2c)z_i - c'h}}{(1+e^{-cz_i})^2(1+e^{c'(z_i-h)})^2}$.

By (i) of Theorem2 we know when $c[1 + (k+1)e^{-c'h} + ke^{-2c'h}]\sum_{i=1}^{4} w_{ii} = 2c[1 + (k+1)e^{-c'h} + ke^{-2c'h}] > 2$, i.e., $c[1 + (k+1)e^{-c'h} + ke^{-2c'h}] > 1$, the system will never be almost globally stable. Especially, let $c = c', k = -1$ and c_0 is the value such that $c_0 - c_0 e^{-2c_0 h} - 1 = 0$, then by (i) of Theorem2when $c > c_0$, the system will never be almost globally stable.

Example 3. Consider the network

$$\begin{pmatrix} x_1'(t) \\ x_2'(t) \end{pmatrix} = \begin{pmatrix} -1 & 0 \\ 0 & -\frac{1}{4} \end{pmatrix} \begin{pmatrix} x_1(t) \\ x_2(t) \end{pmatrix} + \begin{pmatrix} 3 & -2 \\ -\frac{9}{4} & 5 \end{pmatrix} \begin{pmatrix} f(x_1(t)) \\ f(x_2(t)) \end{pmatrix}$$

where $f(x) = \frac{1-e^{-x}}{1+e^{-x}}$. Because $f'(0) = \frac{1}{2}, b = 3, a_1 + a_2 = \frac{5}{4}, bf'(0) > a_1 + a_2$, by Corollary1, the equilibrium $x = 0$ can't be almost globally stable.

5 Conclusion

In this paper, the Hopfield-type neural networks has been analyzed. Lebesgue measure is used to evaluate the 'enough' many in the condition of almost stability. When the complete of stable set of the considered equilibrium has zero measure in R^n, this equilibrium is said to be almost globally stable with respect to this measure. Necessary conditions and sufficient conditions of almost globally stability are derived. Examples are presented to test the conditions.

Since most of the practical neural networks have time delay, we'll try to apply this 'dual' idea to time-delayed hopfeild-type neural network and Cohen-Grossberg neural networks in the future research.

Acknowledgments. The author would like to express her gratitude to the referee for his very helpful and detailed comments.

References

1. Liu, Y., Wang, Z., Liu, X.: Asymptotic stability for neural networks with mixed time-delays: The discrete-time case. Neural Networks 22, 67–74 (2009)
2. Zhang, X.-M., Han, Q.-L.: New Lyapunov-Krasovski functionals for global asymptotic stability of delayed neural networks. IEEE T. Automat. Contr. 20(2), 533–539 (2009)
3. Qiao, H., Peng, J., Xu, Z.: Nonlinear measures: a new approach to exponential stability analysis for Hopfield-type neural networks. IEEE T. Neural Networ. 12(2), 360–370 (2001)
4. Mak, K.L., Peng, J.G., Xu, Z.B., Yiu, K.F.C.: A new stability criterion for discrete-time neural networks: Noniear spectral radius. Chaos Soliton Fract. 31, 424-436-1190 (2007)
5. Rantzer, A.: A dual to Lyapunov's stability theorem. Syst. Control Lett. 42(3), 161–168 (2001)
6. Prajna, S., Parrilo, P.A., Rantzer, A.: Nonlinear control synthesis by convex optimization, IEEE T. Automat. Contr. 49(2), 117–128 (2004)
7. Vaidya, U., Mehta, P.G.: Lyapunov measure for almost everywhere stability. IEEE T. Automat. Contr. 30(1), 307–323 (2008)
8. Monzon, P., Potrje, R.: Local implication of almost global stability. Dynam. Syst. 24(1), 109–115 (2009)

A Double Layer Dementia Diagnosis System Using Machine Learning Techniques

Po-Chuan Cho and Wen-Hui Chen[*]

Graduate Institute of Automation Technology,
National Taipei University of Technology, Taipei, Taiwan
whchen@ntut.edu.tw

Abstract. Studies show that dementia is highly age-associated. Early diagnosis can help patient to receive timely treatment and slow down the deterioration. This paper proposed a hierarchical double layer structure with multi-machine learning algorithms for early stage dementia diagnosis. Fuzzy cognitive map (FCM) and probability neural networks (PNNs) were adopted to give initial diagnosis at based-layer, and then Bayesian networks (BNs) was used to make a final diagnosis at top-layer. Diagnosis results, "proposed treatment" and "no treatment required" can be used to provide self-testing or secondary dementia diagnosis to medical institutions. To demonstrate the reliability of the proposed system, a clinical data provided by the Cheng Kung University Hospital was examined. The accuracy of this system was as high as 83%, which showed that the proposed system was reliable and flexible.

Keywords: Dementia diagnosis, machine learning, probability neural networks, fuzzy cognitive map, Bayesian networks.

1 Introduction

Since the structure of the society is becoming aging, financial and social problems derived from the increasing number dementia patients are becoming more and more significant. Carrying issues have become one of the important problems in the future. Dementia is resulting in the decline of brain aging. It is an inevitable illness for an aging society. Its early symptoms are irreversible, and there is no cure to the illness while medication can only be used to delay its deterioration. If treated in the early stages of dementia, the better efficacy it is. Thus, early diagnosis for dementia has become an important issue.

Current early stage dementia diagnosis relies on the results based on pathology characteristic or cognitive diagnosis test. Pathology characteristic can be detected by neuroimaging. Magnetic resonance imaging (MRI) was used to examine the change of neuron-structure in [1], [2]. The electroencephalography (EEG) was used to analyze event-related potentials (ERPs) and detected early stages of dementia in [3], [4].

---

[*] Corresponding author.

C. Jayne, S. Yue, and L. Iliadis (Eds.): EANN 2012, CCIS 311, pp. 402–412, 2012.

Patel et al. combined both EEG and MRI imaging to better the accuracy to detect early stages of dementia [5]. However, those instruments are not adequate for detecting dementia because the cost of testing is very expensive, the testing process is too long and uncomfortable as they are invasive.

Along with diagnosis method, good treatment has to be accurate and easy to perform. The advantages of using cognitive tests to diagnose early stage of dementia are fast and easy performance; yet in reality, it is difficult for paramedics to be in contact with patients and promote the tests because of general old people's dislike for visiting hospitals. The mere access to perform tests is through untrained relatives who do not fully understand the scales. This has caused inaccuracy of test results. Machine learning algorithms provide a new solution to this problem. Through information technology, paramedics have better access to approach the patients' life and can discover the abnormal cognitive in early stage. Furthermore, machine learning approaches can provide professional medical knowledge. The combination with online system can provide a high accuracy and easy used way of early stage dementia diagnosis.

At present, there are many machine learning approaches used to detect dementia: Sun et al. adopted Bayesian networks to classify the dementia and solve problems of data loss [6]. Mazzocco and Hussain used logic regression model to improve the accuracy of Bayesian networks [7]. Support vector machine was used to classify three different cognitive test data in [8], while Sun et al. chose to use Bayesian belief networks to diagnose dementia [9]. Moreover, an online assessment tool [10] for patients and families to validate dementia was built by John Hopkins University. Patients will receive results and suggestions after taking the test.

However dementia diagnosis involves evaluation of various cognitive abilities, which causes physicians have different interpretation about the test results. It is difficult to reach good accuracy with the use of only one machine learning algorithm. Therefore, how to combine different algorithms benefits and make a suitable decision is quite important.

Different from above mentioned studies, this study proposed a hierarchical double layer structure system for early stage dementia detection. The goal of this system is try to combine different machine learning method's benefits to give more reliable and accuracy diagnosis. The initial diagnosis was obtained through two based-layer approaches: fuzzy cognitive map (FCM) and probability neural networks (PNNs), and then Bayesian networks algorithm was adopted at top-layer to make the final diagnosis. Through this structure, diagnosis results are extensive and intuitional. The system also gives better performance than traditional interviews.

2   Introduce of Cognitive Tests

MMSE [11] is mainly used to detect dementia in Taiwan. It's also one of the references for health insurance. The test includes: disorientation, attention, memory language, spoken language, understanding of spoken language, and construction. There are 30 questions in total. Each question is 1 point. That's 30 points in total.

Although MMSE is widely used, its test results are easily influenced by education and cultural background.

CASI [12] combines the commonly used Japanese Hasegawa dementia screen scale (HDSS), MMSE, and modified mini-mental state test (3Ms). It can be applied cross nationally. There have been English, Japanese, Chinese, Spanish, and Vietnamese versions. The test includes 25 questions of 6 cognitive aspects. There are 100 points in total. The flow of this test is that it takes too long to take the test and it needs a trained person to hold the test.

CERAD [13] is used in 16 hospitals in the USA to determine dementia. It includes two part, clinical performance and neuropsychological evaluation. Neuropsychological performance is commonly used on psychiatric patients. It improves the accuracy for detecting dementia.

In this study, ten cognitive meters are select as system input features. Those features contain MMSE scores, CASI scores and eight scores of CERAD which are verbal fluency, object naming, visual ability, visual recognition, verbal registration, verbal recall, verbal recognition, and visual spatial ability. Please note that the total cognitive meters are ten features, and FCM and PNNs select different features from these ten features.

3   Methodology

This section will explain the hierarchical double layer structure as shown in Fig. 1. Ten attributes are combined as a vector and input to FCM and PNNs. Three diagnosis result: normal (NM), mild cognitive impairment (MCI), and Alzheimer's disease (AD) will be diagnosis in both algorithms. According to the diagnosis result of FCM and PNNs, the final diagnosis (FD) result will be made by Bayesian networks.

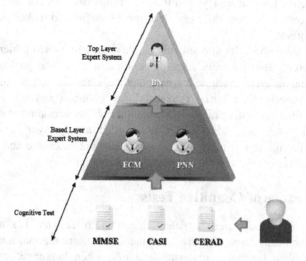

Fig. 1. The proposed double layer structures

3.1   Fuzzy Cognitive Map

FCM [14] is derived from graph theory. Plus fuzzy cognitive information, FCM has expanded its application toward to disease diagnosis region. FCM is an unsupervised learning algorithm, which means it doesn't need training procedure; therefore, it is suit to apply to online system. Inherited from graph theory, each node in FCM is regarded as a concept in the professional field denoted as A. Each links between nodes is represented as weight, denoted as w. The value of weight shows the relations between connected concepts, where w_{ij} denote the influence value form A_i to A_j. Through constructing all concepts as a relation map, FCM can interpretation the professional knowledge of the specific field. In this study the proposed dementia diagnosis FCM is shown in Fig. 2. The weight value is corresponding to initial weight listed in table. 1.

To build a well-designed FCM, we have to visit the doctors to transform the doctor's dementia diagnosis knowledge with fuzzy linguistic variable. Three concept membership functions: NM, MCI, and AD, three weight membership functions: influenced, probably influenced, and not influenced and three fuzzy rules listed followed are adopted to compute the initial conditions of FCM.

1) If node A changes, then node B changes, and the inference to the degree of influence of C. Where A, B and C are linguistic variables.
2) If there is a minor change in A, then there will be a minor change for node B. This suggests less influence between A and B.
3) If there is a big change in node A, then there will be a big change for node B. This suggests bigger influence between node A and B.

Once we transform the doctor knowledge as FCM initials values, the input cognitive meters vector can be inference through fuzzification, fuzzy inference and defuzzification. Additionally, Kannappan et al. proposed the training weights methods by using non-linear Hebbian learning algorithm and the combination with FCM. NHL adjusts with non-zero weights [15]. Its advantage is the weights aren't decided in the beginning but change with time. This will improve the efficiency and flexibility of FCM. Once the FCM initial value is decided, NHL algorithm will update the weights and get new concept values. There won't be a result until the concept value meets the convergence conditions. Its final output value and other cut-off scores will be compared and the diagnosis results will be there. The step of FCM and NHL algorithm are described as follow:

1) Set the initial conditions of FCM which includes initial concept value $A_i^{(k=1)}$ and initial weight values $w_{ij}^{(k=1)}$. Where i is the concept number, k is the iteration number. Moreover, there are two initial parameters: the learning rate $\eta = 0.001$, and updating rate $\gamma = 0.98$.
2) Repeat step 3~5 until it meets convergence.
3) Update FCM concept value with (1), where denoted the updated concept value and f(x) is defined as (2).

4) Adopt NHL algorithm to update the weights with (3) and (4).
5) We defined desired output concept (DOC) as the decision concept. With (5), if the continues DOC difference meets the stop conditions or less than the threshold value e=0.0001, then the procedures is stopped.

The constrained DOC value represents the diagnosis result of FCM. At last, 0.9725 is the threshold of MCI and AD classes, and 0.9828 is the threshold of MCI and NM classes.

$$A_i^{(k+1)} = f\left[A_i^{(k)} + \sum_{j=i}^{N} A_i^{(k)} \cdot w_{ij}^{(k)} \right] \tag{1}$$

$$f(x) = \frac{1}{1+e^{-\lambda x}} \tag{2}$$

$$w_{ji}^{(k)} = \gamma \cdot w_{ji}^{(k-1)} + \eta_k A_i^{(k-1)} \left(A_j^{(k-1)} - \left(w_{ji}^{(k-1)} \right) w_{ji}^{(k-1)} A_j^{(k-1)} \right) \tag{3}$$

$$\Delta w_{ji} = \eta_k A_i^{(k-1)} A_j^{(k-1)} - \Delta w_{ji}^{(k-1)} A_i^{(k-1)} \tag{4}$$

$$\left| DOC_i^{(k-1)} - DOC_i^{(k)} \right| < e \tag{5}$$

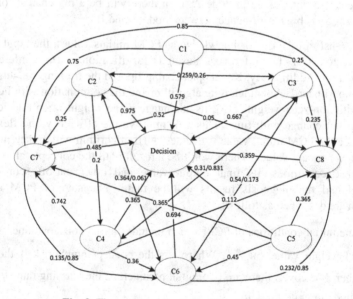

Fig. 2. The proposed dementia diagnosis FCM

Table 1. Proposed FCM initial weights

	C1	C2	C3	C4	C5	C6	C7	C8	C9
C1	0	0	0	0	0	0	0.75	0.25	0.579
C2	0	0	0.259	0.2	0	0.364	0.25	0.05	0.975
C3	0	0.26	0	0.31	0	0.54	0.52	0.235	0.667
C4	0	0	0.831	0	0	0	0.36	0.742	0.365
C5	0	0	0	0	0	0	0.45	0.365	0.112
C6	0	0.061	0.173	0	0	0	0.135	0.232	0.694
C7	0	0	0	0	0	0	0	0.85	0.458
C8	0	0	0	0	0	0	0.85	0	0.359
C9	0	0	0	0	0	0	0	0	0

3.2   Probability Neural Networks

In 1990, Donald Specht introduced probabilistic neural networks, a solution to problem classification and it is a decision making neural networks that has the ability to learn [16]. Unlike back-propagation neural networks that need multiple iterations learning to reach convergence, probabilistic networks uses one-pass learning system and it has the advantage of short calculating time. In addition, the PNNs are based on the idea that the winner has the best probabilistic rates in the alternative networks. With such strong classification accuracy, it has made probabilistic neural networks a reliable tool for classification in the field of medication.

The PNNs this study uses is shown as Fig. 3. It is divided to four layers. The input layer links training samples through radial basis. The pattern layer operates supervised learning training through (6), where X is the input vector and W_i denotes the ith training samples. It finds similarity in inner products. Then, through a non-linear function (7), values are converted between 0~1 before output to the summation layer. The parameter σ is the spread value of PNNs model. In PNNs, each sample is considered a kernel. Different categories of probability density can be evaluated through Parzen window estimator. The accumulation layer can be calculated by (8).

Since PNNs is a supervised learning algorithm, it needs existing cases as samples for training experts PNNs model. The only parameter of PNNs model is the spread parameter. In training process, different spread parameters will be tested and the one with highest accuracy will be selected. New samples will be normalized to values between 0 and 1, and then compared with training models in the PNNs model. Then it will calculate the value of the nodes in hidden layers before it completes diagnosis.

$$D_i^2 = (W_i - X)^t (W_i - X) \tag{6}$$

$$f(x) = \exp\left[-\frac{x}{2\sigma^2}\right] \tag{7}$$

$$f(x) = \frac{1}{(2\pi)^{(p+1)/2}\sigma^{(p+1)}} \cdot \frac{1}{n}\sum_{i=1}^{n}\exp\left[-\frac{(x-x^i)^T(x-x^i)}{2\sigma^2}\right] \tag{8}$$

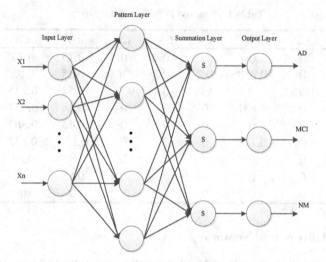

Fig. 3. The proposed PNNs model

3.3   Bayesian Networks

We adopt Bayesian networks as our top-layer system. Bayesian networks belong to the undirected acyclic graph in the graphic theory. It inherits the benefits of graphic model with intuition represent way of representing uncertainty of problems with probabilities. Additionally, Bayesian model can help reduce the complexity of conditional probability and let user easily design the structure of models. The Bayesian model is defined as $\Theta = \{\chi, \varepsilon\}$, where $\chi = \{X_1, \dots X_n\}$ represents n nodes, and ε represents the links. Each node represents a random variable which can be continuous or discrete. Each link $\varepsilon = \{X_i \rightarrow X_j, X_i, X_j \in \chi, i \neq j\}$ represents the relation between nodes. The arrow points form parent nodes to child nodes, which are recorded as the condition probability table. Through the chain rule, Bayesian theory can break down the joint probability to different conditional probability factor with multiplication, as shown in (9). When $i=1$ the node's prior is $P(X_1)$. To infer the conditional probability, calculate with (10), where $Pa(X_i)$ are the parent nodes of X_i. It is known that if the variables are independent, calculate with (11).

The proposed Bayesian networks structure as shown in Fig. 4, and the corresponding condition probability is calculated with (12).

$$P(X_1, \dots, X_n) = \prod_{i=1}^{n} P(X_i \mid X_{i-1}, \dots, X_1) \qquad (9)$$

$$(X_i \mid X_{i-1}, \dots, X_1) = P(X_i \mid Pa(X_i)) \qquad (10)$$

$$P(X_1, \dots, X_n) = \prod_{i=1}^{n} P(X_i \mid Pa(X_i)) \qquad (11)$$

$$P(X_{FD} \mid X_{FCM}, X_{PNN}) =$$
$$\frac{P(X_{FD}, X_{FCM}, X_{PNN})}{P(X_{FCM}, X_{PNN})} = \tag{12}$$
$$\frac{P(X_{FD})P(X_{FCM} \mid X_{FD})P(X_{PNN} \mid X_{FD}, X_{FCM})}{P(X_{FCM})P(X_{PNN} \mid X_{FCM})}$$

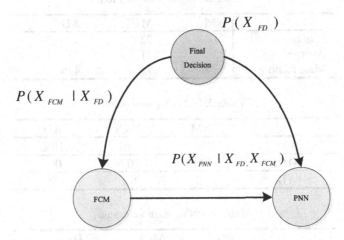

Fig. 4. The proposed Bayesian networks structure

4   Experiment Result

To verify the accuracy of the proposed systems, the data set form Cheng Kung University Hospital [17] are adopted. The data set contained 81 subjects, and subject's basic information is listed in table 2. The number of subject in NM, MCI and AD are 21, 22, and 38 respectively. The average age is around 71-73. The dataset especially avoid the gender balance problem, the male and female ratio is controlled near to 50%. Furthermore, both MCI and AD subjects are adopted SPECT image examination, and s diagnosed by professional doctors.

Due to the PNNs is a supervised learning method, the PNNs model has to be trained before testing. Leave one out validation skill was adopted to calculate the PNNs accuracy. The classified result of FCM and PNNs are listed in table 3 and table 4, where the diagonal numbers represented the positive true accuracy.

The accuracy of single approach FCM or PNNs are 74% and 69% respectively which are acceptable compared to the 40% with classified using MMSE threshold value. It shows that single algorithm can achieve better performance than common threshold based method. However the accuracy in MCI class is 50% and 45% in FCM and PNNs. To increase the total accuracy, the proposed system is used, where the accuracy is listed in table 5. The accuracy of MCI is up to 57% and the accuracy of NM and AD classes are also increased. This result proves the double layer algorithms can fusion the benefit of different learning strategy algorithms. Furthermore, if we set a rule to make the final decision with comparing the accuracy between FCM and PNNs, the accuracy is 69%, which is still worse than Bayesian networks.

In addition, if the system diagnosis the AD and MCI class as "no treatment require", and the AD class into "proposed treatment", then the accuracy will be up to 85%.

Table 2. Subject's information

	NM	MCI	AD
Number	21	22	38
Average Age	72	71	73
Male Ratio	52%	77%	42%

Table 3. FCM diagnosis result

	NM	MCI	AD
NM	76%	14%	10%
MCI	50%	50%	0%
AD	5%	8%	87%

Table 4. PNNs diagnosis result

	NM	MCI	AD
NM	71%	19%	10%
MCI	32%	45%	23%
AD	0%	18%	82%

Table 5. Accuracy rate of different methods

	FCM	PNNs	Rule Based	BNs
Total Accuracy	74%	69%	69%	83%
NM	76%	71%	80%	86%
MCI	50%	45%	9%	57%
AD	86%	82%	97%	94%

5   Conclusion

This study proposed a hierarchical double layer structure of multi machine learning approaches to diagnose early stage dementia. Two popular machine learning approaches: FCM and PNNs are used in based-layer, and the final diagnosis is made with Bayesian networks. The data provided by National Cheng Kung University Medical Center is adopted for actual test. The test samples suggest the accuracy is up to 83%. It is proved that this system can surely help to detect and diagnose. In addition, this system can be operated online and is good for domestic use. It does help to detect dementia for early treatment. The contributions of this study are:

1) Provide an online assisting system to help evaluate dementia earlier and easier.
2) The proposed system is highly accurate and stable since it combines with three popular approaches.
3) Diagnosis is made through computer inference. Tested data is not only limited in certain cognitive test. It is easy to promote and build the database.

We will try to add more machine learning algorithms at based-layer to increase the system flexibility to deal with variety user cases. Furthermore, we will cooperate with doctors to apply to real dementia cases in the future.

Acknowledgment. The authors would like to thank Dr. Shen-Ing Liu at the Department of Psychiatry, Mackay Memorial Hospital, Taiwan for helpful discussions concerning this study, and the National Science Council of Taiwan for the financial support. This work was partially supported by the projects sponsored by the National Science Council, Taiwan, under the Grant No. 100-2221-E-027-014 and 99-2628-E-027-007.

References

1. Studholme, C., Cardenas, V., Song, E., Ezekiel, F., Maudsley, A., Weiner, M.: Accurate Template-based Correction of Brain MRI Intensity Distortion with Application to Dementia and Aging. IEEE Trans. Med. Imag. 23, 99–110 (2004)
2. Duchesne, S., Caroli, A., Geroldi, C., Barillot, C., Frisoni, G.B., Collins, D.L.: MRI-Based Automated Computer Classification of Probable AD Versus Normal Controls. IEEE Trans. Med. Imag. 27, 509–520 (2008)
3. Ahiskali, M., Polikar, R., Kounios, J., Green, D., Clark, C.M.: Combining Multichannel ERP for Early Diagnosis of Alzheimer's Disease. In: 4th International IEEE/EMBS Conference on Neural Engineering, pp. 522–525. IEEE Press, Antalya (2009)
4. Lai, C.L., Lin, R.T., Liou, L.M., Liu, C.K.: The Role of Event-related Potentials in Cognitive Decline in Alzeimer's Disease. Clin. Neur. 121, 194–199 (2010)
5. Patel, T., Polikar, R., Davatzikos, C., Clark, C.M.: EEG and MRI Data Fusion for Early Diagnosis of Alzimer's. In: 30th Annual International Conference of the IEEE Engineering in Medicine and Biology Society, Vancouver, pp. 1757–1760 (2008)
6. Sun, Y., Tang, Y., Ding, S., Lv, S., Cui, Y.: Diagnose the Mild Cognitive Impairment by Constructing Bayesian Network with Missing Data. Exp. Syst. with Appl. 38, 442–449 (2011)
7. Mazzocco, T., Hussain, A.: Novel Logistic Regression Models to Aid the Diagnosis of Dementia. Exp. Syst. with Appl. 39, 3356–3361 (2012)
8. Fung, G., Stoeckel, J.: SVM Feature Selection for Classification of SPECT Images of Alzheimer's Disease Using Spatial Information. Knowle. and Info. Syst. 11, 243–358 (2007)

9. Wei, W., Visweswaran, S., Cooper, G.F.: The Application of Naive Bayes Model Averaging to Predict Alzheimer's Disease from Genome-wide Data. J. Amer. Med. Info. Associ. 18, 370–375 (2011)
10. Dementia Risk Assessment, http://alzcast.org/memorysurvey/
11. MMSE, http://ntuh.sicu.org.tw/upload/sicu_protocol/f115513.doc
12. CASI, http://inspired-aging.nursing.ncku.edu.tw/eer/flash/boldman/a/A4scale-2other-CASI.html
13. CERAD, http://cerad.mc.duke.edu/Default.html
14. Kosko, B.: Fuzzy Cognitive Map. Int. J. Man-Mach. Stud. 24, 65–75 (1986)
15. Kannappan, A., Tamilarasi, A., Papageorgiou, E.I.: Analyzing the Performance of Fuzzy Cognitive Maps with Non-linear Hebbian Learning Algorithm in Predicting Autistic Disorder. Exp. Syst. with Appl. 38, 1282–1292 (2011)
16. Specht, F.D.: Probability Neural Networks and the Polynomial Adaline as Complementary Techniques for Classification. IEEE Trans. Neur. Net. 1, 111–121 (1990)
17. Pa, M.C., Hsiao, S.S.: Navigation Ability in Advancing Alzheimer's Disease Patients. Doctoral thesis (2002), http://ndltd.ncl.edu.tw/

A Hybrid Radial Basis Function and Particle Swarm Optimization Neural Network Approach in Forecasting the EUR/GBP Exchange Rates Returns

Georgios Sermpinis[1], Konstantinos Theofilatos[2], Andreas Karathanasopoulos[3], Efstratios Georgopoulos[4], and Christian Dunis[5]

[1] University of Glasgow Business School, University of Glasgow, Adam Smith Building, Glasgow, G12 8QQ, UK
[2] Department of Computer Engineering and Informatics, University of Patras, Greece
[3] London Metropolitan Business School, London Metropolitan University, 31 Jewry Street, London, EC3N 2EY, UK
[4] Technological Educational Institute of Kalamata, 24100, Kalamata, Greece
[5] Liverpool Business School, JMU, John Foster Building, 98 Mount Pleasant, Liverpool L3 5UZ
georgios.sermpinis@glasgow.ac.uk, theofilk@ceid.upatras.gr,
a.karathanasopoulos@londonmet.ac.uk, sfg@teikal.gr,
christian.dunis@hpwmg.com

Abstract. The motivation for this paper is to introduce in Finance a hybrid Neural Network architecture of Adaptive Particle Swarm Optimization and Radial Basis Function (ARBF-PSO) and a Neural Network fitness function for financial forecasting purposes. This is done by benchmarking the ARBF-PSO results with those of three different Neural Networks architectures and three statistical/technical models. As it turns out, the ARBF-PSO architecture outperforms all other models in terms of statistical accuracy and trading efficiency in the examined forecasting task.

Keywords: PSO, RBF Neural Networks, Financial Forecasting, Trading, Leverage, Transaction Costs.

1   Introduction

Neural networks (NN) are considered an emergent technology with an increasing number of real-world applications including finance [1]. Their numerous limitations and contradictory empirical evidence around their forecasting power have created scepticism about their use among practitioners. The motivation for this paper is present a hybrid Neural Network architecture of Particle Swarm Optimization and Adaptive Radial Basis Function (ARBF-PSO) and its application in forecasting the EUR/GPB exchange rates returns. This method tries to overcome some of the classical Neural Networks limitations. More specifically our proposed architecture is fully adaptive something that decreases the numbers of parameters that the practitioner needs to experiment while on the other hand it increases the forecasting

C. Jayne, S. Yue, and L. Iliadis (Eds.): EANN 2012, CCIS 311, pp. 413–422, 2012.
© Springer-Verlag Berlin Heidelberg 2012

ability of the network. In our study we benchmark our proposed algorithm with a Multi-Layer Percepton (MLP), a Recurrent Neural Network (RNN), a Psi Sigma Neural Network (PSI), an autoregressive moving average model (ARMA), a moving average concergence/divergence model (MACD) plus a naïve strategy in a forecasting and trading simulation of the EUR/GBP European Central Bank (ECB) daily fixing.

Moreover, we introduce a fitness function for our NNs that not only minimizes the MSE of our forecasts but also increase their profitability. This is crucial in financial applications where statistical accuracy is not always synonymous with financial profitability of the derived forecasts.

Our proposed methodology, which is an extension of the algorithm proposed by Ding et. al. [6] for financial forecasting purposes has not been applied to Finance yet. However, several scientists have applied other NNs algorithms to the task of forecasting the exchange rates with ambiguous empirical evidence. In [2] they apply Higher Order Neural Networks in forecasting the AUD/USD exchange rate with a 90% accuracy while Hussain et al. [3] forecast with the same class of NNs the EUR/USD, the EUR/GBP and the EUR/JPY exchange rates with a similarly good statistical accuracy. Panda and Narasimhan [4] use a single hidden layer feedforward NN to produce statistical accurate forecasts of the INR/USD exchange rate having several linear autoregressive models'as benchmarks. On the other hand, Kiani and Kastens [5] forecast the GBP/USD, the CAD/USD and the JPY/USD exchange rates with feedforward and recurrent NNs having as benchmarks several ARMA models. In their application, NNs outperform in statistical terms their ARMA benchmarks in forecasting the GBP/USD and USD/JPY but not in forecasting the USD/CAD exchange rate. More recently, Khashei et. al. [6] propose a hybrid NN-ARIMA model which demonstrates promising empirical results in forecasting the IRR/USD exchange rate and Dunis et. al. [7] compared several NNs and autoregressive models in trading the EUR/USD exchange rate. Their results demonstrate the forecasting superiority of a class of NNs, the Psi Sigma, which is able to capture higher order correlation within their dataset.

As it turns out the ARBF-PSO algorithm does remarkably well and outperforms all other models in terms of statistical accuracy and trading efficiency. It seems that its adaptability and flexibility allows it to outperform in our forecasting competition compared with the more 'traditional' MLP, RNN and PSI models. These results provide the first empirical evidence around the utility of the ARBF-PSO in finance and forecasting.

2   The EUR/GBP Exchange Rates and Related Financial Data

The European Central Bank (ECB) publishes a daily fixing for selected EUR exchange rates: these reference mid-rates are based on a daily concentration procedure between central banks within and outside the European System of Central Banks, which normally takes place at 2.15 p.m. ECB time. The reference exchange rates are published both by electronic market information providers and on the ECB's website shortly after the concentration procedure has been completed. Although only a reference rate, many financial institutions are ready to trade at the EUR fixing and it is therefore possible to leave orders with a bank for business to be transacted at this level.

The ECB daily fixings of the EUR exchange rates are therefore tradable levels which makes using them a more realistic alternative to, say, London closing prices and these are the series that we investigate in this paper. We examine the ECB daily fixings of the EUR/GBP since their first trading day on 4 January 1999 until 29 April 2011. The data period is partitioned as follows.

Table 1. The total dataset

Name of period		Beginning	End
Total dataset	3158	4 January 1999	29 April 2011
Training dataset	2645	4 January 1999	30 June 2009
Out-of-sample dataset [Validation set]	513	1 August 2009	29 April 2011

The observed time series are non-normal (the Jarque-Bera statistics [8] confirms this at the 99% confidence level) containing slight skewness and high kurtosis. They are also nonstationary and hence we decided to transform them into a stationary daily series of rates of return using the formula:

$$R_t = \ln\left(\frac{P_t}{P_{t-1}}\right) \qquad (1)$$

where R_t is the rate of return and P_t is the price level at time t. The stationary property of the return series is confirmed at the 1% significance level (ADF and PP test statistics).

The summary statistics of the EUR/GBP returns series reveal positive skewness and high kurtosis. The Jarque-Bera statistic confirms again that the two return series are non-normal at the 99% confidence level. These two return series will be forecasted from our models.

In the absence of any formal theory behind the selection of the inputs of a neural network, we conduct neural networks experiments and a sensitivity analysis on a pool of potential inputs in the training dataset in order to help our decision. Based on these experiments and the sensitivity analysis we select as inputs the sets of variables that provide the higher trading performance for each network in the in-sample period. These sets of inputs under study are presented in Appendix.

3   Proposed Model

A radial basis function neural network (RBFNN) is a feedforward neural network where hidden units do not implement an activation function, but a radial basis function. An RBFNN approximates a desired function by superposition of nonorthogonal, radially symmetric functions. They have been proposed by Broomhead and Lowe [9] as an approach to improve accuracy of artificial neural networks while decreasing training time complexity.

The hybrid methodology proposed in the present paper is an extension of the hybrid algorithm proposed by Ding et. al. [2]. In this algorithm the Particle Swarm Optimization (PSO) methodology was used to locate the parameters C_i of the RBFNN while in parallel locating the optimal number for the hidden layers of the network. This methodology is extended in our proposed algorithm in order to increase accuracy, make it appropriate for predicting financial time series and avoid the time consuming step of optimizing the parameters of PSO.

The PSO algorithm, proposed by Kennedy and Eberhart [10], is a population based heuristic search algorithm based on the simulation of the social behavior of birds within a flock. In PSO, individuals which are referred to as particles are placed initially randomly within the hyper dimensional search space. Changes to the position of particles within the search space are based on the social-psychological tendency of individuals to emulate the success of other individuals.

In our approach PSO searches for the optimal values of the parameters c_i and for the optimal number of hidden neurons which should be used for our network. Each particle i is initialized randomly to have m_i (within a predefined interval starting from the number of inputs until 100) hidden neurons and is represented as shown in equation (2):

$$C^i = \begin{bmatrix} c_{11}^i & c_{12}^i & \dots & c_{1d}^i \\ c_{21}^i & c_{22}^i & \dots & c_{2d}^i \\ \cdot & \cdot & \dots & \cdot \\ \cdot & \cdot & \dots & \cdot \\ \cdot & \cdot & \dots & \cdot \\ c_{m^i 1}^i & c_{m^i 2}^i & \dots & c_{m^i d}^i \\ N & N & N & N \\ \cdot & \cdot & \dots & \cdot \\ \cdot & \cdot & \dots & \cdot \\ \cdot & \cdot & \dots & \cdot \\ N & N & N & N \end{bmatrix} \quad (2)$$

where N is a large number to point that it does not represent an RBF center.

Furthermore each particle is assigned initially with a random velocity matrix to move within the search space. It is this velocity matrix that drives the optimization process, and reflects both the experiential knowledge of the particle and socially exchanged information from the particles neighborhood. The form of the velocity matrix for every particle is described in the equation below:

$$V^i = \begin{bmatrix} v^i_{11} & v^i_{12} & \cdots & v^i_{1d} \\ v^i_{21} & v^i_{22} & \cdots & v^i_{2d} \\ \cdot & \cdot & \cdots & \cdot \\ \cdot & \cdot & \cdots & \cdot \\ \cdot & \cdot & \cdots & \cdot \\ v^i_{m^i 1} & v^i_{m^i 2} & \cdots & v^i_{m^i d} \\ N & N & \cdots & N \\ \cdot & \cdot & \cdots & \cdot \\ \cdot & \cdot & \cdots & \cdot \\ \cdot & \cdot & \cdots & \cdot \\ N & N & \cdots & N \end{bmatrix} \tag{3}$$

From the centers of its particle described in equation (2) we deployed the Moody-Darken approach [11] to compute the RFB's widths.

At this point of the algorithm the centers and the widths of the RBFNN have been computed. The computation of its optimal weights w_i is accomplished by solving the equation:

$$w^i = (H_i^T \cdot H_i)^{-1} \cdot H_i^T \cdot Y \tag{4}$$

where

$$H_i = \begin{bmatrix} \phi^i_1(x_1) & \phi^i_2(x_1) & \cdots & \phi^i_{m^i}(x_1) \\ \phi^i_1(x_2) & \phi^i_2(x_2) & \cdots & \phi^i_{m^i}(x_2) \\ \cdot & \cdot & \cdot & \cdot \\ \cdot & \cdot & \cdot & \cdot \\ \cdot & \cdot & \cdot & \cdot \\ \phi^i_1(x_{n_1}) & \phi^i_2(x_{n_1}) & \cdots & \phi^i_{m^i}(x_{n_1}) \end{bmatrix}$$

where n1 is the number of training samples.

The computation of $(H_i^T \cdot H_i)^{-1}$ is computationally hard when the rows of Hi are highly dependent. In order to solve this problem we filtered the in-sample dataset and when the mean absolute distance of two training samples is less than 10-3 from the mean values of their input values then randomly we do not use one of them in our final training set. By this way, the algorithm becomes faster while maintaining its high accuracy.

Next, our novel multi-objective fitness function (5) was used for evaluating the performance of its particle. The constraint 10^{-2} was used to state that me networks simplicity is a secondary goal compared to its profitability.

Fitness= Annualized_Return − MSE − 10^{-2} * number_of_hidden_neurons (5)

Iteratively, the position of each particle is changed by adding in it its velocity vector and the velocity matrix for each particle is changed using the equation below:

$$V_{i+1} = w * V_i + c_1 * r_1 * (C^i_{pbest} - C_i) + c_2 * r_2 * (C^i_{gbest} - C_i) \tag{6}$$

where w is a positive-valued parameter showing the ability of each particle to maintain its own velocity, C^i_{pbest} is the best solution found by this specific particle so far, C^i_{gbest} is the best solution found by every particle so far, c_1 and c_2 are used to balance the impact of the best solution found so far for a specific particle and the best solution found by every particle so far in the velocity of a particle. Finally, r_1, r_2 are random values in the range of [0,1] sampled from a uniform distribution.

Ideally, PSO should explore the search space thoroughly in the first iterations and so the values for the variables w and c_1 should be kept high. For the final iterations the swarm should converge to an optimal solution and the area around the best solution should be explored thoroughly. Thus, c_2 should be valued with a relatively high value and w, c_1 with low values. In order to achieve the described behavior for our PSO implementation and to avoid getting trapped in local optima when being in an early stage of the algorithm's execution we developed a PSO implementation using adaptive values for the parameters w, c_1 and c_2. Equations (7), (8) and (9) mathematically describe how the values for these parameters are changed through PSO's iterations helping us to endow the desired behaviour in our methodology.

$$w(t) = (0.4/n2) * (t-n)2 + 0.4 \tag{7}$$

$$c_1(t) = -2 * t/n + 2.5 \tag{8}$$

$$c_2(t) = 2 * t/n + 0.5 \tag{9}$$

where t is the present iteration and n is the total number of iterations.

In order to enhance the optimization of the RBFNN structure we applied two more operators in our hybrid method in addition to the classical ones of the PSO. The first operator is used to add a hidden neuron in every particle with a probability equal to 0.1. The second operator reduces the hidden neurons by one with a probability equal to 0.1. The probabilities which were applied for increasing and decreasing the hidden neurons were set as equal to reassure that the algorithm is not further biased towards larger or smaller architectures. Furthermore, a high probability equal to 0.8 was assigned to the fact of not changing the network's architecture in order to enforce the algorithm to explore thoroughly the potential of existing architectures in the population before investigating different ones.

For the initial population of particles we use a small value of 30 particles and the number of iterations used was 100 combined with a convergence criterion. Using this termination criterion the algorithm stops when the population of the particles is deemed as converged. The population of the particles is deemed as converged when the average fitness across the current population is less than 5% away from the best fitness of the current population. Specifically, when the average fitness across the current population is less than 5% away from the best fitness of the population, the diversity of the population is very low and evolving it for more generations is unlikely to produce different and better individuals than the existing ones or the ones already examined by the algorithm in previous generations.

In summary the novelty of our algorithm lies in the following points. First of all, all PSO parameters in our approach are adaptive using equations (7), (8) and (9) making our method appropriate for usage by non experts while at the same time avoiding the risky and time consuming trial and error approach of optimizing the parameters of a NN. Moreover, in order to adapt our methodology to our specific task of financial time series forecasting we use the multi-objective fitness function (5) to select more profitable predictors for our ARBF-PSO and the other three NNs algorithms retained.

4   Empirical Results

4.1   Statistical Performance

The out-of-sample statistical performance of our models is presented in table 2 below for the EUR/GBP exchange rates. For all four error statistics retained (RMSE, MAE, MAPE and Theil-U) the lower the output, the better the forecasting accuracy of the model concerned.

Table 2. Out-of-sample statistical performance – EUR/GBP

	NAIVE	ARMA	MACD	MLP	RNN	PSI	ARBF-PSO
MAE	0.008	0.006	0.006	0.006	0.005	0.005	0.004
MAPE	243.6%	226.9%	254.3%	196%	159%	162.9%	147.3%
RMSE	0.008	0.007	0.008	0.003	0.005	0.004	0.004
Theil-U	0.999	0.773	0.768	0.534	0.419	0.423	0.404

From the results above, we note that ARBF-PSO retains its forecasting superiority in the out-of-sample sub-period for the statistical measures applied. Once more, the RNN and the PSI present the second best performance with MLP having the third best forecasts in terms of statistical accuracy.

In order to test if our best model in terms of statistical measures produces forecasts that are statistically significant and superior to its counterparts, we apply the Diebold-Mariano statistic [12] for predictive accuracy for both MSE and MAE loss. The results of the Diebold-Mariano statistic, comparing the ARBF-PSO network with its benchmarks for the EUR/GBP exchange rates, are summarized in tables 3.

Table 3. Diebold-Mariano Statistics – EUR/GBP

	NAIVE	ARMA	MACD	MLP	RNN	PSI
MSE	-8.443	-7.174	-5.721	-4.112	-2.817	-2.374
MAE	-8.321	-8.662	-6.983	-4.872	-4.983	-3.475

From the above table we note that the null hypothesis of equal predictive accuracy is rejected for all comparisons and for both loss functions at a 5% confidence interval, since all the test statistics are above the critical value of 1.96. Moreover, the statistical superiority of the ARBF-PSO forecasts is confirmed as for both loss functions the realizations of the Diebold-Mariano (1995) statistic are negative.

4.2   Trading Performance

The trading strategy applied is to go or stay 'long' when the forecast return is above zero and go or stay 'short' when the forecast return is below zero. The 'long' and 'short' EUR/GBP position is defined as buying and selling Euros at the current price respectively.

Since we consider the EUR/GBP time series as a series of middle rates, the transaction costs is one spread per round trip. With an average exchange rate for EUR/GBP of 0.887 for the out-of-sample period, a cost of 1 pip is equivalent to an average cost 0.012% per position for the EUR/GBP.

In table 4 we present the results of our method in the out-of-sample period.

Table 4. Out-of-sample trading performance – EUR/GBP

	NAIVE	ARMA	MACD	MLP	RNN	PSI	ARBF-PSO
Maximum Drawdown	-11.22%	-17.99%	-12.65%	-14.54%	-11.85%	-9.87%	-7.56%
Positions Taken (annualized)	134	125	88	125	144	158	92
Transaction Costs (annualized)	1.92%	1.31%	1.09%	1.89%	2.22%	2.42%	1.37%
Annualized Return (including costs)	2.77%	7.53%	-0.67%	12.43%	17.42%	17.99%	26.43%
Information Ratio (including costs)	0.30	0.80	-0.04	1.12	1.74	1.81	2.97

We observe from the last two rows of tables 5 that the ARBF-PSO confirms its trading superiority in the out-of-sample period. This is consistent with the superior statistical performance of our proposed methodology compared to its other NN and statistical benchmarks. A superiority that can be attributed to the proposed network algorithm that seems to excel in recognizing the pattern of the two series under study compared to more traditional models. It is also worth noting that ARBF-PSO forecasts present a remarkably low maximum drawdown of -7.56% in the out-of-sample period. Thus the maximum potential losses of an investor are almost 4 times lower than the annualized profit in the out-of-sample period. Concerning our benchmarks models, we note that PSI and RNN present the second and the third best performance in terms of trading efficiency. The results of the more traditional models are mixed the naïve and ARMA slightly positive annualized returns after transaction costs.

Concerning our proposed fitness function of equation (5), the results from the statistical and trading evaluation of our NNs forecasts seem promising. Firstly we note that all our NNs present significant profits after transaction costs for the financial index under study in the out-of-sample period. Moreover, we did not note any large inconsistencies in our NNs statistical and trading performance for both sub-periods. Large inconsistencies could indicate that the training of our NNs is biased to either statistical accuracy or trading efficiency, something that could possibly lead to profitable in-sample forecasts but disastrous out-of-sample results.

5   Concluding Remarks

In this paper, we introduced a hybrid Neural Network architecture of Particle Swarm Optimization and Adaptive Radial Basis Function (ARBF-PSO) and a neural network fitness function for financial forecasting purposes. We applied the proposed architecture to the task of forecasting the one day ahead return of the EUR/GBP ECB daily fixings and benchmark its results with a   Multi-Layer Perceptron (MLP), a Recurrent Neural Network (RNN), a Psi Sigma Neural Network (PSI), an autoregressive   moving   average   model   (ARMA),   a   moving   average convergence/divergence model (MACD) plus a naïve strategy. More specifically, the trading and statistical performance of all models was investigated in a forecast and trading simulation over the period January 1999 to March 2011 using the last two years for out-of-sample testing.

In terms of results, the ARBF-PSO outperformed all its benchmarks for statistical accuracy and trading efficiency for the out-of-sample period. This superiority is further confirmed by the Diebold-Mariano test which proves that the ARBF-PSO forecasts are statistically different and superior to its benchmarks.

Concerning our proposed fitness function for NNs, we noted that all our networks produce substantial profitability in the out-of-sample period. Moreover, we observed that the ranking of our models is almost the same in statistical and trading terms. This allows us to argue that our NNs were trained in a way that allowed them to increase not only their statistical accuracy but also their trading efficiency.

Finally, we note that all models presented significant positive returns after transaction costs, a result that questions the market efficiency hypothesis for the EUR/GBP exchange rates.

References

1. Lisboa, P., Vellido, A.: Business Applications of Neural Networks. In: Lisboa, P., Edisbury, B., Vellido, A. (eds.) Business Applications of Neural Networks: The State-of-the-Art of Real-World Applications, pp. vii–xxii. World Scientific, Singapore (2000)
2. Ding, H., Xiao, Y., Yue, J.: Adaptive Training of Radial Basis Function Networks Using Particle Swarm Optimization Algorithm. In: Wang, L., Chen, K., Ong, Y.S. (eds.) ICNC 2005, Part I. LNCS, vol. 3610, pp. 119–128. Springer, Heidelberg (2005)

3. Fulcher, J., Zhang, M., Xu, S.: The Application of Higher-Order Neural Networks to Financial Time Series. In: Kamruzzaman, J., Begg, R., Sarker, R. (eds.) Artificial Neural Networks in Finance and Manufacturing, Hershey, PA. Idea Group, London (2006)

4. Panda, C., Narasimhan, V.: Forecasting exchange rate better with artificial neural network. Journal of Policy Modeling 29(2), 227–236 (2007)

5. Kiani, K., Kastens, T.: Testing Forecast Accuracy of Foreign Exchange Rates: Predictions from Feed Forward and Various Recurrent Neural Network Architectures. Computational Economics 4(32), 383–406 (2008)

6. Khashei, M., Bijari, M., Ardali, G.: Improvement of Auto-Regressive Integrated Moving Average models using Fuzzy logic and Artificial Neural Networks (ANNs). Neurocomputing 4(6), 956–967 (2009)

7. Dunis, C., Laws, J., Sermpinis, G.: Modelling and trading the EUR/USD exchange rate at the ECB fixing. The European Journal of Finance 16(6), 541–560 (2010)

8. Jarque, C.M., Bera, A.: A Test for Normality of Observations and Regression Residuals. International Statistical Review 55(2), 163–172 (1987)

9. Broomhead, S., Lowe, D.: Multivariate Functional Interpolation and Adaptive Networks. Complex Systems 2, 321–355 (1988)

10. Kennedy, J., Eberhart, R.C.: Particle Swarm Optimization. In: Proceedings of the IEEE International Conference on Neural Networks, vol. 4, pp. 1942–1948 (1995)

11. Moody, J., Darken, C.J.: Fast learning in networks of locally tuned processing units. Neural Computation 1(2), 281–294 (1989)

12. Diebold, F.X., Mariano, R.S.: Comparing Predictive Accuracy. Journal of Business and Economic Statistics 13, 253–263 (1995)

Forecasting and Trading the High-Low Range of Stocks and ETFs with Neural Networks

Hans-Jörg von Mettenheim and Michael H. Breitner

Information Systems Research, Leibniz Universität Hannover
Königsworther Platz 1, 30167 Hannover, Germany
{mettenheim,breitner}@iwi.uni-hannover.de
http://www.iwi.uni-hannover.de

Abstract. Intraday trading has some appealing characteristics. For example, overnight gap risks are greatly reduced. Intraday trading strategies tend to achieve better risk adjusted returns. However, academic literature on intraday trading strategies is relatively scarce compared to a significant amount of literature based on daily closing data. This may be partly related to the increased difficulty of dealing with intraday data. In the present paper we expand on a novel approach that builds an intraday trading strategy on open-high-low-close (OHLC) data. OHLC data is easily available from most database vendors. We use OHLC data to train neural networks that forecast the day's high and low of liquid US stocks and ETFs. The resulting long-short strategy tries to take advantage of the daily trading range of a security and exits all positions at the close. A volatility filter further improves risk-adjusted returns.

Keywords: Neural networks, intraday trading, open-high-low-close data.

1 Introduction

For a long time open-high-low-close (OHLC) data has been available cheaply or even for free on financial websites. While there is a host of forecasting studies using OHLC data as inputs for close-to-close analysis the (academic) literature on intraday systems trading highs and lows is scarce. At the other end of the spectrum are studies using high-frequency data. Curiously, a literature search reveals more high-frequency studies than OHLC studies, although high-frequency data is arguably more difficult to obtain and to handle.

However, studies which even just use close-open or open-close data show attractive results. For example [1] analyze returns on different S&P indices. In the present study we additionally use high-low data to trade on. We derive our methodology from [2, 3]. They analyze the Brazilian stock market and trade the high-low range resulting in outstanding economic performance.

The question arises what the reason for this performance is. Is it due to the novel method, to the pecularities of the Brazilian stock market or to a combination of both? The authors themselves acknowledge that there is not much literature available on the Brazilian stock market. We can therefore not put

C. Jayne, S. Yue, and L. Iliadis (Eds.): EANN 2012, CCIS 311, pp. 423–432, 2012.

the authors' results in comparison to other studies in Brazil. We can, however, transfer and adapt the methodology to the US stock market and qualitatively compare the results. As it turns out, results on the US stock market are generally very promising, too. The Brazilian stock market studies by [2, 3] use a 3-layer perceptron. In contrast we use a Historically Consistent Neural Network (HCNN), introduced by [4]. HCNNs provide robust forecasts, see [5–7].

We structure the remainder of this paper as follows. Section 2 briefly recalls the HCNN and presents general characteristics of the learning data. Section 3 deals with the peculiarities of building a testable trading system using only OHLC data and introduces several trading ideas. Section 4 presents economical performance results for different US stocks. Section 5 concludes, discusses limitations of our approach and outlines further work.

2   Neural Network Modeling

The goal of our study is to forecast the daily high and low of a security. To achieve this we select HCNN. This advanced neural network models several time series at once and facilitates multi-step forecasts. The following state equation computes new states in the network:

$$s_{t+1} = \tanh(W \cdot s_t) \tag{1}$$

W is the weight matrix which we optimize. As usual in the context of neural networks we use the abbreviated notation $\tanh(\ldots)$ to mean a component-wise computation of tanh on the resulting vector $W \cdot s_t$. The upper part of s contains our observable time-series. These are the inputs and outputs although we do not make this distinction here, because we consider every time-series equal. The lower part of s contains hidden states. We can think of these hidden states as representing the unknown dynamics of the system under investigation.

It is our experience that the results are best when matrix W is sparsely populated. Sparsity is one meta-parameter in the design of a HCNN. For the present study we use a sparsity of 0.2 corresponding to a memory of $1/0.2$ that is five days. Another parameter is the dimension of s. We generally choose this dimension to be much larger than the number of time-series that we learn. Experimentation shows that choosing a factor of 20 leads to good results. This applies for all HCNNs in the paper.

When designing a neural network the observable time-series are very important. They contain the information that is available for learning the behavior of the dynamical system. Ultimately, the choice of observables leads to good or bad forecasts. To stick with the original paper of [2] we first used the (preprocessed) raw time-series as inputs to the network. This seemed like an interesting option when the time-series is not too long and meandering in a range. However, we soon noticed that for all but the most simple test runs network convergence was very slow due to the lack of stationarity in our time-series. Therefore this paper uses simple returns as basis for the observable time-series. With simple returns all convergence problems disappear. Making a time-series (almost) stationary by computing returns is a very common routine in time-series analysis.

However, using returns leaves us with the problem from where to compute returns for daily highs and lows. Two possibilities arise: first, we can compute returns from the previous day's, individually for high and low. Second, we can use the current day's open and compute the return from there. We tried both options on part of the in-sample dataset and decided to take the second variant because it leads to more robust result. That means that the performance on the validation set better matches the performance on the training set. The remaining time-series are similar to those of [2]:

- returns of the lows
- returns of the highs
- five-period exponential moving average of the lows
- five-period exponential moving average of the highs
- five-period lower bollinger band of the closes
- five-period upper bollinger band of the closes
- returns of the open
- same-day return open to low (to be forecast)
- same-day return open to high (to be forecast).

We have to keep in mind that the first six time-series are only available with a lag of one day when we consider the open of the current day our starting point. For example, we cannot possibly know the current day's close. For this reason we shift these time-series by one day. In the context of HCNN the same applies to the two forecast time-series. Therefore, when forecasting we fill up state vector s with the available information (that is everything except the current day's open) and on the open we add this last information and forecast. We exclude exchange holidays from the dataset.

Our out-of-sample dataset ranges from 2010-12-09 until 2012-03-02 (310 trading days). As is common in HCNN application we use a rolling window forecast. In this work we stick to the parameters of the original study and use 128 trading days of training data to forecast the next ten days. We then move the training data forward by ten days, train and forecast again. The first training interval for the final model therefore begins 128 trading days before 2010-12-09. We also used the trading year before 2010-12-09 as combined training and validation set for preliminary experiments. The length of the rolling window and the length of the forecast horizon is open to debate.

Typically for HCNN the learning rate is set to a small number, in our case $r = 0.001$. Different values for r in that order of magnitude did not change the results substantially in preliminary tests. The goal of HCNN is to reproduce history consistently. That means that we strive to achieve a training error of almost zero. We add diversity by computing an ensemble of 200 HCNNs. The average value is our forecast. Due to the comparatively short training span (128 days) and an overall small model 1000 iterations are enough to achieve a satisfactory error. This allows to train an entire ensemble in a few minutes on an eight-core computer. Remember that training (that is: a model update) is only necessary every ten days. This can easily be done when the exchange is closed or at the weekend. Model eveluation is almost instanteneous and takes less than one second. We can therefore forecast at the open without any noticeable lag.

3 Trading System

Each trading day, at the open, our system delivers a forecast for low and high of the day. We want to exploit this information in a viable trading system using only OHLC historical data for reasons of practicability. This imposes the limitation that we cannot tell in which order high and low will occur. We also do not know whether prices near high and low will be hit only once or several times. In other words: we can check if the range forecast is correct, but we cannot see, if we would have been able to exploit several swings within that range. On the other hand the times of open and close are well defined, obviously.

Nevertheless we can still use OHLC data to realistically backtest historical performance along two general trading ideas. The first involves just trading the range once (if the range forecast is correct) or exiting at the close (if the other side of the forecast range is not hit). This system leads to the following rules:

- if $l < l_f$ and $h_f < h$ then book the difference $h_f - l_f$ to the PnL (profitable long or short trade on correct range)
- if $l < l_f$ and $h < h_f$ then book the difference $c - l_f$ to the PnL (long trade at l_f but h_f not hit, exit at c)
- if $l_f < l$ and $h_f < h$ then book the difference $h_f - c$ to the PnL (short trade at h_f but l_f not hit, exit at c)

where o, h, l, c represent the realized open, high, low, close values, and l_f, h_f the corresponding forecasts for l and h.

We add some obvious consistency rules:

- only trade if $l_f < o < h_f$ (forecast of low and high should be consistent with the open price)
- no trade is possible when $h < l_f$ or $h_f < l$ (forecasts are entirely outside the range)

The second system involves trading into the direction of the more promising extreme (low or high) at the open. When the price hits the presumed extreme we reverse the position and exit at the close. We also exit at the close if we do not reach the forecast extreme. For this scheme we first have to determine the forecast returns from the open:

$$l_r = \left| \frac{l_f}{o} - 1 \right| \tag{2}$$

and

$$h_r = \left| \frac{h_f}{o} - 1 \right|. \tag{3}$$

The same consistency rules as above apply. The more promising direction is the greater of l_r and h_r. For the purpose of illustration let's say that h_r is more promising. We would therefore go long at the open and eventually reverse our position if we reach h_r. For the case $h_r > l_r$ the following rules apply:

- if $h_f < h$ then book $h_f - o$ and $h_f - c$ to the PnL (long at the open, reverse at h_f, buy to cover at c)
- if $h < h_f$ then book $c - o$ to the PnL (long at the open, h_f not hit, exit at c)

The reverse rules apply for $l_r > h_r$. We would short at the open, eventually reverse at l_r and exit at c in any case.

We enhance both systems by introducing a parameter α to encourage or discourage trading. If we narrow the forecast range we encourage trading, because we make it more likely that price will hit our range. If we make the forecast range wider we discourage trading, because we make it more unlikely that price will hit our range. Assuming that the forecasts are consistent ($l_f < h_f$) we denote the (positive) range by $r = h_f - l_f$. We can then modify our forecasts according to

$$\tilde{l}_f = l_f + \alpha \cdot r \tag{4}$$

and

$$\tilde{h}_f = h_f - \alpha \cdot r. \tag{5}$$

$\alpha > 0$ corresponds to narrowing the range and makes trading more likely. We set α heuristically by preliminary experiments on the in-sample data set. We do not perform a strict optimization for α.

Preliminary tests also showed that the second system seems more promising (with and without the use of α). This is not surprising. If we assume that trading on the range forecast has positive expectation then we can expect to be better off being in the market during the entire trading day from open to close (like in the second system) than just being in the market for the shorter time span of the first system. The first system also suffers from the fact that we cannot exploit the range more than once.

Optionally, the system uses one additional rule: a volatility filter. This works as follows: the system simply does not trade in unusually volatile periods. As we consider intraday volatility it is useful to define volatility as (relative) width of the high-low range $r = h - l$. We smooth r with a simple 10-day moving average and compute the rolling standard deviation σ for the past 128 days (corresponding to the training period). The current day's r is not available at the open. Therefore we use the lagged values and compare the smoothed r to it's rolling average $\bar{r}$ and σ. This leads to one additional rule:

- do not trade if smoothed $r > \bar{r} + 2\sigma$.

The rationale of this rule is to avoid the most volatile periods because these are periods that are only seldom present in the training data. We therefore do not expect the networks to work particularly well in these periods and prefer to avoid them. Note that the same filter applies to all following results. We did not specifically tune the volatilty filter to different assets.

4   Results

The core of our trading system is trading the high-low intraday range. For this to work, we need some amount of intraday volatility. We expect that volatile stocks will yield better results than comparatively "duller" stocks. At the same time the stocks should be liquid in the sense that our transaction does not affect the price significantly. For the purpose of our analysis we will consider S&P500 stocks as liquid. We also expect that we will have to encourage trading ($\alpha > 0$) on low volatility stocks and discourage trading on high volatility stocks. Furthermore we also include two liquid ETFs on stock indices. Table 1 shows the securities in our sample.

Table 1. Observable time-series for the HCNN model

Ticker	Company Name	Exchange
XOM	Exxon Mobil Corporation	NYSE
MSFT	Microsoft Corporation	NASDAQ
DMND	Diamond Foods, Inc.	NASDAQ
FSLR	First Solar, Inc.	NASDAQ
SHLD	Sears Holdings Corporation	NASDAQ
SPY	SPDR S&P 500	NYSEArca
IWM	iShares Russell 2000 Index	NYSEArca

The reasons for inclusion in our analysis are as follows. XOM is one of the largest S&P500 companies by market capitalization. It is also quite volatile (for the time immediately preceding the out-of-sample period) for a large cap stock. This is in contrast to MSFT which is also a very large S&P500 company but not volatile and (especially) with low intraday dollar volatility. We expect a reduced performance trading MSFT. The next three stocks are included because they are very volatile. DMND is the most volatile stock of the S&P1500 while FSLR and SHLD are the most volatile stocks of the S&P500. DMND therefore stands for a stock which is not in the prime stock index but still in the upper half of market capitalization. It should still be tradable but the strategy might not scale well to high volumes. FSLR and SHLD should be easier to trade in large quantities. FSLR stands for a volatile company in an uptrend while SHLD is a volatile company in a downtrend. SPY and IWM are the most liquid ETFs on the S&P500 and the Russell 2000. We include these to gauge whether the system also works on closely watched broad indices.

Our system trades quite often: at least twice a day (open to close) or three times a day (open to extreme, extreme to close). We could expect that transaction costs severely impact our returns. As a conservative estimation we take USD 0.005 per share traded. This is a generally available flat rate at Interactive Brokers. However, with higher volumes of shares traded, transaction costs will go down to approximately USD 0.001. If we add liquidity we may even get a

Table 2. Values of α

	XOM	MSFT	DMND	FSLR	SHLD	SPY	IWM
α	0.2	0.2	0.0	-1.0	-0.5	0.0	-0.2

rebate and transaction costs will positively impact the PnL. The trade at l_f or h_f is a limit order and could therefore increase liquidity. As we will see it turns out that transaction costs do not impact the results by much. We use the conservative estimate knowing that we might be able to improve the results.

Trading systems using market orders have to deal with slippage. However, our system does not use market orders. Entering the trade is done using an at-the-open order type. The possible reversion at the extreme uses a limit order. And exiting at the close again uses the corresponding order type. The situation would change if we considered a stop-loss, that would trigger a market order. Without tick data we cannot test this. Strictly speaking the at-the-open order could be problematic. As the neural network needs the open price as input we can only trade a split second later (and would probably use a market order). In preliminary experiments we simulated the behavior of the forecast when varying the open price. In turns out that the forecast is quite stable. We could therefore use pre-open market prices to gauge the probable open price and place our at-the-open order accordingly.

Table 2 shows the chosen values for α and table 3 shows the out-of-sample results of the undisturbed system. The first row shows the number of trades. With decreasing α the number of trades also tends to decrease, because the system reaches the forecast extreme less often. Decreasing α therefore lets the system act more like a trend rider than like a swing trader. The second row shows the Profit and Loss of the system and absolute terms and in the following the PnL from buy-and-hold. The next row shows the realized return over the entire trading period of 310 days. For example, XOM realized a return of 64.4%. This is also compared to the buy-and-hold return. The next row is the annualized return assuming 252 trading days per year, also compared to buy-and-hold. Max DD, refers to the maximum (absolute and relative) encountered drawdown and the row after that is a important risk measure. It can be interpreted as a reward to risk ratio. A (negatively) higher ratio is better because it means that for the same amount of risk (as measured by drawdown) a higher reward is reached. Then the table shows the number of winning trades and the corresponding percentage of winning trades. Finally we compute transaction costs and the annualized return after transaction costs.

We see from the result table that the volatile stocks generally performed well, with good annualized returns and also mostly attractive reward to risk ratios. On the other hand, MSFT included as a large cap stock with low volatility does not perform well and even leads to a loss of 0.5% after transaction costs. In the case of SHLD the very large drawdown would have consumed more than the invested capital before the strategy eventually recovered. In the case of DMND and FSLR the drawdowns are also unacceptably high, but using a leverage of

Table 3. Out-of-sample results: undisturbed strategy

	XOM	MSFT	DMND	FSLR	SHLD	SPY	IWM
Trades	815	790	748	647	672	742	710
PnL	44.86	3.79	35.08	125.83	152.29	64.83	23.32
PnL bh	16.21	5.03	37.88	-104.43	4.46	14.64	3.22
Total return	0.644	0.139	0.912	0.934	2.359	0.517	0.295
Total return bh	0.233	0.184	0.985	-0.775	0.069	0.117	0.041
Ann. return	0.498	0.111	0.698	0.709	1.678	0.403	0.234
Ann. return bh	0.185	0.147	0.75	-0.702	0.056	0.094	0.033
Max DD	-8.25	-3.77	-19.78	-44.68	-94.37	-8.33	-11.47
Max DD rel.	-0.119	-0.138	-0.514	-0.331	-1.462	-0.066	-0.145
Pnl to max DD	-5.436	-1.005	-1.774	-2.817	-1.614	-7.785	-2.034
Nr winning	514	461	436	390	384	446	371
Winning ratio	0.631	0.584	0.583	0.603	0.571	0.601	0.523
TC	4.075	3.95	3.74	3.235	3.36	3.71	3.55
Ann. ret. after tc	0.455	-0.005	0.623	0.692	1.644	0.381	0.199

0.5 would produce more acceptable drawdowns while still conserving attractive returns. A closer look at the trade history of DMND, FSLR, and SHLD reveals that the loss producing trades are mostly clustered at the beginning of the out-of-sample period. It is obvious that a simple no-trade filter that prevents trading when a certain monthly loss has been surpassed would have greatly reduced the seriousness of the drawdowns. Likewise we leave it to further research to implement proper money management rules as stop losses also seem to have a mitigating effect on drawdowns. In total XOM produces the most attractive results with good annualized return and a very attractive reward to risk ratio. The two ETFs also show attractive returns and reward to risk ratios. Especially, SPY manages to achieve a respectable return with a very low drawdown.

Table 4 shows the same measures as above but after applying the volatility filter. This filter essentially acts as a no-trade filter and prevents trading in very volatile periods. In has the side effect of reducing the total number of trades by approximately ten percent for the different assets. The effects of the volatility filter are mixed but often improve risk measures of securities which already show good risk measures. This effect is especially strong with XOM and SPY where the volatility filter enhances the attractive risk to reward ratios.

Considering the performance compared to buy-and-hold we note that the system is not able to beat this benchmark in the case of MSFT and DMND. However, we note that the performance is generally independent from the buy-and-hold performance. For example, in the out-of-sample period FSLR looses more than half its value and SHLD and IWM are slightly positive but essentially flat. Still the strategy produces good returns.

Technically the performance measure should also include overnight and short-term interest rates. As the system is flat throughout the non-trading hours and on week-ends the deposit could earn the overnight rate. With the volatility filter the amount earned by interest rate payments further increases because the system

Table 4. Out-of-sample results: with volatility filter

	XOM	MSFT	DMND	FSLR	SHLD	SPY	IWM
Trades	733	733	660	593	612	695	665
PnL	44.0	2.82	18.44	113.45	123.51	74.99	25.64
PnL bh	16.21	5.03	37.88	-104.43	4.46	14.64	3.22
Total return	0.632	0.103	0.479	0.842	1.913	0.598	0.324
Total return bh	0.233	0.184	0.985	-0.775	0.069	0.117	0.041
Ann. return	0.489	0.083	0.377	0.643	1.385	0.464	0.257
Ann. return bh	0.185	0.147	0.75	-0.702	0.056	0.093	0.033
Max DD	-5.56	-3.77	-21.24	-45.31	-82.76	-7.77	-9.5
Max DD rel.	-0.08	-0.138	-0.552	-0.336	-1.282	-0.062	-0.12
Pnl to max DD	-7.912	-0.747	-0.868	-2.504	-1.492	-9.655	-2.698
Nr winning	470	425	375	352	347	426	348
Winning ratio	0.641	0.58	0.568	0.594	0.567	0.613	0.523
TC	3.665	3.665	3.3	2.965	3.06	3.475	3.325
Ann. ret. after tc	0.45	-0.025	0.31	0.627	1.353	0.443	0.224

stays out of the market for even more time. Interest rate payments on unused funds would obviously only improve performance measures and it does not bias the results to leave them out. Short-term USD interest rates have been very low in recent times and including interest rate payments would only result in a minimal improvement of performance measures anyway.

5   Conclusion, Limitations, and Future Research

Our analysis shows that it is possible to successfully model the intraday dynamics of liquid US securities with only a few (technical) indicators as inputs to a neural network. The trading strategy always goes flat at the close of the day. This eliminates the overnight gap risk present in all daily strategies. A volatility filter acting as a no-trade filter in volatile periods can further improve risk measures. This paper confirms the good results that [2] obtained for the Brazilian stock market with very similar methodology. Our strategy only uses easily available OHLC data. It does not need real intraday tick data for backtesting.

This is also a limitation of our work. It is quite probable that we could enhance the risk-reward ratio of the strategy with tick data. For example, we could imagine trading the range more than once a day if the occasion presents itself. With OHLC data we cannot backtest this, because we don't know the order of possibly recurring daily highs and lows. Also, without tick data it is not possible to evaluate the effect of money management strategies, because, again, we cannot tell the order in which prices arrive. However, considering the serious drawdowns that can occur in certain securities, we would not trade the strategy without at least a proper stop loss on every trade.

Another limitation of our work is the lack of diversified input factors. One could argue that we use four different price series (open, high, low, close). But these price series are strongly correlated. It would be interesting to see if the results improve if we add different time series, for example from securities in the same industry group. Also general market indices could prove a useful addition.

The present study should be extended to include more securities. We motivated the choice of securities qualitatively. However, this cannot replace a broad market analysis. For every trade we know at least the approximate entry price, the exit target and the probable range. We can therefore rank every trade by its forecast reward to risk ratio.This also leads to the idea of creating a (possibly dynamically) combined portfolio of the best performing assets. This could improve risk measures.

Additionally, investigating other markets could prove enlightening. How does the same methodology fare when forecasting foreign exchange, commodities, or other assets?

References

1. Dunis, C., Laws, J., Rudy, J.: Profitable mean reversion after large price drops: A story of day and night in the s&p 500, 400 midcap and 600 smallcap indices. Journal of Asset Management 12, 185–202 (2011)
2. Martinez, L.C., da Hora, D.N., de M. Palotti, J.R., Meira Jr., W., Pappa, G.L.: From an artificial neural network to a stock market day-trading system: A case study on the bmf bovespa. In: Proceedings of International Joint Conference on Neural Networks, Atlanta, Georgia, USA, June 14-19 (2009)
3. Gomide, P., Milidiu, R.: Assessing stock market time series predictors quality through a pairs trading system. In: 2010 Eleventh Brazilian Symposium on Neural Networks (SBRN), pp. 133–139 (2010)
4. Zimmermann, H.G.: Forecasting the Dow Jones with historical consistent neural networks. In: Dunis, C., Dempster, M., Terraza, V. (eds.) Proceedings of the 16th International Conference on Forecasting Financial Markets: Advances for Exchange Rates, Interest Rates and Asset Management, Luxembourg, May 27-29 (2009)
5. Zimmermann, H.G.: Advanced forecasting with neural networks. In: Dunis, C., Dempster, M., Breitner, M.H., Rösch, D., von Mettenheim, H.J. (eds.) Proceedings of the 17th International Conference on Forecasting Financial Markets: Advances for Exchange Rates, Interest Rates and Asset Management, Hannover, May 26-28 (2010)
6. von Mettenheim, H.J., Breitner, M.H.: Robust forecasts with shared layer perceptrons. In: Dunis, C., Dempster, M., Breitner, M.H., Rösch, D., von Mettenheim, H.J. (eds.) Proceedings of the 17th International Conference on Forecasting Financial Markets: Advances for Exchange Rates, Interest Rates and Asset Management, Hannover, May 26-28 (2010)
7. von Mettenheim, H.J., Breitner, M.H.: Neural network model building: A practical approach. In: Dunis, C., Dempster, M., Girardin, E., Péguin-Feissolle, A. (eds.) Proceedings of the 18th International Conference on Forecasting Financial Markets: Advances for Exchange Rates, Interest Rates and Asset Management, Marseille, May 25-27 (2011)

Kalman Filters and Neural Networks in Forecasting and Trading

Georgios Sermpinis[1], Christian Dunis[2], Jason Laws[3], and Charalampos Stasinakis[4]

[1] University of Glasgow Business School
georgios.sermpinis@glasgow.ac.uk
[2] Liverpool Business School
cdunis@tiscali.co.uk
[3] University of Liverpool Management School
J.Laws@liverpool.ac.uk
[4] University of Glasgow Business School
c.stasinakis.1@research.gla.ac.uk

Abstract. The motivation of this paper is to investigate the use of a Neural Network (NN) architecture, the Psi Sigma Neural Network, when applied to the task of forecasting and trading the Euro/Dollar exchange rate and to explore the utility of Kalman Filters in combining NN forecasts. This is done by benchmarking the statistical and trading performance of PSN with a Naive Strategy and two different NN architectures, a Multi-Layer Perceptron and a Recurrent Network. We combine our NN forecasts with Kalman Filter, a traditional Simple Average and the Granger- Ramanathan's Regression Approach. The statistical and trading performance of our models is estimated throughout the period of 2002-2010, using the last two years for out-of-sample testing. The PSN outperforms all models' individual performances in terms of statistical accuracy and trading performance. The forecast combinations also present improved empirical evidence, with Kalman Filters outperforming by far its benchmarks.

Keywords: Psi Sigma Network, Recurrent Network, Forecast Combinations, Kalman Filter.

1   Introduction

The term of Neural Network (NN) originates from the biological neuron connections of human brain. The artificial NNs are computation models that embody data-adaptive learning and clustering abilities, deriving from parallel processing procedures [34]. The NNs are considered a relatively new technology in Finance, but with high potential and an increasing number of applications. However, their practical limitations and contradictory empirical evidence lead to skepticism on whether they can outperform existing traditional models.

   The motivation of this paper is to investigate the use of a Neural Network (NN) architecture, the Psi Sigma Neural Network (PSN), when applied to the task of forecasting and trading the Euro/Dollar (EUR/USD) exchange rate and to explore the

C. Jayne, S. Yue, and L. Iliadis (Eds.): EANN 2012, CCIS 311, pp. 433–442, 2012.

utility of Kalman Filters in combining NN forecasts. This is done by benchmarking the statistical and trading performance of PSN with a Naive Strategy and two different NN architectures, a Multi-Layer Perceptron (MLP) and a Recurrent Network (RNN). We combine our NN forecasts with Kalman Filter, a traditional Simple Average and the Granger- Ramanathan's Regression Approach (GRR). The statistical and trading performance of our models is estimated throughout the period of 2002-2010, using the last two years for out-of-sample testing. In terms of our results, the PSN outperforms all models' individual performances in terms of statistical accuracy and trading performance. The forecast combinations also present improved empirical evidence, with Kalman Filters outperforming by far its benchmarks.

Section 2 is a short literature review and Section 3 follows with the description of the EUR/USD ECB fixing series, used as our dataset. Sections 4 and 5 give an overview of the forecasting models and the forecast combination methods we implemented respectively. The statistical and trading performance of our models is presented in Section 6. Finally, some concluding remarks are summarized in Section 7.

2   Literature Review

The most common NN architecture is the MLP and seems to perform well at time-series financial forecasting [16], although the empirical evidence can be contradictory in many cases. Ince and Trafalis [14] forecast the EUR/USD, GBP/USD, JPY/USD and AUD/USD exchange rates with MLP and Support Vector Regression and their results show that MLP achieves less accurate forecasts. On the other hand, Tenti [21] and Dunis and Huang [3] achieved encouraging results also by using RNNs to forecast the exchange rates. PSNs were first introduced by Ghosh and Shin [7] as architectures able to capture high-order correlations. Ghosh and Shin [7, 8] also present results on their forecasting superiority in function approximation, when compared with a MLP network and a Higher Order Neural Network (HONN). Satisfactory forecasting results of PSN were presented by Hussain et al. [13] on the EUR/USD, the EUR/GBP and the EUR/JPY exchange rates using univariate series as inputs in their networks. Bates and Granger [1] and Newbold and Granger [17] suggested combining rules based on variances-covariances of the individual forecasts, while Granger and Ramanathan [10] presented a regression combination forecast framework with encouraging results. According to Palm and Zellner [18], it is sensible to use simple average for combination forecasting, while Deutsch et al. [2] achieved substantially smaller squared forecasts errors combining forecasts with changing weights. Time-series analysis is often based on the assumption that the parameters are fixed. However, Harvey [12] and Hamilton [11] both suggest using state space modeling, such as Kalman Filter, for representing dynamic systems where unobserved variables (so-called 'state' variables) can be integrated within an 'observable' model. According to Goh and Mandic [9] the recursive Kalman Filter is suitable for processing complex-valued nonlinear, non-stationary signals and bivariate signals with strong component correlations. Kalman Filter is also considered an optimal time-varying financial forecast for financial markets [4]. Terui and van Dijk [22] also suggest that the combined forecasts perform well, especially with time varying coefficients.

3   The EUR/USD Exchange Rate and Related Financial Data

The European Central Bank (ECB) publishes a daily fixing for selected EUR exchange rates: these reference mid-rates are based on a daily concentration procedure between central banks within and outside the European System of Central Banks, which normally takes place at 2.15 p.m. ECB time. The reference exchange rates are published both by electronic market information providers and on the ECB's website shortly after the concentration procedure has been completed. Although only a reference rate, many financial institutions are ready to trade at the EUR fixing and it is therefore possible to leave orders with a bank for business to be transacted at this level.

In this paper, we examine the EUR/USD over period 2002 -2010, using the last two years for out-of-sample.

Table 1. The EUR/USD Dataset - Neural Networks' Training Dataset

PERIODS	TRADING DAYS	START DATE	END DATE
Total Dataset	2295	3/01/2002	31/12/2010
Training Dataset *(In-sample)*	1270	3/01/2002	29/12/2006
Test Dataset *(In-sample)*	511	02/01/2007	31/12/2008
Validation Dataset *(Out-of-sample)*	514	02/01/2009	31/12/2010

The graph below shows the total dataset for the EUR/USD and its volatile trend since early 2008.

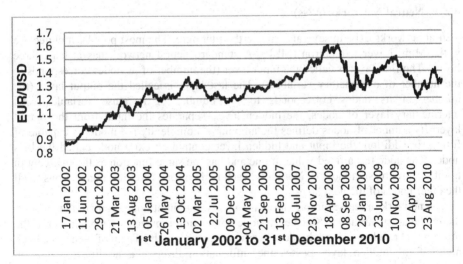

Fig. 1. EUR/USD Frankfurt daily fixing prices

To overcome the non-stationary issue, the EUR/USD series is transformed into a daily series of rate returns. So given the price level $P_1, P_2,\ldots, P_t$, the return at time t is calculated as:

$$R_t = \left(\frac{P_t}{P_{t-1}}\right) - 1 \tag{1}$$

In the absence of any formal theory behind the selection of the inputs of a neural network, we conduct some neural networks experiments and a sensitivity analysis on a pool of potential inputs in the training dataset in order to help our decision. Our aim is to select the set of inputs for each network which is the more likely to lead to the best trading performance in the out-of-sample dataset. In our application, we select as inputs sets of autoregressive terms of the EUR/USD, EUR/GBP and the EUR/JPY exchange rates, based on the higher trading performance for each network in the test sub-period.

4   Forecasting Models

4.1   Naive Strategy

In this paper we use the Naive Strategy in order to benchmark the efficiency of the NNs' trading performance. The Naive Strategy is considered to be the simplest strategy to predict the future. That is to accept as a forecast for time $t+1$, the value of time t, assuming that the best prediction is the most recent period change. Thus, the model takes the form: $\hat{Y}_{t+1} = Y_t$ (2) where Y_t is the actual rate of return at time t and $\hat{Y}_{t+1}$ is the forecast rate of return at time $t+1$.

4.2   Neural Networks (NNs)

Neural networks exist in several forms in the literature. The most popular architecture is the Multi-Layer Perceptron (MLP). A standard neural network has at least three layers. The first layer is called the input layer (the number of its nodes corresponds to the number of explanatory variables). The last layer is called the output layer (the number of its nodes corresponds to the number of response variables). An intermediary layer of nodes, the hidden layer, separates the input from the output layer. Its number of nodes defines the amount of complexity the model is capable of fitting. In addition, the input and hidden layer contain an extra node called the bias node. This node has a fixed value of one and has the same function as the intercept in traditional regression models. Normally, each node of one layer has connections to all the other nodes of the next layer.

The network processes information as follows: the input nodes contain the value of the explanatory variables. Since each node connection represents a weight factor, the information reaches a single hidden layer node as the weighted sum of its inputs. Each node of the hidden layer passes the information through a nonlinear activation function and passes it on to the output layer if the calculated value is above a threshold.

The training of the network (which is the adjustment of its weights in the way that the network maps the input value of the training data to the corresponding output value) starts with randomly chosen weights and proceeds by applying a learning algorithm called backpropagation of errors [19].The learning algorithm simply tries to find those weights which minimize an error function (normally the sum of all squared differences between target and actual values). Since networks with sufficient hidden nodes are able to learn the training data (as well as their outliers and their noise) by heart, it is crucial to stop the training procedure at the right time to prevent overfitting (this is called 'early stopping'). This can be achieved by dividing the dataset into 3 subsets respectively called the training and test sets used for simulating the data currently available to fit and tune the model and the validation set used for simulating future values. The training of a network is stopped when the mean squared forecasted error is at minimum in the test-sub period. The network parameters are then estimated by fitting the training data using the above mentioned iterative procedure (backpropagation of errors). The iteration length is optimised by maximising the forecasting accuracy for the test dataset. Then the predictive value of the model is evaluated applying it to the validation dataset (out-of-sample dataset).

4.2.1   The Multi-Layer Perceptron Model (MLP)
MLPs are feed-forward layered NN, trained with a back-propagation algorithm. According to Kaastra and Boyd [15], they are the most commonly used types of artificial networks in financial time-series forecasting. The training of the MLP network is processed on a three-layered architecture, as described above.

4.2.2   The Recurrent Neural Network (RNN)
The next NN architecture used in this paper is the RNN. For an exact specification of recurrent networks, see Elman [6]. A simple recurrent network has an activation feedback which embodies short-term memory. The advantages of using recurrent networks over feed-forward networks for modeling non-linear time series have been well documented in the past. However, as mentioned by Tenti [21], "the main disadvantage of RNNs is that they require substantially more connections, and more memory in simulation than standard back-propagation networks" (p. 569), thus resulting in a substantial increase in computational time.

4.2.3   The Psi-Sigma Neural Network (PSN)
The PSNs are a class of Higher Order Neural Networks with a fully connected feed-forward structure. Ghosh and Shin [7] were the first to introduce the PSN, trying to reduce the numbers of weights and connections of a Higher Order Neural Network. Their goal was to combine the fast learning property of single-layer networks with the mapping ability of Higher Order Neural Networks and avoid increasing the required number of weights. The price for the flexibility and speed of Psi Sigma networks is that they are not universal approximators. We need to choose a suitable order of approximation (or else the number of hidden units) by considering the estimated function complexity, amount of data and amount of noise present. To overcome this, our code runs simulations for orders two to six and then it presents the best network. The evaluation of the PSN model selected comes in terms of trading performance.[1]

---

[1] For a complete description of all the neural network models we used and their complete specifications see Sermpinis et al. [20].

5   Forecasting Combination Techniques

In this section we present the techniques that we used to combine our NNs forecasts. It is important to outline that a forecast combination targets either to follow the trend of the best individual forecast (*'combining for adaptation'*) or to significantly outperform each one of them (*'combining for improvement'*) [23].

5.1   Simple Average

The first forecasting combination technique used in this paper is Simple Average, which can be considered a benchmark forecast combination model. Given the three NNs' forecasts $f_{MLP}^t, f_{RNN}^t, f_{PSN}^t$ at time t, the combination forecast at time t is calculated as:

$$f_{c_{NNs}}^t = (f_{MLP}^t + f_{RNN}^t + f_{PSN}^t)/3 \tag{3}$$

5.2   Granger and Ramanathan Regression Approach (GRR)

According to Bates and Granger [1], a combining set of forecasts outperforms the individual forecasts that the set consists of. Based on Granger and Ramanathan [10] we combine our forecasts as follows:

$$f_c = a_0 + \sum_{i=1}^{n} a_i f_i + \varepsilon_1 \tag{GRR}$$

Where:

* f_i, $i=1,...,n$ are the individual one-step-ahead forecasts,
* f_{c1}, f_{c2}, f_{c3} are the combination forecast of each model,
* a_0 is the constant term of the regression
* a_i are the regression coefficients of each model
* ε_1, ε_2, ε_3 are the error terms of each regression model

The GRR model at time t used in this paper is specified as shown below:

$$f_{c_{NNs}}^t = 0.0422 + 35.023 f_{MLP}^t + 13.461 f_{RNN}^t + 56.132 f_{PSN}^t + \varepsilon_t \tag{4}$$

From (4) it is obvious that GRR favors PSN forecasts.

5.3   Kalman Filter

Kalman Filter is an efficient recursive filter that estimates the state of a dynamic system from a series of incomplete and noisy measurements. The time-varying coefficient combination forecast suggested in this paper is shown below:

$$\text{Measurement Equation: } f_{c_{NNs}}^t = \sum_{i=1}^{3} a_i^t f_i^t + \varepsilon_t, \quad \varepsilon_t \sim NID\left(0, \sigma_\varepsilon^2\right) \tag{5}$$

$$\text{State Equation: } a_i^t = a_i^{t-1} + n_t, \quad n_t \sim NID(0, \sigma_n^2) \tag{6}$$

Where:

- $f^t_{c_{NNs}}$ is the dependent variable (combination forecast) at time t

- f^t_i $(i = 1, 2, 3)$ are the independent variables (individual forecasts) at time t

- a^t_i $(i = 1, 2, 3)$ are the time-varying coefficients at time t for each NN

- ε_t, n_t are the uncorrelated error terms (noise)

The alphas are calculated by a simple random walk and we initialized $\varepsilon_1 = 0$. Based on the above, our Kalman Filter model has as a final state the following:

$$f^t_{c_{NNs}} = 5.80 f^t_{MLP} + 1.16 f^t_{RNN} + 75.89 f^t_{PSN} + \varepsilon_t \tag{7}$$

From the above equation we note that the Kalman filtering process also favors the PSN model. This is what one would expect, since it is the model that performs best individually.

6   Statistical and Trading Performance

As it is standard in literature, in order to evaluate statistically our forecasts, the RMSE, the MAE, the MAPE and the Theil-U statistics are computed (see Dunis and Williams [5]). For all four of the error statistics retained the lower the output, the better the forecasting accuracy of the model concerned. In Table 2 we present the statistical performance of all our models in the out-of-sample period.

Table 2. Summary of the Out-of-sample Statistical Performance

	NAIVE	MLP	RNN	PSN	Simple Average	GRR	Kalman Filter
MAE	0.0084	0.0058	0.0056	0.0048	0.0048	0.0047	0.0044
MAPE	405.62%	112.37%	105.97%	97.88%	94.07%	92.83%	88.37%
RMSE	0.0107	0.0061	0.0060	0.0054	0.0053	0.0049	0.0043
Theil-U	0.7958	0.7301	0.6001	0.4770	0.5672	0.5297	0.5212

We note that from our individual forecasts, the PSN statistically outperformed all other models. Similarly, for our forecast combinations methodologies the Kalman Filter beat its benchmarks for the four statistical criteria retained in the out-of-sample period.

The trading strategy applied in this paper is to go or stay 'long' when the forecast return is above zero and go or stay 'short' when the forecast return is below zero.[2] In Table 3 below we present the out-of-sample trading performance of our models before and after transaction costs.

---

[2] The transaction costs for a tradable amount, say USD 5-10 million, are about 1 pip (0.0001 EUR/USD) per trade (one way) between market makers. But since we consider the EUR/USD time series as a series of middle rates, the transaction costs is one spread per round trip. With an average exchange rate of EUR/USD of 1.369 for the out-of-sample period, a cost of 1 pip is equivalent to an average cost of 0.007% per position.

Table 3. Summary of Out-of-Sample Trading Performance

	NAIVE	MLP	RNN	PSN	Simple Average	GRR	Kalman Filter
Annualised Return (excluding costs)	-4.80%	14.80%	16.07%	18.37%	16.37%	16.99%	28.79%
Annualised Volatility	12.03%	11.83%	11.02%	10.89%	10.85%	11.02%	10.92%
Information Ratio (excluding costs)	-0.4	1.25	1.46	1.69	1.51	1.54	2.64
Maximum Drawdown	-6.41%	-6.23%	-6.23%	-6.31%	-6.31%	-6.31%	-6.31%
Annualized Transactions	77	71	71	76	70	63	73
Transaction Costs	0.54%	0.50%	0.50%	0.53%	0.49%	0.44%	0.51%
Annualised Return (including costs)	-5.34%	14.30%	15.57%	17.84%	15.88%	16.55%	28.28%
Information Ratio (including costs)	-0.44	1.21	1.41	1.64	1.46	1.50	2.59

From the last two rows of Table 3, we note that the PSN continues to outperform all other single forecasts in terms of trading performance, coinciding with its statistical superiority. From our forecast combinations, only the Kalman Filter beats our best single forecast. The Simple Average and GRR methods seem unable to outperform PSN in the out-of-sample period. On the other hand, the GRR strategy still outperforms the MLP and the RNN models in terms of annualised return and information ratio. That could be thought as a trend to adapt to the best individual performance (*'combining for adaptation'* [23]). It seems that the ability of Kalman Filter to provide efficient computational recursive means to estimate the state of our process, gives it a considerable advantage compared to our fixed parameters combination models.[3]

7   Concluding Remarks

The motivation of this paper is to investigate the use of a Neural Network (NN) architecture, the Psi Sigma Neural Network (PSN), when applied to the task of forecasting and trading the Euro/Dollar (EUR/USD) exchange rate and to explore the utility of Kalman Filters in combining NN forecasts. This is done by benchmarking the statistical and trading performance of PSN with a Naive Strategy and two different NN architectures, a Multi-Layer Perceptron (MLP) and a Recurrent Network (RNN). We combine our NN forecasts with Kalman Filter, a traditional Simple Average and the Granger- Ramanathan's Regression Approach (GRR). The statistical and trading performance of our models is estimated throughout the period of 2002-2010, using the last two years for out-of-sample testing.

As it turns out, the PSN outperforms its benchmarks models in terms of statistical accuracy and trading performance. It is also shown that all the forecast combinations outperform in the out-of-sample period all our single models, except the PSN, for the

---

[3] The in-sample statistical and trading performances of our models are not presented for the sake of space. Nonetheless, the ranking of the models does not change considerably compared with the out-of-sample period.

statistical and trading terms retained. Simple Average and GRR do not beat PSNs' best individual performance, but are better than MLP and RNN, while Kalman Filter presents the best results. It seems that the ability of Kalman Filter to provide efficient computational recursive means to estimate the state of our process gives it a considerable advantage compared to our fixed parameters combination models. The remarkable trading performance of Kalman Filter allows us to conclude that it can be considered as an optimal forecast combination for the models and time-series under study. Our results should also go some way towards convincing a growing number of quantitative fund managers to experiment beyond the bounds of traditional models and trading strategies.

References

[1] Bates, J.M., Granger, C.W.J.: The Combination of Forecasts. Operational Research Society 20(4), 451–468 (1969)
[2] Deutsch, M., Granger, C.W.J., Teräsvirta, T.: The combination of forecasts using changing weights. International Journal of Forecasting 10(1), 47–57 (1994)
[3] Dunis, C.L., Huang, X.: Forecasting and trading currency volatility: an application of recurrent neural regression and model combination. Journal of Forecasting 21(5), 317–354 (2002)
[4] Dunis, C.L., Shannon, G.: Emerging Markets of South-East and Central Asia: Do they Still Offer a Diversification Benefit? Journal of Asset Management 6(3), 168–190 (2005)
[5] Dunis, C.L., Williams, M.: Modelling and Trading the EUR/USD Exchange Rate: Do Neural Network Models Perform Better? Derivatives Use, Trading and Regulation 8, 211–239 (2002)
[6] Elman, J.L.: Finding Structure in Time. Cognitive Science 14(2), 179–211 (1990)
[7] Ghosh, J., Shin, Y.: The Pi-Sigma Network: An efficient Higher-order Neural Networks for Pattern Classification and Function Approximation. In: Proceedings of International Joint Conference of Neural Networks, vol. 1, pp. 13–18 (1991)
[8] Ghosh, J., Shin, Y.: Efficient Higher-Order Neural Networks for Classification and Function Approximation. International Journal of Neural Systems 3(4), 323–350 (1992)
[9] Goh, S.L., Mandic, D.P.: An Augmented Extended Kalman Filter Algorithm for Complex-Valued Recurrent Neural Networks. Neural Computation 19(4), 1039–1055 (2007)
[10] Granger, C.W.J., Ramanathan, R.: Improved methods of combining forecasts. Journal of Forecasting 3(2), 197–204 (1984)
[11] Hamilton, J.D.: Time series analysis. Princeton University Press, Princeton (1994)
[12] Harvey, A.C.: Forecasting, structural time series models and the Kalman filter. Cambridge University Press, Cambridge (1990)
[13] Hussain, A.J., Ghazali, R., Al-Jumeily, D., Merabti, M.: Dynamic Ridge Polynomial Neural Network for Financial Time Series Prediction. In: IEEE International Conference on Innovation in Information Technology, Dubai, pp. 1–5 (2006)
[14] Ince, H., Trafalis, T.B.: A hybrid model for exchange rate prediction. Decision Support Systems 42(2), 1054–1062 (2006)
[15] Kaastra, I., Boyd, M.: Designing a Neural Network for Forecasting Financial and Economic Time Series. Neurocomputing 10(3), 215–236 (1996)

[16] Makridakis, S., Andersen, A., Carbone, R., Fildes, R., Hibon, M., Lewandowski, R., Newton, J., Parzen, E., Winkler, R.: The accuracy of extrapolation (time series) methods: Results of a forecasting competition. Journal of Forecasting 1(2), 111–153 (1982)

[17] Newbold, P., Granger, C.W.J.: Experience with Forecasting Univariate Time Series and the Combination of Forecasts. Journal of the Royal Statistical Society 137(2), 131–165 (1974)

[18] Palm, F.C., Zellner, A.: To combine or not to combine? issues of combining forecasts. Journal of Forecasting 11(8), 687–701 (1992)

[19] Shapiro, A.F.: A Hitchhiker's guide to the techniques of adaptive nonlinear models. Insurance: Mathematics and Economics 26(2-3), 119–132 (2000)

[20] Sermpinis, G., Laws, J., Dunis, C.L.: Modelling and trading the realised volatility of the FTSE100 futures with higher order neural networks. European Journal of Finance, 1–15 (2012)

[21] Tenti, P.: Forecasting foreign exchange rates using recurrent neural networks. Applied Artificial Intelligence 10(6), 567–581 (1996)

[22] Terui, N., Van Dijk, H.K.: Combined forecasts from linear and nonlinear time series models. International Journal of Forecasting 18(3), 421–438 (2002)

[23] Yang, Y.: Combining Forecasting Procedures: Some theoretical results. Econometric Theory 20(1), 176–222 (2004)

Short-Term Trading Performance of Spot Freight Rates and Derivatives in the Tanker Shipping Market: Do Neural Networks Provide Suitable Results?

Christian von Spreckelsen, Hans-Jörg von Mettenheim,
and Michael H. Breitner

Leibniz Universitaet Hannover, Koenigsworther Platz 1, 30167 Hannover
{spreckelsen,mettenheim,breitner}@iwi.uni-hannover.de
http://www.iwi.uni-hannover.de

Abstract. In this paper we investigate the forecasting and trading performance of linear and non-linear methods, in order to generate short-term forecasts in the dirty tanker shipping market. We attempt to uncover the benefits of using several time series models and the potential of neural networks. Maritime forecasting studies using neural networks are rare and only focus on spot rates. We build on this kind of investigation, but we extend our study on freight rates derivatives or Forward Freight Agreements (FFA) in a simple trading simulation. Our conclusion is, that non-linear methods like neural networks are suitable for short-term forecasting and trading freight rates, as their results match or improve on those of other models. Nevertheless, we think that further research with freight rates and corresponding derivatives is developable for decision and trading applications with enhanced forecasting models.

Keywords: Shipping Freight Market, Neural Network, Forecasting Performance, Trading Performance.

1   Introduction

In this paper we investigate the forecasting and trading performance of non-linear forecasting models, to generate short-term forecasts of spot rates and corresponding freight forwards respectively Forward Freight Agreements (FFA) in the dirty tanker shipping market. Freight rates exhibit certain characteristics in the class of commodities: The freight rate represents as an underlying asset a transport service and can be classified as non-storable commodity like electricity. In contrast to other established markets, only a small number of actors operate in the freight market and it is not sure that all relevant information is contained in the forward price.

In recent time freight derivatives become interesting in the maritime market, due to the fact, that freight rates are very volatile. Derivative markets provide a

C. Jayne, S. Yue, and L. Iliadis (Eds.): EANN 2012, CCIS 311, pp. 443–452, 2012.
© Springer-Verlag Berlin Heidelberg 2012

way in which these risks may be transferred to other individuals who are willing to bear them, through hedging. Therefore, actors in the shipping market are forced to use forecasting techniques for the purpose of risk management.

Most studies on forecasting freight rates use traditional time series models and focus on statistical forecast performance measures (e.g. Culliane et al. [4], Veenstra and Franses [9]). Batchelor et al. [3] compared a range of time series models in forecasting spot freight rates and freight forward contracts respectively FFA rates. They concluded, that freight forward contracts are suitable to detect the tendency of future spot freight rate, but FFA rates do not seem to be helpful in predicting the direction of future spot rates. However, the use of linear time series models for freight rates is sometimes criticized, due to the fact that most financial time series show non-linear patterns (see Goulielmos and Psifia [5] and Adland and Culliane [1]). As a representative of non-linear methods, neural networks could be implemented for several financial applications. Li and Parsons[6] and Lyridis et al. [7] attempted to investigate the advantages of neural networks in predicting spot freight rates. They pointed out that neural networks can significantly outperform simple time series models for longer-term forecasting of monthly tanker spot freight rates. A more recent study of Mettenheim and Breitner [8] showed that neural networks achieve good forecasting and trading results in predicting the Baltic Dry Index (BDI), which measures the cost to haul dry freight over the world's oceans. According to trading investigations of freight rates and derivatives some examples and details are given by Alizadeh and Nomikos [2].

Nevertheless, we find a lack of jointly spot and forward forecasting and trading investigations with neural networks. We build on these investigations, but we extend our study on freight derivatives and a wider range of time series models. The main objective of this paper is to investigate neural networks' prediction ability for maritime business short-term forecasting and provide a practical framework for actual trading applications of neural networks. We therefore implement a simple trading simulation to investigate the economic evaluation of the predicted spot freight rates and FFA prices.

The paper is organized as follows. Section 2 gives a short introduction about the methodology of neural networks and alternative statistical time series models. Section 3 describes the data and data preparation. Section 4 gives a brief introduction about our forecasting and trading strategy and shows the performance measures of the neural network and alternative models. We evaluate the statistical forecasting performance via a simple trading simulation. Finally, Section 5 summarizes our conclusions.

2   Methodology of Neural Networks

Neural networks (NN) can be described as non-linear input-output models. They provide the basis for an entirely different approach to the analysis of time series. The connections between inputs and outputs are typically made via one or more hidden layers of neurons, sometimes alternatively called processing units or nodes. NN also appear to have potential application in time series modelling and forecasting. Nevertheless, the success of NN modelling depends on a suitable topology or architecture. This includes determining the number of layers, the number of neurons in each layer and which variables to choose as inputs and outputs. The number of hidden layers is often taken to be one, while the number of hidden neurons is found heuristically. In the case of time series prediction, feedforward NN use the past lagged observations as inputs to conduct one or multi-step ahead forecasts. They do not require any assumptions relating to the underlying data-generating process. Figure 1 shows an example of a neural network topology for time series forecasting purposes.

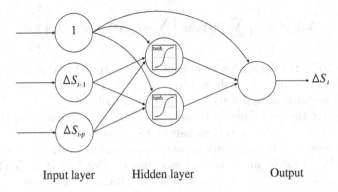

Input layer   Hidden layer   Output

Fig. 1. Topology of a typical NN for time series forecasting
Example with one hidden layer of two neurons. The output, e.g. the forecast variable, depends on the lagged input values at times $t - 1, \ldots, t - p$.

In our case, a one-step ahead forecast of spot freight rates returns $\Delta \hat{S}_t$ is computed using lagged input variables $(\Delta S_{t-1}, \Delta S_{t-2}, \ldots, \Delta S_{t-p})$ as follows (for FFA prices in the same manner):

$$\Delta \hat{S}_t = f(\Delta S_{t-1}, \Delta S_{t-2}, \ldots, \Delta S_{t-p}) \tag{1}$$

f denotes the function determined by the network. Thus the NN is equivalent to the nonlinear autoregressive model for time series forecasting problems. One of the input variables will usually be a constant. The neural network attempts to find the best possible approximation of the function f as a complex combination of elementary non-linear functions. This approximation is coded in the neurons of the network using weights that are connected with each neuron. These weights effectively measure the 'strength' of the different connections and are parameters that need to be estimated from the given data. We further assume there are H neurons in one hidden layer and then attach the weight w_{ij} to the connection between input S_{t-i} and the jth neuron in the hidden level. Given values for the weights, the value to be attached to each neuron may then be found in two stages. First, a linear function of the inputs is found:

$$net_j = \hat{w}_o + \sum_i^p w_{ij} \Delta S_{t-i} \tag{2}$$

For $j = 1, 2, \ldots, H$. Second, the quantity net_j is converted to the final value for the jth neuron by applying an activation function - in our case we use the hyperbolic tangent, $\tanh(net_j)$. Having calculated values for each neuron, a similar pair of operations can then be used to get the predicted value for the output using the values at the H neurons. This requires a further set of weights $\hat{w}_j$ to be attached to the links between the neurons and the output. Overall the output $\Delta \hat{S}_t$, is related to the inputs by the following expression:

$$\Delta \hat{S}_t = f_o \left[\left(\sum_j \hat{w}_j \tanh \left(\sum_i^p w_{ij} \Delta \hat{S}_{t-i} \right) + \hat{w}_o \right) \right] \tag{3}$$

where f_o denotes the activation functions at the output layer. It is also easy to incorporate further input variables into NN model. In this case, we are able to extend such an univariate NN to a multivariate topology.

In addition to the neural network model, we apply traditional linear time series models like the univariate Auto-Regressive Integrated Moving Average (ARIMA) model. Correspondingly to univariate models, statistical multivariate time series methods include the Vector Auto-Regressive process (VAR) and the Vector Error Correction model (VECM). The potential advantage of the multivariate VAR and VECM according to the univariate ARIMA model is that it takes into account the information content in the spot price movement in determining the forward price and vice versa.

3   Description of Data and Data Preparation

A forward freight agreement (FFA) is an agreement between two counterparties to settle a freight rate or hire rate at a certain date in the future. Tanker routes are centralized around the biggest physical routes for shipments of crude oil, known as trade dirty (TD) or trade clean (TC) followed by a numeral to designate the vessel size and cargo. We sample daily prices of most liquid International Maritime Exchange (Imarex) TD3 and TD5 freight forward contracts. These contracts are written on daily spot rates for TD3 and TD5 published by the Baltic Exchange. The spot and FFA data is available from 5 April 2004 to 1 April 2011 (1748 observations).

To avoid expiry effects, we calculate "perpetual" forward contract for one month (22 trading days; FFA 1M) and two month (44 trading days; FFA 2M) as a weighted average of a near and distant futures contracts, weighted according to their respective number of days from maturity. This procedure generates a series of futures prices with constant maturity and avoids the problem of price-jumps caused by the expiration of a particular futures contract. Figure 2 shows data points of TD3 contracts.

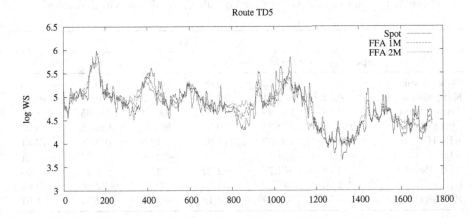

Fig. 2. Spot and forward prices for TD3

All prices are transformed to natural logarithms. Summary statistics of logarithmic first-differences ("log returns") of daily spot and FFA prices are presented in table 1 for the whole period in the two dirty tanker routes. The result's excess kurtosis in all series, and the skewness does not necessarily imply a symmetric distribution. The Jarque-Bera tests indicate departures from normality for both spot and FFA prices in all routes. This seems to be more acute for the spot freight rates. The Ljung-Box $Q(12)$ statistic on the first 12 lags of the sample autocorrelation function and Engles ARCH test indicate significant serial correlation and existence of heteroscedasticity, respectively. In contrast to storable commodities, such as stocks, there is no reason to expect changes in spot freight to be serially uncorrelated. Demand and supply for freight services are determined by the needs of trade. Augmented Dickey Fuller (ADF) and Phillips-Peron (PP) unit root tests indicate that all variables are log first-difference stationary, but the levels indicate, that most price series follow unit root processes.

4   Forecasting and Trading Performance Test

For purpose of forecasting and trading, each data set is divided into two subsets: the first subset runs from 5 April 2004 to 16 February 2010, the second subset from 17 February 2010 to 1 April 2011. The first subset is used to estimate the statistical models and identify the neural network structure while the second is used only for out-of-sample prediction comparison. This implies that we get a sample of 1466 daily observations for the estimation period and a sample of 282 daily observations for the forecasting and trading period – a ratio of 5.25 to 1. All models seem to be well specified, as indicated by relevant diagnostic tests.

We separate all models in univariate and multivariate model classes: The univariate models consist of an ARIMA and a NN model, where we include only the relevant single spot or FFA time series. For the multivariate models VAR,

Table 1. Descriptive statistics for the log differences of spot and FFA rates[a]

	TD3			TD5		
	Spot	FFA 1M	FFA 2M	Spot	FFA 1M	FFA 2M
Skew (levels)	0.38	0.13	−0.06	−0.06	−0.12	−0.13
Kurtosis (levels)	−0.12	−0.01	−0.13	−0.24	−0.09	−0.22
Jarque-Bera	10,212.88	1146.48	1482.50	9560.54	1966.65	1581.69
Ljung-Box	974.58	179.33	120.73	465.72	140.06	90.27
ARCH	115.64	126.99	176.95	21.90	105.88	73.31
ADF	−18.32	−24.77	−26.33	−22.09	−25.52	−26.11
ADF (levels)	−3.78	−2.72	−2.39	−4.10	−2.25	−1.65
PP	−22.56	−31.79	−33.73	−25.81	−32.56	−34.80
PP (levels)	−3.53	−2.59	−2.19	−3.57	−2.13	−1.67

[a] Critical 5% values in brackets: Jarque-Bera [5.99], Ljung-Box $Q(12)$ [51.48], ARCH(12) [1.81], ADF/PP [-2.89].

VECM and a multivariate neural network, namely NN+, we include both spot and all FFA rates of each route.

4.1   Statistical Forecasting Performance Results

We generate one-step ahead out-of-sample forecasts of each model, estimated over the initial estimation period. The forecasting performance of each model is presented in matrix form in table 2 for all contracts. Forecasts made using first-differences will be transformed back to levels to ensure that the measures presented above are comparable for all models. The forecast performance of each model is assessed using the conventional root mean square error metric (RMSE) and Theil's U statistic. The latter allows a relative comparison of formal forecasting methods with a naïve model, a no-change random walk (RW1).

All models outperform their naïve benchmark, except the ARIMA model in predicting the TD5 FFA 2M. Some regularities stand out from the table. First, the FFA rates are much harder to forecast than the spot rates. This phenomenon is not unusual for freight rates and confirms prior studies in the tanker market. Second, in most cases the multivariate models are superior against the univariate representatives. We can find this advantage especially for spot freight rates. But this error difference or advantage declines in the FFA contracts. Furthermore, the VECM, which has an equilibrium correction feature, perform better than VAR models for forecasts of spot rates, but not for forecasts of FFA rates. The neural network results are comparable to those of the other models. It is interesting that the univariate NN achieve relatively good results, but the multivariate NN+ is not able to reinforce this advantage significantly. It seems, that the neural network as a non-linear approximator is already able to extract sufficient information out of the univariate time-series. The additional information contained in other time-series is therefore not needed.

Table 2. One-step ahead forecast performance for Route TD3 and TD5

				Univariate		Multivariate		
Route	Contract	Measure	RW1	ARIMA	NN	VAR	VECM	NN+
TD3	Spot	$\bar{R}^2$	0.9659	0.9723	0.9748	0.9752	0.9756	0.9752
		RMSE	0.0468	0.0425	0.0406	0.0408	0.0397	0.0403
		Theils U	1.0000	0.9087	0.8738	0.8726	0.8591	0.8710
	FFA 1M	$\bar{R}^2$	0.9740	0.9764	0.9771	0.9761	0.9759	0.9769
		RMSE	0.0307	0.0293	0.0289	0.0295	0.0295	0.0291
		Theils U	1.0000	0.9530	0.9406	0.9609	0.9614	0.9494
	FFA 2M	$\bar{R}^2$	0.9577	0.9593	0.9595	0.9594	0.9590	0.9601
		RMSE	0.0284	0.0279	0.0277	0.0279	0.0279	0.0276
		Theils U	1.0000	0.9826	0.9786	0.9839	0.9825	0.9744
TD5	Spot	$\bar{R}^2$	0.9397	0.9475	0.9521	0.9578	0.9595	0.9564
		RMSE	0.0578	0.0540	0.0516	0.0485	0.0471	0.0496
		Theils U	1.0000	0.9345	0.8996	0.8401	0.8185	0.8649
	FFA 1M	$\bar{R}^2$	0.9595	0.9627	0.9625	0.9623	0.9621	0.9628
		RMSE	0.0298	0.0287	0.0287	0.0290	0.0288	0.0287
		Theils U	1.0000	0.9622	0.9619	0.9701	0.9626	0.9612
	FFA 2M	$\bar{R}^2$	0.9617	0.9616	0.9628	0.9637	0.9639	0.9623
		RMSE	0.0222	0.0223	0.0219	0.0216	0.0216	0.0220
		Theils U	1.0000	1.0043	0.9873	0.9751	0.9725	0.9936

4.2 Trading Strategy and Experiment

Statistical performance measures are often inappropriate for financial applications. Trading strategies guided by forecasts on the direction of price change may be more effective and generate higher profits. Therefore, predicting the direction is a practical issue which usually affects a financial trader's decision to buy or sell a freight rate contract. Based on the generated results we provide a simple trading simulation to evaluate our forecasting results in this section.

The trading simulation assumes that, at the beginning of each trading day, the investor makes an asset allocation decision. Consider a freight rate contract whose prices fluctuate from day to day and the mid price on the tth day ($t = 0, 1, 2, \ldots, n$) is q_t. Let $p_t = \ln q_t$ be the log price and $r_t = p_t - p_{t-1}$ be the continuously compounded return on day t. We can generate trading signals now by the following rule:

$$\begin{cases} \text{long, if } \hat{p}_{t+1} > p_t \\ \text{short, if } \hat{p}_{t+1} < p_t \end{cases}$$

A long signal is to buy contracts at the current price, while a short signal is to sell contracts at the current price. This approach has been widely used in the literature.

Except for the straightforward naïve strategy, a random walk, all benchmark models were estimated on our in-sample period. The naïve benchmark strategy

Table 3. Trading performance for Route TD3 and TD5

		TD3						TD5					
		Univariate		Multivariate				Univariate		Multivariate			
Contract	Measure	RW2	ARIMA	NN	VAR	VECM	NN+	RW2	ARIMA	NN	VAR	VECM	NN+
Spot	# trades	282	282	282	282	282	282	282	282	282	282	282	282
	net gain	4.76	5.52	6.07	5.72	5.48	4.64	5.73	6.45	6.95	7.13	7.15	7.32
	log-returns %	1.69	1.96	2.15	2.03	1.94	1.65	2.03	2.29	2.47	2.53	2.54	2.60
	Sharpe ratio	0.39	0.46	0.52	0.48	0.46	0.38	0.38	0.43	0.51	0.49	0.49	0.50
	max profit	0.17	0.33	0.33	0.33	0.33	0.17	0.23	0.40	0.40	0.40	0.40	0.40
	max loss	−0.33	−0.20	−0.08	−0.20	−0.20	−0.33	−0.40	−0.11	−0.11	−0.11	−0.11	−0.11
	winning trades %	0.73	0.73	0.71	0.73	0.72	0.71	0.73	0.74	0.74	0.73	0.72	0.74
FFA 1M	# trades	282	282	282	282	282	282	282	282	282	282	282	282
	net gain	2.01	2.12	2.35	1.79	1.81	2.05	2.01	2.12	2.35	1.81	1.81	2.24
	log-returns %	0.71	0.75	0.83	0.64	0.64	0.73	0.71	0.75	0.83	0.64	0.64	0.79
	Sharpe ratio	0.24	0.25	0.28	0.21	0.21	0.24	0.20	0.22	0.27	0.22	0.23	0.28
	max profit	0.14	0.14	0.14	0.14	0.14	0.14	0.14	0.11	0.11	0.11	0.11	0.11
	max loss	−0.07	−0.09	−0.07	−0.07	−0.09	−0.09	−0.07	−0.10	−0.08	−0.09	−0.09	−0.07
	winning trades %	0.55	0.57	0.58	0.56	0.58	0.59	0.55	0.59	0.62	0.59	0.59	0.60
FFA 2M	# trades	282	282	282	282	282	282	282	282	282	282	282	282
	net gain	1.01	1.12	1.23	0.61	1.14	1.72	1.01	0.58	1.28	1.40	1.52	1.08
	log-returns %	0.36	0.40	0.44	0.22	0.40	0.61	0.36	0.21	0.45	0.49	0.54	0.38
	Sharpe ratio	0.13	0.14	0.16	0.08	0.14	0.22	0.16	0.09	0.21	0.23	0.25	0.17
	max profit	0.10	0.10	0.10	0.10	0.10	0.10	0.10	0.10	0.10	0.10	0.10	0.10
	max loss	−0.08	−0.08	−0.08	−0.07	−0.08	−0.08	−0.08	−0.07	−0.07	−0.06	−0.06	−0.07
	winning trades %	0.51	0.54	0.53	0.50	0.57	0.54	0.51	0.52	0.56	0.57	0.59	0.56

is defined by $\hat{r}_{t+1} = r_t$, where r_t is the actual rate of return at period t and $\hat{r}_{t+1}$ is the forecast rate of return for the next period. So, we switch from the no-change random walk to a constant (last) change random walk model (RW2).

In the trading experiment, it is assumed that during the initiation period, an investor will invest 1 monetary unit at the beginning of each contract period. So far, our results have been presented without accounting for transaction costs during the trading simulation.

4.3 Trading Results and Analysis

The net gain in assets, number of trades executed, and the rate of return over the out-of-sample forecast horizon are shown in table 3 for TD3 and TD5. The initial investments are identical in all models, due to the buy-and-hold strategy. Therefore, all measures are comparable. We see some implications: First, all models earn a positive trading result in case of no transaction costs. Furthermore, it is obvious, that the trading results in spot rates are more profitable than those for FFA contracts. This is also valid for the directional measure "winning trades per %". In most cases all models outperform the naïve RW2 model, except some time series models in predicting FFA prices. The multivariate NN+ undermatch the RW2 benchmark trading results for TD3 spot freight rates.

Additional observations are worth pointing out. The results generated by NN are encouraging in comparison to the other models – for every predicted asset the univariate NN shows the best performance across all models (univariate and multivariate) with respect to the important measures of net gain and risk-adjusted return as measured by the Sharpe ratio. ARIMA and VAR results show no unambiguous picture. Both models outperform the RW2 in case of spot rates. But ARIMA does not perform for TD5 FFA 2M contracts and VAR get worse results for TD3 FFA contracts. The multivariate VECM shows relatively good and stable results, except for the FFA 1M contracts. The multivariate NN+ achieves only in some cases preeminent trading results, e.g. for the TD5 spot freight rates. As mentioned above, additional time series do not improve the neural network performance. We conclude, that both VECM and univariate NN may generate more robust trading results for this time series and perform better than the other forecasting models.

5 Conclusions and Recommendations

In this paper, we have examined the forecasting and trading performance of various standard linear time series models and a non-linear neural network to jointly predict spot and forward freight rates (FFA prices). We have focused on short-term forecasting. To our knowledge there is a lack in the literature of joint predictions of freight rates and derivatives with neural networks and traditional time series models.

We conclude, that neural networks are suitable for short-term forecasting and trading of tanker freight rates and derivatives. For the two most liquid tanker routes TD3 and TD5 we implicate that short-term forecasting with neural networks leads to better results than other traditional time series models. Our forecasting results confirm prior studies concerning time series models. However, out-of-sample forecasting with multivariate forecasting models show that spot freight rates are not helpful in predicting FFA prices, but FFA prices do help predict spot freight rates. The results of neural networks are in line with these findings.

We have implemented a simple trading simulation to evaluate the forecasting performance with economical criteria. Trading strategies guided by forecasts on the direction of price change may be more effective and generate higher profits. In our evidence, both VECM and univariate neural networks may generate more robust trading results for the analyzed time series than the other forecasting models. Therefore, neural networks could be a starting point for building a decision support model for spot freight rate and FFA trading purposes. Several extensions for further research are also thinkable, e.g. longer investment horizons and inclusion of further exogenous input variables in multivariate models like crude oil prices, maritime data or any other variables.

References

1. Adland, R., Cullinane, K.: The non-linear dynamics of spot freight rates in tanker markets. Transportations Research Part E 42, 211–224 (2006)
2. Alizadeh, A.H., Nomikos, N.K.: Shipping Derivatives and Risk Management. Palgrave-McMillan, London (2009)
3. Batchelor, R., Alizadeh, A.H., Visvikis, I.D.: Forecasting spot and forward prices in the international freight market. International Journal of Forecasting 23(1), 101–114 (2007)
4. Culliane, K., Mason, K.J., Cape, M.B.: A comparison of models for forecasting the Baltic Freight Index: Box-Jenkins revisited. International Journal of Maritime Economics 2(1), 15–39 (1999)
5. Goulielmos, A.M., Psifia, M.E.: Forecasting weekly freight rates for oneyear time charter 65 000 dwt bulk carrier, 1989-2008, using nonlinear methods. Maritime Policy & Management 36(5), 411–436 (2009)
6. Li, J., Parsons, M.: Forecasting tanker freight rate using neural networks. Maritime Policy & Management 24(1), 9–30 (1997)
7. Lyridis, D.V., Zacharioudakis, P., Mitrou, P., Mylonas, A.: Forecasting tanker market using artificial neural networks. Maritime Economics and Logistics 6, 93–108 (2004)
8. Mettenheim, H.-J., Breitner, M.H.: Robust decision support systems with matrix forecasts and shared layer perceptrons for finance and other applications. In: ICIS 2010 Proceedings 83 (2010)
9. Veenstra, A.W., Franses, P.H.: A co-integration approach to forecasting freight rates in the dry bulk shipping sector. Transportation Research Part A 31(6), 447–458 (1997)

Modelling and Trading the DJIA Financial Index Using Neural Networks Optimized with Adaptive Evolutionary Algorithms

Konstantinos Theofilatos[1], Andreas Karathanasopoulos[2], Georgios Sermpinis[3], Thomas Amorgianiotis[1], Efstratios Georgopoulos[4], and Spiros Likothanassis[1]

[1] Department of Computer Engineering and Informatics, University of Patras, Greece
[2] London Metropolitan Business School, London Metropolitan University, 31 Jewry Street, London, EC3N 2EY, UK
[3] University of Glasgow Business School, University of Glasgow, Adam Smith Building, Glasgow, G12 8QQ, UK
[4] Technological Educational Institute of Kalamata, 24100, Kalamata, Greece
{theofilk,amorgianio,likothan}@ceid.upatras.gr,
a.karathanasopoulos@londonmet.ac.uk,
georgios.sermpinis@glasgow.ac.uk, sfg@teikal.gr

Abstract. In the current paper we study an evolutionary framework for the optimization of various types Neural Networks' structure and parameters. Two different adaptive evolutionary algorithms, named as adaptive Genetic Algorithms (aGA) and adaptive Differential Evolution (aDE), were developed to optimize the structure and the parameters of two different types of Neural Networks: Multilayer Perceptron (MLPs) and Wavelet Neural Networks (WNN). Wavelet neural networks have been introduced as an alternative to MLPs to overcome their shortcomings presenting more compact architecture and higher learning speed. Furthermore, the evolutionary algorithms, which were implemented, are both adaptive in terms that their most important parameters (Mutation and Crossover probabilities) are assigned using a self adaptive scheme. The motivation of this paper is to uncover novel hybrid methodologies for the task of forecasting and trading DJIE financial index. This is done by benchmarking the forecasting performance the four proposed hybrid methodologies (aGA-MLP, aGA-WNN, aDE-MLP and aDE-WNN) with some traditional techniques, either statistical such as a an autoregressive moving average model (ARMA), or technical such as a moving average covcergence/divergence model (MACD). The trading performance of all models is investigated in a forecast and trading simulation on our time series over the period 1997-2012. As it turns out, the aDE-WNN hybrid methodology does remarkably well and outperforms all other models in simple trading simulation exercises. (This paper is submitted for the ACIFF workshop).

Keywords: Trading Strategies, Financial Forecasting, Transaction costs, Multi-Layer Perceptron, Wavelet Neural Networks, Genetic Algorithms, Differential Evolution, Hybrid forecasting methodologies.

C. Jayne, S. Yue, and L. Iliadis (Eds.): EANN 2012, CCIS 311, pp. 453–462, 2012.
© Springer-Verlag Berlin Heidelberg 2012

1   Introduction

Stock market analysis is an area of growing quantitative financial applications. Modeling and trading financial indices is a very challenging open problem for the scientific community due to their chaotic non-linear nature. Thus, forecasting financial time series is a difficult task because of their complexity and their nonlinear, dynamic and noisy behavior. Traditional methods such as ARMA and MACD models fail to capture the complexity and the nonlinearities that exist in financial time series. Neural network approaches have given satisfactory results but they suffer from certain limitations. In the present paper we used Multi Layer Perceptron (MLP) Neural Networks (NN), as being the most famous among them, and Wavelet Neural Networks (WNN) as being a modern architecture which is believed to surpass the limitations of MLP such as slow convergence and training difficulties.

Both MLPs and WNN have already been used for the modeling of financial time series. In specific Dunis et al. [1] in 2010 applied the MLP NN to the problem of modeling and trading the ASE 20 index. Compared to classical models, such as ARMA models, MLPs were proved to outperform them in terms of annualized return. Moreover, WNN have also been applied to financial forecasting problems [2] and they have been prove to outperform simple MPL NNs.

Despite their high performance in many practical problems including financial forecasting, NNs present some drawbacks that should be surpassed in order to establish them as reliable trading tools. In specific, their performance is highly related to their structure and parameter values. The optimization of the structure and parameter values of NNs for financial forecasting and trading is an open problem even if some initial approaches have been developed [3, 4]. All these approaches appear to be able to optimize the NNs structure and parameters but include some extra parameters of the optimization technique which should in turn be optimized through experimentation.

In the present paper, we introduce two novel evolutionary approaches (adaptive Genetic Algorithms, adaptive Differential Evolution) which adapt their parameters during the evolutionary process to achieve better convergence characteristics and to enable not experienced users to use them. These evolutionary techniques were used to optimize the structure and the parameters of MLP NNs and WNNs. The produced hybrid techniques were applied to the problem of modeling and trading the DJIA index. Experimental results demonstrated that they present better trading performance than statistical trading techniques. From the proposed hybrid techniques, the hybrid technique which combines the adaptive Differential Evolution method with WNNs outperformed all other examined models in terms of annualized returns even when transactions costs are taken into account.

2   The DJIA Index and Related Financial Data

The Dow Jones Industrial Average (DJIA) also called the Industrial Average, the Dow Jones, the Dow 30, or simply the Dow, is a stock market index, and one of several indices created by Wall Street Journal editor and Dow Jones & Company co-founder Charles Dow. It was founded on May 26, 1896, and is now owned by Dow Jones Indexes, which is majority owned by the CME Group. It is an index that shows how 30 large, publicly owned companies based in the United States have traded during a standard trading session in the stock market.

In Table 1 we present the dataset and the dataset's subsets which were used for training the examined predictors and validating their performance.

Table 1. Total dataset

Name of Period	Trading Days	Beginning	End
Total Dataset	3597	27 October 1997	6 February 2012
Training set	2157	27 October 1997	23 May 2006
Test Set	720	24 May 2006	31 March 2009
Validation Set	720	1 April 2009	6 February 2012

As inputs to our algorithms and our networks, we selected 14 autoregressive inputs described in detail in Table 2 below.

Table 2. Explanatory variables

Number	Variable	Lag
1	DJIA all share return	1
2	DJIA all share return	2
3	DJIA all share return	3
4	DJIA all share return	4
5	DJIA all share return	5
6	DJIA all share return	6
7	DJIA all share return	7
8	DJIA all share return	8
9	DJIA all share return	9
10	DJIA all share return	10
11	DJIA all share return	11
12	DJIA all share return	12
13	DJIA all share return	13
14	DJIA all share return	14

3   Forecasting Models

3.1   Benchmark Models

For comparison reasons two benchmark models were applied to the problem of modeling the DJIA financial index. These are an ARMA and a MACD models.

3.1.1 ARMA Model

Autoregressive moving average models (ARMA) assume that the value of a time series depends on its previous values (the autoregressive component) and on previous residual values (the moving average component)[1] . The ARMA model takes the form of equation (1)

$$Y_t = \phi_0 + \phi_1 Y_{t-1} + \phi_2 Y_{t-2} + \ldots + \phi_p Y_{t-p} + \varepsilon_t - w_1 \varepsilon_{t-1} - w_2 \varepsilon_{t-2} - \ldots - w_q \varepsilon_{t-q} \quad (1)$$

where Y_t is the dependent variable at time t

Y_{t-1}, Y_{t-2}, and Y_{t-p} are the lagged dependent variable

ϕ_0, ϕ_1, ϕ_2, and ϕ_p are regression coefficients

ε_t is the residual term

ε_{t-1}, ε_{t-2}, and ε_{t-p} are previous values of the residual

w_1, w_2, and w_q are weights.

Using as a guide the correlogram in the training and the test sub periods we have chosen a restricted ARMA (6, 6) model. All of its coefficients are significant at the 99% confidence interval. The null hypothesis that all coefficients (except the constant) are not significantly different from zero is rejected at the 99% confidence interval.

The selected ARMA model takes the form:

$$Y_t = 3.20 \cdot 10^{-4} + 0.276 Y_{t-1} - 0.446 Y_{t-3} - 0.399 Y_{t-6} + 0.264 \varepsilon_{t-1} - 0.170 \varepsilon_{t-3} - 0.387 \varepsilon_{t-6} \quad (2)$$

The model selected was retained for out-of-sample estimation. The performance of the strategy is evaluated in terms of traditional forecasting accuracy and in terms of trading performance.

3.1.2 MACD

The moving average model is defined as:

$$M_t = \frac{(Y_t + Y_{t-1} + Y_{t-2} + \ldots + Y_{t-n+1})}{n}$$

Where M_t is the moving average at time t

n is the number of terms in the moving average

Y_t is the actual rate of return at period t

The MACD strategy used is quite simple. Two moving average series are created with different moving average lengths. The decision rule for taking positions in the market is straightforward. Positions are taken if the moving averages intersect. If the short-term moving average intersects the long-term moving average from below a 'long' position is taken. Conversely, if the long-term moving average is intersected from above a 'short' position is taken.

The forecaster must use judgement when determining the number of periods n on which to base the moving averages. The combination that performed best over the in-sample sub-period was retained for out-of-sample evaluation. The model selected was a combination of the Djie and its 8-day moving average, namely n = 1 and 8 respectively or a (1, 8) combination. The performance of this strategy is evaluated solely in terms of trading performance.

3.2   Proposed Hybrid Techniques

Neural networks exist in several forms in the literature. In the present study, we used Multi Layer Perceptron (MLP) which is one of the most famous architectures and Wavelet Neural Networks (WNN) which are an alternative neural network architecture designed to overcome convergence limitations of the MLP Neural Networks.

A standard MLP neural network has three layers and this setting is adapted in the present paper. The first layer is called the input layer (the number of its nodes corresponds to the number of explanatory variables). The last layer is called the output layer (the number of its nodes corresponds to the number of response variables). An intermediary layer of nodes, the hidden layer, separates the input from the output layer. Its number of nodes defines the amount of complexity the model is capable of fitting. In addition, the input and hidden layer contain an extra node, called the bias node. This node has a fixed value of one and has the same function as the intercept in traditional regression models. Normally, each node of one layer has connections to all the other nodes of the next layer. The network processes information as follows: the input nodes contain the value of the explanatory variables. Since each node connection represents a weight factor, the information reaches a single hidden layer node as the weighted sum of its inputs. Each node of the hidden layer passes the information through a nonlinear activation function and passes it on to the output layer if the calculated value is above a threshold. The training of the network (which is the adjustment of its weights in the way that the network maps the input value of the training data to the corresponding output value) starts with randomly chosen weights and proceeds by applying a learning algorithm called backpropagation of errors. The backpropagation algorithm simply tries to find those weights which minimize an error function.

WNNs are a generalized form of radial basis function feed forward neural networks. Similar to simple MLPs they are three layered architectures having only one hidden layer. In contrast to simple MLPs, WNNs use radial wavelets as activation functions to the hidden layer, while using the linear activation function in the output layer. The information processing of a WNN is performed as follows. Suppose $x=[x_1,...x_d]$ to be the input signal, where d is the inputs dimensionality, then the output of its j-th hidden neuron is estimated using equation (3)

$$\psi_j(x) = \prod_{i=1}^{d} \phi_{d_{ij}, t_{i,j}}(x_i) \tag{3}$$

where $\phi_{d_{i,j}, t_{i,j}}$ is the wavelet activation function (one among Mexican Hat, Morlet and Gaussian wavelet) of the j-th hidden node and $d_{i,j}$ and $t_{i,j}$ are the scaling and translational vectors respectively. Then the output of the WNN is computed by estimating the weighted sum of the outputs of each hidden neuron using the weights that connect them with the output layer. The learning process involves the approximation of the scaling and translational vectors which should be used for the hidden layer and of the weights that connect the hidden layer with the output. For the approximation of the scaling vector we used the methodology proposed by Zhang and Benveniste [5], and for the approximation of the translational vectors we used the k-means algorithm [6] with k being the number of hidden nodes. The weight vector W can easily be computed analytically by computing the following equation

$$W=(\Psi^T\Psi)^{-1}\Psi^TY \tag{4}$$

where

$$\Psi = \begin{pmatrix} \psi(x_1,d_1,t_1) & \psi(x_1,d_2,t_2) & ... & \psi(x_1,d_m,t_m) \\ \psi(x_2,d_1,t_1) & ... & ... & ... \\ ... & ... & ... & ... \\ \psi(x_d,d_1,t_1) & \psi(x_d,d_2,t_2) & ... & \psi(x_d,d_m,t_m) \end{pmatrix} \tag{5}$$

The parameters of the MLP that need to be tuned in every modeling problem are the inputs that the MLP neural networks should have, the size of the hidden layer, the activation functions which should be used for the hidden and output layers, the learning rate and the momentum parameter. The parameters of the WNN that need to be tuned in every modeling problem are the inputs that is should use, the size of the hidden layer and the activation function that should be used in the hidden layer. The appropriate tuning of all these parameters requires a hard time consuming step of trial and error procedure. This trial error procedure may lead to overfitting of the algorithms and to the data snooping effects. Both these phenomenon may lead us in overestimating our results. To overcome these difficulties in the present paper we propose the application of evolutionary optimization algorithms to optimize the structure and aforementioned the parameters of MLP Neural Networks and WNNs. The evolutionary algorithms which were developed for this purpose are designed to be adaptive. Specifically, in order to achieve better convergence behavior, we attempted to adapt their parameter values during the evolutionary process. These algorithms are described in detail in the following sections.

For comparison reasons, we used the same size of the initial population (20 candidate solutions) and the same termination criterion for both evolutionary approaches. The termination criterion which was applied is a combination of the maximum number of iterations and a convergence criterion. The maximum number of iterations was set to 100 and the convergence criterion is satisfied when the mean population fitness is less than 5% away from the best population fitness for more than five consecutive iterations.

As fitness function both evolutionary methods use the one described in equation (6) in order to force them towards more profitable strategies.

$$\text{Fitness = annualized return - MSE -0.001*\#selected \_inputs} \tag{6}$$

3.2.1 Adaptive Genetic Algorithms (aGA)
GAs [7] are search algorithms inspired by the principle of natural selection. They are useful and efficient if the search space is big and complicated or there is not any available mathematical analysis of the problem. A population of candidate solutions, called chromosomes, is optimized via a number of evolutionary cycles and genetic

operations, such as crossovers or mutations. Chromosomes consist of genes, which are the optimizing parameters. At each iteration (generation), a fitness function is used to evaluate each chromosome, measuring the quality of the corresponding solution, and the fittest chromosomes are selected to survive. This evolutionary process is continued until some termination criteria are met. It has been shown that GAs can deal with large search spaces and do not get trapped in local optimal solutions like other search algorithms [7].

In the present paper we deployed an adaptive GA to optimize the structure and the parameters of MLP neural networks and WNNs. The chromosomes are encoded as binary strings. When, continuous values are needed the string genes are transformed to their corresponding decimal value. As for the selection operator, roulette wheel selection was deployed. In order to raise the evolutionary pressure and speed up convergence, the fitness functions were scaled using the exponential function.

The two main genetic operators of a Genetic Algorithm are crossover and mutation. For the crossover operator, two-point crossover was used to create two offsprings from every two selected parents. The parents are selected at random, two crossover points are selected at random and two offsprings are made by exchanging genetic material between the two crossover points of the two parents. The crossover probability was set equal to 0.9 to leave some part of the population to survive unchanged to the next generation.

Most studies on the selection of the optimal mutation rate parameter coincide that a time-variable mutation rate scheme is usually preferable than a fixed mutation rate [8]. Accordingly, we propose the dynamic control of the mutation parameter using equation (7):

$$Pm(n) = 0.2 - n \cdot \frac{0.2 - \frac{1}{P_S}}{MAX_G} \tag{7}$$

where n is the current generation, PS is the size of the population and MAXG is the maximum generation specified by the termination criteria. Using equation (7), we start with a high mutation rate for the first generations and then gradually decrease it over the number of generations. In this manner global search characteristics are adopted in the beginning and are gradually switched to local search characteristics for the final iterations. The mutation rate is reduced with a smaller step when a small population size is used in order to avoid stagnation. For bigger population sizes the mutation rate is reduced with a larger step size since a quicker convergence to the global optimum is expected.

3.2.2   Adaptive Differential Evolution (aDE)

The DE algorithm is currently one of the most powerful and promising stochastic real parameter optimization algorithms [9]. As an evolutionary algorithm, it iteratively applies selection, mutation and crossover operators until some termination criteria are reached. In opposite to GAs, DE is mainly based on a specific mutation operator which perturbs population individuals with the scaled differences of randomly selected and distinct population members. DE is able to handle continuous gene representation. Thus, candidate solutions are represented as strings of continuous variables comprising of feature and parameter variables. In order to compute discrete values which are to be optimized like the size of the hidden layer rounding of the continuous values was used.

The mutation operator which was used in our proposed wrapper method, for every population member Xi, initially selects three random distinct members of the population $(X_{1,i}, X_{2,i}, X_{3,i})$ and produces a donor vector using the equation (8):

$$V_i = X_{1,i} + F*(X_{2,i} - X_{3,i})$$ (8)

where F is called mutation scale factor.

The crossover operator applied was the binomial one. This operator combines every member of the population xi with its corresponding donor vector Vi to produce the trial vector U_i using the equation (9).

$$U_i(j) = \begin{cases} V_i(j) & , \quad if\ (rand_{i,j}[0,1] \le Cr) \\ X_i(j) & , \quad otherwise \end{cases}$$ (9)

where $rand_{i,j}[0,1]$ is a uniformly distributed random number and Cr is the crossover rate.

Next, every trial vector U_i is evaluated and if it suppresses the corresponding member of the population X_i it takes its position in the population.

The most important control parameters of a DE algorithm are the mutation scale factor F and the crossover rate Cr. Parameter F controls the size of the differentiation quantity which is going to be applied to a candidate solution from the mutation operator. Parameter Cr determines the number of genes which are expected to change in a population member. Many approaches have been developed to control these parameters during the evolutionary process of the DE algorithm [9]. In our adaptive DE version, we deployed one of the most recent promising approaches [10]. This approach uses in every iteration a random value for the F parameter selected from a uniform distribution with mean value 0.5 and standard deviation 0.3 and a random value for the parameter Cr from a uniform distribution with mean value Crm and standard deviation 0.1. Crm is initially set to 0.5. The Crm is replaced during the evolutionary process with values that have generated successful trial vectors. Thus, this approach replaces the sensitive user defined parameters F and Cr with less sensitive parameters like their mean values and their standard deviation.

4   Empirical Trading Results

In this section we present the results of the studied models in the problem of trading the DJIA index. The trading performance of all the models considered in the out-of-sample subset is presented in the table below. Our trading strategy is simple and identical for all of them: go or stay long when the forecast return is above zero and go or stay short when the forecast return is below zero. Because of the stochastic nature of the proposed methodologies a simple run is not enough to measure their performance. This is the reason why ten runs where executed and the mean results are presented in the next tables.

Table 3. Out of sample trading performance results

	ARMA	MACD	aGA-MLP	aDE-MLP	aGA-WNN	aDE-WNN
Information Ratio (excluding costs)	0,12	0,37	0,98	0,99	1,07	1,44
Annualized Volatility (excluding costs)	18,04%	18,04%	18,01%	18,01%	18,00%	17,97%
Annualized Return (excluding costs)	2,16%	6,61%	17,74%	17,79%	19,30%	25,96%
Maximum Drawdown (excluding costs)	-28,01%	-14,37%	-17,19%	-16,91%	-13,75%	-16,05%
Positions Taken (annualized)	34	27	33	6	64	104
Transaction costs	0,30%	0,24%	0,10%	0,05%	0,57%	0,94%
Annualized Return (including costs)	1,86%	6,37%	17,64%	17,74%	18,73%	25,02%

It is easily observed from Table 3 that the hybrid evolutionary - neural network methods clearly outperform the classical ARMA and MACD models. Furthermore, among the hybrid proposed techniques aDE-WNN is the one that prevails. The robustness of WNNs when combined with the supreme global search characteristics of the proposed adaptive differential evolution approach, is able to produce highly profitable trading strategies.

4.1   Transaction Costs

The transaction costs for a tradable amount are about 1 pip per trade (one way) between market makers. But as the DJIE time series considered here is a series of middle rates, the transaction cost is one spread per round trip therefore the cost of 1 pip is equivalent to an average cost of 0.009% per position.

From Table 3 one can easily see that even when considering transaction costs, the aDE-WNN predictor still significantly outperforms all other examined trading strategies in terms of annualized return.

5   Concluding Remarks

In the present paper we introduced a computational framework for the combination of evolutionary algorithms and neural networks for the forecasting and trading of financial indices. The evolutionary algorithms are used to optimize the neural networks structure and parameters. Two novel adaptive evolutionary algorithms were developed named as adaptive Genetic Algorithms and adaptive Differential Evolution. Both of them use parameters whose values are being adapted during the evolutionary process either to fasten the evolutionary process or to avoid getting trapped in local optimal solutions. These algorithms were deployed to optimize the structure and parameters of MLP and wavelet neural networks. The derived hybrid methodologies were applied in the problem of modeling and short term trading of the DJIA financial index using a problem specific fitness function.

Experimental results proved that the proposed hybrid techniques clearly outperformed statistical techniques in terms of sharp ratio and annualized return. Furthermore, as expected, WNNs produced more profitable trading strategies than MLP NNs. The optimal hybrid technique, as proved experimentally, is the aDE-WNN method. This technique outperformed the other examined models in the applied trading simulations.

References

1. Dunis, C., Laws, J., Karathanasopoulos, A.: Modeling and Trading the Greek Stock Market with Mixed Neural Network Models. Refereed at Applied Financial Economics (2010)
2. Bozic, J., Vukotic, S., Babic, D.: Prediction of the RSD exchange rate by using wavelets and neural networks. In: Proceedings of the 19th Telecommunication Forum TELFOR, pp. 703–706 (2011)
3. Karathanasopoulos, A.S., Theofilatos, K.A., Leloudas, P.M., Likothanassis, S.D.: Modeling the Ase 20 Greek Index Using Artificial Neural Nerworks Combined with Genetic Algorithms. In: Diamantaras, K., Duch, W., Iliadis, L.S. (eds.) ICANN 2010, Part I. LNCS, vol. 6352, pp. 428–435. Springer, Heidelberg (2010)
4. Tao, H.: A Wavelet Neural Network Model for Forecasting Exchange Rate Integrated with Genetic Algorithms. IJCSNS International Journal of Computer Science and Network Security 6(8), 60–63 (2006)
5. Zhang, Q., Benveniste, A.: Wavelet networks. IEEE Transactions on Neural Network 3, 889–898 (1982)
6. Zainuddin, Z., Pauline, O.: Improved Wavelet Neural Networks for Early Cancer Diagnosis Using Clustering Algorithms. International Journal of Information and Mathematics Sciences 6(1), 30–36 (2010)
7. Holland, J.: Adaptation in Natural and Artificial Systems: An Introductory Analysis with Applications to Biology, Control and Artificial Intelligence. MIT Press, Cambridge (1995)
8. Thierens, D.: Adaptive Mutation Rate Control Schemes in Genetic Algorithms. In: Proceedings of the 2002 IEEE World Congress on Computational Intelligence: Congress on Evolutionary Computation, pp. 980–985 (2002)
9. Das, S., Suganthan, P.: Differential Evolution: A survey of the State-of-the-Art. IEEE Transactions on Evolutionary Computation 15(1), 4–30 (2011)
10. Qin, A., Huang, V., Suganthan, P.: Differential evolution algorithm with strategy adaptation for global numerical optimization. IEEE Transactions of Evolutionary Computation 13(2), 398–417 (2009)

Applying Kernel Methods on Protein Complexes Detection Problem

Charalampos Moschopoulos[1,2], Griet Laenen[1,2], George Kritikos[3,4], and Yves Moreau[1,2]

[1] Department of Electrical Engineering-ESAT, SCD-SISTA, Katholieke Universiteit Leuven,
Kasteelpark Arenberg 10, box 2446, 3001, Leuven, Belgium
[2] IBBT Future Health Department, Katholieke Universiteit Leuven, Kasteelpark Arenberg 10,
box 2446, 3001, Leuven, Belgium
[3] Bioinformatics & Medical Informatics Team, Biomedical Research Foundation,
Academy of Athens, Soranou Efessiou 4, 11527 Athens, Greece
[4] EMBL - European Molecular Biology Laboratory, Department of Genome Biology,
Meyerhofstr. 1, D-69117 Heidelberg, Germany
{Charalampos.Moschopoulos,Griet.Laenen,
Yves.Moreau}@esat.kuleuven.be, kritikos@embl.de

Abstract. During the last years, various methodologies have made possible the detection of large parts of the protein interaction network of various organisms. However, these networks are containing highly noisy data, degrading the quality of information they carry. Various weighting schemes have been applied in order to eliminate noise from interaction data and help bioinformaticians to extract valuable information such as the detection of protein complexes. In this contribution, we propose the addition of an extra step on these weighting schemes by using kernel methods to better assess the reliability of each pairwise interaction. Our experimental results prove that kernel methods clearly help the elimination of noise by producing improved results on the protein complexes detection problem.

Keywords: kernel methods, protein-protein interactions, protein interaction graphs, protein complexes.

1   Introduction

Protein-protein interaction (PPI) data are generated by a variety of experimental methodologies which either produce thousands of PPIs introducing a significant amount of noise in the form of false positives or false negatives, either they produce highly reliable interaction data suffering from poor coverage of the complete interactome graph. The aggregation of these data is stored in online repositories alongside with information concerning the methods used for their detection.

During the last years, various computational methods have been used in order to extract information from PPI data such as protein complex detection [1, 2] and definition of the functionality of unknown proteins [3]. It has been proved that the results of these methods depend vastly on the quality of the input data, so there is a

C. Jayne, S. Yue, and L. Iliadis (Eds.): EANN 2012, CCIS 311, pp. 463–471, 2012.
© Springer-Verlag Berlin Heidelberg 2012

great interest from the bioinformatics community in the reliability evaluation of each recorded PPI.

Different weighting schemes have been developed that assign confidence scores which reflect the reliability and biological significance of each protein interaction pair [4]. Several schemes use methods derived from graph theory (such as the Czekanowski-Dice distance) [5]. Other approaches compute confidence scores with the help of other biological data such as expression data, number of experiments, protein functionality, protein localization etc. [6, 7].

In this manuscript we propose the use of kernel methods on the results of any weighting scheme in order to further improve the quality of weighted PPI datasets by revealing the importance of indirect connections between proteins in the PPI networks. We prove the efficiency of this methodology by applying kernel methods on three different weighting schemes (Adjust-CD [8], Functional similarity [9] and MV scoring [10]). Afterwards, we apply the clustering algorithms Markov clustering (MCL) [11] and EMC [12] and we prove that better results are derived on the protein complex detection problem due to the kernel methods.

2   Materials and Methods

In this study, the protein interaction data of the yeast subset of the IrefIndex database [13] has been used (version 9.0, release date: November 2011). IrefIndex database merges thirteen different interactome databases after removal of the redundant entries.

Three different weighting schemes, widely used in the protein interaction evaluation problem [10],were used in our study: Adjust-CD, Functional similarity and MV scoring. On the results of these methods we applied exponential and Laplacian exponential diffusion kernel methods, as these kernel methods are the most popular and widely used in the literature. Then, the MCL algorithm was applied in order to result in the final protein complex candidates. The selection of MCL was made based on the fact that other similar algorithms like RNSC [2] or COACH [14] algorithms could not be applied to a weighted PPI graph, since they do not take edge weights into account. In order to further filter the derived results, we chose EMC algorithm which use the MCL algorithm in the first step of its methodology. Finally, after applying a clustering algorithm, the derived results were validated against the golden standard of recorded protein complexes of yeast, which is available in the MIPS database [15]. All these methods are discussed in the following subparagraphs.

2.1   Weighting Schemes

Adjust-CD Weighting Scheme. The Adjust-CD method [8] is derived from the PRODISTIN weighting method that uses the Czekanowski-Dice distance (CD-Distance) [5] and consists of an iterative procedure that relies solely on the network topology to calculate the reliability of a binary protein interaction. More specifically, the CD-Distance estimates the degree of functional similarity between two proteins from the number of neighbors they have in common. While PRODISTIN takes into consideration only the 1st degree neighbors, Adjust-CD calculates also the CD-distances of the 2nd degree neighbors. Interestingly, the Adjust-CD scoring method may also be used to discover protein interactions that do not originally exist in the protein interaction network.

Functional Similarity Weighting Scheme. The functional similarity weighting scheme (FS-weight) [9] is based on both topology and reliability of interactions estimated from the frequency and sources of physical evidence to estimate functional similarity between 1st and 2nd degree neighbors. Similarly to Adjust-CD, FS-weight considers the indirect functional association between two proteins important in case these proteins share a common neighborhood. However, it gives lower weight values to common neighbors by penalizing the pairwise interactions where any of these proteins has few 1st degree neighbors.

MV Scoring Weighting Scheme. Instead of using graph properties in its function, like the previous mentioned methods do, the MV scoring scheme [10] assigns higher weights to those interactions that are reported from a high number of experiments or have been recorded during an experiment with low plurality. This kind of information is hosted in IrefIndex database, so there is no bias at the integration of this information in the tested dataset. However, this weighting scheme assigns high confidence interactions derived from a single experiment with low plurality.

2.2 Graph Kernel Methods

Graph kernel methods can help assess the importance of direct as well as indirect connections between pairs of proteins in PPI graphs by assigning a weight to each pair. In this manuscript, two different kernel methods have been applied on already weighted PPI graphs and as a result pairwise weighted interactions between almost all proteins are generated.

The Adjacency and Laplacian Matrix of a Graph. Given a weighted, undirected graph G, with symmetric weights w_{ij} between nodes i and j, the adjacency matrix A is defined to have entries

$$A_{ij} = \begin{cases} w_{ij}, & \text{if } i \sim j \\ 0, & \text{otherwise.} \end{cases} \tag{1}$$

The combinatorial Laplacian matrix L of G is defined as $L = D - A$ where $D = \text{diag}(d_{ii}) = \text{diag}\left(\sum_{j=1}^{n} a_{ij}\right)$ such that L has entries

$$L_{ij} = \begin{cases} -w_{ij}, & \text{if } i \sim j \\ d_{ii}, & \text{if } i = j \\ 0, & \text{otherwise.} \end{cases} \tag{2}$$

The Exponential and Laplacian Exponential Graph Kernel. A kernel function $\kappa \colon \mathcal{X} \times \mathcal{X} \mapsto \mathbb{R}$ provides a similarity measure on the input space $\mathcal{X}$. Now let $\mathcal{X}$ be the vertices of a weighted, undirected graph G, then κ can be represented as a matrix K with each element $K_{ij} = K_{ji}$ capturing a global relationship between node i and node j that takes indirect paths between the nodes into account.

We consider two different graph kernels to calculate these relationships, namely the exponential and the Laplacian exponential kernel.

The ***exponential kernel*** was introduced by Kandola *et al.* [16] as:

$$K_E = e^{\alpha A} = \lim_{n \to \infty} \left(I + \frac{\alpha A}{n} \right)^n \tag{3}$$

, where A is the adjacency matrix, I is the identity matrix, α determines the degree of diffusion, and n is the number of iterations. This kernel integrates a contribution from all paths between node i and node j, while discounting these paths according to the number of steps.

A meaningful alternative to the exponential kernel can be obtained by simply substituting the adjacency matrix A in (3) with the negated Laplacian matrix L, resulting in the ***Laplacian exponential kernel*** defined by Kondor and Lafferty [17] as:

$$K_{LED} = e^{-\alpha L} = \lim_{n \to \infty} \left(I - \frac{\alpha L}{n} \right)^n. \tag{4}$$

This kernel is also known as the Laplacian exponential diffusion or heat kernel, since it originates from the differential equation describing heat diffusing in material. It can be seen as a random walk, starting from a node i and transitioning to a neighboring node j with probability αw_{ij} or staying at node i with probability $1 - \alpha d_{ii}$.

2.3   Clustering Algorithms

In order to prove the assets of the proposed methodology, we performed clustering on the weighted PPI graphs. We chose to use the Markov clustering algorithm (MCL) as its efficiency has been proved in various review articles about the protein complex detection problem [18, 19]. Moreover, we filtered the results of MCL in order to obtain better quality results by applying the EMC algorithm.

MCL Algorithm. The MCL algorithm [11] is a fast and scalable unsupervised clustering algorithm. It is one of the most widely used algorithms on the protein complexes detection problem and it is based on simulating stochastic flows in networks. The MCL algorithm can detect cluster structures in graphs by taking advantage of a mathematical bootstrapping procedure. The process is trying to perform random walks through a graph and deterministically compute their probabilities to find the best paths. It does so by using stochastic Markov matrices and applying iteratively the inflation and the expansion parameters. By pruning "weak" edges in the graph and simultaneously promoting "strong" edges, the algorithm discovers the cluster structure in the graph.

EMC Algorithm. The EMC algorithm [12] is constituted by a two-step procedure. In the first step the PPI graph is clustered by the MCL algorithm and in the second step the results are filtered based on either individual or a combination of 4 different methods. These are density, haircut operation, best neighbor and cutting edge. Contrary to the MCL algorithm, which assigns each protein of the initial PPI graph to a cluster, this two-step approach preserves only those clusters that have high probability to be real biological complexes.

2.4   Evaluation Procedure

To prove the efficiency of each weighting scheme, we compare all results with the recorded yeast protein complexes of the MIPS database, which are widely used as a golden standard dataset [15] and are constituted (when the redundancy entries are removed) by 220 protein complexes. To determine whether a derived cluster represents a protein complex or not, we used the evaluation metric called geometrical similarity index (ω) presented in [1] and defined as follows:

$$\frac{I^2}{A * B} \geq 0.2 \tag{5}$$

, where I is the number of common proteins, A the number of proteins in the predicted complex and B the number of proteins in the recorded complex.

Moreover, we used the geometrical accuracy metric (Acc_g), which is measured through the geometrical mean of the sensitivity and the positive predictive value. It has the advantage that it gives a more "objective" picture of the quality of the results as it obtains high values only if the values of both sensitivity and positive predictive value are high.

3   Results and Discussion

In our experiments we tested two different kernels, the exponential and the Laplacian exponential diffusion kernel, we tried different numbers of iterations and diffusion rates. The best results were obtained for the exponential kernel with 3 iterations and a diffusion rate of 0.7. All our experiments can be found at the supplementary material at: http://homes.esat.kuleuven.be/~cmoschop/Supplementary_ciab2012.pdf. Figures 1, 2 and 3 show our results for the different weighting schemes. That is the performance of these methodologies with and without the intermediate kernel step.

It has to be noted that when the kernel methods were applied to the already weighted datasets and before the clustering procedure, where almost 2 million interaction weights were recorded. As a consequence, the derived clustering results were extremely poor since the enormous number of weighted interactions just adds noise in the clustering procedure. So we decided to have the same number of weighted interactions as the initial weighted datasets, before application of the kernel methods (that is approximately 80.000 interactions) by keeping the higher generated weights.

As it can be seen in Figure 1, the use of the kernel method clearly helped all tested weighted schemes. Especially, the Adjust-CD method seems to have benefited more in its prediction rates. However, the absolute number of valid predictions has been decreased significantly from 72 to 44. On the other two occasions (MV scoring and FS weighting) there was a smaller improvement but the absolute number of valid predictions suffered a smaller decrease.

Another quality component that shows the beneficial effect of the kernel methods is the mean score of the valid predicted complexes (those for which the mean geometric similarity index of the predicted clusters surpasses the threshold of 0.2), shown in Figure 2. All three cases reach approximately the value of 0.5, which implies that the derived predictions have high quality as they manage to identify more than satisfactory the recorded protein complexes of the MIPS database.

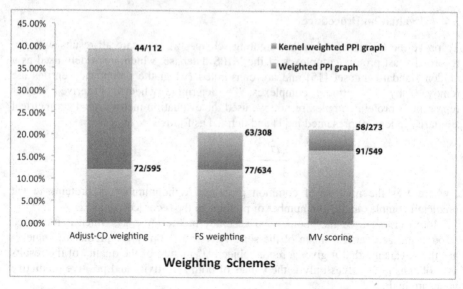

Fig. 1. The percentage of successful predictions of each weighting scheme dataset

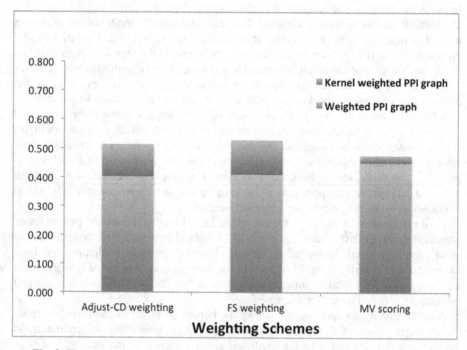

Fig. 2. The mean score of valid predicted complexes of each weighting scheme dataset

On the other hand, as it is shown in Figure 3, there is a small reduction concerning the geometrical accuracy metric. However, this reduction is infinitesimal and it won't be an exaggeration to claim that the Acc_g metric remains stable after application of the kernel method.

Some conclusions can be extracted concerning the use of the kernel methods, derived from our experiments shown in the Supplementary material. First of all, it seems that the higher the iteration step parameter of these method, the fewer and bigger clusters are generated at the final results. Despite that, it was proved that using kernel methods on unweighted graphs does not lead to better results comparing to the other weighted schemes. Obviously, the kernel methods cannot surpass the other weighted schemes as they are using additional data sources in their PPI evaluation.

As mentioned before, the MCL algorithm assigns each protein of the weighted graph to a generated cluster. For this reason, the number of final clusters is enormous in all three cases. In order to propose a comprehensive method, which will reproduce high quality results on protein complex detection, we used in the clustering step the EMC algorithm.

As it is shown in Figures 4, 5 and 6, this leads to a huge improvement in the final results of the three different weighted schemes. The percentage of successful predictions increased approximately 10% in all cases. Furthermore, the Acc_g of all methods reached or surpassed the value of 80%. If this result is combined with the high mean score of valid predicted protein complex values, it is clear that the derived results can be considered as high quality ones and obviously superior comparing to the ones derived using the MCL clustering algorithm. However, the absolute number of final valid predicted clusters is significantly reduced, which is a known drawback of the EMC algorithm, as has been shown in [12].

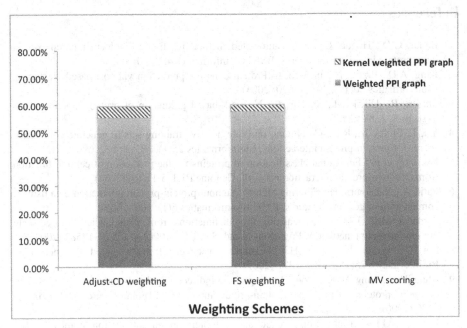

Fig. 3. The geometrical accuracy (Acc_g) of each weighting scheme dataset

4   Conclusions

In this manuscript, a new methodology is proposed for dealing with the problem of PPI reliability by adding a step where a kernel method is applied on already weighted PPI datasets. In the performed experiments, three different weighting schemes were used on the yeast PPI dataset of the IrefIndex database. The derived results prove that kernel methods helped these methods to reach better quality solutions concerning the protein complex detection problem. A future perspective is the investigation of the impact the kernel methods may have on the evaluation of human PPI datasets. A Supplementary file containing a complete list of all the performed experiments of this manuscript can be found at:

http://homes.esat.kuleuven.be/~cmoschop/Supplementary_ciab2012.pdf.

Acknowledgments. Research supported by: Research Council KUL: ProMeta, GOA MaNet, KUL PFV/10/016 SymBioSys, START 1, OT 09/052 Biomarker, several PhD/postdoc & fellow grants. Flemish Government: IOF: IOF/HB/10/039 Logic Insulin, FWO: PhD/postdoc grants, projects: G.0871.12N (Neural circuits) research community MLDM; G.0733.09 (3UTR); G.0824.09 (EGFR), IWT: PhD Grants; TBM-IOTA3, TBM-Logic Insulin, FOD: Cancer plans, Hercules Stichting: Hercules III PacBio RS. EU-RTD: ERNSI: European Research Network on System Identification; FP7-HEALTH CHeartED. COST: Action BM1104: Mass Spectrometry Imaging, Action BM1006: NGS Data analysis network.

References

1. Bader, G.D., Hogue, C.W.: An automated method for finding molecular complexes in large protein interaction networks. BMC Bioinformatics 4, 2 (2003)
2. King, A.D., Przulj, N., Jurisica, I.: Protein complex prediction via cost-based clustering. Bioinformatics 20(17), 3013–3020 (2004)
3. Sharan, R., Ulitsky, I., Shamir, R.: Network-based prediction of protein function. Mol. Syst. Biol. 3, 88 (2007)
4. Yu, J., Finley Jr., R.L.: Combining multiple positive training sets to generate confidence scores for protein-protein interactions. Bioinformatics 25(1), 105–111 (2009)
5. Brun, C., et al.: Functional classification of proteins for the prediction of cellular function from a protein-protein interaction network. Genome Biol. 5(1), R6 (2003)
6. Patil, A., Nakamura, H.: Filtering high-throughput protein-protein interaction data using a combination of genomic features. BMC Bioinformatics 6(1), 100 (2005)
7. Samanta, M.P., Liang, S.: Predicting protein functions from redundancies in large-scale protein interaction networks. Proc. Natl. Acad. Sci. USA 100(22), 12579–12583 (2003)
8. Liu, G., Wong, L., Chua, H.N.: Complex discovery from weighted PPI networks. Bioinformatics 25(15), 1891–1897 (2009)
9. Chua, H.N., Sung, W.K., Wong, L.: Exploiting indirect neighbours and topological weight to predict protein function from protein-protein interactions. Bioinformatics 22(13), 1623–1630 (2006)
10. Kritikos, G.D., et al.: Noise reduction in protein-protein interaction graphs by the implementation of a novel weighting scheme. BMC Bioinformatics 12, 239 (2011)

11. Enright, A.J., Van Dongen, S., Ouzounis, C.A.: An efficient algorithm for large-scale detection of protein families. Nucleic Acids Res. 30(7), 1575–1584 (2002)
12. Moschopoulos, C.N., et al.: An enchanced Markov clustering method for detecting protein complexes. In: 8th IEEE International Conference on BioInformatics and BioEngineering (BIBE 2008), Athens (2008)
13. Razick, S., Magklaras, G., Donaldson, I.M.: iRefIndex: a consolidated protein interaction database with provenance. BMC Bioinformatics 9, 405 (2008)
14. Wu, M., et al.: A core-attachment based method to detect protein complexes in PPI networks. BMC Bioinformatics 10, 169 (2009)
15. Mewes, H.W., et al.: MIPS: analysis and annotation of proteins from whole genomes in 2005. Nucleic Acids Res. 34(Database issue), D169–D172 (2006)
16. Kandola, N., Cristianini, N., Shawe-Taylor, J.: Learning semantic similarity. In: Advances in Neural Information Processing Systems, pp. 657–664 (2002)
17. Kondor, R., Lafferty, J.: Diffusion kernels on graphs and other discrete structures. In: Proceedings of the Nineteenth International Conference on Machine Learning (2002)
18. Brohee, S., van Helden, J.: Evaluation of clustering algorithms for protein-protein interaction networks. BMC Bioinformatics 7, 488 (2006)
19. Moschopoulos, C., et al.: Which clustering algorithm is better for predicting protein complexes? BMC Research Notes 4(1), 549 (2011)

Efficient Computational Prediction and Scoring of Human Protein-Protein Interactions Using a Novel Gene Expression Programming Methodology

Konstantinos Theofilatos[1], Christos Dimitrakopoulos[1], Maria Antoniou[1], Efstratios Georgopoulos[2], Stergios Papadimitriou[3], Spiros Likothanassis[1], and Seferina Mavroudi[1,4]

[1] Department of Computer Engineering and Informatics, University of Patras, Greece
[2] Technological Educational Institute of Kalamata, 24100, Kalamata, Greece
[3] Department of Information Management, Technological Institute of Kavala, Greece
[4] Department of Social Work, School of Sciences of Health and Care, Technological Educational Institute of Patras, Greece
{theofilk,dimitrakop,antonium,likothan,mavroudi}@ceid.upatras.gr,
sfg@teikal.gr, sterg@teikav.edu.gr

Abstract. Proteins and their interactions have been proven to play a central role in many cellular processes. Thus, many experimental methods have been developed for their prediction. These experimental methods are uneconomic and time consuming in the case of low throughput methods or inaccurate in the case of high throughput methods. To overcome these limitations, many computational methods have been developed to predict and score Protein-Protein Interactions (PPIs) using a variety of functional, sequential and structural data for each protein pair. Existing computational methods can still be enhanced in terms of classification performance and interpretability. In the present paper we present a novel Gene Expression Programming (GEP) algorithm, named as jGEPModelling 2.0, and apply it to the problem of PPI prediction and scoring. jGEPModelling2.0 is a variation of the classic GEP algorithm to make it suitable for the problem of PPI prediction and enhance its classification performance. To test its efficiency, we applied it to a public available dataset and compared it to two other state-of-the-art PPI prediction models. Experimental results proved that jGEPModelling2.0 outperformed existing methodologies in terms of classification performance and interpretability. (This paper is submitted for the CIAB2012 workshop).

Keywords: Protein Protein Interactions, Human, PPI scoring methods, Gene Expression Programming, Genetic Programming.

1 Introduction

Proteins are nowadays considered to be the most important participants in molecular interactions. Specifically, proteins play a significant role in almost all the cellular functions such as regulatory signals transmission in the cell and they catalyze a huge number of chemical reactions. The total number of possible interactions within the cell is astronomically large and the full identification of all true PPIs is a very

C. Jayne, S. Yue, and L. Iliadis (Eds.): EANN 2012, CCIS 311, pp. 472–481, 2012.
© Springer-Verlag Berlin Heidelberg 2012

challenging task. Moreover, PPIs range from weak ones that can be formed only under certain circumstances to stronger ones that are formed in various cases.

Many high throughput methodologies have been developed for the experimental prediction of PPIs with the prevailing ones among them being the yeast two-hybrid (Y2H) system, mass spectrometry (MS), protein microarrays, and Tandem Affinity purification (TAP) [1]. These methodologies overcome the time and cost limitations of low throughput experimental methods as they can predict many PPIs in a single experiment. Thus, they improved the coverage of the known PPIs. However, the improvement in terms of coverage introduced many false positives and some researchers believe that the false positive rate among experimental PPI databases is over 50% [1, 2].

In order to further improve PPI coverage and overcome the false positive rate problem of experimental techniques, many computational approaches have been developed. All these computational methods use protein and protein-interaction data which are located in public databases and most of them are supervised machine learning classifiers. These classifiers use as inputs a variety of functional, sequential and structural features. The main machine learning methods have a prominent role among computational PPI prediction methods. Their most important representatives are Bayesian classifiers, Artificial Neural Networks, Support Vector Machines and Random Forests [1, 2]. All these approaches fail to achieve both high classification performance and interpretability. Moreover, all existing computational approaches are facing the class imbalance problem, failing to balance the tradeoff between the metrics of sensitivity and specificity. Furthermore, most existing computational techniques do not incorporate a feature selection step and select the features which are going to be used empirically. This fact, is a restrictive factor for the extraction of high performance interpretable classifiers.

In the present paper we introduce a novel computational PPI prediction tool, called jGEPmodel2.0, which is a Gene Expression Programming variation that extends the basic algorithm introducing a novel case specific fitness function, a new local search operator for the models' constants optimization and an adaptive mutation operator to enhance its convergence behavior. The proposed tool was applied to a dataset extracted using HINT-KB (http://150.140.142:84) which is a publicly available database for Human PPI data. In order to test its efficiency we compared it with two modern PPI prediction methods: A Random Forest method [3-6] and the wrapper methodology combining Genetic Algorithms and SVM Classifiers [7, 8].

Experimental results proved that the proposed methodology outperformed existing methods in terms of classification performance and interpretability. Specifically, a simple equation for the prediction and scoring of human PPIs was built. By analyzing this equation, some important conclusions have been made about the intra cellular mechanisms that decide if a pair of proteins interacts or not.

2   jGEPModelling2.0

Gene Expression Programming is a modern Evolutionary Algorithm proposed by Ferreira [9] as an alternative method to overcome the drawbacks of Genetic Algorithms (GAs) and Genetic Programming (GP) [10]. Similar to GA and GP, GEP follows the Darwinian Theory of natural selection and survival of the fittest individuals.

The main difference between the three algorithms is that, in GEP there is a distinct discrimination between the genotype and the phenotype of an individual. This difference resides in the nature of individuals, namely in the way individuals are represented: in GAs individuals are symbolic strings of fixed length (chromosomes); in GP individuals are non-linear entities of different sizes and shapes (parse trees); finally in GEP individuals are also symbolic strings of fixed length representing an organism's genome (chromosomes/genotype), but these simple entities are encoded as non-linear entities of different sizes and shapes, determining the models fitness (expression trees/phenotype).

Each GEP chromosome is composed of a head part and a tail part. The head contains genes-symbols that represent both functions and terminals, whereas the tail contains only terminal genes. The set of functions usually incorporates a subset of mathematical or Boolean user specified functions. The set of terminals is composed of the constants and the independent variables of the problem. The head length (denoted h) is chosen by the user, whereas the tail length (denoted t) is evaluated by:

$$t = (n-1)h+1 \tag{1}$$

where, n is the number of arguments of the function with most arguments. Despite its fixed length, each gene has the potential to encode Expression Trees (ETs) of different sizes and shapes, ranging from the simplest composed of only one node (when the first element of a gene is a terminal) and the largest composed of as many nodes as the length of the gene (when all the elements of the head are functions with maximum arity). One of the advantages of GEP is that the chromosomes will always produce valid expression trees, regardless of modification, and this means that no time needs to be spent on rejecting invalid organisms, as in case of GP.

The initial eversion of our implementation [11, 12] named as jGEPmodel1.0 was applied to the problem of modeling the fatigue of composite materials and the problem of prediction financial time series and the results were satisfactory. In the present paper, we extended our initial implementation by introducing a novel local constant optimization operator, a problem specific fitness function and an adaptive mutation operator. Next, we briefly outline the algorithm's steps:

1. **Creation of initial population:** Initially a population of random chromosomes is produced using the user defined set of functions and head's size.
2. **Express chromosomes:** For the evaluation of the performance for each individual of the population the expression trees (ET) are built. This process is also very simple and straightforward. For the complete expression, the rules governing the spatial distribution of functions and terminals must be followed. First, the start of a gene corresponds to the root of the ET, forming this node the first line of the ET. Second, depending on the number of arguments of each element (functions may have a different number of arguments, whereas terminals have an arity of zero), in the next line are placed as many nodes as there are arguments to the elements in the previous line. Third, from left to right, the new nodes are filled, in the same order, with the elements of the gene. This process is repeated until a line containing only terminals is formed.
3. **Execute each program:** Using the post order traversal, each expression tree is transformed to a mathematical expression.

4. **Evaluate Fitness:** Using the training set, the equation 2 and the mathematical expression produced for every individual, we compute its fitness. The second term in equation 2 is the Fisher's Discriminant Rate for classification problem of two classes. Because we are encountering a classification problem we not only want an accurate prediction for the classes but we need the averages of the assigned values for each class to be as far as possible, while their covariances to be as small as possible. The third term is used to handle the bloat effect which is present in Genetic Programming approaches. Specifically, we are interested in selecting the smallest model that achieves the optimal classification. The models complexity is a secondary goal and thus this term is multiplied with a weight equal to 10^{-3} in order to reduce its significance in the overall fitness function.

$$Fitness = -MSE + \frac{(\mu_{pos} - \mu_{neg})^2}{\sigma_{pos}^2 + \sigma_{neg}^2} - 10^{-3} * \# \exp ression\_tree\_nodes \quad (2)$$

where, MSE is the mean square error.

5. **Selection:** Tournament selection is applied. Tournament selection involves running several "tournaments" among a few individuals chosen randomly from the population. The winner of each tournament (the one with the best fitness) is selected for genetic modification. Selection pressure is easily adjusted by changing the tournament size. If the tournament size is larger, weak individuals have a smaller chance to be selected.

6. **Reproduction:** At this step we apply the genetic operators of recombination, mutation and constant local search on the winners of the tournaments.
   a. **Recombination:** the parent chromosomes are recombined using two point recombination to produce two new solutions. This operator is applied with a probability named as recombination rate.
   b. **Adaptive mutation:** This operator is applied to a chromosome to randomly vary one part of it. It is applied with a user defined probability named as mutation probability. In the first generation the 30% of a chromosome is randomized if it is selected to be mutated. As the generations pass the randomization percentage of a mutated chromosome is linearly decreased until it alternates only 10% of the chromosome's genes. This adaptive behavior enables the algorithm to explore a larger area of the search space initially and to exploit more promising areas of the search space in the final phases of the algorithm. When the mean fitness of the population is less that 5% away from the fitness of the best individual the percentage of randomization that mutation operator applies is risen by a 5% in order to enable the algorithm to get unstuck from a possible local optimal.
   c. **Local search operator:** This operator is applied to a chromosome to optimize locally one of its constant values. It is applied with a user defined probability named as local search rate. The constants in a model play a significant role in its performance. The classical GEP operators accomplish a global constant optimization. However, even small variations in the constants of a model may improve its performance. This is the reason why this operator is very crucial for the overall algorithm.

7. **Prepare new programs for the next generation:** At this step, we replace the tournament losers with the new individuals created by the variation operators in the population.

8. **Termination criterion:** We check if the termination criterion is fulfilled and if not we return to step 2. We used as a termination criterion the maximum number of 100.000 generations that GEP was left to run combined with a convergence criterion to avoid overfitting. Specifically the performance of the individuals in each iteration is measured in the validation set. The algorithm terminates if for 10 subsequent best individuals variations their performance in the validation set is decreased.

9. **Results:** As a result we return the best individual ever found during the evolution process. The classification threshold is then optimized for this individual using training and validation sets and the classification metrics of accuracy, sensitivity and specificity are estimated.

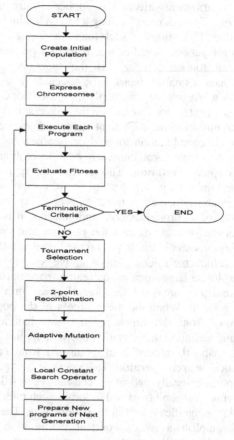

Fig. 1. Block Diagram of jGEPModel2.0 algorithm

The jGEPmodel2.0 tool is standalone tool implemented in java. It provides a user friendly interface which enables users to define the method's parameters, upload their data and browse the results. In Figure 1, the jGEPmodel2.0 interface is demonstrated. Tool's implementation support multi-threading experimentation to execute the algorithm in parallel.

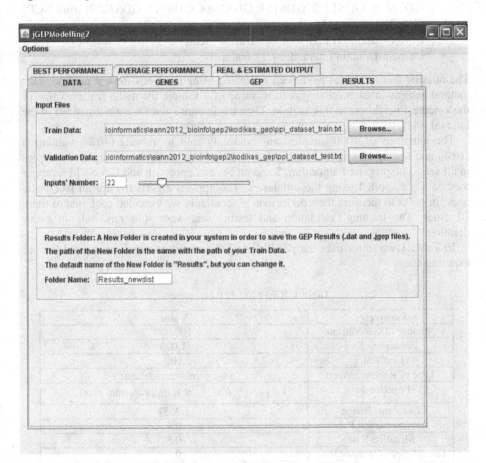

Fig. 2. jGEPmodel2.0 Tool's User Interface

3   Experimental Results

In order to test the performance of the algorithm we applied it to a public available human PPI dataset downloaded from HINT-KB (http:150.140.142.24:84/). From this database we downloaded 2000 positive interactions which are also stored in HPRD database [13] and 2000 negative random interactions which have not been mentioned as true interactions in any other public PPI database.

A list of 22 informative features is downloaded from HINT-KB in both positive and negative sets. These features are:

- Gene Ontology [14] (Co-function, Co-process, Co-localization) (3 features)
- Sequence Similarity BlastP e-value. (1 feature)
- Yeast Homology (1 feature)
- Gene Expression Profile Pearson Similarity (datasets (GDS531, GDS534, GDS596, GDS651, GDS806, GDS807, GDS843, GDS987, GDS1085, GDS2855, GDS1402, GDS181, GDS1088, GDS841, GDS3257) from NCBI Gene Expression Omnibus [15]) (15 features)
- Co-localization (PLST tool localization predictions [16]) (1 feature)
- Domain-Domain Interactions. (1 feature)

The missing feature values have been estimated using the KNN-impute method. This method assigns in each missing feature value of a sample the mean feature value of the k-nearest neighbors to this sample. Then the feature values were normalized in the interval [0-1].

The initial set of 4000 samples was randomly split to training (40%), validation (10%) and testing (50%) set keeping 1:1 rate between positive and negative samples in all sets. The proposed algorithm, Random Forests approach and GA-SVM wrapper method were applied using these datasets. The applied algorithms are all stochastic ones. In order to measure their performance accurately we executed each one of them 10 times. The training, validation and testing sets were split randomly in every iteration.

In Table 1 we summarize the jGEPModel2.0 parameters which were used for our experiments.

Table 1. jGEPModel2.0 parameters

Parameter	Value
Number of Generations	100.000
Population Size	1.000
Head's Size	100
Type of Recombination	Two points recombination
Function Set	$\{+, -, *, /, \wedge, \sqrt{\ }, abs, log, exp, min, max\}$
Constants Range	[-3, 3]
Recombination Rate	0.9
Mutation Rate	0.5
Local Search Operator Rate	0.1

In table 2 we present the classification performance of every algorithm.

Table 2. Classification Performance of computational methods for the prediction of human PPIs

Computational Method	Accuracy	Sensitivity	Specificity
jGEPmodel2.0	0.8267 ($\pm$ 0.031)	0.8328 ($\pm$ 0.015)	0.8206 ($\pm$ 0.033)
Random Forests	0.8183 ($\pm$ 0.003)	0.8145 ($\pm$ 0.002)	0.8220 ($\pm$ 0.004)
GA-SVM	0.7919($\pm$ 0.029)	0,7434 ($\pm$ 0.029)	0,8404 ($\pm$ 0.021)

It is easily observed that the proposed model outperforms the other classifiers in terms of classification performance. In equation 3 we present the mathematical equation derived from the best execution of the jGEPmodel2.0. The optimal classification threshold derived for this classification model was 0.5092.

Output= (Domain_Domain_Interactions ^ (log((max((Gene_EXPR8 + (Gene_EXPR8^2)), (Sequence_Similarity + (((abs(Yeast_Homology + (exp(Sequence_Similarity^1/2)))) * (GO_process ^ Gene_EXPR1)) ^1/2)))) - (3.0))))^1/2   (3)

In Figure 3 we demonstrate the predicted versus the real values of our PPI dataset. It can be easily observed that there do not only exist values near the real output but the two classes are clearly distinguished.

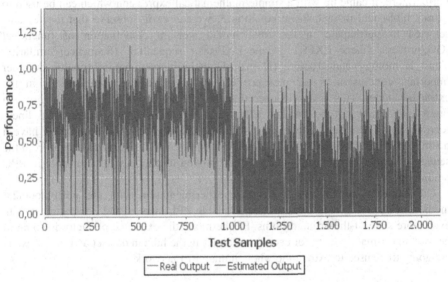

Fig. 3. Real and Estimated by equation 3 output for the test dataset

4   Conclusions

GEP algorithm has been applied to a variety of problems so far, and it has been proven to be a very useful tool for modeling and classification . In contrast to other machine learning techniques, it searches for a mathematical expression which will then be used to accomplish the prediction or classification task. The high degrees of freedom in this mathematical expression enable it to find a solution even in very complex non-linear problems such as the PPI prediction problem. Its main drawbacks are the bloat effect, overfitting and slow convergence.

jGEPmodel2.0 is a novel GEP variation which tries to overcome the aforementioned drawbacks of the classical GEP. The bloat effect is limited by incorporating in the fitness function the term of the number of nodes of each expression tree in order to force the optimization procedure to prefer shorter solutions than longer ones. Overfitting was faced deploying a complex termination criterion. Using the performances in the validation set, the algorithm diagnoses when overfitting starts and terminates the evolutionary process. In order to overcome the slow convergence of the classical GEP, two basic modifications were implemented: the adaptive mutation rate which enhances the convergence behavior of the algorithm and the constant local search operator which provides the algorithm local search features.

For the problem of PPI prediction we designed a novel fitness function which enables the algorithm to be efficient for this classification task. When jGEPmodel2.0 was applied to the problem of predicting Human PPIs using a public available dataset it outperformed existing methodologies in terms of classification performance. Furthermore, it came up with a simple mathematical expression which can be used to extract biological insight. From equation 3, we can easily observe that the features selected to participate in the final model were a combination of functional (GO_process, Gene_EXPR1, Gene_EXPR8), sequential (Sequence_Similarity, Yeast_Homology) and structural (Domain-Domain Interactions) information. Another important conclusion is the presence of non-linear feature combinations in the extracted model. This is an evidence that the PPI prediction problem is highly non linear and could not be handled effectively by using linear models like linear regression etc. Our final conclusion is that all selected terms in equation 3 have a positive effect in predictor's output. This was expected for all features except from the sequence similarity which has not yet been proven to play a significant role in PPIs prediction.

Our future plans involve the further experimentation with the jGEPmodel2.0 method by applying it to other public available PPI datasets and comparing it with even more state-of-the-art algorithms. Furthermore, the extracted predictor is going to be used to estimate a score for each protein pair in the human dataset and these results are going to be used to extract the whole human PPI network.

Acknowledgments. This research has been co-financed by the European Union (European Social Fund - ESF) and Greek national funds through the Operational Program "Education and Lifelong Learning" of the National Strategic Reference Framework (NSRF) - Research Funding Program: Heracleitus II. Investing in knowledge society through the European Social Fund.

References

1. Theofilatos, K.A., Dimitrakopoulos, C.M., Tsakalidis, A.K., Likothanassis, S.D., Papadimitriou, S.T., Mavroudi, S.P.: Computational Approaches for the Prediction of Protein-Protein Interactions: A Survey. Current Bioinformatics 6(4), 398–414 (2011)

2. Rivas, J., Fortanillo, C.: Protein-Protein Interactions Essentials: Key Concepts to Building and Analyzing Interactome Networks. PLoS Computational Biololy 6(6), e1000807 (2010)
3. Breiman, L.: Random forests. Machine Learning J. 45, 5–32 (2001)
4. Liu, Y., Kim, I., Zhao, H.: Protein interaction predictions from diverse sources. Drug Discov. Today 13, 409–416 (2008)
5. Chen, X., Liu, M.: Prediction of protein–protein interactions using random decision forest framework. Bioinformatics 21, 4394–4400 (2005)
6. Thahir, M., Jaime, C., Madhavi, G.: Active learning for human protein-protein interaction prediction. BMC Bioinformatics 11(1), S57 (2010)
7. Wang, B.: Prediction of protein interactions by combining genetic algorithm with SVM method. In: IEEE Congress on Evolutionary Computation, pp. 320–325 (2007)
8. Wang, B., Chen, P., Zhang, J., et al.: Inferring Protein-Protein Interactions Using a Hybrid Genetic Algorithm/Support Vector Machine Method. Protein & Peptide Letters 17, 1079–1084 (2010)
9. Ferreira, C.: Gene Expression Programming: A New Adaptive Algorithm for Solving Problems. Complex Systems 13(2), 87–129 (2001)
10. Koza, J.R.: Genetic programming: on the programming of computers by means of natural selection. MIT Press, Cambridge (1992)
11. Antoniou, M.A., Georgopoulos, E.F., Theofilatos, K.A., Vassilopoulos, A.P., Likothanassis, S.D.: A Gene Expression Programming Environment for Fatigue Modeling of Composite Materials. In: Konstantopoulos, S., Perantonis, S., Karkaletsis, V., Spyropoulos, C.D., Vouros, G. (eds.) SETN 2010. LNCS (LNAI), vol. 6040, pp. 297–302. Springer, Heidelberg (2010)
12. Antoniou, M.A., Georgopoulos, E.F., Theofilatos, K.A., Likothanassis, S.D.: Forecasting Euro – United States Dollar Exchange Rate with Gene Expression Programming. In: Papadopoulos, H., Andreou, A.S., Bramer, M. (eds.) AIAI 2010. IFIP AICT, vol. 339, pp. 78–85. Springer, Heidelberg (2010)
13. Keshava, T., Goel, R., Kandasamy, K., et al.: Human Protein Reference Database–2009 update. Nucleic Acids Res. 37, D767–D772 (2009)
14. Ashburner, M., Ball, C., Blake, J., et al.: Gene ontology: tool for the unification of biology. The Gene Ontology Consortium. Nat. Genet. 25, 25–29 (2000)
15. Barrett, T., Troup, D., Wilhite, S., et al.: NCBI GEO: archive for functional genomics data sets -10 years on. Nucleic Acids Research 39, D1005–D1010 (2012)
16. Scott, M., Thomas, D., Hallet, M.: Predicting subcellular localization via protein motif co-occurrence. Genome Res. 14(10A), 1957–1966 (2004)

Biomedical Color Image Segmentation through Precise Seed Selection in Fuzzy Clustering

Byomkesh Mandal and Balaram Bhattacharyya[*]

Department of Computer & System Sciences,
Visva-Bharati University, Santiniketan-731 235, West Bengal, India
balaramb@gmail.com

Abstract. Biomedical color images play major role in medical diagnosis. Often a change of state is identified through minute variations in color at tiny parts. Fuzzy C-means (FCM) clustering is suitable for pixel classification to isolate those parts but its success is heavily dependent on the selection of seed clusters. This paper presents a simple but effective technique to generate seed clusters resembling the image features. The HSI color model is selected for near-zero correlation among components. The approach has been tested on several cell images having low contrast at adjacent parts. Results of segmentation show its effectiveness.

Keywords: Color image segmentation, blood cell images, histogram, pixel classification, fuzzy C-means.

1   Introduction

Segmentation of blood cell images has prevalent interest in medical research and is a key component in diagnosis and treatment planning. Several diseases can be determined by the count, size and shape of different blood cells. Detection of cells that are normally absent in peripheral blood, but released in some diseases, is another key for diagnosis [1]. Segmentation of such cells images will facilitate further processing to classify them as belonging to a particular class, or declaring them to be either healthy or diseased [2]. The manual segmentation is not only tiresome and time-consuming; it may suffer from inter-observer variability [3]. Computerized processing of this job will help both in minimizing time consumption and human error. Feature-rich high-resolution images with fast processing are instrumental in diagnosis. However, enhanced accuracy in image segmentation invites multi-dimensional space-time complexity.

Like many other biomedical images, the main difficulty with segmentation of blood cell images is its inherent low signal-to-noise ratio [4]. Such microscopic images often suffer from staining and illumination inconsistencies [5]. Image clarity is often hampered due to background influence and cell overlapping [6]. All these uncertainties together make segmentation of blood cells a challenging task. Common segmentation algorithms often fall short of exhibiting even the minimum acceptable results [7].

---

[*] Corresponding author.

C. Jayne, S. Yue, and L. Iliadis (Eds.): EANN 2012, CCIS 311, pp. 482–491, 2012.
© Springer-Verlag Berlin Heidelberg 2012

The primary objective of the current study is to develop a segmentation algorithm, which will be robust enough to cope with the uncertainties without significant enhancement in space-time complexity.

2   Pixel Classification Using Clustering

Clustering is a powerful unsupervised learning technique that has successfully been used to unearth the similarities and dissimilarities within a set of objects represented by vectors of measurements. Algorithms follow the basic principle that objects within a cluster show a high degree of similarity among them but exhibit very low affinity to the objects belonging to other clusters. Thus, cluster analysis allows for the discovery of similarities or dissimilarities hidden within data patterns as well as for drawing conclusions from those patterns. Clusters can be formed from images based on pixel intensity, color, texture, location or combinations of them. The clusters serve as foundation for detecting segments of the image through similarity of adjacent pixels.

The cluster building process may be started with some initial cluster centers, called seeds, on ad hoc basis or based on some prior information. Pixels closest to a center are attached to the respective cluster and the center is updated from the distances of the pixels belonging to it. All clusters are treated accordingly. Redistribution of pixels is done upon distances from the new centers. The process continues until no center moves any further.

In clustering, there always remains a chance of loss of information during the classification process. Minimizing this information loss is earnestly required especially in biomedical cell image segmentation where normal and pathological samples are often separated by only subtle differences in visual features. In general, the more the information incorporated into the image segments, the better the quality of recognition.

Crisp clustering algorithms, e.g. K-means, define hard-decision membership function to map each pixel to its respective cluster [8], that is, each pixel belongs to one and only one cluster. Such algorithms have an inherent problem of overlooking small localized variations of image components. This limitation results in loss of information in segmented images. The information loss may impose a serious burden in recognition especially in biomedical images where various image parts are typically separated by minute variation in pixel colors.

Contrary to crisp clustering, FCM assigns a soft-decision membership between pixels and clusters and is thus less prone to falling into local optima [9]. A pixel can belong to several clusters at the same time with varying degree of membership. The uncertainty in making decision is preserved until the final conclusions are made. Clusters formed by FCM retain more information compared to those created by crisp counterparts, thus minimizes the risk of loss of information during the classification process. In addition, it plays an important role in resolving classification ambiguities due to overlapping regions, identifying low-contrast regions and imprecise boundaries, which are very common phenomena in biomedical images. All these are strong points to employ the FCM algorithm in the present work with due care on computational complexities.

3   Related Work

FCM fares better in preserving image characteristics due to its flexibility in pixel classification. However, computational complexity is a concern for FCM algorithm in color image segmentation. Several variants are found in the literature.

Ahmed et al [10] propose a modification to the objective function so that labelling of a pixel is influenced by the labels of immediate neighbors and they term it FCM_S. The neighborhood term acts as a regularizing agent and promotes homogeneous labelling. Although this modification helps to handle intensity inhomogeneities, computation of the neighborhood term of each pixel in successive iterations involves significant processing time.

Computational complexity of the neighborhood term of FCM_S is resolved by Chen and Zhang [11] by introducing two variants of FCM_S namely, FCM_S1 and FCM_S2. An extra mean-filtered image is incorporated in FCM_S1, whereas median-filtered image is introduced in FCM_S2. As both the mean- and median-filtered images can be constructed in advance, the neighborhood complexity is reduced. But both the algorithms still suffer from a serious problem of parameter choice.

Liew [12] presents a new image segmentation algorithm based on adaptive fuzzy c-means (AFCM) clustering. He introduces a novel dissimilarity index in the modified FCM objective function, and exploits the high inter-pixel correlation inherent in most real-world images. But it cannot ensure generation of appropriate regions.

A mean-shift based fuzzy c-means is reported in [13]. The authors introduce a new mean field term within the objective function of conventional FCM. Since mean shift can find cluster centers quickly and reliably, the approach is capable of generating optimal clusters for the test images. More and more approaches are proposed by using multiple information fusion [14], [15].

4   Problem Statement and Proposed Approach

Classification of pixels based on cluster formation is sensitive upon fine-tuning of clusters. Positions of seed clusters have profound influence in convergence of clustering [16]. Moreover, arbitrary seeds may lead the algorithm falling into local minima [17] and the algorithm is far away from generating acceptable results. Deciding the number of seed clusters is yet another challenging task. Starting with a large number of seeds may result in over segmentation, whereas too few of them can lead to under segmentation. Although, significant and continuous attempts have been made towards the solution of this problem, it still remains challenging [18].

In cluster-based segmentation, the final clusters serve as foundation for region development and it is thus a primary requirement that the clusters are consistent enough with the color distribution of pixels, which, in turn, requires as much as image information to be incorporated into the classification procedure. Cluster formation is closely related to distribution of pixels in an image and hence, brute force approach for clustering may result in wide variation of segmentation qualities from image to image.

We thus attempt to associate the image with the process of selecting the seeds for segmentation. Histogram of an image provides a global description of image information and serves as an important basis of statistical approaches in image processing. In HSI color space, each pixel of an image can be viewed as a 3D vector representing the three primary components namely, hue (H), saturation (S) and intensity (I). Hence, information extracted from each of the H, S and I histograms as a whole can provide global description of color distribution over the entire image and can be utilized to find the seed clusters. If we can associate color variation at pixel level with the seed clusters, the final clusters obtained via successive refinements will be able to encompass image features at higher levels. Such clusters are much prone to image at hand and hence regions obtained through segmentation on the basis of those clusters are expected to be more reliable.

5   FCM Algorithm

In FCM [19], each pixel has certain membership values associated to each cluster center. Each membership value lies within the range [0, 1] and indicates the degree of association between a pixel and a particular cluster.

The algorithm is summarized in the following. Let, $X = \{x_1, x_2, ..., x_n\}$ is a set of n data points, where each data point x_i ($i = 1, 2, ..., n$) is a vector in a real d-dimensional space R^d, U_{cn} is a set of real matrices, each of dimension $c \times n$, and c ($2 \leq c < n$) is an integer. Then the classical FCM algorithm aims to minimize the following objective function

$$J_{FCM}(U, V) = \sum_{i=1}^{n} \sum_{j=1}^{c} u_{ij}^m d_{ij}^2, \quad \text{subject to} \quad \sum_{j=1}^{c} u_{ij} = 1 \text{ and } \sum_{i=1}^{n} u_{ij} > 0 \qquad (1)$$

where $U (= [u_{ij}]) \in U_{cn}$, u_{ij} is the degree of membership of x_i in the jth cluster, $V = \{v_1, v_2, ..., v_c\}$ is a cluster center set, $v_i \in R^d$ and $m \in [1, \infty)$, known as fuzzy exponent, is a weighting exponent on each fuzzy membership which indicates the amount of fuzziness in the entire classification process. d_{ij} is the distant norm which indexes the vector distance between the data point x_i and the center of the jth cluster, v_j. Usually, the Euclidean distance is taken, i.e.

$$d_{ij} = \left\| x_i - v_j \right\| \qquad (2)$$

The objective function J_{FCM} controls the uniformity of cluster centers and the degree of compactness. In general, the smaller the value of objective function, the higher is the degree of compactness and uniformity.

Minimization of the cost function J_{FCM} is a nonlinear optimization problem that can be implemented by using the following iterative process:

Step 1: Choose appropriate value for c and initialize the cluster center set V randomly. Also select a very small positive number ε and set the step variable t to 0.

Step 2: Calculate (at $t = 0$) or update (at $t > 0$) the membership matrix $U = [u_{ij}]$ by

$$u_{ij}^{(t+1)} = \frac{1}{\sum_{k=1}^{c} \left(\frac{\left\| x_i - v_j^{(t)} \right\|}{\left\| x_i - v_k^{(t)} \right\|} \right)^{\frac{2}{(m-1)}}} \quad \text{for } j = 1, 2, \ldots, c \tag{3}$$

Step 3: Update the cluster center set V by

$$v_j^{(t+1)} = \frac{\sum_{i=1}^{n} (u_{ij}^{(t+1)})^m \cdot x_i}{\sum_{i=1}^{n} (u_{ij}^{(t+1)})^m} \quad \text{for } j = 1, 2, \ldots, c \tag{4}$$

Step 4: Repeat steps 2 and 3 until

$$\left| U^{(t+1)} - U^{(t)} \right| \le \varepsilon \tag{5}$$

6   Selection of Color Model

Selecting color model is crucial in segmentation since it leads the process long way towards successful categorization of pixels into color clusters. RGB is a default model for digital images but is not well suited for image processing because of high correlation among components. Moreover, color perception does not depend on distance in RGB space.

The HSI color model, that resembles human visual perception, seems to be a better alternative. Hue (H) represents basic color tones without any nuance. Saturation (S) is a measure of purity of color and signifies the amount of white light mixed with hue. Finally, intensity (I), which describes brightness of an image, is determined by the amount of light reflected from source. The near-zero correlation among the H, S and I components [20] is helpful in component-wise peak selection and generating initial cluster centers. Moreover, the component hue, which is invariant to highlighting, shading, or inhomogeneous illumination [21], makes the HSI model appropriate for biomedical image segmentation where such artifacts are very common. Experiments on sample images (Fig. 4) substantiate better applicability of HSI in color image segmentation. Geometrically the model can be represented as a cone as in Fig. 1.

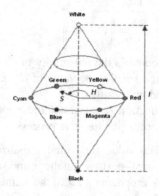

Fig. 1. Schematic representation of the HSI color model

7   Our Approach

Histogram of a digital image is simple yet a powerful measure for providing global description of image information. Each color in HSI model can be viewed as three-dimensional vector containing three primary components, namely hue, saturation and intensity. Information extracted from each of the histograms is utilized to provide overall description of color distribution among the image pixels. A typical segmentation algorithm based on histogram analysis can be carried out only if dominant peaks are recognized correctly [22].

The entire segmentation approach is done in five steps.

Step 1: Selection of significant peaks
Let $h(i)$ represents histogram in a color channel. (i is an integer and $0 \leq i \leq ch\_max$; $ch\_max = 360$ for H, 100 for S and 255 for I). Significant peaks [23] are found in the following way:

(1) Find all possible peaks: Inspect each component histogram and find all such points which correspond to local maximums:

$$P_0 = \{(i, h(i)) \mid h(i) > h(i-1) \wedge h(i) > h(i+1), 1 \leq i \leq ch\_max - 1\} \quad (6)$$

(2) Select the significant peaks: Form a new set P_1 using the following formula:

$$P_1 = \{(p_i, h(p_i)) \mid h(p_i) > h(p_{i-1}) \wedge h(p_i) > h(p_{i+1}), p_i \in P_0\} \quad (7)$$

Points in the set P_1 are much more significant than the points in P_0 in determining the peaks from a component histogram.

Step 2: Merging of close peaks
The above approach may pick too many close peaks. Difference between two nearest peaks may not even be perceivable. Two close peaks in a channel are merged to a single one until the distance between them is greater than a predefined distance. Experimentally, the threshold distance is taken to be 3% of the entire range of that particular channel.

Step 3: Formation of seed clusters
Suppose m, n and p numbers of peaks are thus found in H, S and I channels, respectively. The set of seed clusters, K, contains $m \times n \times p$ members. A seed is constituted with one value from each of m, n and p in order. That is, if α_i ($i = 1, 2, ...,$ m), β_j ($j = 1, 2, ..., n$) and γ_k ($k = 1, 2, ..., p$) are the positions of such peaks in H, S, and I histograms, respectively, K is defined as

$$K = \{(\alpha_i, \beta_j, \gamma_k)\}; \quad i = 1, 2, ..., m; j = 1, 2, ..., n; k = 1, 2, ..., p \quad (8)$$

The set K is given input to FCM. After convergence of FCM, each pixel is assigned to that cluster with which it has the highest degree of membership for belonging.

Step 4: Elimination of stray clusters
Depending upon cardinality of the set K, which is $m \times n \times p$, FCM may generate so many void clusters or small clusters containing too few image pixels. Applying region

growing upon small clusters results in over-segmentation. So it is better to treat such clusters as non-contributing and pixels within them are assigned to the next closest cluster. Based on analysis done using numerous images, a cluster is not considered for further processing if it contains less than 0.05% of total image pixels.

Step 5: Region growing

Following the classification of pixels into valid clusters, the process of region growing categorizes the 8-connected pixels into a region if they belong to the same cluster; otherwise a new region is initiated.

Computational complexity of the proposed approach is analyzed as follows. The process starts with construction of component histograms. It scans the entire image, complexity of which is $O(N)$, where N is the dimension of the image. Finding all possible peaks from each of H, S and I histograms requires ch_max operations whereas selection of significant peaks requires another p number of operations, where p denotes the number of candidate peaks. So total complexity of this phase is $O(3(ch\_max + p))$. Again, if s be the count of close peaks to be merged in a channel, total complexity of merging in three channels is $O(3s)$. Finally, complexity of FCM clustering is $O(3Nc^2l)$, where c denotes the number of cluster centers and l is the required number of iterations for convergence. Hence, total complexity of the proposed approach is $O(3Nc^2l) + O(N) + O(3(ch\_max + p)) + O(3s) \approx O(3Nc^2l)$, i.e., complexity of the FCM algorithm itself.

8   Experiment

Preciseness in clustering brings accuracy in capturing segments. The algorithm is thus tested on a set of color images of human blood cell of varying types with gradually decreasing contrast. While Fig. 2 contains images having several parts of moderately low contrast, Fig. 3 contains those of very low contrast. In both the figures, first row contains original images and the second and the third present segmentation results. Results of segmentation using usual FCM with arbitrary selection of seeds are in the second row and those using the proposed approach are in the third row. It is apparent in Fig. 2 that majority of the regions are identified by proposed as well the usual approached but the very low contrast parts (shown with outline) within the same images are identified only by the proposed approach. This indicates the strength of the proposed approach. This is corroborated when the algorithm is applied on images having parts at very low contrast (Fig. 3). It is apparent from Fig. 2 and Fig. 3 that the proposed approach is quite capable of identifying regions even at very low contrast which is not possible with the usual FCM. Regions that are identified only by the proposed approach are outlined with boundary. The tests establish effectiveness of the precise seed selection technique over their random selection.

The HSI model is preferable over the RGB for color image segmentation due to near-zero correlation among components. A set of experiments is conducted with both the models to reveal it. The results are presented in Fig. 4. Successive rows contain original images, segmentation result using proposed algorithm in RGB and the same in HSI. Due to high correlation among components, the RGB model affects segmentation (row 2) in two ways: under-segmentation at very low contrast parts and over-segmentation occurring with increase in contrast among parts in the image.

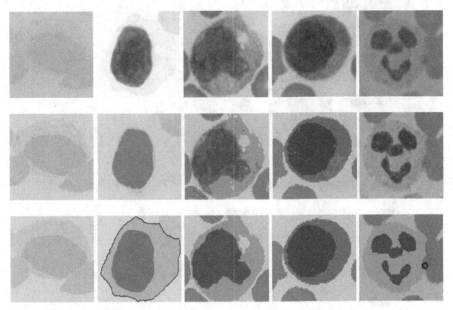

Fig. 2. Segmentation of moderately low contrast cell images. Row 1: Original image, row 2: Segmentation using usual FCM, row 3: Segmentation using proposed approach.

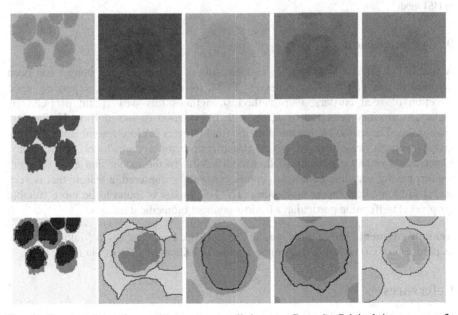

Fig. 3. Segmentation of very low contrast cell images. Row 1: Original image, row 2: Segmentation using usual FCM, row 3: Segmentation using proposed approach.

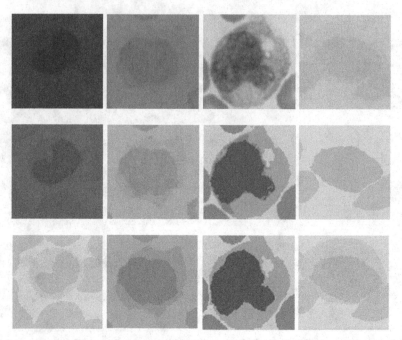

Fig. 4. Segmentation of blood cell images in RGB model. Row 1: Original test images, row 2: Segmentation using proposed FCM in RGB model, row 3: Segmentation using proposed FCM in HSI model.

9   Conclusion

A simple but effective method for segmentation of biomedical cell images has been reported in this paper. Effectiveness of FCM, to a large extent, is dependent upon selection of seed clusters. Our method strengthens this weak point of FCM by replacing arbitrary selection of seeds with information extracted from H, S and I histograms. The method helps building the final clusters in close resemblance with the image features, resulting in pixel classification to a high level of accuracy. Experiments reflect efficiency of this modification to the extent that even adjacent regions having very low contrast in colors can well be captured, a feature that is very useful in biomedical image processing. The HSI model is found to be more suitable for pixel classification particularly for low contrast biomedical images.

Acknowledgment. The authors would like to thank Prof. Tathagato Mukhopadhyay, Department of Computer & System Sciences, for critical reading of the manuscript.

References

1. Adollah, R., et al.: Blood Cell Image Segmentation: A Review. In: 4th Kuala Lumpur International Conference on Biomedical Engineering, vol. 21, pp. 141–144 (2008)
2. Sinha, N., Ramakrishnan, A.: Blood Cell Segmentation using EM Algorithm. In: 3rd Indian Conf. on Comp. Vision, Graphics & Image Proc. (ICVGIP), pp. 445–450 (2002)

3. Ma, Z., et al.: A Review on the Current Segmentation Algorithms for Medical Images. In: 1st International Conference on Imaging Theory and Applications (IMAGAPP), pp. 135–140. INSTICC Press (2009)
4. Maulik, U.: Medical Image Segmentation Using Genetic Algorithms. IEEE Trans. Information Technology in Biomedicine 13(2), 166–173 (2009)
5. Montseny, E., Sobrevilla, E., Romani, S.: A Fuzzy Approach to White Blood Cells Segmentation in Color Bone Marrow Images. In: IEEE International Conference on Fuzzy Systems, pp. 173–178 (2004)
6. Vromen, J., McCane, B.: Red Blood Cell Segmentation using Guided Contour Tracing. In: 18th Annual Colloquium of the Spatial Information Research Centre, pp. 25–29 (2006)
7. Pham, D.L., Xu, C., Prince, J.L.: Current Methods in Medical Image Segmentation. Anu. Rev. Biomed. Eng. 2, 315–337 (2000)
8. Bhattacharyya, B., Mandal, B., Mukhopadhyay, T.: Robust Segmentation of Color Image for Wireless Applications. In: Fourth International Conference on Machine Vision (ICMV 2011). Proc. of SPIE, vol. 8349, 83490R (2011)
9. Kang, B.-Y., Kim, D.-W., Li, Q.: Spatial Homogeneity-Based Fuzzy c-Means Algorithm for Image Segmentation. In: Wang, L., Jin, Y. (eds.) FSKD 2005, Part I. LNCS (LNAI), vol. 3613, pp. 462–469. Springer, Heidelberg (2005)
10. Ahmed, M.N., et al.: A Modified Fuzzy c-Means Algorithm for Bias Field Estimation and Segmentation of MRI Data. IEEE Trans. Medical Imaging 21, 193–199 (2002)
11. Chen, S.C., Zhang, D.Q.: Robust Image Segmentation using FCM with Spatial Constraints based on New Kernel-induced Distance Measure. IEEE Trans. Systems Man Cybernetics, B 34(4), 1907–1916 (2004)
12. Liew, A.W.-C., Yan, H., Law, N.F.: Image Segmentation Based on Adaptive Cluster Prototype Estimation. IEEE Trans. on Fuzzy Systems 13(4), 444–453 (2005)
13. Zhou, H., Schaefer, G., Shi, C.: A Mean Shift based Fuzzy C-Means Algorithm for Image Segmentation. In: 30th Annual International Conference of the IEEE Engineering in Medicine and Biology Society (EMBS 2008), pp. 3091–3094 (2008)
14. Yu, Z., et al.: An Adaptive Unsupervised Approach toward Pixel Clustering and Color Image Segmentation. Pattern Recognition 43(5), 1889–1906 (2010)
15. Wang, X.-Y., Bu, J.: A Fast and Robust Image Segmentation using FCM with Spatial Information. Digital Signal Processing 20, 1173–1182 (2010)
16. Jain, A.K., Murty, M.N., Flynn, P.J.: Data Clustering: A Review. ACM Computing Surveys 31(3), 264–323 (1999)
17. Asultan, K.S., Selim, S.: A Global Algorithm for the Fuzzy Clustering Problem. Pattern Recognition 26(9), 1357–1361 (1993)
18. Mandal, B., Bhattacharyya, B.: A Rough Set Integrated Fuzzy C-Means Algorithm for Color Image Segmentation. In: Das, V.V., Vijaykumar, R. (eds.) ICT 2010. CCIS, vol. 101, pp. 339–343. Springer, Heidelberg (2010)
19. Dunn, J.: A Fuzzy Relative of the ISODATA Process and its Use in Detecting Compact Well-separated Clusters. Journal of Cybernetics 3, 32–57 (1973)
20. Gonzalez, R.C., Woods, R.E.: Digital Image Processing. Prentice Hall, Englewood Cliffs (2002)
21. Valavanis, K.P., Zhang, J., Paschos, G.: A Total Color Difference Measure for Segmentation in Color Images. Journal of Intelligent and Robotic Systems 16, 269–313 (1996)
22. Tan, K.S., Isa, N.A.M.: Color Image Segmentation using Histogtam Thresholding – Fuzzy C-Means Hybrid Approach. Pattern Recognition 44, 1–15 (2011)
23. Cheng, H.D., Sun, Y.: A Hierarchical Approach to Color Image Segmentation Using Homogeneity. IEEE Trans. Image Processing 9(12), 2071–2082 (2000)

Author Index